STUDENT STUDY GUIDE

James Zubricky
University of Toledo

CHEMISTRY

McMURRY • FAY • ROBINSON

SEVENTH EDITION

PEARSON

Editor-in-Chief: Jeanne Zalesky
Acquisitions Editor: Chris Hess
Project Manager: Crissy Dudonis
Program Manager: Lisa Pierce
Program Management Team Lead: Kristen Flatham
Project Management Team Lead: David Zielonka
Compositor: Lumina Datamatics Ltd.
Cover Designer: Elise Lansdon
Executive Marketing Manager: Will Moore
Cover Photo Credit: Dr. Keith Wheeler / Science Source

2 16

www.pearsonhighered.com

ISBN 10: 0-133-88881-9
ISBN 13: 978-0-13-388881-2

Table of Contents

Preface

This study guide is intended to provide a bare-bones version of the material presented in *Chemistry*, Seventh edition, by McMurry/Fay/Robinson. The reasoning behind this book is that it is often helpful for you, the student, to have a somewhat less-intimidating introduction to what can be an overwhelming subject. Sometimes, when you're in class or studying on your own, it's very easy not to see the forest for the trees; you might get so caught up in the details that you forget to take a step back and see the overall picture. My goal in writing this book was to help you get a sense for the "forest" of chemistry and not get lost in the "trees" of the minute details.

Each chapter of this study guide has been completely revised and rewritten to correspond directly, section-by-section, with the textbook. Each study guide chapter contains the following sections:

- **Learning Objectives**—These were taken from the end of each chapter of the text. The Learning Objectives are the goals that the authors of your textbook (and your professor!) want you to accomplish as you study that chapter.
- **Chapter Summary**—This gives you a very brief overview of what the chapter is about.
- **The Chapter in Detail**—Each section corresponds section-by-section to your textbook and contains a condensed version of the chapter contents, along with example problems that are completely worked out for you and problems that you need to complete. Space is provided for you to write the solutions in the book. Keep in mind: chemistry is not a spectator sport, and giving you problems to solve that are relevant to the text is my way of making sure that you're not just passively reading it. (For instructors: the workbook problems are written in such a way that I would use these in recitation/discussion sections.)
- **Putting It Together**—Most of the later chapters contain a Putting It Together problem that brings the topics studied in that particular chapter together with topics from previous chapters.
- **Self-Test**—At the end of each chapter, I've included a sample test covering that chapter's material. Each Self-Test contains multiple choice, true–false, fill-in-the-blank, matching, and word problems to make sure that you're on top of your game. There is also a Challenge Problem for each chapter that pushes you to the next level. As an instructor who teaches from this book, I believe these are pretty fair questions. They are the kind of questions that I would set in an hour-long exam based on the book.

Finally, a Solutions Key is provided at the end of the study guide that provides worked-out solutions for all the problems in the book. The biggest change from the sixth edition to the seventh edition of this study guide is that I've compressed the four different solutions appendices into one streamlined appendix.

A writing project like this is never a one-person project; in fact, there is a team of people that are responsible for the book you hold in your hands (or are reading online): I would like to thank Terry Haugen and Chris Hess for trusting me with this project and Crissy Dudonis for dealing with me when I went on writing spurts and droughts. I must also thank both Jennifer Edwards and David Berg for insisting that I get going on this project. I am also indebted to the authors of the previous six editions of this study guide, who also developed the format: DonnaJean Freeden and Julie Klare. I've more or less modeled this edition of the study guide based on their work. Finally, a big thank you to both David Boatright (University of West Georgia) for proofreading and correcting the errors of the text and Jill Robinson (Indiana University) for her help and guidance in this project, especially during the last few weeks of writing.

As well as the excellent people at Pearson, I must thank my colleagues at the University of Toledo for their constant support, advice, opinions, political discussions, jokes, laughs, and inappropriate comments—all of which I badly needed during the writing of this study guide: Jon Kirchhoff, Andy Jorgensen, Pam Samples, and Ruth Hottell.

Finally, I want to thank Jessica Willinger, who was incredibly patient and understanding while I was writing the vast bulk of this book during snow days, early weekend mornings, and vacations. She listened to me when I freaked out about getting everything done and she calmed me down. This book is dedicated to her.

James Zubricky
The University of Toledo
March 2015

CHAPTER ONE

Chemical Tools: Experimentation and Measurement

Learning Objectives

As a result of reading and studying this chapter, you should be able to

Section 1.1	**The Scientific Method**	
	1.	Identify the steps in the scientific method.
	2.	Differentiate between a qualitative and quantitative measurement.
Section 1.2	**Experimentation and Measurement**	
	3.	Write numbers in scientific notation and use prefixes for multiples of SI units.
Section 1.3	**Mass and Its Measurement**	
	4.	Describe the difference between mass and weight.
	5.	Convert between different prefixes used in mass measurement.
Section 1.4	**Length and Its Measurement**	
	6.	Convert between different prefixes used in length measurements.
Section 1.5	**Temperature and Its Measurement**	
	7.	Convert between common units of temperature measurements.
Section 1.6	**Derived Units: Volume and Its Measurement**	
	8.	Convert between SI and metric units of volume.
	9.	Convert between different prefixes used in volume measurements.
Section 1.7	**Derived Units: Density and Its Measurement**	
	10.	Calculate mass, volume, or density using the formula for density.
	11.	Based on density, predict whether a substance will float or sink in another substance.
Section 1.8	**Derived Units: Energy and Its Measurement**	
	12.	Calculate kinetic energy of a moving object.
	13.	Convert between common energy units.
Section 1.9	**Accuracy, Precision, and Significant Figures in Measurement**	
	14.	Specify the number of significant figures in a measurement.
	15.	Evaluate the level of accuracy and precision in a data set.
	16.	Report a measurement to the appropriate number of significant figures.
Section 1.10	**Rounding Numbers**	
	17.	Report the answer of mathematical calculations to the correct number of significant figures.
	18.	Round a measurement to a specified number of significant figures.

Section 1.11   Unit Conversions
>   19.   Change a measurement into different units using appropriate conversion factors.

Chapter Summary

Chemistry is the study of the composition, properties, and transformations of matter. The basis for all experiments performed in science is the scientific method.

Chemistry knowledge proceeds through experimentation, and experimentation requires both measurement and data analysis. The seven base units of the International System of Units (abbreviated as SI) are introduced in this chapter, and the most immediately useful are discussed in some detail. Units derived from these basic units are also introduced, as well as prefixes used to enhance the utility of these basic units for very small and very large measurements. Scientific notation, and the rules for significant figures are explained, and the process of dimensional analysis demonstrated for converting from one unit to another.

The Chapter in Detail

Section 1.1   The Scientific Method in a Chemical Context: Improved Pharmaceutical Insulin

Chemistry is a not a spectator science; we learn more about ourselves, our environment, and our universe through experimentation. The method or thought-process that scientists use to investigate and solve problems is called the **scientific method**.

There are four stages to the scientific method:

- **Observation** – a systematic recording of natural phenomena that may be qualitative (descriptive) or quantitative (involving measurements).
- Formation of a **hypothesis** – a possible explanation for the observed phenomena based upon facts collected from previous experiments as well as scientific knowledge and intuition.
- Testing of the hypothesis through **experimentation.**
- Development of a **theory** from the hypothesis that is consistent with experimental data. It is also a unifying principle that explains the experimental results. In science, theories can be shown to be sound through repeated experimentation, but a theory can never be absolutely proven. If new experiments discount the theory, it must be changed or set aside.

Section 1.2   Experimentation and Measurement

To make results easier to use worldwide, in 1960 a system of measurements was adopted known as the International System of Units, or SI system (from the French, Système International d'Unités). It is based on the metric system and has seven fundamental units—the kilogram, which measures mass; the meter, which measures length; the kelvin, which measures temperature; the mole, which measures the amount of a substance; seconds, used to measure time; the ampere, which is a unit of electrical current; and the candela, which is a unit of luminous intensity. Not all of these are used in the introductory study of chemistry, and there are units, such as those for volume or energy, that appear to be missing, but all units used in science can be derived from these seven.

To make the units more broadly applicable to many different fields, from atomic chemistry to astrophysics, a system of prefixes is available. *It is absolutely essential that you memorize these prefixes!*

These prefixes are used for large units:

- giga (G) $= 10^9$
- mega (M) $= 10^6$
- kilo (k) $= 10^3$

These prefixes are used for small units:

- deci (d) $= 10^{-1}$
- centi (c) $= 10^{-2}$
- milli (m) $= 10^{-3}$
- micro (μ) $= 10^{-6}$
- nano (n) $= 10^{-9}$

To make numbers easier to read, scientific notation can also be used. Review Appendix A for a more in-depth review, but in brief, scientific notation takes a number and puts it in an exponential format in which there is one digit to the left of the decimal place and an exponent to indicate where the decimal belongs. So 306 becomes 3.06×10^2, and 0.00306 becomes 3.06×10^{-3}.

If a number represents a measurement, it must have a unit. The difference between 20 feet, 20 meters, and 20 miles is important! Without the unit, a measured number is meaningless.

Workbook Example 1.1

PROBLEM:
Express the following length measurements in scientific notation: a) 3652 m and b) 0.000 054 95 m. What unit prefixes could reasonably be used with these numbers?

SOLUTION:
We will use the rules outlined in Appendix A in your textbook.

For part a), we are asked to write the number 3652 in scientific notation. We need to move the decimal point three places to the left so that we get the number 3.652, which is between 1 and 10. Because we moved the decimal point to the left three places, we will use the number 3 as our exponent for the power of 10. That gives us the answer 3.652×10^3 m. *Let's take this one step further*: looking at the list of possible prefixes, 10^3 corresponds with *kilo*. Therefore, this number could also be expressed as 3.652 kilometers.

Part b) deals with the number 0.000 054 95. In order to write this in scientific notation, we need to move the decimal point five places to the right to obtain the number 5.495, which is between 1 and 10. Because we moved the decimal point to the right, the exponent will be -5. That gives us the answer 5.495×10^{-5}.

Workbook Example 1.2

PROBLEM:
State the SI unit and abbreviation used for a) length, b) temperature, c) amount of substance, and d) mass.

SOLUTION:
From Table 1.2 in your textbook, we find that the SI unit for a) length is meter (m);
b) temperature is kelvin (K); c) amount of substance is mole (mol); and d) mass is kilogram (kg).

Section 1.3   Mass and Its Measurement

Mass is the amount of matter in an object. Matter, in its turn, is everything that has a physical presence. The SI unit of mass is the kilogram. This is much too big for most uses in chemistry, so grams (g), milligrams (mg), and micrograms (μg) are commonly used.

Mass is frequently confused with weight. Weight has to do with gravitational pull, and it will vary depending on the strength of gravity; mass is a measurement of the amount of matter in an object. An object will have the same *mass* on Earth and on the moon, whereas it will have much less *weight* on the moon. The confusion is rarely a problem in chemistry, as it is unusual to be in a gravitational field other than that of the Earth.

Section 1.4   Length and Its Measurement

The meter is the SI unit for length. The meter is equal to 39.37 inches and is too large for most measurements in chemistry. Chemists use centimeters and millimeters to deal with most measurements.

Section 1.5   Temperature and Its Measurement

Although the official SI unit for the measurement of temperature is the kelvin, degrees Celsius are more commonly used. The two scales are easily interconverted, because both have 100 degrees between the freezing point of water and the boiling point of water. The difference in the scales is the temperature assigned the value zero: on the Celsius scale, 0 °C is the freezing point of water. On the Kelvin scale, 0 K is known as *absolute zero*, the coldest temperature possible.

To convert a temperature from degrees Celsius to kelvins, we use the following equation:

$$\text{Temperature in K} = \text{temperature in °C} + 273.15$$

To convert from kelvins to degrees Celsius, we use the following equation:

$$\text{Temperature in °C} = \text{temperature in K} - 273.15$$

The Fahrenheit scale is still used predominantly in the United States. The Fahrenheit scale sets 32 °F as the freezing point of water and 212 °F as the boiling point of water. So whereas the Celsius and Kelvin scales have 100 degrees between the freezing and boiling points of water, the Fahrenheit scale has 180°. An additional complication is the shifting up of the freezing point of water from 0 °C to 32 °F.

To convert between degrees Fahrenheit and degrees Celsius requires two adjustments: one to adjust for the difference in degree size and the other to adjust for the difference in zero point.

To convert from Celsius to Fahrenheit, we use this equation:

$$°F = \left(\frac{9}{5} \times °C\right) + 32$$

To convert from Fahrenheit to Celsius, we use this equation:

$$°C = \left(\frac{5}{9} \times °F\right) - 32$$

Workbook Example 1.3

PROBLEM:
Gallium, a metal whose salts are used to produce Blu-ray lasers, melts at 302.9 K. What is this temperature in °F?

SOLUTION:
This is a good time to use the "thinking approach." Not having been given a conversion from kelvins to degrees Fahrenheit, the easiest approach is to do the quick and easy conversion from kelvins to degrees Celsius, then do the more complex conversion from Celsius to Fahrenheit.

Zero on the Celsius scale is much higher than zero in the Kelvin scale, so to convert from kelvins to degrees Celsius, subtract 273.15:

$$302.9 \text{ K} - 273.15 = 29.75 °C$$

Now convert to Fahrenheit.

A Fahrenheit degree is smaller than a Celsius degree. Just think about the freezing point of water. Water freezes at 0° on the Celsius scale and at 32° on the Fahrenheit scale. Therefore, when converting from Celsius to Fahrenheit, the temperature should be higher.

Using the temperature conversion equation, we get

$$°F = \left(\frac{9}{5} \times 29.75\right) + 32 = 85.55°F$$

Gallium is actually really awesome; the melting point is lower than the temperature of the human body (98.6 °F), so gallium metal will actually melt in your hand if you hold it long enough!

Workbook Problem 1.1

You wish to impress upon a friend in Russia, which uses the Celsius scale, the misery of the heat wave you are enduring in Detroit. What is the Celsius equivalent of a balmy 102 °F? If you really want to feel hot, figure out the temperature in kelvins!

Strategy: Consider the differences in the degree size and zero point adjustments for the Fahrenheit and Celsius scales.

Step 1: Based on the discussion of the Fahrenheit scale and the Celsius scale, determine if °C should be higher or lower than °F.

Step 2: Apply the formula for the conversion of °F to °C.

Step 3: Does your answer in step 2 agree with your answer in step 1?

Step 4: Apply the formula for the conversion of °C to K.

Section 1.6   Derived Units: Volume and Its Measurement

Derived units are those made by combining the seven basic units. A simple example is speed, which would be expressed in SI units as meters per second, or m/s.

Volume is also a derived unit that comes from the unit of length: remember from geometry that volume is equal to length x width x height. If this is done with centimeters as the lengths, the result is cm^3. This is exactly equivalent to a milliliter (1×10^{-3} L). One liter is equivalent to a dm^3 – a cubic decimeter.

In the chemistry lab, volume is measured using graduated cylinders, syringes, volumetric flasks, and burets.

Section 1.7   Derived Units: Density and Its Measurement

Density is measured by dividing the mass of a substance by its volume. For liquids and solids, density is expressed in g/mL or g/cm^3. (Recall that these units are identical: 1 mL = 1 cm^3.) Density is an intrinsic physical property that can be used to identify a substance. Density can also be used to convert back and forth between mass and volume.

Density is temperature-dependent. Most substances increase in volume as they are heated and decrease in volume as they are cooled, so when reporting a density, the temperature must also be noted.

The properties of water make it unique: down to a temperature of 4 °C water shrinks; then it begins to expand. Solid water (ice) is about 10% larger than the same mass of liquid water. This is why ice floats. More generally, substances with a lower density will float on top of substances with higher density.

Workbook Example 1.4

PROBLEM:
A student determined that a metal cylinder having a mass of 25.478 g has a volume of 9.597 mL. What is the density of this metal cylinder?

SOLUTION:
Knowing both the mass and volume of the metal cylinder, we can determine the density by simply dividing the mass by the volume. Make sure to keep the units with the numbers, as this will prevent many careless errors.

$$Density = \frac{25.478\,g}{9.597\,mL} = 2.655\,\frac{g}{mL}$$

Given a chart of metal densities, it would be possible to identify the metal directly from its density, because density is an intrinsic property. The density of aluminum (Al) is about 2.7 g/cm^3 at room temperature, making it likely that the cylinder is made of aluminum.

Workbook Example 1.5

PROBLEM:
A student is given containers of isopropyl alcohol (approximate density 0.79 g/mL) and vegetable oil (approximate density 0.90 g/mL). Vegetable oil will not mix with isopropyl alcohol (rubbing alcohol). If both liquids are poured into a graduated cylinder, which will sink? Which will float? If an ice cube (approximate density 0.91 g/mL) is dropped into the cylinder, where will it settle?

SOLUTION:
It is helpful to understand how density works: a substance that is more dense (has a higher number of grams per milliliter) will sink in something that is less dense. So when the liquids are put into the same container, the more dense liquid (vegetable oil) will sink, and the isopropyl alcohol will float, because the substances will not mix. The ice cube has an intermediate density: it will sink in the alcohol, but float on the oil. Based on the given densities the ice would sink to bottom.

Workbook Problem 1.2

The density of lead at 25 °C is 11.34 g/mL. How many grams of lead are found in a volume of 53.43 mL?

Strategy: Using the information provided about density, set up a mathematical equation that allows you to solve for the mass of lead.

Step 1:   Substitute the information given in the problem into the mathematical equation you set up.

Step 2:   Solve for the mass of lead.

Step 3:   Does your answer make physical sense?

Section 1.8   Derived Units: Energy and Its Measurement

Energy is defined as the ability to do work or to supply heat. There are several different forms of energy (which we'll talk about further in Chapter 9), but the two biggest forms of energy that we come across in general chemistry is kinetic energy and potential energy.

Kinetic energy (E_K) is the energy of motion and can be calculated by the following equation:

$$E_K = \frac{1}{2}mv^2$$

where m is the mass of an object (measured in kg) and v is the velocity of the object (measured in m/s). The final answer for energy is expressed with units $kg\left(\frac{m}{s}\right)^2$, where $1\ kg\left(\frac{m}{s}\right)^2$, is equal to 1 joule (J), which is the SI unit for energy.

While most modern chemists use joules to describe energy, some old-school chemists and biochemists still use calories to describe energy. To convert from joules to calories, 1 calorie is equal to 4.184 joules. When we talk about calories in terms of the energy we get from food, we have another unit to deal with: Calories (where the capital C refers to food calories); 1 Calorie is equal to 1000 calories, or 1 kilocalorie.

Workbook Example 1.6

PROBLEM:
I went to a Chicago Cubs game and caught a fly ball! Using a radar speed gun, I was able to clock the ball coming towards me at 45.15 meters per second. If the mass of the baseball was 0.143 kilograms, how much kinetic energy did that baseball have?

SOLUTION:
Knowing the mass of the baseball (0.143 kg) and the speed of the baseball (45.15 m/s) will make solving this problem much easier: we can put our values into the kinetic energy equation and solve for E_K:

$$E_K = \frac{1}{2}mv^2$$

$$E_K = \frac{1}{2}(0.413\ kg)(45.15\ m/s)^2$$

$$E_K = 146\ J$$

Section 1.9   Accuracy, Precision, and Significant Figures in Measurement

Accuracy and **precision** are concepts related to the reliability of a measurement. Accuracy refers to the closeness of a number to its accepted value, while precision refers to the closeness of a set of measurements to one another. If a piece of metal is agreed to weigh 4.5 grams, and you weigh it three times, getting answers of 4.787 g, 4.763 g, and 4.775 g, you have measurements that are close to one another—they are *precise*—but they do not agree with the accepted value: they are not *accurate*.

Significant figures are the number of digits recorded for a measurement. For the example masses in the previous paragraph, four significant figures are given. When reading a digital instrument, write down all the numbers: if the instrument reports it, it is significant. When reading an analog scale, such as a ruler, record every number of which you are certain, and record the last one as a good estimate.

When trying to determine the number of significant figures in a measurement, refer to the four rules in your textbook on Page 18. When writing in scientific notation, only significant figures are recorded. A special case is *exact numbers*. These are counting numbers that can have no fractional part: the number of students in a class, for example, will be an exact number, as there will never be a fractional student present. These numbers are said to have an infinite number of significant figures.

Workbook Example 1.7

PROBLEM:
How many significant figures do the following measurements have?
10.002 g,   0.00672 mL,   5.3000 m,   83,500 s

SOLUTION:
10.002 g — five significant figures (Rule 1)

0.006 72 mL — three significant figures (Rule 2); if you write the number as 6.72×10^{-3} mL, it is easier to see the number of significant figures. This could also be written as 6.72 μL.

5.3000 m — three significant figures (Rule 3)

83,500 s — uncertain (Rule 4); if the number is 8.3500×10^4, there are five significant figures. If the number is 8.350×10^4, there are four significant figures. If the number is 8.35×10^4, there are three significant figures.

Section 1.10   Rounding Numbers

When working with numbers, often you will get an answer from a calculator that has far more digits in it than are reasonably useful. For example, if you are calculating a density from measurements that are 1 gram and 3 cm^3, a calculator will give you an answer of 0.333 333 333 333 g/cm^3. A few simple rules help deal with these situations:

1. When multiplying or dividing, the answer must have the same number of significant figures as the smallest number used. In the density example in the above paragraph, the

answer can have no more than one significant figure because the smallest number, 1, has only one significant figure. Therefore, the density must be reported as 0.3 g/cm³.

2. When adding or subtracting, the answer must not have more digits to the right of the decimal point than either of the original numbers. So if you added the following: 0.3, 5.75, and 7.267, the answer could have no more than one digit on the right of the decimal point: 13.317 would become 13.3.

3. When truncating a number, round up if the number past the last possible digit is a 5 with one or more additional numbers higher than 5 (e.g., 13.356 becomes 13.4), round down if the number is a 5 with nothing following or numbers lower than 5 (e.g., 13.3500 becomes 13.3).

4. Most importantly, **do not round or truncate any numbers until the calculation is complete**.

Workbook Example 1.8

PROBLEM:
If a high-speed train in Japan covers 532 miles in 3.7 hours, what is the average speed of the train?

SOLUTION:

three significant figures $\qquad \dfrac{532 \text{ miles}}{3.7 \text{ hours}} = 140\dfrac{\text{miles}}{\text{hour}}$ *answer needs two significant figures*

If this problem is plugged in to a calculator, the answer provided will have more digits than can possibly be justified: 143.783 783 8 miles/hour suggests far more precision than the numbers permit.

Workbook Example 1.9

PROBLEM:
What is the final answer to the following problem, using the correct number of significant figures?

103.7835 g
−1.52   g

SOLUTION:

103.7835 g	*ends four places past decimal point*
− 1.52 g	*ends two places past decimal point*
102.26 g	*ends two places past decimal point*

The final answer cannot have more digits to the right of the decimal than the smallest original number. However, in doing calculations, use all figures until the end of the problem. Only round off the final answer.

Workbook Example 1.10

PROBLEM:
The mass of a sample of granite is found to be 15.896 g. If the volume is determined to be 5.8 mL, what can you report for the density of the granite?

SOLUTION:

$$\text{Density} = \frac{15.896 \text{ g}}{5.8 \text{ mL}} = 2.740\ 689\ 65 \frac{\text{g}}{\text{mL}}$$

Decide how many significant figures should be in your answer. The denominator has only two significant figures; therefore, the answer must also have only two significant figures.

Round off your answer. The first digit to be dropped is less than 5. 2.740 689 65 g/mL becomes 2.7 g/mL. This situation is not unusual: the determination of the mass of a granite sample can be done far more precisely than the determination of the volume of an irregular sample.

Workbook Example 1.11

PROBLEM:
Round off the numbers to three significant figures:
(a) 83.567 g   (b) 0.000 358 500 g   (c) 28,572 g

SOLUTION:
(a) Because the last digit is a 5 with following digits greater than 5, the mass rounds to 83.6 g.

(b) Because the last digit is a 5 with nothing following, the number is truncated to 0.000 358 g.

(c) The rounding of this number to 28,600 g introduces ambiguity about the number of significant figures. To prevent this, the number can be reported as 2.86×10^4 g, or as 28.6 kg.

Workbook Problem 1.3

A quantity of carbon dioxide gas, which has a density of 1.977 g/L, is found to have a mass of 4.843 65 g. Determine the volume of the gas, and report this volume with the correct number of significant figures.

Strategy: Set up a mathematical equation to determine the volume of the gas. Apply the rules found above to determine the correct number of significant figures.

Step 1:   Solve for the volume of the gas.

Step 2:   Apply the rules for multiplication and division to determine the number of significant figures in your answer.

Step 3:   If necessary, use the rules for rounding off numbers.

Section 1.11   Calculations: Converting from One Unit to Another

To convert from one unit to another requires the use of a conversion factor. The **dimensional-analysis method** clarifies the use of these factors by setting up multiplications such that units cancel out. Conversion factors can be written as fractions that can be interchanged:

$$1 \text{ in.} = 2.54 \text{ cm}$$

This tells you that there is one inch per 2.54 cm ("per" always implies division) OR 2.54 centimeters per inch.

$$\text{Conversion factors: } \frac{1\,in}{2.54\,cm} \text{ or } \frac{2.54\,cm}{1\,in}.$$

The quantity in the numerator is equal to the quantity in the denominator, making this equivalent to multiplying by 1.

Choose the form of the conversion factor that will eliminate the unwanted units and leave the desired units. If the units are correct, the numbers will necessarily follow. Make sure that the units are multiplied and divided as the problem is worked: bizarre final units (g^2/cm^3, for example) are an indication that an error was made.

Workbook Example 1.12

PROBLEM:
On her first trip to the United States, a student rents the least expensive car available. When she gets on the highway, she discovers that the car's top speed is 43 miles per hour. When she calls home, what speed in kilometers per hour does she report to her family?

SOLUTION:
Using the conversion table on the inside back cover, we learn that 1 mi = 1.6093 km. Set up the equation beginning with the information given (43 mi/h), and write the conversion factor so that the unit *mi* cancels.

$$\frac{43\,mi}{1\,hr} \times \frac{1.6093\,km}{1\,mi} = 69\frac{km}{hr}$$

Ballpark Check: There are more kilometers than miles in the same length, but less than double. So the answer in kilometers will be greater than the number of miles, but less than twice as much.

Workbook Example 1.13

PROBLEM:
In the United Kingdom and Ireland, the unit of "stone," though forbidden for use in trade, is still commonly used to report the weights of people. One stone is equal to 14 pounds. If a friend in Ireland, complaining about having put on the "freshman stone," tells you that he now weighs 11 stone, how much does he weigh in kilograms?

SOLUTION:
Given that 14 lbs = 1 stone and 2.205 lbs = 1 kg, set up the equation knowing that you need to convert 11 stone to pounds, then pounds to kg. The units stone and pounds should cancel each other out.

$$11\,stone \times \frac{14\,lbs}{1\,stone} \times \frac{1\,kg}{2.205\,lbs} = 70\,kg$$

Ballpark Check: A stone is 14 pounds, and there are about 2 pounds per kilogram, so the original number of stone will be multiplied by about 7 to get the equivalent quantity in kilograms. So, the answer will be close to the original number of stone, multiplied by about 7. In this case, the number of significant figures in the original number determines the number of significant figures.

WARNING: It's easy to get the right answer using dimensional analysis without really understanding what you're doing.

Workbook Problem 1.4

If a runner completes a 5-mile race in 27 minutes 46 seconds, how long can she expect it will take her to run a 10-kilometer race?

Strategy: Using the conversion table on the inside back cover of your textbook, determine the conversion factor for meters to miles, and convert time to one unit. Use the time and distance for the 5-mile race to determine the runner's speed, convert that speed to kilometers per minute, and determine the time expected for the distance in kilometers.

Step 1:   Calculate the amount of time it will take for the athlete to run that distance.

Self–Test

This section is intended to test your knowledge of the material covered in this chapter. Think through these problems, and make certain you understand what the problem is about—ask yourself if your answer makes sense. Many of these questions are linked to the chapter learning goals. Therefore, successful completion of these problems indicates that you have mastered the learning goals for this chapter. You will receive the greatest benefit from this section if you use it as a mock exam. You will then discover which topics you have learned thoroughly and which topics you need to study in more detail.

True–False
1. One of the key steps of the scientific method involves forming a theory.

2. The number 0.002 05 contains three significant figures.

3. _Weight_ and _mass_ are the same thing.

4. The unit most commonly used for laboratory volume measurements is the m^3.

5. 1×10^6 g is equal to 1 µg.

6. The number 4000 can be expressed in scientific notation as 4×10^3.

7. Kinetic energy is the energy of movement.

8. The SI unit for temperature is the degree Fahrenheit.

9. The temperature "absolute zero" describes 0 K.

10. Density describes an object's mass and its temperature.

Multiple Choice
11. Volume is
    a. anything physically real
    b. the pull of gravity on an object by the earth or other celestial body
    c. how much space an object takes up
    d. the amount of matter in an object

12. The following measurements were made by a student during a laboratory exercise:
    23.05 cm, 22.93 cm, 35.02 cm.
    The reference value in this experiment is 30.00 cm. These numbers represent data that are
    a. precise and accurate   b. precise but not accurate
    c. accurate but not precise   d. neither precise nor accurate

13. If 3.73 was multiplied by 153.2, the answer would contain
    a. three significant figures   b. four significant figures
    c. five significant figures   d. six significant figures

14. The SI unit for mass is
    a. oz   b. lb   c. g   d. kg

15. The SI prefix _nano_ corresponds to the multiplier
    a. 10^{-3} b. 10^{-6} c. 10^{-9} d. 10^{-12}

16. A Celsius degree is
    a. larger than a Fahrenheit degree, but the same as a Kelvin.
    b. smaller than a Kelvin and the same as a Fahrenheit degree.
    c. larger than a Kelvin.
    d. smaller than a Fahrenheit degree.

Matching

17. Scientific Method

18. Theory

19. Mass

20. Precision

21. Accuracy

22. Dimensional analysis

a. the amount of matter contained in an object

b. how close a measurement is to an accepted value

c. how well a number of independent measurements agree with one another

d. an experiment-based method for increasing understanding of the natural world

e. a way to carry out calculations involving different units

f. a consistent explanation of known observations based on many independent experiments; useful, but cannot be proven

Fill-in-the-Blank

23. A _____ is always open to modification by new experimental data.

24. The SI units for mass, temperature, and length are the _____, the

    _____, and the _____.

25. There are three commonly used temperature scales. _____ and

    _____ have degrees of the same size, while _____ has degrees

    that are smaller.

26. The intrinsic property of a substance _____ can be determined by dividing a

    substance's mass by its volume.

27. If an instrument gives you measurements that average-out to the accepted reference value but

    are not close to each other, you would say that your data are _____, but not

    _____.

28. To write numbers that are very small or very large, it is helpful to use _____.

29. When finished working a problem, it is often necessary to _____ the answer to

    the correct number of _____.

30. 1 inch = 2.54 cm is an example of a _____ _____.

31. When adding numbers together, the answer cannot have more significant figures on the right of the decimal point than _____.

32. The most appropriate prefix to use for 360,000,000 watts is _____, abbreviated _____.

33. If you know the density of a liquid and its volume, you can calculate its _____.

34. A graduated cylinder and a syringe can both be used to measure _____.

Problems

35. 246 cm = _____ mm = _____ m = _____ km

36. Perform the following calculations and report your answer with the correct number of significant figures:

    a. $\dfrac{8.3}{2.793} =$ b. $23.945\ 62 \times 8.3421 =$

    c. $3.2 + 5.3875 =$ d. $9434.21 - 4.199 =$

37. The freezing point of paraffin wax, commonly used in making preserves, is 56 °C. What is this in degrees Fahrenheit?

38. The land area of Zambia is 290,587 mi^2. Convert this to km^2.

39. How many nanometers are there in 6.73×10^{-7} m?

40. The boiling point of ethanol is 78.5 °C. Convert this temperature to °F and K.

41. In 2009, the world's record for the 100-meter dash was 9.58 seconds, and the world's record for the 100-meter freestyle swim was 47.85 seconds. How fast were the record holders moving in miles per hour?

42. The mass of an amorphous chunk of a yellow element is found to be 9.634 g. When it is placed in a graduated cylinder containing 11.5 mL of water, the water level rises to 16.2 mL. What is the density of the substance in g/cm^3?

43 As of July 8 2014, the *Voyager 2* Spacecraft had completed 13,741 days of continuous operation and reached a distance of 19.1 billion kilometers from the Earth. If communications travel at the speed of light (3.00×10^8 m/s), how long does it take for data to get from the spacecraft to ground control?

44. A pycnometer is a device used to make precise density determinations. A pycnometer weighs 23.75 g when empty and 37.82 g when filled with water at 20 °C. The density of water at 20 °C is 0.9982 g/mL. When 6.345 g of beryllium are placed in the pycnometer, a total mass of 40.72 g

is obtained when the pycnometer is again filled with water at 20 °C. What is the density of beryllium?

45. If the density of titanium is 4.55 g/mL, and a titanium metal plate used to repair a broken skull weighs 10.67 g, what is the volume of the plate?

46. Olive oil has a density of 0.918 g/mL. It will mix with trichloroethane, a solvent with a density of 1.3492 g/mL. In what proportions would 100 total grams of these substances have to be mixed to obtain the same density as water, 0.997 g/mL?

47. What is the physical basis of 0 °C? How is this different from 0 K?

48. When working through calculations, at what step are the numbers rounded or truncated to the correct number of significant figures? Why is it done this way?

Challenge Problem
49. Gallium has a melting point of 29.78 °C and a boiling point of 2403 °C. This very large liquid range coupled with the fact that gallium is a liquid near room temperature makes this metal ideal for the construction of a high-temperature thermometer.

You need to measure the melting points of some metals and have decided to construct a thermometer using gallium as the liquid. However, you don't want the thermometer to be too awkward to handle, so you have decided to define the melting point of gallium as 0°Ga and the boiling point as 1000°Ga. To ensure the accuracy of your thermometer, you measure the melting point of copper. The melting point of copper on your thermometer is 447°Ga. The reported melting point for copper is 1085°C. Is your thermometer accurate?

HINT: Think about how to convert from Fahrenheit to Celsius. How would you convert from Celsius to your new temperature system? You can then write the equation for conversion between °Ga and °C.

CHAPTER TWO

Atoms, Molecules, and Ions

Learning Objectives

As a result of reading and studying this chapter, you should be able to

Section 2.1   **Chemistry and the Elements**
1. Use symbols to represent element names.

Section 2.2   **Elements and the Periodic Table**
2. Identify the location of metals, nonmetals, and semimetals on the periodic table.
3. Indicate the atomic number, group number, and period number for an element whose position on the periodic table is given.
4. Identify groups as main group, transition metal group, or inner transition metal group.

Section 2.3   **Some Common Groups of Elements and Their Properties**
5. Specify the location and give examples of elements in the alkali metal, alkaline-earth, transition metal, halogen, and noble gas groups.
6. Use the properties of an element to classify it as metal, nonmetal, or semimetal and give its location in the periodic table.

Section 2.4   **The Law of Conservation of Mass and the Law of Definite Proportions**
7. Determine the mass of the products in a reaction using the law of mass conservation.

Section 2.5   **The Law of Multiple Proportions and Dalton's Atomic Theory**
8. Demonstrate the law of multiple proportions using mass composition of two compounds of the same element.
9. Determine the formula of a compound given mass-composition data for two compounds and the formula of one compound.

Section 2.6   **Atomic Structure: Electrons**
10. Describe Thomson's cathode ray experiment and what it contributed to the current model of atomic structure.
11. Describe Millikan's oil-drop experiment and what it contributed to the current model of atomic structure.

Section 2.7   **Atomic Structure: Protons and Neutrons**
12. Describe Rutherford's gold foil experiment and what it contributed to the current model of atomic structure.
13. Describe the structure and size of the atom.
14. Calculate the number of atoms in sample given the size of the atom.

Section 2.8   **Atomic Numbers**
15. Determine the mass number, atomic number, and number of protons, neutrons, and electrons from an isotope symbol.
16. Write isotope symbols for elements.

Chapter Summary

In order to understand the behavior of chemical reactions, we first need to understand how chemistry as a science has developed from the ancient Greeks to our present time. This chapter begins with a brief description of the ways in which matter is categorized and how various properties, both chemical and physical, can be used in its identification. This discussion is followed by an introduction to the periodic table and how the elements are organized within it. Next, we begin to understand chemistry by examining Dalton's atomic theory and how his theory has developed and expanded since the early nineteenth century. Atoms will be described, and the three basic subatomic particles—protons, neutrons, and electrons— are introduced, along with the ways in which these vary in numbers of ions and isotopes. After a quick introduction to the mole and how chemists weigh atoms, we then turn to looking at compounds, mixtures, chemical-bond types, acids and bases, and chemical nomenclature.

The Chapter in Detail

Section 2.1   Chemistry and the Elements

An element is a fundamental substance that can't be chemically changed or broken down. There are 118 known elements, 90 of which occur naturally, and the rest have been produced through nuclear reactions.

Elements are given one- or two-letter symbols; the first letter is capitalized, and if there is a second letter, it is lowercased. The symbols usually come from the English name of the element, but sometimes they are derived from other languages. For instance, the symbol for sodium is Na; this comes from the Latin word *natrium*, an obsolete word for sodium.

Workbook Example 2.1

PROBLEM:
Write out the full names of the following elements: F, Mg, I, Te, P.

SOLUTION:
Using the periodic table inside the front cover of the textbook, we find that the full names of the symbols are

F – fluorine	Mg – magnesium	I – iodine
Te – tellurium	P – phosphorus	

Workbook Example 2.2

PROBLEM:
Find the symbols for the following elements: iron, oxygen, potassium, chlorine, sodium.

SOLUTION:
Using the periodic table inside the front cover of the textbook, we find that the symbols for these elements are

iron – Fe	oxygen – O	potassium – K
chlorine – Cl	sodium – Na	

You may find it helpful to make flashcards with the name of the element on one side and its symbol on the other. You will generally have access to a periodic table, but it may have only the symbols of the elements, and not their names. Focus first on the main group elements (groups IA through 8A), because these are the ones you will use the most, particularly at the beginning of this course.

Section 2.2   Elements and the Periodic Table

Ten elements have been known since antiquity, but as the pace of chemical discovery increased in the 1700s and 1800s, efforts were made to categorize the elements by their properties. The most successful of these efforts was the **periodic table,** published in 1869 by Dmitri Mendeleev, a Russian chemist.

The periodic table divides the elements into 18 vertical columns called **groups** and 7 horizontal rows called **periods**. The elements in each group have similar chemical properties. The number of elements in each period—two in period one, eight in period two, and so on—is an indication of increasing atomic size and the structure of atoms.

Groups 1A and 2A on the left, and 3A through 8A on the right of the periodic table, are the **main group elements**. Their chemical properties are easy to predict based on their location on the periodic table. Groups 1B through 8B are known as the **transition metals**. The 14 unnumbered groups shown separately on the bottom of the periodic table are the **inner transition metals**.

The periodic table of the elements is the most important organizing principle of chemistry. It is worth printing out a black-and-white periodic table and coloring in these groups to get a solid understanding of where they lie on the table: use different colors for the main group, transition metals, and inner transition metals.

Section 2.3   Some Common Groups of Elements and Their Properties

Elements can be described in terms of their **properties**. Properties are themselves classified in a number of ways:

- Intensive properties do not depend on the size of the sample. An example is melting point: ice melts at 0 °C, whether it is a snowflake or an iceberg.
- Extensive properties do depend on the amount of the sample. To continue the previous example, the amount of energy needed to melt a snowflake or an iceberg is very different and depends on the sample size.
- Physical properties can be observed without changing the chemical makeup of the sample: melting is a physical property, as is boiling. Whether ice, liquid, or steam, the sample is still water. Physical changes tend to be readily reversible.
- Chemical properties cannot be observed without a change in the composition of the sample. If hydrogen and oxygen are brought together and a spark is applied, water will result. Something has been learned about hydrogen (that it combines with oxygen to make water), but you no longer have hydrogen alone, because the sample has changed.

Workbook Example 2.3

PROBLEM:
There is overlap in the kinds of properties that are observed. Classify these as intensive or extensive, and also as physical or chemical.

Color	Combustibility	Smell	Length	Rusting

SOLUTION:
Using the descriptions of these properties above:
- Color is an *intensive physical* property.
- Combustibility is an *intensive chemical* property.
- Smell is an *intensive physical* property.
- Length is an *extensive physical* property.
- Rust is an *intensive chemical* property.

Four groups in the periodic table are described using both physical and chemical properties:

- **Group 1A: Alkali metals**. These are silvery, soft metals. All react strongly with water to form very alkaline solutions, which gives the group its name and also makes it impossible to find these metals in pure form naturally because of the ubiquitous presence of water.
- **Group 2A: Alkaline earth metals**. These are also lustrous, silvery metals that are not found free in nature. They also react with water, though not as strongly as the alkali metals.
- **Group 7A: Halogens**. These are colorful, corrosive nonmetals. They are found in nature only in combination with other elements, including the alkali metals and the alkaline earth metals. These compounds are known as salts: the world "halogen" is from the ancient Greek word for salt.

- **Group 8A: Noble gases.** As the name suggests, these elements are almost completely unreactive.

As above, it is a good idea to print out a black-and-white periodic table and color in these categories to help you remember them.

The elements are also divided into three major categories:

- **Metals.** These are found on the left side of the periodic table, bounded by a zigzag line running from boron at the top to astatine at the bottom. They are generally shiny, malleable, ductile (able to be drawn into wire), and conduct electricity and heat. All except mercury are solid.
- **Nonmetals.** These are found to the right of the zigzag line. They are typically dull in appearance (when solid), brittle, and poor conductors of heat and electricity. They have a variety of colors, and are divided between solids, liquids (bromine), and gases.
- **Metalloids.** These nine elements are found sitting above the zigzag line, and also include germanium and antimony underneath the line.

Here again, making yourself a periodic table with the metals, nonmetals, and semimetals colored in with different colors will go a long way in helping you remember exactly where these groups lie.

Workbook Example 2.4

PROBLEM:
Determine the type of the following elements (metal, nonmetal, or semimetal):

iridium   antimony   iodine

SOLUTION:
By determining the position of the element in the periodic table, you can determine the class of the element. (It is also helpful to know the symbol for each of the elements.)

Iridium – Ir; iridium is to the left of the zigzag line in the periodic table and is a metal.

Antimony – Sb; antimony is adjacent to the zigzag line in the periodic table and is a semimetal.

Iodine – I; iodine is found to the right of the zigzag line in the periodic table and is a nonmetal.

Workbook Example 2.5

PROBLEM:
Determine which class of elements each of these descriptions matches:
1. Corrosive yellow gas;
2. Shiny, malleable, conducts electricity

SOLUTIONS:
1. Gases are all nonmetals.
2. If something is malleable, that is, easily shaped without breaking, it is solid and a metal.

Section 2.4   Observations Supporting Atomic Theory: The Conservation of Mass and the Law of Definite Proportions

By carefully examining the behavior of gases, Englishman Robert Boyle (1627–1691) was able to propose an atomic theory, as well as to postulate the existence of a number of elements. From his work came two fundamental laws of chemistry:

- **Law of mass conservation** — Mass is neither created nor destroyed in chemical reactions. This law is a cornerstone of chemical science and makes it possible to perform reliable calculations about chemical reactions.
- **Law of definite proportions** — Different samples of a pure chemical substance always contain the same proportion of elements by mass. This means that specific mass ratios of elements will form specific compounds. For example, if hydrogen and water are combined to make water, 8 grams of oxygen will always combine with 1 gram of hydrogen. Where this gets tricky is when compounds can form more than one compound with each other. Carbon monoxide and carbon dioxide, for example, both only contain carbon and oxygen, but they will have different mass ratios.

Workbook Example 2.6

PROBLEM:
5.63 grams of methane are burned in oxygen to form carbon dioxide and water. If the reaction produces 15.5 grams of carbon dioxide and 12.7 grams of water, how much oxygen is required for the reaction to be completed?

SOLUTION:
According to the law of mass conservation, mass is neither created nor destroyed in chemical reactions. Therefore, the total mass of the reactants (the starting material) must equal the total mass of the products (the substances produced after the reaction has occurred). In this case:

$$g \text{ methane} + g \text{ oxygen} = g \text{ carbon dioxide} + g \text{ water}$$
$$(5.63 \text{ g methane}) + (x \text{ g oxygen}) = (15.5 \text{ g carbon dioxide}) + (12.7 \text{ g water})$$
$$x \text{ g oxygen} = (15.5 \text{ g} + 12.7 \text{ g}) - 5.6 \text{ g}$$
$$x \text{ g oxygen} = 22.6 \text{ g}$$

Workbook Example 2.7

PROBLEM:
If 23 grams of sodium combine with 37.5 grams of chlorine to form sodium chloride, what is the mass ratio of the two elements in the compound? How much chlorine will be required to react with 8.5 grams of sodium?

SOLUTION:
The mass ratio will simply be the mass of either element divided by the mass of the other. So, the mass ratio can either be expressed as 1.63 grams of chlorine per gram of sodium ($37.5 \div 23$), or as 0.613 grams of sodium per gram of chlorine ($23 \div 37.5$). These ratios are best expressed as 1.63 g Cl / 1 g Na (1.63 grams of chlorine per 1 gram of sodium), or 0.613 g Na/g Cl.

To figure out how much chlorine will be needed, either of the ratios can be used, but care must be taken to make sure the units cancel:

$$8.5 \, g \, Na \times \frac{1.63 \, g \, Cl}{1 \, g \, Na} = 13.85 \, g \, Cl$$

This answer has more digits than it should: significant figure rules reduce the answer to 14 grams of chlorine.

Section 2.5   The Law of Multiple Proportions and Dalton's Atomic Theory

John Dalton (1766–1844) proposed a new atomic theory in 1808, which contains a number of precepts that have survived essentially unchanged to this day:

- Elements are made of tiny particles called atoms.
- Each element is characterized by the mass of its atoms: atoms of the same element have the same mass, while atoms of different elements have different masses.
- When atoms join together to form compounds, they do so in small, whole-number ratios. It is never possible to have a fraction of an atom participating in a reaction.
- Chemical reactions rearrange the combinations of atoms, but the atoms themselves are not changed.
- **The law of multiple proportions**: If two elements combine in different ways to form different substances, the mass ratios are small, whole-number multiples of each other.

Workbook Example 2.8

PROBLEM:
Nitrogen and oxygen form a large number of different compounds with each other. In one such compound, 2.8 grams of nitrogen combines with 3.2 grams of oxygen. In a different compound, 2.3 grams of nitrogen combines with 5.3 grams of oxygen. Do these compounds obey the law of multiple proportions?

SOLUTION:
Find the O:N mass ratio in each compound. First compound: O:N mass ratio = $\dfrac{3.2 \, g \, O}{2.8 \, g \, N} = 1.14$

Second compound: O:N mass ratio = $\dfrac{5.3 \, g \, O}{2.3 \, g \, N} = 2.3$

Determine if the two C:O mass ratios are small, whole-number multiples of each other.

$$\frac{O : N \, mass \, ratio \, in \, second \, compound}{O : N \, mass \, ratio \, in \, first \, compound} = \frac{2.30}{1.14} = 2.00$$

Note: the ratios can be done in any way, and give the same information. The ratio of the oxygen to nitrogen in the second compound is twice as high as in the first compound. If this was done the other way around, the answer obtained would state that the ratio in the first compound is half that of the second compound. This is the same answer. With a problem of this type, the important thing is to show that the ratios are small: 1:2 and 2:1 are both small, whole-number ratios.

Workbook Problem 2.1

Carbon and hydrogen form a large number of compounds. In one such compound, 3.43 g of carbon combines with 0.857 g of hydrogen. In another, 4.80 g of carbon combines with 0.400 g of hydrogen. Show that the law of multiple proportions is being obeyed.

Strategy: Determine the carbon-to-hydrogen mass ratios for both compounds, and compare these ratios to each other.

Step 1: Determine the carbon-to-hydrogen ratio for the first compound.

Step 2: Determine the carbon-to-hydrogen ratio for the second compound.

Step 3: Divide the ratio found for the first compound by the ratio for the second compound.

Step 4: Can your answer be converted to a small, whole-number ratio?

Section 2.6   Atomic Structure: Electrons

Dalton had no way to investigate what atoms are made of, but the invention of **cathode ray tubes** made it possible for J. J. Thomson (1856–1940) to unlock the mystery of the electron. He discovered that:

- High voltage causes electric current in the form of electrons to be emitted from electrodes made of two thin pieces of metal.
- The electrons travel toward the positive electrode, so they must have a negative charge, and they travel like a stream of infinitesimally small particles.
- Many different metals may be used to make electrodes that will emit electrons: electrons are a fundamental part of matter.
- Cathode rays can be deflected by bringing either a magnet or an electrically charged plate near the tube.

Thomson reasoned that this deflection must depend on the strength of the deflecting magnetic or electric field, the size of the negative charge on the electron, and the mass of the electron. By making a large number of very careful measurements, Thomson determined that the charge-to-mass ratio, e/m, of the electron = $1.758\ 819 \times 10^8$ C/g (C = coulombs, a measure of electrical current; we will talk about this in Chapter 17, Electrochemistry).

Robert Millikan (1868–1953) built on Thomson's work by constructing an apparatus that would enable him to coat tiny drops of oil with electrons and then determine the charge on the drop. By the use of repeated measurements, he was finally able to determine that the charges on the drops were always multiples of $1.602\ 177 \times 10^{-19}$ C (the value is negative, because the charge on an electron is negative).

With the values for the charge of an electron, and the charge-to-mass ratio, it was then possible to calculate the mass of an electron. These infinitesimally tiny particles weigh a mere $9.109\ 390 \times 10^{-28}$ g. This calculation is shown in your textbook: note that it was determined using nothing more than simple unit analysis.

Section 2.7 Atomic Structure: Protons and Neutrons

If electrons are negatively charged, there must be other particles that provide positive charge to maintain electrical neutrality of matter. Ernest Rutherford (1871–1937) successfully investigated this with the use of alpha (α) particles. These have the following characteristics:

- They are 7000 times more massive than electrons.
- They have a positive charge that is twice the magnitude of, but opposite in sign to, the charge on an electron (electrons have a charge of -1, alpha particles have a charge of +2).
- They are naturally emitted by some radioactive elements.

In his experiment, Rutherford directed a beam of alpha particles at a sheet of gold foil only a few atoms thick. Most passed through the foil, but a very few (0.005%) were deflected, and a tiny number bounced back at the source.

Rutherford explained this by proposing that atoms consist mostly of empty space, have a positive charge, and that most of their mass is concentrated in a central core, which Rutherford called the **nucleus**. Most of the alpha particles passed through the empty space of the atoms, but those that were deflected were affected both by the mass and the charge of the particles within the atoms.

Further experiments showed that the nucleus itself contains not only positively charged protons but also uncharged neutrons.

To outline the nuclear model:

- Atoms contain three subatomic particles: electrons, protons, and neutrons
- Electrons and protons have equal but opposite charges, with electrons carrying negative charges and protons carrying positive charges.
- Protons and neutrons have almost the same mass, which is much larger than the mass of an electron.
- The nucleus of the atom (protons and neutrons) provides most of the mass of the atom but very little of the volume.
- The electrons provide almost none of the mass of an atom but almost all of the volume.

Section 2.8 Atomic Numbers

Elements differ from each other according to the number of protons their nuclei contain. This is also known as their **atomic number**, or "Z". The elements in the periodic table are arranged in order of increasing atomic number, which resolved a few discrepancies that had occurred when they were arranged by mass. This was determined by Henry Mosely in 1913, based on results from Rutherford's gold foil experiment. We will actually discuss Mosely's work in more detail later on in the textbook.

For example, tellurium ($Z = 52$, mass $= 127.6$) and iodine ($Z = 53$, mass $= 126.9$): iodine behaves very much like a halogen, but if the periodic table is arranged by mass, it will not fall in the halogen category. The discovery of atomic numbers put this to rights.

In a neutral (uncharged) atom, the number of protons will be the same as the number of electrons.

Most atomic nuclei also contain neutrons. The **mass number** of an atom, or "A," is found by adding the atomic number Z to the number of neutrons, N as shown in the following equation. The mass of electrons is so small compared to the mass of the other particles that it can be ignored.

$$A = Z + N$$

As we said above, elements are defined by the amount of protons: any variation in the number of protons and you will have a different element. However, the number of neutrons in the nucleus can change without changing the identity of the element. Atoms with different numbers of neutrons are called **isotopes**. The number of neutrons has very little effect on the chemical properties of an atom.

Isotope symbols are written using the following convention, where "X" indicates the elemental symbol:

$$^A_Z X$$

From this, it is possible to calculate the number of neutrons ($A - Z$). It is not difficult to remember which of these numbers is which: the smaller will always be the atomic number, the larger will always be the mass number. Sometimes the atomic number is omitted, because it can be found in the periodic table and so does not vary.

Workbook Example 2.9

PROBLEM:
Determine the number of protons, neutrons, and electrons in the following isotopes:

$$^{71}_{31}Ga \qquad ^{191}_{77}Ir \qquad ^{88}_{39}Y$$

SOLUTION:
The subscript is the atomic number (Z), and the superscript is the mass number (A). Remember that $Z =$ the number of protons $=$ the number of electrons, and $A - Z =$ the number of neutrons.

$^{71}_{31}Ga$: $Z = 31$; number of protons = number of electrons = 31
$\qquad A - Z = 71 - 31 = 40$; number of neutrons = 40

$^{191}_{77}Ir$: $Z = 77$; number of protons = number of electrons = 77
$\qquad A - Z = 191 - 77 = 114$; number of neutrons = 114

$^{88}_{39}Y$: $Z = 39$; number of protons = number of electrons = 39
$\qquad A - Z = 88 - 39 = 49$; number of neutrons = 49

Workbook Problem 2.2

Oxygen is the most abundant element by mass in the biosphere. All classes of biomolecules, proteins, sugars, fats, and nucleic acids, contain oxygen, and oxygen makes up almost 90% of the mass of the world's oceans. Oxygen is made up of three stable isotopes, with mass numbers 16, 17, and 18. Write the standard symbols for these isotopes.

Strategy: Use the periodic table to determine the atomic number and the number of neutrons in each isotope. Give the standard symbol for each.

Step 1: Find the atomic number for oxygen.

Step 2: Use the mass numbers given in the periodic table to determine the number of neutrons in each isotope.

Step 3: Write the standard symbol with the mass number as a superscript and the atomic number as a subscript, both to the left of the symbol.

Workbook Problem 2.3

The metalloid X has been used as a cosmetic since at least 3000 B.C. When bound to sulfur atoms in a 2:3 ratio, it is known as "kohl." An atom of the metal alloid contains 51 protons, and its most stable isotope contains 70 neutrons. Identify the metal, and write the symbol in standard format.

Strategy: Use the periodic table to determine the atomic number and identify the metal. Give the standard symbol for the metal.

Step 1: Determine the atomic number from the information given.

Step 2: Knowing the atomic number, identify the metal by using the periodic table.

Step 3: Write the standard symbol with the mass number as a superscript and the atomic number as a subscript, both to the left of the symbol.

Section 2.9   Atomic Weights and the Mole

Because the masses of individual atoms and their components are so tiny, grams are an inconvenient unit. As a result, **atomic mass units** are used instead. These are abbreviated *amu* in chemistry and are known as daltons (Da) in biochemistry. Atomic mass units are defined as exactly half the mass of an atom of $^{12}_{6}C$; therefore, 1 amu = $1.660\,54 \times 10^{-24}$ g.

Most elements occur as a mixture of isotopes, so the atomic mass reported on the periodic table is a weighted average of the naturally occurring isotopes.

The atomic mass of an element = Σ(mass of each isotope × the abundance of the isotope). The Greek letter Σ (sigma) means "the sum of."

Workbook Example 2.10

PROBLEM:
Calculate the mass in grams of a single atom of cadmium.

SOLUTION:
From the mass given in the periodic table, we know that 1 atom of Cd has a mass of 112.411 amu. We also know that 1 amu = 1.6605×10^{-24} g. The final unit we want to end up with is g/atom of Cd. Therefore, we should set up our problem in the following manner:

$$\frac{112.4\,amu}{1\,atom\,Cd} \times \frac{1.6605 \times 10^{-24}\,g}{1\,amu} = 1.867 \times 10^{-22}\,g$$

Workbook Example 2.11

PROBLEM:
Boron has two stable isotopes: B-10 and B-11. Boron-10 is 19.9% of a sample of boron and has an atomic mass of 10.0129 amu. The remaining 80.1% is boron-11, with an atomic mass of 11.0093 amu. Calculate the average atomic mass of boron.

SOLUTION:
To work out the average atomic mass of boron, multiply the individual masses by their abundance, and then add them together:

$$\Sigma(0.199 \times 10.0129\,amu) + (0.801 \times 11.0093\,amu) = 10.81\,amu$$

This agrees with the value on the periodic table, which is always a good sign.

Workbook Problem 2.4

A 14-carat gold ring is 14/24 gold. If a 14-carat gold ring weighs 3.65 g, how many gold atoms does it contain?

Strategy: Figure out how many grams of gold are present, then determine the conversion factors needed to convert from grams of sample to number of atoms.

Step 1:   Set up a mathematical equation such that all of the units except number of atoms cancel.

<u>Workbook Problem 2.5</u>

Lead has four naturally occurring isotopes. ^{208}Pb has an atomic mass of 207.977 amu and an abundance 52.4% in an average lead sample. ^{207}Pb has an atomic mass of 206.976 and an abundance of 22.1%. ^{206}Pb has an atomic mass of 205.974, and an abundance of 24.1%. The remaining lead will be ^{204}Pb with an atomic mass of 203.973 amu. Calculate the average atomic mass of Pb.

Strategy: Remember, the atomic mass of an element = Σ(mass of each isotope × the abundance of the isotope), and all abundances must add up to 100%.

Step 1: Use the information given to determine the abundance of ^{204}Pb.

Step 2: Use the information given to determine the average atomic mass.

Section 2.10 Mixtures and Chemical Compounds; Molecules and Covalent Bonds

It is very unusual to find pure elements: there are many more types of substances than there are elements on the periodic table. These substances can all be classified as either **pure substances** or **mixtures**.

Pure substances are either **elements** or **chemical compounds**. In general, elements cannot be changed into another substance, but we will see that in nuclear reactions that that is not always the case (Chapter 22). Chemical compounds are made up of atoms that are chemically bonded together. They are formed by chemical reactions and cannot be separated except by chemical means. They also have properties that are very different from their constituent elements.

Mixtures, in contrast, are formed when two or more substances are mixed together without any chemical interaction. The individual substances maintain their properties and can be separated relatively easily.

Heterogeneous mixtures are not uniformly mixed: dirt in water will not be evenly distributed, and will tend to settle out. If it is mixed in, the liquid will be cloudy.

Homogeneous mixtures are uniform throughout, and they can be solids (metal alloys), liquids (solutions), or gases (air). As an example, sugar dissolved in water will have the same amount of sugar at the top of the water as at the bottom. The liquid will also be clear, because the sugar is perfectly dispersed in the liquid.

Workbook Example 2.12

PROBLEM:
Identify the following household items as elements, compounds, homogeneous mixtures, or heterogeneous mixtures:

        salad   tea   metal utensils   water   soda can   air

SOLUTION:
Think about the composition of these things that you see in everyday life. Salad is a heterogeneous mixture, because it has areas of variable composition: two separate spoonfuls will have different components. Tea is a clear liquid, and does not have constituents that settle: it is a homogeneous mixture. Metal utensils can be made of any number of things from stainless steel to sterling silver, but all are mixtures of metals in unvarying proportions, so these are homogeneous mixtures as well. Water contains elements bound together to form something with very different properties: it cannot easily be separated back into its constituent parts, so it is a compound. A soda can, barring the paint on the outside, is made of aluminum, which is an element. Air is a mixture of elements not bonded to one another. Air contains nitrogen, oxygen, hydrogen, and many other components: it is a homogeneous mixture.

Let's get back to chemical compounds. To form carbon dioxide, a compound, a solid, and a gas react to form a gas. Carbon and oxygen gas combine in a *specific ratio*—one carbon to two oxygens—every time.

To form carbon dioxide, the electrons in the carbon and oxygen atoms form connections to each other, called **chemical bonds**. One type of chemical bond is called a **covalent bond**; this happens when two atoms share two (or more) electrons. When two or more atoms are joined together through covalent bonding, like carbon dioxide, we call that a **molecule**.

Covalent bonds are represented in a number of ways (see Figure 2.9 in the textbook), including *ball-and-stick models*, which show the bonds as "sticks" connecting the atoms, and *space-filling models* that show the atoms stuck together without explicitly showing the bonds.
Covalent bonds are commonly shown using **structural formulas**: these use lines between atoms to show how the atoms are connected. This is particularly important in organic chemistry, which is the study of carbon compounds. In these compounds, structure determines behavior, even if the atoms are the same.

Workbook Example 2.13

PROBLEM:
Ethanol and dimethyl ether have the same chemical formula: C_2H_6O. Because the atoms in the molecules are connected in different ways, one is a gas used as a propellant in aerosol cans, and the other is a liquid used in mixed drinks. In dimethyl ether, the carbon atoms are linked with the oxygen in the middle, and in ethanol, the carbons are attached to each other with the oxygen attached to one end. Each carbon has four bonds, each oxygen has two bonds, and each hydrogen has one bond. Draw the structures.

SOLUTION:

         dimethyl ether   ethanol

Workbook Example 2.14

PROBLEM:
1-butanol has the structural formula shown below. Write the chemical formula.

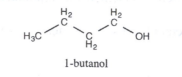

1-butanol

SOLUTION:
Chemical formulas are written from structural formulas by putting the constituents in alphabetical order. This structure has four carbon atoms, ten hydrogen atoms, and one oxygen atom. Its chemical structure is $C_4H_{10}O$.

Section 2.11 Ions and Ionic Bonds

When atoms approach one another, the electrons interact to form **chemical bonds**. Two main types are possible: **covalent bonds**, described above, occur between nonmetals, and **ionic bonds** occur between metals and nonmetals.

Ionic bonds are not formed by shared electrons, as covalent bonds are. Ionic bonds are the result of electrons being *transferred* from one atom to another, resulting in atoms with opposite charges that serve to hold the atoms together.

Metals give up electrons easily to form **cations**—positively charged particles. It is easy to remember that metals give up electrons if you recall that metals conduct electricity because they are not holding on to their electrons very tightly. Nonmetals readily gain electrons to form **anions**—negatively charged particles.

When anions and cations interact, they do so generally on the basis of opposite charges. They do not form specific bonds, but they do all they can to neutralize their opposing charges. A cation will surround itself with anions, and *vice versa*. As a result, ionic bonds result in crystals (like those of common table salt) called **ionic solids**, rather than forming molecules. When a formula is written for an ionic solid, it is written so as to show the neutralized charges, but it does not indicate that a specific anion is bonded exclusively to a specific cation.

Polyatomic ions also occur: these are charged (usually negatively, because they consist of nonmetals), covalently bonded groups of atoms that carry a charge. There are covalent bonds within polyatomic ions; as units, they can form ionic bonds.

Workbook Example 2.15

PROBLEM:
Identify the following as covalent molecules or ionic compounds:
a) CH_4,   b) MgO,   c) KBr,   d) NO_2

SOLUTION:
The easiest way to differentiate between molecules and ionic compounds is to determine whether the formula contains both metals and nonmetals or only nonmetals. If only nonmetals are present, a covalent molecule has formed. If metals and nonmetals are both present, ionic bonding is occurring. CH_4 and NO_2 contain only nonmetals, so they are molecules. MgO and KBr are both combinations of metals and nonmetals. These are ionic compounds.

Section 2.12 Naming Chemical Compounds

Special rules have been developed to name chemical compounds. The simplest compounds are **binary ionic compounds**: as the name suggests, these have only two elements and are ionic. They are named cation first, then anion with the suffix *–ide*. For all practical purposes, these compounds can be named as *metal nonmetalide*. No information about the ratio of one element to the other is given in the name, but must be given in the formula.

The formula of a binary ionic compound must be written to give the smallest ratio of anion to cation that provides for electrical neutrality. What this means in plain English is that if you are balancing charges, make sure you don't include more than you need to: for example, Mg_3N_2 is correct, not Mg_6N_4. If all your subscripts have a common factor, you'll need to fix it.

To name chemical compounds, you will need to know the charges on the ions. For main group metals and nonmetals, this information can be taken straight from the periodic table.

- Main group metals carry a positive charge equal to their group number. Alkali metals, for example, are found in group 1A, so they all carry charges of +1.
- Nonmetals carry negative charges equal to their group number minus 8. So, oxygen, which falls in group 6A, will carry a charge of $6 - 8 = -2$.

When transition metals are involved as cations, it is not possible to take the charges from the periodic table. Because of this, the charge must be given when the name is written: for example, iron(II) chloride indicates that the iron atoms in the compound carry a charge of +2. When the formula is written, the charges on the iron atom must be balanced by negative charges. Because chlorine carries a charge of –1 when it is an ion (group $7A - 8 = -1$), there will need to be two chlorines. The formula will be $FeCl_2$.

Workbook Example 2.16

PROBLEM:
Name the following compounds: MgO $CuCl_3$ RbCl SnO_2

SOLUTION:

MgO: magnesium oxide | No Roman numeral is needed: magnesium is a group 2A metal and forms only Mg^{2+}.

$CuCl_3$:   copper (III) chloride | The charge on Cl is –1, and there are three of them. To have electrical neutrality, the charge on copper, which is a transition metal, must be +3. The Roman numeral (III) is used to indicate this charge.

RbCl: rubidium chloride | Rubidium is a group 1A metal and forms only Rb^+.

SnO_2:   tin(IV) oxide | There are 2 O^{2-} ions with a total charge of –4. To have electrical neutrality, the charge on tin must be +4. Tin can form more than one kind of ion; therefore the charge must be indicated by a Roman numeral.

Workbook Example 2.17

PROBLEM:
Write formulas for the following compounds:
a) cesium oxide   b) magnesium nitride   c) vanadium(V) bromide

SOLUTION:

Cs_2O Cesium is a group 1A metal and forms only Cs^+. Oxygen forms O^{2-} oxide ions, so two cesium ions are needed for each oxygen.

Mg_3N_2 Magnesium is a main group, group 2A metal that has a +2 charge, while nitride ions have a –3 charge (group $5 - 8 = -3$). To have electrical neutrality, you need two Mg^{2+} and two N^{3-}. $[(3 \times +2) + (2 \times -3)] = 0$. There will need to be six positive charges and six negative charges to achieve neutrality.

VBr_5 Vanadium(V) has a +5 charge, while bromine has a –1 charge. To have electrical neutrality, you need 1 V^{5+} and 5 Br^-.

Workbook Problem 2.6

Name or write formulas for the following:

MgF_2 sodium chloride   chromium(III) oxide   CaO TiI_4
bismuth(II) nitride   aluminum sulfide CoF_3 CuSe

Strategy: Follow the rules outlined here and in your textbook.

Step 1:   When naming compounds, determine if the metal is a main group metal or a transition metal. If the metal is a transition metal, you must indicate the charge on the metal when naming the compound. The charge on the metal is determined from the number and charge on the anion. (Remember, the charges must all add up to zero.)

Step 2:   When writing formulas, use the periodic table or the name of the compound to determine the charge on the metal. Use the periodic table to determine the charge on the anion. (Remember, the charges must all add up to zero.)

Step 3:   Make sure the formula contains the smallest whole-number ratio of cation to anion.

Binary molecular compounds contain only two nonmetals. They are named by first naming the element closest to the metals on the periodic table, followed by the element farthest from the metals with an *–ide*.

Because these are electron-sharing interactions, they can occur in many different ways (remember the law of multiple proportions?). Numerical prefixes must be used to identify the number of each element in the molecule. *Mono-* is not generally used for the first element. These prefixes should already be familiar to you, and you should memorize them.

Careful! If a binary molecular compound contains hydrogen, it will be named using these conventions if it is a gas, but it will be named as an acid if it is dissolved in water (aqueous). This will be indicated in parentheses after the formula: HCl (*g*) is hydrogen chloride, but HCl (*aq*) is hydrochloric acid!

Workbook Example 2.18

PROBLEM:
Name the following compounds: SF_6 N_2O_5 P_4S_{10} NI_3

SOLUTION:
  SF_6:   sulfur hexafluoride
  N_2O_5:   dinitrogen pentoxide
  P_4S_{10}:   tetraphosphorus decasulfide
  NI_3:   nitrogen triiodide

Workbook Example 2.19

PROBLEM:
Write formulas for the following compounds:
        hydrogen chloride (gas)   xenon hexafluoride   diphosphorus trioxide

SOLUTION:
hydrogen chloride (gas): $HCl(g)$ gaseous, so it is named as a binary molecule

xenon hexafluoride: XeF_6

diphosphorus trioxide: P_2O_3

Workbook Problem 2.7

Name or write formulas for the following:
a) PF_6 b) dinitrogen tetroxide   c) selenium dioxide   d) NBr_3 e) $HF(g)$

Strategy: Follow the rules given here and in your textbook.

Step 1:   When naming the molecules, remember that the element that is more anion-like uses the -*ide* suffix.

Step 2:   Add numerical prefixes.

Ionic compounds containing polyatomic ions are named by following the rules for naming binary ionic compounds: cation, then anion. A number of polyatomic ions are listed in a table in your book: these should be memorized. Flashcards are a great way to do this!

Oxoanions have special naming conventions that may make their names easier to learn: first learn the name and formula of the ion whose name ends with *-ate* (including the charge on the anion). Other similar ions will be named using the following conventions:

- Add one O; add prefix *-per*.
- Remove one O; change ending to *-ite*.
- Remove two Os; add prefix *-hypo* and change *-ate* to *-ite*.

Workbook Example 2.20

PROBLEM:
Name the following compounds: a) $NaClO_4$ b) MgS_2O_3 c) $Fe_2(CO_3)_3$ d) $Mn(NO_3)_2$

SOLUTION:

$NaClO_4$:　　　sodium perchlorate

MgS_2O_3:　　　magnesium thiosulfate

$Fe_2(CO_3)_3$:　　iron(III) carbonate. The carbonate anion has a –2 charge, and there are three of them, for a total charge of -6. Therefore, there must be a +6 charge spread over the two iron cations, so the charge on the iron cation must be +3. Each iron cation has a charge of +3 (+6/2). The charge on the iron is indicated with Roman numerals because iron is a transition metal.

$Mn(NO_3)_2$:　　manganese nitrate.

Workbook Example 2.21

PROBLEM:
Write formulas for the following compounds:
　　　iron(III) sulfate　　　　　　cesium nitrite　　　　　　sodium acetate

SOLUTION:

| iron(III) sulfate: | $Fe_2(SO_4)_3$ | The sulfate anion has a –2 charge. Iron(III) by definition has a charge of +3, so the charges must be balanced. Three sulfate ions are needed to give a charge of –6, which will balance two Fe^{3+} ions, with a charge of +6. |

| cesium nitrite: | $CsNO_2$ | Cesium is a group 1A metal, so carries a charge of +1, as does the nitrite ion. |

| sodium acetate: | $NaCH_3CO_2$ | The acetate anion is one of the most complex of the polyatomic ions. It has a –1 charge, and is thus perfectly balanced by the +1 charge of the group 1A sodium ion. |

Workbook Problem 2.8

Name or write formulas for the following:

$Fe(NO_3)_3$ potassium chromate   magnesium phosphate $Al_2(SO_4)_3$ ammonium nitrate

Strategy: Follow the rules listed above and in your textbook.

Step 1:   Identify the cation and the anion. Write the cation first.

Step 2:   Determine the charge on the cation, and write the formula to make sure all charges are neutralized.

Self–Test

This section is intended to test your knowledge of the material covered in this chapter. Think through these problems, and make certain you understand what is going on. Ask yourself if your answer makes sense. Many of these questions are linked to the chapter learning goals. Therefore, successful completion of these problems indicates you have mastered the learning goals for this chapter. You will receive the greatest benefit from this section if you use it as a mock exam. You will then discover which topics you have mastered and which topics you need to study in more detail.

True–False

1. Isotopes are atoms of the same element that carry a different charge.

2. The mass of the products of a chemical reaction will always be the same as the mass of the reactants.

3. The smallest unit of matter that retains its identity as an element is an atom.

4. Two nonmetals will only ever form one type of compound.

5. Electrons and neutrons are found in the nucleus of an atom, protons in a diffuse cloud outside the nucleus.

6. The atomic number – the number of protons in an element – is what gives an element its identity.

7. All atoms of the same element have the same infinitesimally tiny mass.

8. Ionic compounds are named using prefixes to denote the number of each element.

9. Main-group metals such as sodium can carry a variety of different charges.

Multiple Choice

10. The isotope $^{70}_{31}Ga$ contains:
    a. 71 protons, 31 neutrons, and 71 electrons
    b. 31 protons, 71 neutrons, and 31 electrons
    c. 31 protons, 40 neutrons, and 31 electrons
    d. 31 protons, 40 neutrons, and 40 electrons

11. The compound copper(I) selenide is
    a. covalent, with formula CuS
    b. covalent, with formula CuSe
    c. ionic, with formula CuSe
    d. ionic, with formula Cu_2Se

12. The proper configuration of an atom is
    a. neutrons and protons in the nucleus, electrons outside the nucleus
    b. neutrons and electrons in the nucleus, protons outside the nucleus
    c. protons and electrons in the nucleus, neutrons outside the nucleus
    d. none of the above

13. The compound Na_2SO_4 is
    a. covalent, called sodiosulfuric acid
    b. ionic, called sodium sulfate
    c. covalent, called sodium sulfide
    d. ionic, called sodium sulfoxide

14. Covalent bonding involves:
    a. interactions between electrons
    b. formation of molecules
    c. nonmetals bonding to nonmetals
    d. all of the above

15. Ionic bonding requires:
    a. similarly sized atoms
    b. metal and nonmetal atoms
    c. oppositely charged particles
    d. isotopes

16. In a homogenous mixture:
    a. the properties of the mixture are completely different from the properties of the components
    b. the composition is uniform throughout, and the components retain their separate identities
    c. the composition is variable, with different proportions in different locations
    d. only liquids can be involved

17. Which of the following correctly identifies the subatomic particles?
    a. proton, positive charge, mass of 1 amu
    b. neutron, negative charge, mass of 1 amu
    c. neutron, no charge, negligible mass
    d. electron, negative charge, mass of 1 amu

18. The correct name of N_2O_5 is
    a. nitrogen oxide
    b. oxygen nitrate
    c. dinitrogen pentoxide
    d. nitropentoxygen

19. The correct name of $NaC_2H_3O_2$ is
    a. sodiumcarbohydrooxide
    b. sodium acetate
    c. monosodium monoacetate
    d. none of the above

Fill-in-the-Blank

20. Elements are made up of _____, which contain three fundamental particles:

    _____.

21. The identity of an element is defined by the number of _____ it contains, also called the _____; the other particles can vary.

22. When the number of _____ changes, the mass of the element changes. The result is called a(n) _____. When the number of _____ changes, the charge changes: the result is known as a(n) _____.

23. Very few pure elements exist in the world around us; mostly we are surrounded by _____, which are also considered pure substances, and _____.

24. When nonmetals combine, the result is a(n)_____ bond. In this type of bond, electrons are _____.

25. When a(n) _____ and a(n)_____ combine, the result is a(n) _____ bond. In this type of bond atoms are held together by opposite _____.

26. When a covalently bonded group of atoms also carries a charge, it is known as a(n) _____. Some of the most common of these are the _____ ion, NO_3^-, the _____ ion, NH_4^+, and the _____ ion, SO_4^{2-}.

27. In molecules, the number of each element is denoted using a series of prefixes. One of an element is denoted mono, two is _____, three is _____, five is _____, and so on.

28. When chemical reactions occur, the mass of the _____ will always be the same as the mass of the products. This is known as the _____.

Matching

29. Ionic compound

30. Molecules

31. Atomic number

32. Polyatomic ions

33. Binary compounds

34. Heterogeneous mixture

a. compounds containing two elements

b. number of protons in an element

c. negatively charged subatomic particle

d. discoverer of atomic nucleus

e. bond formed by sharing electrons

f. compound formed by opposite charges

35. Structural formula

36. Electrons

37. Proton

38. Anion

39. Atoms

40. Neutron

41. Ernest Rutherford

42. Isotopes

43. J. J. Thomson

44. Covalent bond

45. Cation

46. Atomic mass unit

g. 1/12 the mass of a C-12 atom

h. positively charged ion

i. covalently bonded units

j. provide information about molecular connectivity

k. uncharged subatomic particle

l. positively charged subatomic particle

m. covalently bonded units that carry a charge

n. discoverer of the electron

o. negatively charged ion

p. same element, varying masses

q. proportions of components vary within mixture.

r. smallest particle of any element

Problems

47. Hydrogen gas reacts explosively with oxygen to form water vapor. If an unknown quantity of hydrogen gas is reacted with 3.2 L of oxygen (density 1.43 g/L), and 5.15 g of water are formed, what volume of hydrogen gas (density 0.0893 g/L) was in the initial mixture?

48. Phosphorus and oxygen form two compounds, one of which is a white crystalline solid, the other of which is a waxy substance that smells like garlic. In the first compound, 4.3 g phosphorus combined with 11.4 g oxygen. In the second, 3.34 g phosphorus combined with 5.33 g oxygen. Demonstrate that these compounds obey the law of multiple proportions.

49. How many protons, neutrons, and electrons are present in the following elements?
    a. $^{97}_{42}Mo$ b. $^{235}_{92}U^{6+}$ c. $^{235}_{93}Np$ d. $^{202}_{80}Hg^{4+}$

50. Neon has three stable isotopes and an average atomic mass of 20.18 amu. If Neon-20 has an atomic mass of 19.992 and an abundance of 90.48%, and Neon-21 has an atomic mass of 20.993 amu and an abundance of 0.27%, what is the abundance and atomic mass of Neon-22?

51. Magnesium, Mg, has three naturally occurring isotopes: ^{24}Mg has an abundance of 78.99% and an atomic mass of 23.985 amu, ^{25}Mg with an abundance of 10.00% and an atomic mass of 24.986 amu, and ^{26}Mg with an abundance of 11.01% and an atomic mass of 25.983 amu. Calculate the average atomic mass of magnesium.

52. Naphthalene is the compound that gives mothballs their distinctive smell. Given its structural formula, below, what is its chemical formula?

naphthalene

53. Write the formula and name for the compound that forms between each of the following pairs of elements:
    a. Li, Cl   b. Na, S   c. Ca, O   d. Mg, N

54. Name the following compounds, and identify them as ionic or covalent:
    a. CO   b. $SrCl_2$ c. CCl_4 d. PtO_2

55. What is the mass (in grams) of 9500 atoms of gold?

56. Formulas for chemical compounds can be worked out using the percent masses of each component. For the organic compound dichloromethane (CCl_2H_2), what percentage of the total mass is provided by each component (use the atomic masses in amu)?

57. Write a formula for the compound that would form between calcium and the following ions:
    a. fluoride   b. phosphate   c. nitrate   d. carbonate

58. Name the following compounds:
    a. $SnCl_4$ b. $H_3PO_4(aq)$ c. $Mg(C_2H_3O_2)_2$ d. HCl
    e. Rb_2S f. CO_2 g. $CrPO_4$ h. F_2O
    i. P_2O_5 j. $NiCl_2$ k. $Fe(OH)_3$ l. HCl(aq)
    m. PbI_2 n. $HNO_3(aq)$ o. $BaBr_2$ p. $Al(NO_3)_3$

59. Write formulas for the following compounds:
    a. calcium nitrite   b. sodium chloride   c. dichlorine monoxide
    d. dihydrogen monoxide   e. nickel(II) hydride   f. boron tribromide
    g. copper(II) chloride   h. xenon tetrafluoride   i. nitrogen dioxide
    j. barium nitride   k. phosphoric acid   l. hydrofluoric acid
    m. lithium permanganate   n. carbonic acid   o. phosphorous trichloride
    p. carbon tetrabromide

Challenge Problem
60. The smallest object that can be resolved by the human eye is about 0.2 mm across. A sphere of osmium metal (density = $22.61 g/cm^3$) with a diameter of 0.2 mm would contain how many atoms? How many atoms would a sphere of aluminum (density $2.7 g/cm^3$) contain? If these atoms were cubes with edge length 0.1 mm, what would the volume of these atoms be?

CHAPTER THREE

Mass Relationships in Chemical Reactions

Learning Objectives

As a result of reading and studying this chapter, you should be able to

Sections 3.1 - 3.2 **Balancing Chemical Equations**
1. Visualize bonds broken and formed in a chemical reaction and relate numbers of molecules/atoms to the balanced reaction.
2. Balance a chemical reaction given the formulas of reactants and products.

Section 3.3 **Chemical Arithmetic: Stoichiometry**
3. Calculate formula weight, molecular weight, and molar mass given a chemical formula or structure.
4. Interconvert between mass, moles, and molecules/atoms of a substance.
5. Relate the amount (moles or mass) of reactants and products in a balanced equation using stoichiometry.

Section 3.4 **Yields of Chemical Reactions**
6. Calculate percent yield given amounts of reactants and products.

Section 3.5 **Reactions with Limiting Amounts of Reactants**
7. Visualize the relative amount of atoms or molecules in the reactants and products of a balanced reaction.
8. Determine which reactant is limiting and calculate the theoretical yield of the product and the amount of excess reagent.
9. Calculate percent yield when one reactant is limiting.

Sections 3.6 - 3.7 **Percent Composition, Empirical Formulas, and Combustion Analysis**
10. Calculate the percent composition given a chemical formula or structure.
11. Determine the empirical and molecular formula given the mass percent composition and molecular weight of a compound.
12. Determine the empirical and molecular formula given combustion analysis data and molecular weight.

Section 3.8 **Determining Molecular Weight: Mass Spectrometry**
13. Determine the molecular weight of a substance given a mass spectrum. Determine empirical and molecular formula using both mass spectral and combustion analysis data.
14. Identify a compound using a molecular weight measured with a high-mass accuracy mass spectrometer.

Chapter Summary

The central concern of chemistry is the change of one substance into another—chemical reactions are at the heart of the science. In this chapter, you begin your study of chemical reactions by learning how to balance chemical equations. Using balanced chemical equations, along with the concepts of the mole and stoichiometry, you will learn how to calculate the amounts of reactants needed for a reaction and the theoretical amount of products produced in a reaction. These calculations are carried out using molar masses for gram $\leftrightarrow$ mole conversions. You will also learn how to determine an empirical formula from the percent composition of a compound and vice versa. Finally, you will see how combustion analysis, a type of elemental analysis, along with mass spectrometry can be used to determine both the empirical and molecular formula of a compound.

The Chapter in Detail

Section 3.1   Representing Chemistry on Different Levels

This section sets the stage for the discussion that will occur in this chapter: how do chemists interpret a chemical equation? We can use a chemical equation to describe what's going on at the macroscopic level by talking about moles (which we'll talk about later) of compounds, or we can analyze a reaction at the microscopic level to tell a story about atoms and molecules.

Section 3.2   Balancing Chemical Equations

The law of mass conservation states that the total mass of products and reactants in a chemical reaction must be equal: matter is neither created nor destroyed. Dalton's atomic theory states that in reactions, the atoms are simply rearranged, not altered. According to these rules, the number and kind of atoms must be the same on both sides of a chemical reaction.

Equations must also be balanced by **formula unit**; that is, the unit that corresponds to any given formula. Therefore, as well as individual atoms, covalent molecules and ionic units must also be balanced.

To balance an equation:

- From the information you are given, write the equation with the correct formula units on each side of the reaction. This equation will be unbalanced.
- Find suitable **coefficients** for the formula units: the coefficients are placed in front of the equations. You cannot balance equations by changing the subscripts within the molecules, or you will be changing the compounds.
- Reduce coefficients to their smallest whole-number value, dividing by a common factor if necessary
- Check that the same number of each kind of atom appears on each side of the reaction. If this is the case, the equation is balanced.

Workbook Example 3.1

PROBLEM:
Methane (CH_4) burns in oxygen to produce carbon dioxide and water. Write and balance the equation.

SOLUTION:
Using the rules we established above:

Step 1:   Write the unbalanced equation.

$$CH_4 + O_2 \rightarrow CO_2 + H_2O$$

Step 2:   Use coefficients to balance the equation. Think about one element at a time. (HINT: It usually helps to save oxygen for last. The molecular nature of oxygen, which requires that it be added two at a time to reactions, can make balancing equations a little tricky.)

Notice that there are four hydrogen atoms on the reactant side, but only two hydrogens on the product side. Begin by placing a 2 in front of H_2O.

$$CH_4 + O_2 \rightarrow CO_2 + 2\,H_2O$$

You now have four hydrogens on each side and one carbon on each side, leaving oxygen still unbalanced: there are two oxygens on the reactant side and four on the product side.

$$CH_4 + 2\,O_2 \rightarrow CO_2 + 2\,H_2O$$

Step 3:   The coefficients are already reduced to their smallest whole-number ratio.

Step 4:   Check your answer.

Reactant side	Product side
1 C	1 C
4 O	4 O
4 H	4 H

Workbook Example 3.2

PROBLEM:
Write a balanced equation for the combustion of 1-pentanol ($C_5H_{11}OH$).

SOLUTION:
The term *combustion* simply means reaction with oxygen. When hydrocarbons (compounds containing primarily C and H) such as methane and 1-pentanol undergo combustion, carbon dioxide and water are produced.

Step 1:   Write the unbalanced equation.

$$C_5H_{11}OH + O_2 \rightarrow CO_2 + H_2O$$

Step 2:   Use coefficients to balance the equation. Begin with carbon. There are five carbon atoms on the reactant side, but only one carbon atom on the product side. Begin by placing a 5 in front of CO_2.

$$C_5H_{11}OH + O_2 \rightarrow 5\,CO_2 + H_2O$$

There are 12 hydrogens on the reactant side—don't be confused by the fact that they are broken up—but only two hydrogens on the product side. Place a 6 in front of H_2O.

$$C_5H_{11}OH + O_2 \rightarrow 5\,CO_2 + 6\,H_2O$$

Now balance the oxygen. There are 3 oxygens on the reactant side—don't forget the methanol oxygen—and 16 oxygens on the product side. To get an even number of oxygens will require doubling the number of pentanol molecules and all of the coefficients that were taken from it:

$$2\,C_5H_{11}OH + O_2 \rightarrow 10\,CO_2 + 12\,H_2O$$

There are now 4 oxygens on the reactant side and a whopping 32 oxygens on the product side. To balance this number of oxygens on the product side, subtract out the 2 oxygens that are associated with the methanol molecules, leaving 30 oxygens that need balancing; put a 15 in front of the molecular oxygen:

$$2\,C_5H_{11}OH + 15\,O_2 \rightarrow 10\,CO_2 + 12\,H_2O$$

Step 3:   Reduce the coefficients to their smallest whole-number ratio.

The ratio is 2:15:10:12, which is the smallest whole-number ratio.

Step 4:   Check your answer.

Reactant side	Product side
10 C	10 C
24 H	24 H
32 O	32 O

Workbook Problem 3.1

Write a balanced chemical equation for the combustion of hexane (C_6H_{14}).

Strategy: Remember, the term *combustion* is used to indicate reaction with oxygen. When hydrocarbons (compounds containing primarily C and H) undergo a combustion reaction, carbon dioxide and water are the products.

Step 1:   Write the unbalanced chemical equation.

Step 2:   Use coefficients to balance the equation. (Remember, it helps to save oxygen for last.)

Step 3:   Reduce the coefficients to their smallest whole-number ratio.

Step 4:   Check your answer.

Section 3.3   Chemical Arithmetic: Stoichiometry

Individual molecules can't be weighed, so in a balanced reaction, a conversion must be made between the ratios of molecules and the masses of reactants.

The conversion factor for this is an SI unit called the **mole**. A mole of any substance is equal to 6.022×10^{23} formula units, also known as **Avogadro's number**. One mole of the substance is equal to its molecular mass (from the periodic table) in grams.

Workbook Example 3.3

PROBLEM:
What is the mass of one mole of hydrogen?

SOLUTION:
The definition of the mass of one mole of a substance is the amount of that substance contained in the molecular mass (from the periodic table) in grams. But is that the actual mass?

Hydrogen exists as a diatomic molecule: H_2. The molecular mass of hydrogen from the periodic table is 1.008 amu, so the mass of a molecule of hydrogen would be 1.008 × 2 or 2.016 amu. By definition, this means that one mole of hydrogen will have a mass of 2.016 grams.

Just this once, let's do the conversion to see if it comes out the same:

$$\frac{2.016\,amu}{1\,molecule} \times \frac{6.022 \times 10^{23}\,molecules}{1\,mole} \times \frac{1.660539 \times 10^{-24}\,g}{1\,amu} = \frac{2.016\,g}{1\,mole}$$

Good news: the lengthy conversion from amu to moles, with a conversion factor in it that no one wants to memorize, is completely unnecessary. When you need to know the mass of one mole of a substance, you can pull it straight off the periodic table, no conversions necessary.

Moles make it possible to use macroscopic quantities of substances for microscopic purposes: one mole of hydrogen molecules will have exactly the same number of particles as a mole of water, or oxygen, or table salt. If you need to react substances together, you will have the same number of particles if you have the same number of moles, even if the masses are wildly different.

Workbook Example 3.4

PROBLEM:
How many grams of each substance will be required/produced if one mole of hydrogen reacts with one mole of chlorine to produce hydrogen chloride gas by the following unbalanced equation?

$$H_2(g) + Cl_2(g) \rightarrow HCl(g)$$

SOLUTION:
First, the equation must be balanced. This is simply done by putting a 2 in front of the hydrogen chloride gas produced.

$$H_2(g) + Cl_2(g) \rightarrow 2\ HCl(g)$$

One mole of any substance has a mass equal to its molecular mass in grams, so the one mole of hydrogen will have a mass of 2.016 g. One mole of chlorine gas will be equal to 2×35.45 g/mol, or 70.9 g. Despite this huge mass difference, the number of particles of each substance—hydrogen and chlorine—is the same. The chlorine molecules are just *much* heavier.

To determine how much hydrogen chloride gas is produced, there is a coefficient to be considered. The coefficient tells you that for every mole of hydrogen gas or chlorine gas, two moles of hydrogen chloride gas are produced. This makes sense, because when you break apart one molecule of hydrogen and one molecule of chlorine, you have four atoms. Redistribute them, and you have made two hydrogen chloride molecules.

So, when we added together one mole of hydrogen and one mole of chlorine, we made two moles of hydrogen chloride. This has a molecular mass of $1.008 + 35.45$, or 36.46 g/mol. If there are two moles, there will be 72.9 grams of hydrogen chloride gas.

Take note that once again, the law of mass conservation is at work. On the reactants side, 2.016 g + 70.9 g = 72.9 grams of reactants. On the product side, 72.9 grams of hydrogen chloride gas is produced.

Workbook Problem 3.2

When hydrogen peroxide (H_2O_2) solutions are left exposed to light, the hydrogen peroxide breaks down to form oxygen and water. If 2 liters of a 3% hydrogen peroxide solution contain 60 grams of hydrogen peroxide, and the solution is left in the light until the reaction is complete, how many moles of oxygen are produced? How many moles of water?

Strategy: To determine mole ratios, you first need a balanced equation. After obtaining the mole ratio from the balanced equation, you can use this knowledge to obtain the conversion factors needed to solve the problem.

Part A

Step 1:   Write an unbalanced chemical equation based on the information given in the problem.

Step 2:   Use coefficients to balance the equation. (Remember to save oxygen for last.)

Step 3:   Reduce the coefficients to their smallest whole number ratio if necessary.

Step 4:   Check your answer.

Part B

Step 1:   Determine the mole ratio for hydrogen peroxide and oxygen from your balanced chemical equation.

Step 2:   Use the mass and molecular weight of the hydrogen peroxide solution to determine the moles of hydrogen peroxide present.

Step 3:   Use the mole ratio from above as a conversion factor, and calculate the number of moles of oxygen generated by the decomposition reaction.

Step 4:   Use the mole ratio from above as a conversion factor, and calculate the number of moles of water generated by the decomposition reaction.

The mole relationships in a chemical equation can also be used with molecular masses to make mass determinations for a chemical reaction. As moles cannot be measured directly in the lab, moles must be converted back to grams for practical use. *Stoichiometry* is the term for the use of these ratios.

Workbook Example 3.5

PROBLEM:
When concentrated sulfuric acid is added to table sugar, a dehydration reaction occurs according to the following unbalanced equation:

$$C_{12}H_{22}O_{11} + H_2SO_4 \rightarrow C + H_2O + H_2SO_4$$

If 0.50 moles of sucrose are to be reacted with an excess of sulfuric acid, how many grams of sucrose are needed, and how many grams of carbon will result?

SOLUTION:
First, balance the equation:

$$C_{12}H_{22}O_{11} + H_2SO_4 \rightarrow 12\ C + 11\ H_2O + H_2SO_4$$

Molar mass of sucrose:

12 carbon atoms	×	12.01 amu	=	144.12 g/mol
22 hydrogen atoms	×	1.008 amu	=	22.176 g/mol
11 oxygen atoms	×	16.00 amu	=	176.00 g/mol
		Total	=	342.30 g/mol

This molar mass also represents a conversion factor of $\dfrac{342.3\ g}{1\ mol}$ or $\dfrac{1\ mol}{342.3\ g}$.

To determine the grams of sucrose in 0.50 moles:

$$0.50\ mol \times \frac{342.3\ g}{1\ mol} = 170\ g$$

It will take a couple of steps to determine the number of grams of carbon that will be produced. First, determine how many moles of carbon will be produced using the ratios from the balanced equation:

$$0.50\ mol\ sucrose \times \frac{12\ mol\ C}{1\ mol\ sucrose} = 6.0\ mol\ C$$

Now convert that number of moles to grams:

$$6.0\ mol\ C \times \frac{12.01\ g}{1\ mol} = 72\ g\ C$$

72 grams of carbon will be produced from the reaction of 0.5 moles of sucrose with sulfuric acid. With practice, it will become possible for you to do this without breaking it up into separate steps. Once you have a balanced equation, you can set up a problem like this to go straight from moles of sucrose to grams of carbon:

$$0.60\ moles\ sucrose \times \frac{12\ mol\ C}{1\ mol\ sucrose} \times \frac{12.01\ g\ C}{1\ mol\ C} = 72\ g\ C$$

Start out by breaking things up into as many steps as you need to when you are starting out, but you will get faster!

The basic steps needed to complete stoichiometry problems can be summarized like this:

- Balance the equation: an unbalanced or incorrectly balanced equation will make it impossible to solve a problem correctly.
- Convert to moles: if you are given grams of one substance, and need grams of another substance, you will need the formula weights of both.
- Set up a dimensional-analysis problem. Keep both units and identifiers with all numbers. For example, don't just write 12.01 g/mol, write 12.01 g carbon/mol. You will have numerous conversion factors of this type, so you want to make sure that not only the units, but even the substances cancel out. Using "fake units" will continue to be helpful as you move through the study of chemistry!
- Solve, making sure that your units are correct and that your answer makes sense.

Workbook Problem 3.3

When elemental iron rusts, it combines with oxygen to form iron(III) oxide. How many grams of rust will be produced if 14 grams of metal nails are left out on a construction site all winter? How many grams of oxygen will have been consumed in this reaction?

Strategy: You can calculate the grams of iron oxide produced by using the molar mass of iron oxide and the mole ratio of iron-to-iron oxide from the balanced chemical equation as conversion factors.

g iron → moles iron → moles iron(III) oxide → g iron(III) oxide

Step 1:   Write an unbalanced chemical equation from the information given above.

Step 2:   Balance the equation.

Step 3:   From the balanced chemical equation, determine the mole to mole ratio of iron to iron(III) oxide.

Step 4:   Find the atomic mass of iron.

Step 5:   Determine the molecular mass for iron(III) oxide.

Step 6:   From the information in Steps 3, 4, and 5, create conversion factors. Use these conversion factors, and set up a dimensional-analysis problem to convert from grams of iron, to moles of iron, to moles of iron oxide, to grams of iron oxide.

Step 7:   Using the law of mass conservation, how many grams of oxygen were added to the iron?

Section 3.4   Yields of Chemical Reactions

When stoichiometric calculations are done, we get an ideal result, known as the **theoretical yield**. Running an experiment will provide the **actual yield**. The actual yield will always be less than the theoretical yield. In the workbook problem above, for instance, there will probably be little pockets of unrusted iron within the nails, which would make the final mass lower than you calculated by assuming that every atom of elemental iron became iron oxide.

A report of an experiment is often given in terms of reaction efficiency or **percent yield**. This is calculated very simply:

$$percent\ yield = \frac{actual\ yield}{theoretical\ yield} \times 100\%$$

Workbook Example 3.6

PROBLEM:
When 15.0 grams of hydrogen react with oxygen, 120 grams of water result. What is the percent yield of this reaction?

SOLUTION:
Write and balance the equation.

$$2\,H_2 + O_2 \rightarrow 2\,H_2O$$

Figure out the atomic or molecular weights of all the relevant components:

H_2 = 2.02 g/mol
H_2O = 18.02 g/mol

Because you are not given any information of oxygen in the problem, it's assumed that the oxygen is the excess reagent. In that case, you don't need the molecular weight for oxygen.

Set up a dimensional analysis equation to determine the theoretical yield of the reaction:

$$15.0\,g\,H_2 \times \frac{1\,mol\,H_2}{2.02\,g} \times \frac{2\,mol\,H_2O}{2\,mol\,H_2} \times \frac{18.02\,g\,H_2O}{1\,mol\,H_2O} = 134\,g\,H_2O$$

Now solve for the percent yield:

$$percent\;yield = \frac{actual\;yield}{theoretical\;yield} \times 100\% = \frac{120\,g}{134\,g} \times 100\% = 89.6\%$$

Workbook Example 3.7

PROBLEM:
If burning hydrogen in oxygen to form water generally has a 95% yield, how many grams of water can be expected from the combustion of 23 grams of hydrogen?

SOLUTION:
Use the balanced equation from above, $2\,H_2 + O_2 \rightarrow 2\,H_2O$, and use it to determine the theoretical yield of the equation if 23 grams of hydrogen are used:

$$23\,g\,H_2 \times \frac{1\,mol\,H_2}{2.02\,g\,H_2} \times \frac{2\,mol\,H_2O}{2\,mol\,H_2} \times \frac{18.02\,g\,H_2O}{1\,mol\,H_2O} = 205\,g$$

Now the actual yield can be estimated: the actual yield will be 95% of the theoretical yield:

$$205\,g \times 0.95 = 195\,g$$

Workbook Problem 3.4

The last step in ammonia production is a high-pressure reaction that proceeds according to the following unbalanced reaction:

$$H_2 + N_2 \rightarrow NH_3$$

If 18.0 grams of nitrogen gas are reacted, and 14.6 grams of ammonia result, what is the percent yield?

Strategy:　　To determine the percent yield, first balance the equation, then determine the theoretical yield of ammonia in the reaction.

Step 1:　　Balance the equation.

Step 2: Calculate the theoretical yield using dimensional analysis.

Step 3: Calculate the percent yield using the actual yield stated in the problem and the theoretical yield just calculated.

Section 3.5   Reactions with Limiting Amounts of Reactants

To maximize yields, reactions are often carried out with much more of one reactant than necessary based on the balanced equation. In this situation, there is a **limiting reactant** and an excess reactant. The limiting reactant determines the theoretical yield of the reaction: once it is used up, no more reaction is possible.

Problems involving a limiting reactant can be complex: proceed with caution, because no information can be obtained from the masses of the reactants: only their mole ratios contain useful information.

To work these problems, convert all reactant quantities to moles and run the reaction mathematically, once with each reactant. The reactant that gives you the fewest moles of product is your limiting reactant, and that amount is the theoretical yield of the reaction.

Workbook Example 3.8

PROBLEM:
15 grams of hydrogen gas and 37 grams of chlorine gas are reacted together to form hydrogen chloride gas. Which is the limiting reactant? How many grams of hydrogen chloride gas can be produced (theoretical yield)? How many grams of the excess reactant will be left over?

SOLUTION: First, write a balanced chemical equation.

$$H_2 + Cl_2 \rightarrow 2\,HCl$$

Determine the number of moles of each reactant present.

$$15\,g\,H_2 \times \frac{1\,mol\,H_2}{2.02\,g\,H_2} = 7.4\,mol\,H_2$$

$$37\,g\,Cl_2 \times \frac{1\,mol\,Cl_2}{71\,g\,Cl_2} = 0.52\,mol\,Cl_2$$

The reactants are required in a 1:1 mole ratio, so in this example it is possible to see that chlorine will be the limiting reactant and hydrogen the excess reactant, despite the fact that by mass there is a great deal more chlorine than hydrogen.

These ratios will not always be this straightforward: a foolproof way to find the limiting and excess reactants is to run the reaction stoichiometry with each reactant and look at the amount of product produced by each:

$$7.4\,mol\,H_2 \times \frac{2\,mol\,HCl}{1\,mol\,H_2} = 14.8\,mol\,HCl$$

$$0.52\,mol\,Cl_2 \times \frac{2\,mol\,HCl}{1\,mol\,Cl_2} = 1.04\,mol\,HCl$$

This method works equally well for simple and complex reactions: the number of moles of chlorine available can produce far less than the number of moles of hydrogen. Chlorine is your limiting reactant, and the theoretical yield of the reaction is 1.04 mol HCl gas.

$$1.04\,mol\,HCl \times \frac{36.5\,g\,HCl}{1\,mol\,HCl} = 38.0\,g\,HCl$$

How many grams of hydrogen will be left over? There are several ways of getting at this information. The most straightforward is to do a quick calculation using the law of mass conservation: we have obtained 38 grams of products from 15 g + 37 g of reactants: 14 grams of reactants must be left over. As hydrogen is in the excess, there are 14 grams of hydrogen left over.

Another way to get at this is to use mole quantities: 0.52 moles of chlorine reacted, so you can set a stoichiometric equation to get at the number of moles of hydrogen that were used. This is extreme overkill for the simple ratio here, but is very useful when the ratio is less straightforward:

$$0.52\,mol\,Cl_2 \times \frac{1\,mol\,H_2}{1\,mol\,Cl_2} = 0.52\,mol\,H_2$$

Now that the number of moles of hydrogen used up is known, you can simply subtract from the original number of moles of hydrogen and convert to grams:

$$\left(7.4\,mol\,H_2 - 0.52\,mol\,H_2\right) \times \frac{2.02\,g\,H_2}{1\,mol\,H_2} = 14\,g\,H_2$$

Workbook Problem 3.5

Aluminum metal and hydrochloric acid react to make aluminum chloride and hydrogen gas. If 5.3 grams of aluminum are reacted with 9.2 grams of hydrochloric acid in solution, how many grams of aluminum chloride can be obtained? What is the limiting reactant? How much of the excess reactant (in grams) is left over? What is the percent yield of the reaction if 8.3 grams of aluminum chloride are produced?

Strategy: Generate a balanced equation and determine the number of moles available for each reactant. Next, run the reaction mathematically with each of the reactants and see which gives the smallest number of moles of product. This will determine the limiting reactant and the theoretical yield and make it possible to calculate the percent yield and the amount of excess reactant left over. Next, compare the mole ratio of the actual amounts to the mole ratio required according to the *balanced chemical equation*. You can then determine which reactant is in excess and which reactant is present in a limiting amount.

Step 1:   Using the information given, write an unbalanced chemical equation.

Step 2:   Balance the chemical equation.

Step 3:   Determine the number of moles of each reactant present.

Step 4:   Determine which reactant gives the smallest number of moles of product. This gives the limiting reactant and the theoretical yield.

Step 5:   Determine how many moles of the excess reactant were used.

Step 6:   Determine how many moles of the excess reactant remain.

Step 7:   Determine how many grams of the excess reactant remain.

Step 8:   Determine the percent yield of the reaction.

Section 3.6   Percent Composition and Empirical Formulas

Determining the composition of a new compound requires that the elements in the compound and their relative abundances be known.

This is expressed as **percent composition**: how much of each element is present in a compound by mass. This can also be calculated for a known compound if desired.

Percent composition can be converted into mole ratios by figuring out how many moles of each element would be present in 100 grams of the unknown compound. This mole ratio can then be expressed as an **empirical formula**: the simplest ratio of the elements to one another.

If the molecular mass is known, the empirical formula can be converted into a **molecular formula**: the actual numbers of each element in a molecule of the compound. Sometimes, the two are the same, but other times the molecular formula will be a whole-number multiple of the empirical formula.

Workbook Example 3.9

PROBLEM:
A simple hydrocarbon is found to be 75% C by mass and 25% H by mass. What is the empirical formula of the compound? If its molecular mass is determined to be 16 g/mol, what is its molecular formula?

SOLUTION:
Assume that 100 grams of the compound are present. 75% of 100 g is 75 grams of carbon; 25% of 100 grams is 25 g hydrogen. Convert to moles:

$$75\,g\,C \times \frac{1\,mol\,C}{12\,g\,C} = 6.25\,mol\,C$$

$$25\,g\,H \times \frac{1\,mol\,H}{1\,g\,H} = 25\,mol\,H$$

This makes it possible to find the ratio of one element to the other. To determine the empirical formula, divide to find the smallest ratio. Be sure to keep element names or symbols with the numbers to avoid confusion:

$$\frac{25\,mol\,H}{6.25\,mol\,C} = \frac{4\,mol\,H}{1\,mol\,C}$$

This tells us that there are four hydrogens for every one carbon, which gives an empirical formula of CH_4.

If the unknown hydrocarbon has a molecular mass of 16 g/mol, we can also determine the molecular formula of the compound. The empirical formula has a molecular weight of 16 g/mol, so in this case, the empirical formula is the same as the molecular formula.

Workbook Example 3.10

PROBLEM:
A compound is found to be 36.8% N and 63.2% O. What is the empirical formula of the compound?

SOLUTION:
Assume that 100 grams of the compound are present. 36.8% of 100 g is 36.8 grams of nitrogen; 63.2% of 100 grams is 63.2 g oxygen. Convert to moles:

$$63.2\,g\,O \times \frac{1\,mol\,O}{16\,g\,O} = 3.95\,mol\,O$$

$$36.8\,g\,N \times \frac{1\,mol\,N}{14\,g\,N} = 2.63\,mol\,N$$

There is no obvious whole-number ratio here, but continue determining the ratio:

$$\frac{3.95\,mol\,O}{2.63\,mol\,N} = \frac{1.5\,mol\,O}{1\,mol\,N}$$

A formula can be written with this, but it will need a little extra work:

$NO_{1.5}$ is not a proper formula, as partial atoms are not allowed. Multiply through to get rid of the fraction (this sometimes takes a little trial and error) and make sure that the formula still is the smallest allowable: N_2O_3 is an allowable formula that maintains the ratio of one-and-a-half oxygens for each nitrogen.

Workbook Example 3.11

PROBLEM:
A slightly more complex compound is determined to be 27% Na, 16.5% N, and 56.5% O. What is the empirical formula of the compound?

SOLUTION: Assume that 100 grams of the compound are present. This will be 27 g Na, 16.5 g N, and 56.5 g O. Convert to moles:

$$27\,g\,Na \times \frac{1\,mol\,Na}{23\,g\,Na} = 1.174\,mol\,Na$$

$$16.5\,g\,N \times \frac{1\,mol\,N}{14\,g\,N} = 1.179\,mol\,N$$

$$56.5\,g\,O \times \frac{1\,mol\,O}{16\,g\,O} = 3.531\,mol\,O$$

There are essentially equal numbers of moles of Na and N, so there will be the same number of each of these in the product. What remains is to determine the ratio of oxygens to the other elements:

$$\frac{3.531\,mol\,O}{1.179\,mol\,N} = \frac{3\,mol\,O}{1\,mol\,N}$$

So there are three oxygens for every one nitrogen and one sodium for each nitrogen: $NaNO_3$. This is sodium nitrate. The most important thing with these problems is to not confuse the elements with one another.

Workbook Problem 3.6

Determine the empirical formula of a compound that is 18.9% Li, 16.2% C, and 64.9% O.

Strategy:	Assume you have a 100-gram sample, and convert the percentages of the elements into grams.
Step 1:	Convert the grams of each element into moles of each element.
Step 2:	Find the mole ratios.
Step 3:	If necessary, convert the ratios to whole numbers, and write the empirical formula.

Workbook Example 3.12

PROBLEM:
What is the percent composition of sulfuric acid?

SOLUTION:
The formula for sulfuric acid is H_2SO_4. Convert this into a mass percent by determining the mass of each element and the total molecular mass:

$$
\begin{array}{lll}
H \times 2 \times 1.008 \text{ g} & = & 2.0016 \text{ g} \\
S \times 1 \times 32.06 \text{ g} & = & 32.06 \text{ g} \\
\underline{O \times 4 \times 16.00 \text{ g}} & = & \underline{64.00 \text{ g}} \\
\text{Molecular mass} & = & 98.06 \text{ g/mol}
\end{array}
$$

Mass % of each element:

$$\%H = \frac{2.0016\,g}{98.06\,g} \times 100\% = 2.04\%\,H$$

$$\%S = \frac{32.06\,g}{98.06\,g} \times 100\% = 32.7\%\,S$$

$$\%O = \frac{64.00\,g}{98.06\,g} \times 100\% = 65.3\%\,O$$

Workbook Problem 3.7

Determine the percent composition of sodium hypochlorite.

Strategy: Write the formula for sodium hypochlorite and determine the mole ratio of the elements in the compound.

Step 1:   Determine the formula of the compound.

Step 2:   Determine the molecular mass and the mass of each component.

Step 3:   Determine the percent composition of each component.

Section 3.7   Determining Empirical Formulas: Elemental Analysis

As we mentioned before, when a hydrocarbon burns, it produces water and carbon dioxide. This useful fact can be used to determine the ratios of the elements to one another: the moles of carbon dioxide and water produced by such a reaction can be used to determine the moles of hydrogen and carbon in the original compound and thus the empirical formula of the compound.

Determine moles of CO_2 – this will be the same as the number of moles of carbon.

Determine moles of H_2O – this will be one-half the number of moles of hydrogen.

If there is another element present, it can be determined by subtracting the masses of carbon and hydrogen (go from moles to grams, to determine this) from the total mass of the unknown compound.

If the molecular mass is also known, it is possible to determine not only the empirical formula, but the molecular formula as well.

Workbook Example 3.13

PROBLEM:
When a 7.03-gram sample of a hydrocarbon is burned, 21.32 grams of carbon dioxide and 10.82 grams of water are produced. If the compound contains only carbon and hydrogen, and its molecular mass is 58 g/mol, what is the molecular formula of the compound?

SOLUTION:
Convert the grams of carbon dioxide and water into moles of carbon and hydrogen, then use the ratio to generate an empirical formula. From the molecular mass, develop the molecular formula from the empirical formula if necessary.

$$21.32\,g\,CO_2 \times \frac{1\,mol\,CO_2}{44.01\,g\,CO_2} \times \frac{1\,mol\,C}{1\,mol\,CO_2} = 0.484\,mol\,C$$

$$10.82\,g\,H_2O \times \frac{1\,mol\,H_2O}{18.00\,g\,H_2O} \times \frac{2\,mol\,H}{1\,mol\,H_2O} = 1.202\,mol\,H$$

Calculate the mole ration of hydrogen to carbon:

$$\frac{1.202\,mol\,H}{0.484\,mol\,C} = \frac{2.5\,H}{1\,C}$$

The initial molecular formula is $C_1H_{2.5}$, which, if we multiply everything by 2, becomes a proper empirical formula, C_2H_5.

In order to get to the actual molecular formula, we need to get a ratio of the molar masses. In the problem, we were told that the molar mass is 58 g/mol, and the molar mass of the empirical formula we just calculated is 29 g/mol. If we get a ratio of the two molar masses

$$\frac{58\,g/mol}{29\,g/mol} = \frac{2}{1}$$

this tells us that we need to multiply the empirical formula by 2 in order to get to the molecular formula. The molecular formula will be C_4H_{10}.

Workbook Example 3.14

PROBLEM:
A compound known as methyl butanoate is one of the compounds responsible for the distinctive smell of pineapple. It contains carbon, hydrogen, and oxygen. When combusted, a 3.73 g sample of this compound produced 8.03 g carbon dioxide and 3.27 g water. If the molecular weight of the compound is 102 g/mol, what is the molecular formula?

SOLUTION:
Convert the grams of carbon dioxide and water into moles of carbon and hydrogen. Convert these back into grams to determine the weight of oxygen in the compound, then determine the number of moles of oxygen. Use these three quantities to develop mole ratios and an empirical formula.

Compare the empirical formula weight and the molecular formula weight to determine the molecular formula.

$$8.03\,g\,CO_2 \times \frac{1\,mol\,CO_2}{44.01\,g\,CO_2} \times \frac{1\,mol\,C}{1\,mol\,CO_2} \times \frac{12.01\,g\,C}{1\,mol\,C} = 2.19\,g\,C$$

$$3.27\,g\,H_2O \times \frac{1\,mol\,H_2O}{18.00\,g\,H_2O} \times \frac{2\,mol\,H}{1\,mol\,H_2O} \times \frac{1.008\,g\,H}{1\,mol\,H} = 0.366\,g\,H$$

Now determine the amount of O by subtracting the masses of carbon and hydrogen from the mass of the starting sample:

$$3.73\,g\,compound - 2.19\,g\,C - 0.37\,g\,H = 1.17\,g\,O$$

Convert the mass of O to moles of O.

$$1.17\,g\,O \times \frac{1\,mol\,O}{16.00\,g\,O} = 0.0731\,mol\,O$$

In order to calculate the mole ratios, we also need the moles of hydrogen and carbon:

$$0.366\,g\,H \times \frac{1\,mol\,H}{1.008\,g\,H} = 0.363\,mol\,H$$

$$2.19\,g\,C \times \frac{1\,mol\,C}{12.01\,g\,C} = 0.182\,mol\,C$$

We can now determine the mole ratios between carbon, hydrogen, and oxygen. Because oxygen has the least amount of moles, we're going to use that as our denominator, so that we always have one mole of oxygen:

$$\frac{0.363\,mol\,H}{0.0731\,mol\,O} = \frac{5\,mol\,H}{1\,mol\,O}$$

$$\frac{0.182\,mol\,C}{0.0731\,mol\,O} = \frac{2.5\,mol\,C}{1\,mol\,O}$$

This gives a preliminary formula of $C_{2.5}H_5O$ and a real empirical formula of $C_5H_{10}O_2$. The empirical formula has a molecular mass of 102 g/mol, so this is also the molecular formula.

Workbook Problem 3.8

2-butene-1-thiol is one of the compounds that gives skunk spray such a memorable odor. It contains only carbon, hydrogen, and sulfur. When 2.76 grams of this compound were burned (over a weekend, to make sure that all the custodial staff didn't quit), 5.70 grams of carbon dioxide and 2.31 grams of water were produced. If the compound has a molecular mass of 88 g/mol, determine the empirical and molecular formulas of this compound.

Strategy: Convert the grams of carbon dioxide and water into moles of carbon and hydrogen. Convert these back into grams to determine the weight of sulfur in the compound, then determine the number of moles of sulfur. Use these three quantities to develop mole ratios and an empirical formula. Compare the empirical formula weight and the molecular formula weight to determine the molecular formula.

Step 1:   Find the molar amounts of C and H in CO_2 and H_2O.

Step 2:   Carry out mole-to-gram conversions to find the number of grams of C and H in the original sample.

Step 3:   Subtract the masses of C and H from the mass of the starting sample to determine the mass of S.

Step 4:   Convert the mass of S to moles of S.

Step 5:   Find the mole ratios.

Step 6:   Use the ratios above to write the empirical formula of the compound.

Step 7:   Compare the empirical formula mass to the molecular formula weight to see if a conversion is needed.

Step 8:   Write the molecular formula.

Self–Test

This section is intended to test your knowledge of the material covered in this chapter. Think through these problems, and make certain you understand what is going on. Ask yourself if your answer makes sense. Many of these questions are linked to the chapter learning goals. Therefore, successful completion of these problems indicates you have mastered the learning goals for this chapter. You will receive the greatest benefit from this section if you use it as a mock exam. You will then discover which topics you have mastered and which topics you need to study in more detail.

True–False
1. When balancing equations, it is essential that there be the same number of molecules on each side.

2. Chemical equations describe behavior on both the atomic (microscopic) level and the bulk (macroscopic) level.

3. A mole of two different substances contains the same number of particles.

4. The coefficients in a balanced chemical equation show the mass ratios of the reactants and products.

5. Theoretical yields of chemical reactions are determined in the experimental laboratory.

6. When reactants are not present in stoichiometric ratios, the limiting reactant will determine the maximum yield of the reaction.

7. Combustion analysis can be used to determine the molecular formula of an unknown compound.

8. The empirical formula and the molecular formula for a compound are sometimes the same.

Matching

9. Balanced

10. Yield

11. Mole

12. Coefficient

13. Stoichiometry

14. Molecular mass

15. Empirical formula

16. Subscript

17. Combustion

18. Actual yield

a. sum of all the masses of the atoms in a molecule

b. the smallest ratio of elements in a compound

c. the amount of product formed in a reaction

d. the number in front of a reactant or product that shows its mole ratio with the other components

e. reaction with oxygen

f. indication of number of each type of atom in a molecule

g. having the same number of all atom types in both products and reactants

h. experimentally determined amount of product in a reaction.

i. the amount of a substance equal to the molecular or atomic mass in grams

j. the use of mole ratios and molecular weights to carry out chemical calculations

Fill-in-the-Blank

19. To be useful in making calculations, a chemical equation must be _____.

20. The mole ratios of the components in a reaction are indicated by _____ in front of the components. These can be changed if necessary, but the _____ within the components cannot be changed, because it will change the identity of the component.

21. To convert amounts of substances from grams to moles, the conversion factor is

    _____ _____.

22. If reactants are not present in stoichiometric ratios, the amount of product will be determined by the _____ _____.

23. The reactant that does not determine the yield of the reaction is known as the _____

    _____.

24. The _____ _____ of a reaction can be calculated using a balanced equation, but it is rarely, if ever, achieved. In the lab, the _____ _____ will be obtained. The two together can be used to calculate the _____ _____ of the reaction.

25. It is possible to determine the _____ _____ of a compound from the mass percentages of the elements in it, but to determine the _____ _____ of the compound, the _____ _____ must be known.

26. When a hydrocarbon is burned in oxygen, the primary products are _____ _____ and _____.

27. The number of particles in one mole of a substance is known as _____ _____. The number is _____.

Problems

28. Balance the following equations:
    a. $N_2 + O_2 \rightarrow N_2O_5$
    b. $C_3H_8 + O_2 \rightarrow CO_2 + H_2O$
    c. $H_3PO_4 + NaOH \rightarrow Na_3PO_4 + H_2O$
    d. $Ba(NO_3)_2 + NaOH \rightarrow Ba(OH)_2 + NaNO_3$

29. Determine the molecular or formula mass of the following compounds and convert the given amounts to moles:
    a. 45 g calcium chloride
    b. 23 g sucrose ($C_{12}H_{22}O_{11}$)
    c. 67 g phosphorus trifluoride
    d. 5 g ammonium carbonate

30. Using the balanced equation – the answer for problem 28a – for the production of dinitrogen pentoxide: if four moles of nitrogen and 10 moles of oxygen are combined, what is the theoretical yield of the reaction in moles and grams? If 376 grams of product are obtained, what is the percent yield?

31. When sodium hydrogen carbonate is heated, it decomposes according to the following unbalanced equation:

$$NaHCO_3 \rightarrow Na_2CO_3 + CO_2 + H_2O$$

If 20 grams of sodium hydrogen carbonate are heated, what is the theoretical yield of water in grams? If the reaction has a percent yield of 87%, what is the maximum quantity of water that can be obtained?

32. When 6.7 grams of methane (CH_4) are burned in 20 grams of oxygen, what is the limiting reactant? What is the theoretical yield of water and carbon dioxide? How many grams of the excess reactant will remain when the reaction is complete?

33. Determine the percent composition of the following compounds:
    a. Carbon dioxide
    b. Sodium phosphate
    c. CH_3COOH
    d. Adenosine triphosphate: $C_{10}H_{16}N_5O_{13}P_3$

34. A compound containing 40.0% carbon, 6.7% hydrogen, and 53.3% oxygen has a molecular mass of 180 g/mol. What is the molecular formula of this compound?

35. Isopentyl acetate is a compound responsible for the distinctive smell of bananas. It contains only carbon, hydrogen, and oxygen. When 1.57 grams of this compound were subjected to combustion analysis, 3.71 grams of CO_2 and 1.51 grams of H_2O were measured. What is the empirical formula of this compound?

CHAPTER FOUR

Reactions in Aqueous Solution

Learning Objectives

As a result of reading and studying this chapter, you should be able to

Section 4.1   **Concentrations in Solution: Molarity**
1.   Calculate the molarity of a solution given the mass of solute and total volume.
2.   Calculate the amount of solute in a given volume of a solution with a known molarity.
3.   Describe the proper technique for preparing solutions of known molarity.

Section 4.2   **Diluting Concentrated Solutions**
4.   Calculate the concentration of a solution that has been diluted.
5.   Describe the proper technique for diluting solutions.

Section 4.3   **Electrolytes in Aqueous Solution**
6.   Classify a substance as a strong, weak, or nonelectrolyte.
7.   Calculate the concentration of ions in a strong electrolyte solution.

Section 4.4   **Types of Chemical Reactions in Aqueous Solution**
8.   Classify a reaction as a precipitation, acid-base neutralization, or oxidation–reduction (redox) reaction.

Section 4.5   **Aqueous Reactions and Net Ionic Equations**
9.   Write an ionic and net ionic equation and identify spectator ions given the molecular equation.

Section 4.6   **Precipitation Reactions and Solubility Guidelines**
10.   Use the solubility guidelines (Table 4.2) to predict the solubility of an ionic compound in water.
11.   Predict whether a precipitation reaction will occur and write the ionic and net ionic equations.

Section 4.7   **Acids, Bases, and Neutralization Reactions**
12.   Convert between name and formula for an acid.
13.   Classify acids as strong or weak based on the molecular picture of dissociation.
14.   Write the ionic equation and net ionic equation for an acid-base neutralization reaction.

Section 4.8   **Solution Stoichiometry**
15.   Convert between moles and volume using molarity in stoichiometry calculations.

Chapter Summary

This chapter begins by continuing the discussion from Chapter 3. In that chapter, we assumed that the reactions talked about were solids; however, most of the time in chemistry we deal with reactions that are aqueous, meaning that a solid has been dissolved in water. In this chapter, we begin by describing how we measure the concentration of a solution by defining the unit *molarity* and then keep going by applying molarity to stoichiometric problems. We then move from the quantitative world to the qualitative to talk about three different types of chemical reactions in aqueous solution: precipitation reactions, acid–base reactions, and oxidation–reduction reactions. These reactions are often written as net ionic equations. Therefore, it is necessary to know the difference between strong and weak electrolytes and nonelectrolytes. You can predict the outcome of each of these kinds of reactions if you know the solubility rules, you can recognize acids and bases, and you know how to assign oxidation numbers to compounds. You are also shown how to balance oxidation–reduction reactions with either the oxidation–number method or the method of half-reactions. Finally, you will apply the concepts you learned in earlier chapters about stoichiometry to the reactions that are discussed in this chapter.

The Chapter in Detail

Section 4.1   Concentrations in Solution: Molarity

Many chemical reactions are carried out in solution, because this provides far more opportunity for particles of reactants to collide with one another. If reactants are dissolved, it is essential that their concentration be known. Chemical quantities in reactions are given in moles, so the most convenient unit of concentration for the study of chemistry should do the same: molarity is defined as the moles of solute per liter of solution. When written in equation form, it can also be rearranged, as needed, to be used as a conversion factor:

$$Molarity\,(M) = \frac{moles\ of\ solute}{volume\ of\ solution\,(in\,L)}$$

To make up a solution to a particular molarity, the moles of solute needed must be converted to grams. This amount of solute is then placed in a properly sized volumetric flask, and the flask is partially filled with water. Once the solute is dissolved completely, the solution is made up to volume by adding solvent up to a calibration line. The reasons for this procedure will be demonstrated in the following examples.

Workbook Example 4.1

PROBLEM:
How many grams of sucrose (MW = 342.3 g/mol) are needed to make 500 mL of a 0.50 M solution? How would you make this solution?

SOLUTION:
Molarity refers to moles, not grams, so it is necessary to first figure out how many moles of sucrose are needed, then solve for grams:

$$moles = 0.50\,M \times 0.500\,L = 0.25\,mol$$

Notice that the volume must be converted to liters. The reason for this is more obvious if molarity is expanded to moles/liter:

$$moles = \frac{0.50\,mol}{1\,L} \times 0.500\,L = 0.25\,mol$$

Always be *very careful* with units. They matter!

Now convert to grams:

$$0.25\,mol\,sucrose \times \frac{342.3\,g\,sucrose}{1\,mol\,sucrose} = 85.6\,g\,sucrose$$

To make up this solution, put 85.6 grams of sucrose in a 500 mL volumetric flask, add enough water to dissolve the sucrose, and then fill the volumetric flask to 500 mL.

Note: 85.6 grams of sucrose is about a half-cup of sugar, and 500 mL is about two cups of water. If the solution were to be made by dropping the sugar into 500 mL of water, the volume of the final solution would be far more than 500 mL, and the concentration would be uncertain. This is why volumetric glassware is used to make these solutions.

Workbook Example 4.2

PROBLEM:
How many milliliters of 0.40 M salt solution contain 0.112 moles of salt?

SOLUTION:
The identity of the chemical compound here generically referred to as "salt" is not given, nor is it required. Don't let that throw you off!

Since you know the amount of moles and the molarity of the solution, we will use the definition of molarity as a conversion factor:

$$0.112\,mol\,salt \times \frac{1\,L}{0.40\,mol} = 0.28\,L$$

The reason we write the molarity expression with L on top and mol on the bottom is that with the 0.112 moles on top, the moles will cancel out.

To convert liters to milliliters, remember that there are 1000 milliliters in 1 liter of solution:

$$0.28\,L \times \frac{1000\,mL}{1\,L} = 280\,mL$$

Workbook Example 4.3

PROBLEM:
What is the molarity of a solution of barium chloride made by dissolving 15 grams of the salt into 250 mL of solution?

SOLUTION:
First, determine the number of moles of barium chloride in 15 grams:

$$15\,g\,BaCl_2 \times \frac{1\,mol\,BaCl_2}{208.3\,g\,BaCl_2} = 0.072\,mol\,BaCl_2$$

The molarity of the solution can now be calculated:

$$\frac{0.072\,mol\,BaCl_2}{0.250\,L} = 0.29\,M$$

Workbook Problem 4.1

How many grams of sodium chloride are needed to make 500 milliliters of a 0.30 M sodium chloride solution?

Strategy: Using the definition of molarity and the volume of solution, determine moles and grams of solute.

Step 1:   First, determine the number of moles found in 500 mL of a 0.30 M solution.

Step 2:   Determine the number of grams of sodium chloride from the number of moles.

Section 4.2   Diluting Concentrated Solutions

Some chemicals are sold as concentrated solutions (often called *stock solutions*) that must be diluted before use. Fortunately, it is very simple to figure out how to dilute a solution. Remember, when a solution is diluted, the amount of *solute* does not change, only the amount of *solvent* does. This leads to a simple equation:

$$M_i V_i = M_f V_f$$

Written out, this means that the initial molarity times the initial volume equals the final molarity times the final volume.

A word of caution: when performing dilutions in the lab, you need to be cautious to always add strong acids to water, and not the other way around. When acids are diluted, they generate large quantities of heat: if water is added to acid, the heat can cause the water to boil and splatter the acid. Be careful!

Workbook Example 4.4

PROBLEM:
A lab protocol asks for 100 mL of 5.0 M HCl. If the stock solution is 12M, how much is needed for the dilution, and how would you perform it?

SOLUTION: Use the above equation, rearranging to solve for the initial volume:

$$M_i V_i = M_f V_f$$

$$V_i = \frac{M_f V_f}{M_i}$$

$$V_i = \frac{5.0\,M \times 100\,mL}{12\,M} = 42\,mL$$

Note: no conversion was done of the volumes here. When this equation is used, molarity cancels out, leaving behind whatever volume unit is being used. If this makes you uncomfortable, by all means convert everything to liters, but it isn't strictly necessary.

To make this solution, 42 mL of acid would be carefully measured, then poured slowly into a volumetric flask containing about 40 mL of water. Once the acid has been diluted by being added to water, the solution could be made up to volume.

Workbook Problem 4.2

What is the final concentration when 50 mL of 10 M magnesium chloride is diluted to 1.0 L?

Strategy: Remember that the number of moles of solute present do not change when a solution is diluted.

Step 1:   Rearrange the dilution formula to solve for the final concentration.

Section 4.3   Electrolytes in Aqueous Solution

Substances that dissolve in water to form ions are known as **electrolytes**; solutions containing these compounds will conduct electricity. Electrolytes are soluble ionic compounds such as sodium chloride (table salt). Acids and bases are also electrolytes, because they dissociate to form ions—in particular, the ions that distinguish them as being acids (H^+) or bases (OH^-).

Nonelectrolytes may dissolve in water, but they will not conduct electricity when dissolved because they do not dissociate into ions. Covalent molecules such as sugars do not dissociate when they dissolve.

Electrolytes can be either **strong** or **weak**: this is defined simply by how much they dissociate. That is, a **strong electrolyte** dissolves almost completely and a weak electrolyte very little. By extension, a strong acid dissociates completely and a weak acid only partially.

Dissociation is a dynamic process: if a weak acid is dissolved in water, there will be both dissociated and un-dissociated molecules present, and the weak acid will conduct electricity to a certain extent. In the weak acid solution, whole molecules dissociate and dissociated molecules reform constantly in equilibrium between the forward and reverse reactions. The process is represented by a double arrow in the following equation for the dissociation of a weak acid:

$$CH_3COOH \rightleftharpoons CH_3COO^- + H^+$$

Organic acids have formulas that contain $-CO_2H$ or COO^-, and all organic acids are weak electrolytes. Alcohols, on the other hand, have formulas that contain $-OH$ groups: these are nonelectrolytes. **Nonelectrolytes** will not dissociate when dissolved in water.

Workbook Example 4.5

PROBLEM:
$Mg(OH)_2$ is a strong base. What is the concentration of OH^- ions in a 0.450 M solution of $Mg(OH)_2$?

SOLUTION:
A strong electrolyte dissociates completely. So the $Mg(OH)_2$ will not be found as units, but only as dissociated ions:

$$Mg(OH)_2 \rightarrow Mg^{2+} + 2OH^-$$

This makes it possible to calculate the concentration of OH^- ions directly:

$$\frac{0.450\,mol\,Mg(OH)_2}{1L} \times \frac{2\,mol\,OH^-}{1\,mol\,Mg(OH)_2} = \frac{0.900\,mol\,OH^-}{1L} = 0.900\,M\,OH^-$$

Workbook Problem 4.3

Determine the total molar concentration of ions in a 0.75 M solution of aluminum chloride.

Strategy: $AlCl_3$ is a strong electrolyte and completely dissociates in water.

Step 1: Determine the total number of moles of ions formed when $AlCl_3$ completely dissociates in water.

Step 2: Create a conversion factor comparing the total number of moles of ions in solution to 1 mol of $AlCl_3$.

Step 3: Use the conversion factor to calculate the molar concentration of ions in solution.

Section 4.4 Types of Chemical Reactions in Aqueous Solutions

There are three types of reactions that we're concerned with in this chapter: precipitation reactions, acid–base neutralization reactions, and oxidation–reduction reactions.

In **precipitation reactions**, two soluble components interact to form an insoluble product, which then falls out of solution.

In **acid–base neutralization reactions**, an acid and a base react to form water and a salt (*salt* is simply a generic term that refers to an ionic compound formed by an acid and a base). In acid–base reactions, both the acid and basic components of the mixture (H^+ and OH^-) are removed from the reaction when they combine to form a salt in neutral water.

In **oxidation–reduction reactions**, also called **redox** reactions, electrons are transferred between components in a reaction mixture. You can recognize redox reactions by, looking for uncharged elements on one side of the reaction that become part of ionic compounds on the other side of the reaction and *vice versa*.

Section 4.5 Aqueous Reactions and Net Ionic Equations

There are a number of ways of writing equations for reactions that occur in aqueous solutions.

Molecular equations are written as though the components are not dissociated: ions are kept together.

Complete ionic equations take the molecular equations and break down everything into ions. The only thing that doesn't break down is a precipitate (it stays as a solid in solution).

Net ionic equations are written so that only the ions that participate in the reaction are shown. Generally, ionic reactions contain ions that do not participate, but are unchanged as the reaction proceeds. These are known as **spectator ions**: they are removed when a net ionic equation is written.

Workbook Example 4.6

PROBLEM:
When solutions of lead(II) nitrate and sodium chloride are mixed, solid lead chloride and aqueous sodium nitrate are produced. Write molecular, ionic, and net ionic equations for this reaction.

SOLUTION:

Molecular Equation:
$$Pb(NO_3)_2 \ (aq) + NaCl \ (aq) \ \rightarrow \ PbCl_2 \ (s) \ + \ NaNO_3 \ (aq)$$

Balanced Molecular Equation:
$$Pb(NO_3)_2 \ (aq) + 2 \ NaCl \ (aq) \ \rightarrow \ PbCl_2 \ (s) \ + \ 2 \ NaNO_3 \ (aq)$$

Complete Ionic Equation:
$$Pb^{2+} \ (aq) + 2 \ NO_3^- \ (aq) + 2 \ Na^+ \ (aq) + 2 \ Cl^- \ (aq) \rightarrow PbCl_2 \ (s) + 2 \ Na^+ \ (aq) + 2NO_3^- \ (aq)$$

Spectator Ions:
$$Na^+ \ and \ NO_3^-$$

Net Ionic Equation:
$$Pb^{2+} \ (aq) + 2 \ Cl^- \ (aq) \rightarrow PbCl_2 \ (s)$$

Workbook Problem 4.4

When solutions of nitric acid and potassium hydroxide are mixed, a neutralization reaction occurs. Write the net ionic equation for this reaction.

Strategy: From the information given, write a molecular equation and determine if any of the reactants or products is a strong electrolyte.

Step 1:   Write and balance the molecular equation.

Step 2:   Write the complete ionic equation (from step 1).

Step 3:   List the spectator ions in the complete ionic equation.

Step 4:   Write the net ionic equation.

Section 4.6   Precipitation Reactions and Solubility Guidelines

Predicting the results of a precipitation reaction requires an understanding of **solubility**: ionic compounds that are highly soluble will not precipitate, or fall out of solution, while those with low solubility will fall out of solution.

Solubility guidelines are called that because they are only partially predictive; however, these guidelines will get you through most problems having to do with precipitation reactions:

- Anything with an alkali metal (group I) cation will be soluble.
- Anything with an ammonium cation (NH_4^+) will be soluble.
- Anything with a nitrate anion (NO_3^-), a perchlorate ion (ClO_4^-), or an acetate ion ($CH_3CO_2^-$) will be soluble.
- Halides (Cl^-, Br^-, and I^-) and sulfates (SO_4^{2-}) will be soluble, with some exceptions.
- For all other anions: "heavy" metals will fall out of solution, "light" metals will stay in: be careful with lead, silver, and mercury cations. Unless they are attached to a nitrate, they are unlikely to be soluble.

Workbook Example 4.7

PROBLEM:
For which of the following solutions would precipitations occur?
Na_2CO_3 and $Pb(NO_3)_2$, NH_4Cl and $NaCH_3CO_2$, $AgNO_3$ and $MgCl_2$.

SOLUTION:
Exchange the cations and anions and examine the possible products.

Na_2CO_3 and $Pb(NO_3)_2$ can yield $NaNO_3$ and $PbCO_3$.

Anything attached to lead is a suspect for precipitation, and $NaNO_3$ contains two ions identified as always soluble. $PbCO_3$ will precipitate.

NH_4Cl and $NaCH_3CO_2$ can yield $NH_4CH_3CO_2$ and $NaCl$

Anything attached to an ammonium ion will be soluble, as will anything attached to a sodium ion: there is no cation here that can generate a precipitate.

$AgNO_3$ and $MgCl_2$ can yield $AgCl$ and $Mg(NO_3)_2$.

Although chlorides are usually soluble, silver is always suspicious: silver chloride will precipitate out. Although magnesium is not listed as "always soluble," it is attached to a nitrate, which is always soluble; therefore, magnesium will not precipitate in this reaction.

Workbook Problem 4.5

Describe how to produce 1.3 grams of lead(II) chloride from 1 M solutions of two soluble salts.

Strategy: Using the solubility guidelines and solution stoichiometry, determine suitable reactants to use in the preparation of lead chloride and the amount of reactants needed.

Step 1: Determine reactants that are soluble and will produce the insoluble lead(II) chloride and another soluble product.

Step 2: Write a balanced chemical equation.

Step 3: Use solution stoichiometry to determine the volume of each reactant needed.

Section 4.7 Acids, Bases, and Neutralization Reactions

In the chapter so far, we have been using the definitions of acid and base proposed by Svante Arrhenius (1859–1927) in 1887: acids dissociate in water to form hydrogen ions and bases dissociate in water to form hydroxide ions.

H^+, as a bare proton, is an exceptionally strong ion. As a result, it does not truly exist free in solution, but binds itself to a water molecule to form a **hydronium ion** with formula H_3O^+.

Different acids will release different amounts of protons into a solution. Hydrochloric acid, HCl, is called a **monoprotic acid** because it has only one proton to release into solution; however, sulfuric acid, H_2SO_4, is a **diprotic acid**, and phosphoric acid, H_3PO_4, is a **triprotic acid.** In addition, protons differ in their strength depending on conditions. For example, it is more difficult for a proton to be removed from an already-charged unit: H_2SO_4 will dissociate completely to $HSO_4^- + H^+$, but HSO_4^- is a *weak* acid.

Acids and bases can be strong and weak, just as other electrolytes, and the definition means the same thing: it has only to do with the amount of dissociation and not the corrosiveness of a substance. A concentrated solution of a weak acid will cause a more serious burn than a dilute solution of a strong acid.

The following is a list of strong and weak acids and bases (there are many more than shown here):

- Strong acids: perchloric ($HClO_4$), sulfuric (H_2SO_4), hydrochloric (HCl), hydrobromic (HBr), and nitric (HNO_3). These acids fully dissociate.
- Weak acids: hydrofluoric (HF), phosphoric (H_3PO_4), and acetic (CH_3CO_2H).
- Strong bases: hydroxide metal salts are all strong bases.
- Weak bases: ammonia (NH_3), and any other compound that has the ability to remove hydrogen ions from solution.

Workbook Example 4.8

PROBLEM:
Determine if the following are acids, bases, or neutral salts: $Ca(OH)_2$, HCl (aq), NaCl, NaOH, $CaBr_2$

SOLUTION:
An acid will produce H^+ ions, a base will produce OH^- ions, and a neutral salt will produce neither.

$Ca(OH)_2$ – produces OH^- in solution; base   HCl (aq) – produces H^+ in solution; acid
NaCl – produces Na^+ and Cl^- in solution; neutral salt   NaOH – produces OH^- in solution; base
$CaBr_2$ – produces Ca^{2+} and Br^- in solution; neutral salt

---

Workbook Problem 4.6

Determine the reaction type, and write balanced molecular, complete ionic, and net ionic equations for the following reactions:

a. LiOH and H_2SO_4
b. NH_4Cl and $AgNO_3$

Strategy: Determine the reaction type by examining the reactants—are they acids or bases or neutral?

Step 1:   Write the balanced molecular equation for each reaction.

Step 2:   Determine the presence of any strong electrolytes. Write the complete ionic equation, showing the strong electrolytes in terms of their free ions.

Step 3:   Write the net ionic equation by removing spectator ions.

Section 4.8   Solution Stoichiometry

Molarity can be used as a conversion factor in stoichiometry calculations to determine volumes of solutions needed in reactions.

Workbook Example 4.9

PROBLEM:
Silver nitrate and calcium chloride, both soluble, react to form silver chloride, an insoluble solid. How much 0.10 M calcium chloride is needed to completely react with 100 mL of 0.5 M silver nitrate?

SOLUTION:
First, write the balanced equation for the reaction.

$$CaCl_2 + 2\ AgNO_3\ \rightarrow\ 2\ AgCl +\ Ca(NO_3)_2$$

Next, calculate the moles of silver nitrate present:

$$0.100\,L \times \frac{0.50\,mol}{1\,L} = 0.050\,mol\ silver\ nitrate$$

Using the balanced equation, calculate the number of moles of $CaCl_2$ needed to react with 0.050 mol silver nitrate:

$$0.050\,mol\ AgNO_3 \times \frac{1\,mol\ CaCl_2}{2\,mol\ AgNO_3} = 0.025\,mol\ CaCl_2$$

The volume of the $CaCl_2$ solution can now be calculated:

$$0.025\,mol\ CaCl_2 \times \frac{1\,L}{0.10\,mol\ CaCl_2} = 0.250\,L = 250\,mL\ CaCl_2$$

Workbook Problem 4.7

Hydrochloric acid reacts with zinc metal to produce a solution of zinc chloride and hydrogen gas. What volume of 5.0 M hydrochloric acid is needed to react with 4.75 g of zinc? If the density of hydrogen gas is 0.0899 g/L, what volume of hydrogen gas will be produced?

Strategy: To solve any stoichiometry problem, we need to know mole ratios of reactants and products as determined by the balanced chemical equation. Knowing the actual number of moles present of one reactant or product, we can then determine the number of moles needed of all other reactants and products.

Step 1:   Write a balanced chemical equation.

Step 2:   Determine the number of moles of zinc present.

Step 3:   Determine the number of moles of HCl needed.

Step 4:   Calculate the volume of 5.0 M HCl needed for this reaction.

Step 5:   To calculate the volume of H_2 gas produced, we first need to calculate the moles, and then the grams of H_2 gas produced.

Step 6:   Determine the volume of H_2 gas produced, using the grams of H_2 produced and the density of H_2 gas.

Section 4.9   Titrations

To determine the concentration of an unknown solution, a procedure called **titration** can be used. A carefully measured volume of the unknown solution is slowly reacted with a solution of known concentration. An **indicator**—a molecule that changes color when the reaction is complete—is generally used to indicate the endpoint of the reaction.

Workbook Example 4.10

PROBLEM:
Apple juice contains malic acid, a weak acid with formula $C_4H_6O_5$ that contains two acid groups. If 27 mL of 0.10 M sodium hydroxide is needed to neutralize 15 mL of a malic acid solution, what is the concentration of the solution?

SOLUTION:
First, write a balanced equation for the neutralization reaction.

$$C_4H_6O_5\ (aq)\ +\ 2\ NaOH\ (aq)\ \rightarrow\ Na_2C_4H_4O_5\ (aq)\ +\ 2\ H_2O\ (l)$$

From the information given, we can calculate the number of moles of sodium hydroxide used to reach the equivalence point of the titration.

$$0.027\,L\,NaOH \times \frac{0.10\,mol}{1\,L} = 0.0027\,mol\,NaOH$$

From the balanced equation, calculate the moles and molarity of malic acid.

$$0.0027\,mol\,NaOH \times \frac{1\,mol\,malic\,acid}{2\,mol\,NaOH} = 0.0014\,mol\,malic\,acid$$

$$\frac{0.0014\,mol\,malic\,acid}{0.015\,L} = 0.093\,M\,malic\,acid$$

Workbook Problem 4.8

A student determines that 16.37 mL of sulfuric acid is required to neutralize a 100.0 mL solution containing 0.153 g of NaOH. Determine the molarity of the nitric acid solution.

Strategy: Balance the equation to find mole ratios, determine moles of known compounds, then calculate moles of the unknown and determine concentration.

Step 1: Write a balanced chemical equation for the reaction.

Step 2: Calculate the number of moles of NaOH present.

Step 3: Determine the number of moles of sulfuric acid needed to react with that quantity of NaOH.

Step 4: Calculate the molarity of the acid solution.

Section 4.10   Oxidation–Reduction (Redox) Reactions

Originally, **oxidation** was defined as the gaining of oxygen and **reduction** was defined as the removal of oxygen. Although these definitions have been significantly broadened, this is still a useful first step in recognizing redox reactions.

Oxidation is now defined as *the loss of electrons.* When an element combines with oxygen, to use the historical definition, the metal loses electrons to become a cation, and the oxygen gains those same electrons to become an oxide ion:

$$2\,Mg + O_2 \rightarrow 2\,MgO$$

When electrons are lost from a chemical species, the result is a change in its charge *toward positive.* It is described this way because some elements will have more than one charge: a change from −2 to −1 is an *oxidation*, despite the fact that no positively charged species is generated.

Reduction is now defined as *the gaining of electrons.* If the reaction above is reversed, the Mg^{2+} ions gain two electrons to become elemental magnesium.

$$2\,MgO \rightarrow 2\,Mg + O_2$$

In reactions of this kind, the number of electrons lost by one species must be the number gained somewhere else: electrons do not just fly away!

So how do we keep track of all this? We use a method of electron "bookkeeping" that uses oxidation numbers to show where electrons are going in a reaction. Sometimes, the oxidation number indicates a charge, but not always.

Simplified rules for assigning oxidation numbers (make sure to refer also to the full-bore rules in your textbook):

- Elements have oxidation numbers of 0.
- Ions that contain only one atom have an oxidation number identical to their charge.
- In molecules or ionic compounds, atoms generally carry oxidation numbers equal to their usual ionic state
- Oxygen is usually −2
- Hydrogen is usually +1 (when bonded to a metal, it will be −1 because metals are prone to losing electrons)
- Halogens are usually −1
- Neutral compounds will have oxidation numbers on their atoms that sum to 0. Polyatomic ions will have oxidation numbers on their atoms that sum to the charge on the ion.

Workbook Example 4.11

PROBLEM:
Determine the oxidation number of each of the atoms in the following: HCl, NaOH, NH_3, H_3PO_4

SOLUTION:
For HCl, halogens are −1, hydrogen is +1. When these are added together, they give 0, correct for a neutral molecule.

$$H = +1$$
$$Cl = -1$$

For NaOH, Na carries its usual charge as its oxidation number, which is +1. Oxygen is −2, leaving hydrogen. Hydrogen's charge depends on whether it is bonded to a metal or a nonmetal, and here we have both as possibilities. However, this is a neutral compound, so the oxidation numbers must add up to 0, and hydrogen must carry an oxidation state of +1.

$$Na = +1$$
$$O = -2$$
$$H = +1$$

In NH_3, hydrogen is bound to a nonmetal and thus each carries an oxidation number of +1. This is a neutral compound, so N must carry an oxidation number of –3 to balance the three +1 numbers on the hydrogens.

$$N = -3$$
$$H = +1$$

In H_3PO_4, there are a number of things to consider. This compound is all nonmetals, so the hydrogens will carry oxidation numbers of +1. Oxygen carries a charge of –2, and there are four of them. This leaves phosphorus to balance things out. $(4 \times -2) + (3 \times +1) + P = 0$, so phosphorus carries an oxidation number of +5. This is clearly not a real charge; it is results of this kind that remind us that these are useful bookkeeping methods, but they do not necessarily reflect ionic reality.

Workbook Problem 4.9

Determine the oxidation number for each element in VO_4^{3-} and $NaVO_3$.

Strategy: Follow the rules above for determining oxidation states.

Step 1:   Identify the elements that have a fixed oxidation state.

Step 2:   Determine the oxidation number of the remaining elements, keeping in mind that the sum of all oxidation numbers must be equal to the ionic charge or to zero for a neutral molecule.

Section 4.11   Identifying Redox Reactions

A **reducing agent** causes reduction to occur. It is oxidized itself by losing one or more electrons, and its oxidation number increases.

Oxidizing agents cause oxidation to occur by accepting electrons, and is thereby reduced itself. The oxidation number of an oxidizing agent decreases.

When one species in a reaction loses electrons, another gains them. Oxidation and reduction must occur together. When a metal and a nonmetal react with one another, the metal will be the reducing agent, and the nonmetal will be the oxidizing agent (this is easy to remember, as oxygen is a nonmetal). Metals can also oxidize and reduce one another, depending on the strength of their attraction for electrons.

Workbook Example 4.12

PROBLEM:
Identify the species oxidized, the species reduced, the oxidizing agent, and the reducing agent in the following reaction:

$$Zn\ (s) + 2\ H^+\ (aq) + 2\ Cl^-\ (aq) \rightarrow Zn^{2+}\ (aq) + 2\ Cl^-\ (aq) + H_2\ (g)$$

SOLUTION:
To solve this problem, look for atoms that have had a change in oxidation number:

$$Zn \rightarrow Zn^{2+}$$

The oxidation number of zinc goes from 0 to +2. Zinc is oxidized and donates electrons to hydrogen, which is the reducing agent.

$$H^+ \rightarrow H_2$$

The oxidation number of hydrogen goes from +1 to 0. Hydrogen is reduced and accepts electrons from zinc, which is the oxidizing agent.

Workbook Problem 4.10

Identify the species oxidized, the species reduced, the oxidizing agent, and the reducing agent in the following reaction:

$$Cu\ (s) + 2\ Ag^+\ (aq) + 2\ NO_3^-\ (aq) \rightarrow 2\ Ag\ (s) + Cu^{2+}\ (aq) + 2\ NO_3^-\ (aq)$$

Strategy: Determine the oxidation number of all species present.

Step 1:   Identify the species that have a change in oxidation number.

Step 2:   Based on the change in oxidation number, identify the species oxidized, the species reduced, and the oxidizing and reducing agents.

Section 4.12   The Activity Series of the Elements

Predicting which metals will take electrons and which will donate electrons when they are together in aqueous solution can be done by using the periodic table, because metals in the periodic table are arranged in what is known as an *activity series*.

If a metal is above another metal in the activity series, it will give up electrons to that metal. So, if you are looking to see what will happen if you put magnesium ribbon into a copper sulfate solution, you will see that magnesium is above copper; therefore you will wind up with magnesium ions and copper metal.

Also in the activity series is H^+, though it is not a metal. Metals above H^+ will react with acids to form metal ions and hydrogen gas; those below it will not. The metals at the bottom of the activity series are the least-reactive metals (that is, the least likely to give up electrons) and those at the top are the most reactive. Those at the very top are so eager to ionize that they will react with water to form hydrogen gas and hydroxide ions.

Partial Activity Series for Oxidation Reactions

Strongly reducing	$Li \rightarrow Li^+ + e^-$	
	$K \rightarrow K^+ + e^-$	
	$Ba \rightarrow Ba^{2+} + 2\,e^-$	These elements react rapidly with aqueous
	$Ca \rightarrow Ca^{2+} + 2\,e^-$	acid (H^+) or H_2O to release H_2 gas.
	$Na \rightarrow Na^+ + e^-$	
	$Mg \rightarrow Mg^{2+} + 2\,e^-$	
	$Al \rightarrow Al^{3+} + 3\,e^-$	
	$Mn \rightarrow Mn^{2+} + 2\,e^-$	These elements react with aqueous H^+
	$Zn \rightarrow Zn^{2+} + 2\,e^-$	ions or with steam to release H_2 gas.
	$Cr \rightarrow Cr^{3+} + 3\,e^-$	
	$Fe \rightarrow Fe^{2+} + 2\,e^-$	
	$Co \rightarrow Co^{2+} + 2\,e^-$	
	$Ni \rightarrow Ni^{2+} + 2\,e^-$	These elements react with aqueous H^+
	$Sn \rightarrow Sn^{2+} + 2\,e^-$	ions to release H_2 gas.
	$\mathbf{H_2 \rightarrow 2\,H^+ + 2\,e^-}$	
	$Cu \rightarrow Cu^{2+} + 2\,e^-$	
	$Ag \rightarrow Ag^+ + e^-$	These elements do not react with aqueous
	$Hg \rightarrow Hg^{2+} + 2\,e^-$	H^+ ions to release H_2 gas.
Weakly reducing	$Pt \rightarrow Pt^{2+} + 2\,e^-$	
	$Au \rightarrow Au^{3+} + 3\,e^-$	

Workbook Example 4.13

PROBLEM:
Using the activity series, write balanced chemical equations for the following reactions:

$Cu\,(s) + Co^{2+}\,(aq) \rightarrow$ $Mn\,(s) + Zn^{2+}\,(aq) \rightarrow$ $Sn\,(s) + HCl\,(aq) \rightarrow$

$Pt\,(s) + Hg^{2+}\,(aq) \rightarrow$ $Al\,(s) + Zn^{2+}\,(aq) \rightarrow$ $Ni\,(s) + Mn^{2+}\,(aq) \rightarrow$

$Ni\,(s) + HCl\,(aq) \rightarrow$ $Pt\,(s) + HBr\,(aq) \rightarrow$

SOLUTION:
To predict the outcome of these reactions, remember that any element higher in the activity series will react with the ion of any element lower in the activity series. Also remember that metals above the H^+ ion in the activity series will displace the hydrogen ion from an acid to form H_2 gas.

$$Cu\,(s) + Co^{2+}\,(aq) \rightarrow \text{no reaction} \qquad Co^{2+} \text{ lies above Cu in the activity series}$$

$$Mn\,(s) + Zn^{2+}\,(aq) \rightarrow Mn^{2+}\,(aq) + Zn\,(s)$$

$$Sn\,(s) + 2\,HCl\,(aq) \rightarrow SnCl_2\,(aq) + H_2\,(g)$$

$$Pt\,(s) + Hg^{2+}\,(aq) \rightarrow \text{no reaction} \qquad Hg^{2+} \text{ lies above Pt in the activity series}$$

$$2\,Al\,(s) + 3\,Zn^{2+}\,(aq) \rightarrow 2\,Al^{3+}\,(aq) + 3\,Zn\,(s)$$

$$Ni\,(s) + Mn^{2+}\,(aq) \rightarrow \text{no reaction} \qquad Mn^{2+} \text{ lies above Ni in the activity series}$$

$$Ni\,(s) + 2\,HCl\,(aq) \rightarrow NiCl_2\,(aq) + H_2\,(g)$$

$$Pt\,(s) + HBr\,(aq) \rightarrow \text{no reaction} \qquad H^{+} \text{ lies above Pt in the activity series.}$$

Workbook Problem 4.11

Using the activity series, write balanced chemical equations for the following reactions:

    a. $Cu\,(s) + HCl\,(aq) \rightarrow$ b. $Mn\,(s) + ZnCl_2\,(aq) \rightarrow$

    c. $Mg\,(s) + Al(NO_3)_3\,(aq) \rightarrow$

Strategy: Remember that any element higher in the activity series will react with the ion of any element lower in the activity series. Also remember that metals above the H^{+} ion in the activity series will displace the hydrogen ion from an acid to form H_2 gas.

Step 1:   Predict the outcome of these reactions.

Section 4.13   Redox Titrations

Just as the concentrations of acids and bases in solution can be determined by titration using an indicator, the concentration of an oxidizing or reducing agent can be similarly determined if the following conditions are met:

- The reaction must have a 100% yield
- A color change in one of the reactants must signal the endpoint.

Because many redox reagents are strongly colored, this can be a useful method. A balanced equation is of course required, as is a known quantity of one of the reactants.

Workbook Problem 4.12

A 50 mL solution of iodine and starch was titrated with an acidic 0.15 M thiosulfate solution. After 15.7 mL of the thiosulfate solution had been added, the blue color of the iodine:starch complex completely disappeared, indicating that the iodine had been completely reduced to I^- ions. What was the original molar concentration of the iodine in solution?

The unbalanced equation is

$$I_2 + S_2O_3^{2-} \rightarrow I^- + S_4O_6^{2-}$$

Strategy: Balance the equation, proceed as for an acid–base titration.

Step 1:   Determine the number of moles of thiosulfate ion that were required to react with the iodine solution to completely reduce the iodine.

Step 2:   Balance the equation.

Step 3:   Use mole ratios to determine the amount of iodine present in the reaction mixture.

Step 4:   Determine the original concentration of I_2 in solution.

Section 4.14   Some Applications of Redox Reactions

Reactions such as the ones discussed here may seem like they are rare and confined to the laboratory, but in reality, electrons move around in many different types of reactions that occur in daily life. All of the following have a redox component:

- Combustion: the oxidation state of oxygen shifts from 0 when it is an element, to −2 after a reaction occurs. This causes a reduction in whatever is being burned.
- Bleaching: strong oxidizing agents such as peroxides, hypochlorite, or elemental chlorine can destroy colored molecules.
- Batteries: not surprisingly, the ability to generate a current, which is a flow of electrons, relies on this type of reaction.
- Metallurgy: most metals are found in the Earth's crust as oxides, and thus must be reduced to get their elemental forms.

- Corrosion, or rusting, is simply combustion writ small. It is a slower process, but if you set steel wool on fire, or leave it out in the rain, you will end up with the same product: rust.
- Cellular respiration: This is a combustion reaction that operates in an extremely controlled manner, but when reduced to its basics, glucose is combined to form carbon dioxide (which you then breathe out), water, and the heat that maintains your body temperature.

Self–Test

This section is intended to test your knowledge of the material covered in this chapter. Think through these problems, and make certain you understand what is going on. Ask yourself if your answer makes sense. Many of these questions are linked to the chapter learning goals. Therefore, successful completion of these problems indicates you have mastered the learning goals for this chapter. You will receive the greatest benefit from this section if you use it as a mock exam. You will then discover which topics you have mastered and which topics you need to study in more detail.

True–False

1. In acid–base reactions, an acid and a base react to form carbon dioxide and water.

2. Sodium nitrate will precipitate out of aqueous solution.

3. When electrons move from one species to another in chemical reactions, this is known as oxidation (if electrons are lost) and reduction (if electrons are gained).

4. Weak electrolytes do not fully dissociate.

5. Spectator ions do not participate in precipitation reactions.

6. Acids release nitronium ions into solution.

7. NaCl is a strong electrolyte.

8. Oxidations are always equal to ionic charge.

9. An activity series makes it possible to predict the outcome of precipitation reactions.

10. Redox reactions can be broken into half-reactions to make them easier to balance.

Matching

11. Reduction	a. dissociating only slightly
12. Net ionic equation	b. reactions involving transfer of electrons
13. Oxidation number	c. H_3O^+
14. Weak electrolyte	d. ranking of elements by reducing ability

15. Redox reaction   e. substance that gives up electrons

16. Dissociate   f. reaction description without spectators

17. Spectator   g. substance that accepts electrons

18. Hydroxide ion   h. gain of electrons

19. Activity series   i. loss of electrons

20. Electrolyte   j. ion that does not participate in a reaction

21. Reducing agent   k. reaction in which two soluble salts form an insoluble product

22. Precipitation reaction   l. measure of electron richness

23. Oxidation   m. substance that will conduct electricity when dissolved

24. Hydronium ion   n. OH^-

25. Oxidizing agent   o. separate into ions in solution

26. Strong electrolyte   p. substance that dissociates to produce hydroxide ions in solution

27. Base   q. substance that dissociates to produce hydronium ions in solution

28. Acid   r. complete dissociation

Fill-in-the-Blank

29. When _____ reactants yield a(n) _____ product, removing some of the dissolved ions, a(n)_____ reaction has occurred.

30. In neutralization reactions, a(n) _____ and a(n) _____ react to form _____ and _____.

31. In _____ reactions, electrons are transferred between reactants.

32. Acids, bases, or electrolytes that dissociate completely in solution are known as _____, while those that dissociate incompletely are known as _____.

33. A(n) _____ shows all reactants in their undissociated forms, a(n) _____ shows all reactants dissociated, and a(n) _____ shows only those reactants that have undergone a change.

34. In the Arrhenius definition, acids dissociate in aqueous solution to form an anion and a(n)

_____ ion, while bases dissociate to form a cation and a(n) _____ ion.

35. When a substance is oxidized, it _____ electrons; when it is reduced, it _____

electrons.

Problems
36. Classify each reaction as a precipitation, neutralization, or redox reaction, and justify your
    answer.
    a. HNO_3 (aq) + KOH (aq) → KNO_3 (aq) + H_2O (l)
    b. $Cu(NO_3)_2$ (aq) + 2 NaOH (aq) → $Cu(OH)_2$ (s) + 2 $NaNO_3$ (aq)
    c. $Cu(NO_3)_2$ (aq) + Zn (s) → $Zn(NO_3)_2$ (aq) + Cu (s)

37. If 15 g of Na_3PO_4, a strong electrolyte, are dissolved in 250 mL of solution, what is the total
    molar concentration of ions?

38. Write balanced molecular, ionic, and net ionic equations for the following reactions in
    aqueous solution:
    a. lead(II) nitrate + sodium chloride   b. sulfuric acid + lithium hydroxide
    c. potassium chloride + silver acetate

39. Given 1.00 M solutions of $Ca(NO_3)_2$ and KOH, how much of each would need to be
    combined to form 1.8 g of $Ca(OH)_2$?

40. Predict reactions for each of the following. If a reaction occurs, write the net ionic equation.
    If no reaction occurs, list the possible soluble products.
    a. $Cu(CH_3CO_2)_2$ + $Pb(NO_3)_2$ b. $NaNO_3$ + $(NH_4)_2SO_4$
    c. $CaCl_2$ + $AgNO_3$

41. Assign oxidation numbers to all of the atoms in the following:
    a. Cl_2 b. NaCl
    c. $CO_3{}^{2-}$ d. Cr_2O_3
    e. $ClO_3{}^-$

42. For each unbalanced equation given below, identify the oxidizing agent and the reducing
    agent.
    a. C_2H_6 + O_2 → CO_2 + H_2O b. Na + Cl_2 → NaCl
    c. Mg + Fe^{2+} → Mg^{2+} + Fe   d. Ca + H^+ → Ca^{2+} + H_2

43. Given the activity series provided here, predict the outcome of
    the following reactions. When a reaction occurs, balance the
    equation.
    a. chromium and copper(II) nitrate
    b. tin and silver nitrate
    c. copper(II) nitrate and cobalt
    d. silver and hydrochloric acid

Activity Series
Li → Li^+ + e^-
Ba → Ba^{2+} + $2e^-$
Na → Na^+ + e^-
Cr → Cr^{3+} + $3e^-$
Co → Co^{2+} + $2e^-$
Sn → Sn^{2+} + $2e^-$
H_2 → $2H^+$ + $2e^-$
Cu → Cu^{2+} + $2e^-$
Ag → Ag^+ + e^-

44. Calculate the molarity of an Na_2SO_3 solution if 15.0 mL of the solution reacts completely with 23.7 mL of 0.124 M $K_2Cr_2O_7$ in acid solution according to the following unbalanced equation:

$$SO_3^{2-} + Cr_2O_7^{2-} \rightarrow SO_4^{2-} + 2\ Cr^{3+}$$

45. 5.3 g of potassium sulfate are combined with 4.9 g of lead nitrate in aqueous solution.
    a. Write the molecular, complete ionic, and net ionic equations for this reaction.
    b. How many grams of solid will be produced?
    c. How much of the excess reactant will be left over?
    d. If the experimental yield is 4.37 g, what is the percent yield?
    e. What will be the molar concentration of ions remaining in the solution if the reaction mixture is made up to 200 mL?

Challenge Problems
46. When sodium or lithium are added to water, the following rather exciting reaction occurs according to the following unbalanced equation:

$$M\ (s) + H_2O\ (l) \rightarrow MOH\ (aq) + H_2\ (g)$$

Enough heat is typically generated by this reaction that the hydrogen gas burns as it is generated. If a 4 g mixture of lithium and sodium is added to water, and the resulting solution titrated, 59.7 mL of 7.5 M HCl is required to neutralize the solution. How many grams of lithium and sodium were contained in the original mixture?

47. A 5.36 g sample that contains some elemental tin is dissolved in strong acid to give a solution of Sn^{2+} (aq). This is then titrated with a solution that is 0.0563 M in NO_3^-, which is reduced to NO (g). If the equivalence point is reached at 27.6 mL, what is the percentage of tin in the original sample?

CHAPTER FIVE

Periodicity and the Electronic Structure of Atoms

Learning Objectives

As a result of reading and studying this chapter, you should be able to

Section 5.1	**The Nature of Radiant Energy and the Electromagnetic Spectrum**
1.	Label the wavelength, frequency, and amplitude in an electromagnetic wave and understand what they mean (Figure 5.3).
2.	Interconvert between wavelength and frequency of electromagnetic radiation.
Section 5.2	**Particlelike Properties of Radiant Energy: The Photoelectric Effect and Planck's Postulate**
3.	Calculate the energy of electromagnetic radiation in units of J/photon or kJ/mol given the frequency or wavelength.
4.	Describe the photoelectric effect and explain how it supports the theory of particlelike properties of light (see Figures 5.4 and 5.5).
5.	Calculate the frequency or wavelength of radiation needed to produce the photoelectric effect given the work function of a metal.
Section 5.3	**The Interaction of Radiant Energy with Atoms: Line Spectra**
6.	Describe the difference between a continuous spectrum and a line spectrum.
Section 5.4	**The Bohr Model of the Atom: Quantized Energy**
7.	Compare the wavelength and frequency of different electron transitions in the Bohr model of the atom.
8.	Relate wavelengths calculated using the Balmer–Rydberg equation to energy levels in the Bohr model of the atom.
Section 5.5	**Wavelike Properties of Matter: de Broglie's Hypothesis**
9.	Calculate the wavelength of a moving object using the de Broglie equation.
10.	Explain why the wavelength of macroscopic objects is not visible.
Section 5.6	**The Quantum Mechanical Model of the Atom: Heisenberg's Uncertainty Principle**
11.	Calculate the uncertainty in position of moving objects if the velocity is known.

Chapter Summary

The birth of quantum mechanics at the turn of the last century greatly expanded the understanding of the atomic nature of matter. This discovery blurred the lines between matter and energy, demonstrating that at the smallest levels, the matter and energy share properties: matter is wave-like, and light is particle like. The observation of line spectra and the development of the Balmer–Rydberg equation, though it was some years before it was fully explained, was the first step in understanding the quantization of energy: it led to the Bohr model of the hydrogen atom, which was soon greatly expanded by Schrödinger's wave functions. The variety of orbital shapes, and the development of the system of quantum numbers to identify each electron in an element, helped to solidify the remarkable depth of information contained in the periodic table. These discoveries also resolved reactivity and periodic properties of the elements that seemed paradoxical, such as the decrease in atomic radius as the atomic number increases across the periodic table.

The Chapter in Detail

Section 5.1   The Nature of Radiant Energy and the Electromagnetic Spectrum

Although we experience them quite differently, visible light, infrared and ultraviolet radiation, and X-rays are all forms of **electromagnetic energy**. Collectively, they make up the **electromagnetic spectrum**.

Electromagnetic energy travels in waves. Electromagnetic waves can be characterized using four properties:

- *Frequency:* the number of waves that pass by a certain point in one second. The units of measurement are per second, abbreviated s^{-1} or 1/s, or given the name hertz (Hz). Frequency is indicated by the Greek lowercase *nu* (v).
- *Wavelength:* the length of one complete wave, usually measured from peak to peak. The units used are meters or nanometers (1×10^{-9} m) and the wavelength is indicated by the Greek lambda – λ.
- *Amplitude:* The height of a wave measured from the center line between peak and trough and proportional to its intensity.
- *Speed:* when wavelength (m) and frequency (1/s) are multiplied together, the result is m/s. All electromagnetic radiation moves at the same speed: 3.00×10^8 m/s, which is the speed of light, indicated by the letter "c."

Because all electromagnetic radiation has the same speed, it is possible to use this speed of light as a conversion factor to convert between wavelength and frequency:

$$c = \lambda v$$

which can also be rearranged to

$$\lambda = \frac{c}{v}$$

and

$$v = \frac{c}{\lambda}$$

It is also helpful to remember the results of this relationship: a longer wavelength is lower in frequency, and a shorter wavelength is higher in frequency.

Workbook Example 5.1

PROBLEM:
Blu-Ray lasers typically operate at 473 nm. What is their frequency?

SOLUTION:
Although it is not stated in the problem, you have a three-variable equation and two of the variables: wavelength and speed. Remember that c is a constant. Also remember that the speed of light is in units of m/s, which means that the 473 nm *must* be converted to m!

$$v = \frac{3.00 \times 10^8 \; m/s}{4.73 \times 10^{-7} \; m} = 6.3 \times 10^{14} \; Hz$$

Note: you will quickly notice that a frequency in this range (10^{14} Hz) is typical of visible light.

Workbook Problem 5.1

Your college radio station transmits at 88.5 MHz. What is the wavelength associated with this frequency?

Strategy: Use the relationship between speed, wavelength, and frequency.

Step 1:   Choose the appropriate form of the equation, and solve.

Section 5.2   Particlelike Properties of Radiant Energy: The Photoelectric Effect and Planck's Postulate

The *photoelectric effect* refers to the fact that light of a specific frequency can eject electrons from the surface of a clean metal. In 1905, Albert Einstein proposed that this was due to light behaving like streams of small particles or **photons**. Below that specific threshold frequency, there is not enough energy per photon to eject an electron.

The relationship of energy to frequency can be expressed thus:

$$E = hv$$

or, since $c = \lambda v$, we can relate the energy of a photon to its frequency:

$$E = \frac{hc}{\lambda}$$

where "h" is Planck's constant, which has a value of 6.626×10^{-36} J·s. This gives the energy of one photon of a specific frequency or wavelength.

Higher frequencies and shorter wavelength produce higher energy. This is borne out in the way humans experience the electromagnetic spectrum: no one ever got a radio-wave burn, but UV radiation damages skin, and X-rays can do even deeper damage, because they have enough energy to pass right through the body and potentially damage living tissue.

Matter is quantized at the level of molecules and atoms, and electromagnetic energy is quantized in **photons**. Using this information, Niels Bohr developed a model of the hydrogen atom as a proton with one electron that had discrete orbits available to it: the electron could be *here*, or *here*, but not in between. This explained the phenomenon of line spectra, and allowed the numbers n and m in the Balmer–Rydberg equation to be identified as energy levels.

The energy of photons emitted is related to the difference in energy levels in the atom.

Workbook Example 5.2

PROBLEM:
If one mole of photons has an energy of 53.7 kJ, what is the wavelength of the light from which the photons came?

SOLUTION:
We need to rearrange the equation $E = h\nu$ to calculate the wavelength. However, the given energy must first be converted from kJ/mol to J/photon.

$$\frac{53.7\,kJ}{1\,mol} \times \frac{1\,mol}{6.022\times10^{23}\,photons} \times \frac{1000\,J}{1\,kJ} = 8.92\times10^{-20}\,J/photon$$

Now, rearrange the equation to solve for wavelength and put in the known values:

$$\lambda = \frac{hc}{E} = \frac{\left(6.626\times10^{-34}\,J\cdot s\right)\left(3.00\times10^{8}\,m/s\right)}{8.92\times10^{-20}\,J/photon} = 2.23\times10^{-6}\,m = 22.3\,nm$$

Workbook Problem 5.2

The threshold frequency of sodium metal is 4.4×10^{14} Hz. What is the wavelength and color of light to which this corresponds, and what is the energy of one mole of photons at this frequency?

Strategy: Use the equation for the energy of a photon, and multiply by Avogadro's number. Then convert frequency to wavelength.

Step 1:   Determine the energy of one photon, then convert to one mole of photons.

Step 2:   Convert the frequency to wavelength.

Step 3:   Consult your textbook for the color of this wavelength of light.

Section 5.3   The Interaction of Radiant Energy with Atoms: Line Spectra

Normal sunlight has a continuous spectrum. It contains all wavelengths of visible light, and it even extends outside of the visible range to ultraviolet and infrared.

When individual elements are subjected to electricity or heat, they are said to be "excited." When this happens, they give off light only in specific wavelengths unique to that element, visualized as a **line spectrum**.

In 1885, Johannes Balmer was able to mathematically describe the pattern in the visible atomic line spectrum of the hydrogen atom. His equation was later extended by Johannes Rydberg to describe all the lines of the spectrum, including those outside the visible range.

The Balmer–Rydberg equation states that

$$\frac{1}{\lambda} = R\left[\frac{1}{m^2} - \frac{1}{n^2}\right]$$

or

$$v = Rc\left[\frac{1}{m^2} - \frac{1}{n^2}\right]$$

where R is the *Rydberg constant* – 1.097×10^{-2} /nm, and n and m are integers such that n > m.

Workbook Example 5.3

PROBLEM:
Calculate the wavelength of light emitted when an electron falls from the $n = 8$ level to the $n = 3$ level in the hydrogen atom.

SOLUTION:
We solve this problem by using the Balmer–Rydberg equation. Assign the higher energy level to be "n," and the lower level to be "m."

$$\frac{1}{\lambda} = R\left[\frac{1}{m^2} - \frac{1}{n^2}\right]$$

$$\frac{1}{\lambda} = 1.097 \times 10^{-2}\, nm^{-1}\left[\frac{1}{3^2} - \frac{1}{8^2}\right] = 0.00105\, nm^{-1}$$

Don't forget to invert the answer—nm^{-1} is not the desired unit!

$$\frac{1}{\lambda} = \frac{1}{0.00105\, nm^{-1}} = 955\, nm$$

This is in the infrared range.

Workbook Problem 5.3

The Pfund series of spectral bands has m = 5 and n > m. What are the longest and shortest wavelengths in this series?

Strategy: Determine the values of n that will make λ the longest and shortest. Remember that the value of λ is greatest when the value of n is smallest, and the value λ is smallest when the value of n is greatest.

Step 1:   Determine the value of n that will make λ the shortest and longest.

Step 2:   Use the Balmer–Rydberg equation with $m = 5$ and solve for the shortest wavelength.

Step 3:   Solve for λ using the Balmer–Rydberg equation and the values of m that make λ the longest.

Section 5.4   The Bohr Model of the Atom: Quantized Energy

The work of Balmer and Rydberg that gave rise to the mathematical model they developed was done before the need for a theory of atomic structure to explain the way line spectra existed; in fact, their work was done before electrons were officially "discovered" by J. J. Thomson in 1896! Between the years of 1888 (when Balmer and Rydberg did their work) and 1913, Planck put forth his theory of quantized energy (1900), Einstein solved the mystery of the photoelectric effect (1905), and the Rutherford performed his gold foil experiment (1911). Building upon all of these ideas, Niels Bohr proposed a model of the hydrogen atom based on the experimental data obtained from line spectra.

Bohr's theory of the hydrogen atom included:
- The nucleus was in the center of the atom (this agreed with Rutherford's theory)
- The electron circled around the nucleus in an orbit
- The energy levels of the orbits are quantized so that only certain values of energy can correspond to specific orbits

When an electron moves from a higher energy level to a lower one (for example, from n = 3 to n = 2), there is an energy change that's associated with that transition. Taking a page from Planck's playbook, Bohr said that this change in energy has to be equal to

$$\Delta E = E_{final} - E_{initial} = h\nu$$

and that when an electron is moving from a lower energy level to a higher one (where n_{intial} is less than n_{final}), then the value of ΔE should be positive because the electron has absorbed a photon of energy. Likewise, if the electron moves from a higher energy level to a lower one (so $n_{initial}$ is larger than n_{final}), the value of ΔE should be negative, which tells us that a photon of energy has been released from the electron.

The Balmer–Rydberg equation takes this part of the Bohr theory into account by using the variables *m* and *n*; *n* is the value of the higher-energy orbit and *m* is the value of the lower-energy orbit.

Although Bohr did a great job explaining the line spectrum of the hydrogen atom, there were several problems that arose from this theory:
- It only works for atoms who only have one electron (such as He^+, Li^{2+}, and so on).
- Electrons do not move in fixed, defined orbits.

Despite these limitations, Bohr's theory of the hydrogen atom was seminal to the rest of quantum theory; he was awarded the Nobel Prize in Physics in 1922 for his work.

Workbook Example 5.4

PROBLEM:
Use the Balmer–Rydberg equation to calculate the wavelength in nanometers of the spectral line for hydrogen when *n* = 6. What is the energy in kilojoules per mole of the radiation corresponding to this line?

SOLUTION:
Remember that one form of the Balmer–Rydberg equation allows us to solve for the wavelength, λ:

$$\frac{1}{\lambda} = R\left[\frac{1}{m^2} - \frac{1}{n^2}\right]$$

We are given *n* = 6, and since we're dealing with the Balmer series (see Figure 5.8 in your text), we know that *m* = 2. Since R is the Rydberg constant (1.097×10^{-2} nm^{-1}), we can solve for λ, and then, since the Planck relation states that

$$E = \frac{hc}{\lambda}$$

we can solve for the energy (remember that h and c are constants) based on our answer from the Balmer–Rydberg equation.

The first step is to solve the Balmer–Rydberg equation for λ:

$$\frac{1}{\lambda} = 1.097 \times 10^{-2}\, nm^{-1}\left[\frac{1}{2^2} - \frac{1}{6^2}\right]$$

$$\frac{1}{\lambda} = 1.097 \times 10^{-2} \, nm^{-1} \left[\frac{1}{4} - \frac{1}{36} \right]$$

$$\frac{1}{\lambda} = 1.097 \times 10^{-2} \, nm^{-1} \left[\frac{2}{9} \right]$$

$$\frac{1}{\lambda} = 2.438 \times 10^{-3} \, nm^{-1}$$

$$\lambda = \frac{1}{2.437 \times 10^{-3} \, nm^{-1}} = 410.2 \, nm$$

The wavelength of the spectra line is 410.2 nm. The next step is to solve for the energy of the radiation using the Planck relation, but in order to use this equation, we have to convert the wavelength we just found from *nm* to *m*:

$$410.2 \, nm \times \frac{1 \, m}{1 \times 10^9 \, nm} = 4.102 \times 10^{-7} \, m$$

With that conversion done, we can solve for the energy of radiation:

$$E = \frac{hc}{\lambda}$$

$$E = \frac{\left(6.626 \times 10^{-34} \, J \cdot s \right) \left(3.00 \times 10^8 \, m/s \right)}{4.102 \times 10^{-7} \, m} = 4.846 \times 10^{-19} \, J/photon$$

Remember that the question asked us for the answer in units of kJ/mol, so we need to convert:

$$\frac{4.846 \times 10^{-19} \, J}{1 \, photon} \times \frac{1 \, kJ}{1000 \, J} \times \frac{6.022 \times 10^{23} \, photons}{1 \, mole} = 291.8 \, kJ/mol$$

Workbook Problem 5.4

Calculate the energy required (in kJ/mol) to remove an electron from the third shell of a hydrogen atom.

Strategy: Identify *n* and *m*, solve for wavelength, then energy, and convert to kJ/mol.

Step 1:   Identify *n* and *m* for this problem.

Step 2:   Solve the Balmer–Rydberg equation to get the wavelength of the energy needed.

Step 3:   Solve for the energy of one photon of this energy.

Step 4:   Convert to kJ/mol.

Section 5.5   Wavelike Properties of Matter: de Broglie's Hypothesis

In 1925, Louis de Broglie suggested that if light can behave like a particle, it is only logical that particles can also behave like light, so that all matter will have wavelike properties. He supported his proposal with a mathematical derivation.

Einstein proposed this now-famous relationship between mass and energy: $E = mc^2$. Using the known relationship between wavelength and frequency, and substituting simple velocity (v) for the speed of light, Einstein came up with the following equation that can be used to solve for the wavelength of any species of matter:

$$\lambda = \frac{h}{mv}$$

In this equation, h is also rearranged: the unit of Joule is kg•m^2/s or mass times velocity. For problems involving the de Broglie equation, use h = 6.626×10^{-34} kg•m^2/s.

Workbook Example 5.5

PROBLEM:
The muzzle velocity of a 50-gram sniper bullet is 860 m/s. The sniper must consider wind, movement of target, humidity, and altitude. What is the de Broglie wavelength of the ammunition at this speed? Could the de Broglie wavelength contribute to a missed shot?

SOLUTION:
Using the de Broglie equation, we can solve for wavelength:

$$\lambda = \frac{h}{mv} = \frac{6.626 \times 10^{-34}\ kg \cdot m^2/s}{(0.050\,kg)(860\,m/s)} = 1.5 \times 10^{-35}\ m$$

The de Broglie wavelength of the bullet is less than the diameter of an atom. But even if I might miss my shot, I'd still give it a try!

Section 5.6   The Quantum Mechanical Model of the Atom: Heisenberg's Uncertainty Principle

The Bohr model was an essential step in the understanding of the structure of matter, but, like the Balmer–Rydberg equation, it only works for hydrogen. Its most important insight was the idea that the electrons in an atom are confined to specific energy levels.

Erwin Schrödinger proposed the **quantum mechanical model** of the atom in 1926. In this model, the idea of electrons as particles is abandoned in favor of their wavelike properties.

Just one year later, Werner Heisenberg provided support for this idea by demonstrating that it is impossible to know simultaneously both where an electron is and what path it follows. Any attempt to "see" an electron's position would require that it interact with photons of light. As these photons would transfer energy to the electron, the speed of the electron would be altered by the interaction.

This idea is called the Heisenberg uncertainty principle, and it can be expressed mathematically:

$$(\Delta x)(\Delta mv) \geq \frac{h}{4\pi}$$

Expressed in words, it means that the uncertainty in the position of the object (x), multiplied by the uncertainty in its momentum (mass × velocity), is greater than or equal to Planck's constant divided by 4π.

Workbook Example 5.6

PROBLEM:
The mass of an electron is 9.11×10^{-31} kilograms, and its velocity, known to within 15%, is 2.50×10^{6} m/s. What is the uncertainty in its position?

SOLUTION:
First, determine the uncertainty in the velocity:

$$0.15\left(2.5 \times 10^{6} \, m\!\!\diagup\!\!_{s}\right) = 3.75 \times 10^{5} \, m\!\!\diagup\!\!_{s}$$

The Heisenberg uncertainty principle equation can be re-arranged to solve for Δx:

$$(\Delta x)(\Delta mv) \geq \frac{h}{4\pi}$$

$$\Delta x \geq \frac{\dfrac{h}{4\pi}}{(\Delta mv)}$$

$$\Delta x \geq \frac{h}{4\pi(\Delta mv)}$$

$$\Delta x \geq \frac{6.626 \times 10^{-34}\ kg \cdot m^2/s}{4\pi \left(9.11 \times 10^{-31}\ kg\right)\left(3.75 \times 10^5\ m/s\right)}$$

$$\Delta x \geq 1.54 \times 10^{-10}\ m\ or\ 154\ pm$$

Although these numbers are quite small, so are the objects in question. Like de Broglie wavelengths, these calculations lose their relevance in the macroscopic world.

Section 5.7   The Quantum Mechanical Model of the Atom: Orbitals and Quantum Numbers

Although both the position and momentum of an electron cannot be precisely known, its probable location can be calculated using a differential equation known as Schrödinger's wave equation. The solutions to this equation are called *wave functions* or *orbitals*, and they are represented by the Greek letter ψ (psi). ψ^2 defines the probability of finding the electron in a specific area around the nucleus.

Three parameters are required for wave functions: these are called the **quantum numbers**, and they are defined using the following rules:

The **principal quantum number** is designated *n*. It defines the energy level of the electron, and it can be any positive integer. It also defines the *shell* of the electron. Electrons with *n* = 5, for example, are said to be in the fifth shell.

The **angular momentum quantum number** is designated *l*. It defines the shape of the orbital, and it can be any integer (including zero) *less than n*. These numbers also define the *subshell* of the electron by the following system:

l = 0	*s* subshell	*l* = 1	*p* subshell
l = 2	*d* subshell	*l* = 3	*f* subshell

The **magnetic quantum number** is designated "m_l". It defines the orientation of an orbital, and it can be any integer from –*l* to *l*. If the angular momentum quantum number is 2, for example, m_l can be –2, –1, 0, 1, or 2.

The energy levels for hydrogen electrons are dependent only on the principal quantum number. In multielectron atoms, things are complicated by the fact that electrons can interact with one another as well as with the nucleus. In general, as energy is required to separate positive and negative charges, an electron with a larger value of *n* will be higher in energy.

Workbook Example 5.7

PROBLEM:
Give the values of the quantum numbers of a 2p subshell.

SOLUTION:
For a 2p subshell, $n = 2$. The fact that it is a p orbital says that l must be 1. Because m_l can be any number from $-l$ to l, it has possible values of -1, 0, and 1. There are three orbitals in a 2p subshell, because there are three possible orientations (as shown by the three magnetic quantum numbers).

Workbook Problem 5.5

Determine the subshell for an electron having $n = 3$, $l = 0$. How many orbitals are in this subshell?

Strategy: Determine the subshell associated with a value of $l = 0$.

Step 1:   Identify the subshell with the value of n and the letter designation for $l = 0$.

Step 2:   Determine the number of orbitals in this subshell.

Section 5.8   The Shapes of Orbitals

The s orbitals are spherical, with the probability of finding an electron dependent only on the distance from the nucleus. The size increases in successively higher shells, with nodes—areas of zero probability—in between the shells. This corresponds to the wave-like nature of electrons: an area of zero probability corresponds to an area of zero amplitude in a wave. There is only one s orbital per shell ($l = 0$, so $m_l = 0$).

The p orbitals are commonly described as dumbbell-shaped, with a node near the nucleus. As expected from the quantum number rules, there are three p orbitals per shell: $l = 1$, so $m_l = -1$, 0, or 1. These are defined as occurring along the x-axis, y-axis, z-axis of a Cartesian coordinate system. The two lobes of a p orbital have different phases, which is important for covalent bonding: only electrons that are in the same phase can interact. As the principal quantum number increases, the size of the p orbitals also increases.

The d orbitals have a more complex shape. With an angular-momentum quantum number of 2, there are five orbitals in each d subshell: $m_l = -2$, -1, 0, 1, 2. Four of the orbitals are shaped like paired p orbitals, 90 degrees rotated from one another; the fifth is shaped like a p orbital with a doughnut-shaped "skirt" in the xy plane. As in the p orbitals, the alternating lobes have different phases.

The f orbitals have eight lobes and three nodal planes through the nucleus. $l = 3$, so $m_l = -3, -2, -1, 0, 1, 2, 3$. This additional complexity makes them difficult to describe, and elements that contain them will not be much discussed at this level of study.

Section 5.9   Electron Spin and the Pauli Exclusion Principle

The three quantum numbers discussed earlier define the orbitals available in an atom but do not allow for a unique description of the electrons in an orbital.

The property that distinguishes the individual electrons in an orbital is their "spin." Electrons will spin either clockwise or counterclockwise, which causes them to have a tiny magnetic field. Each orbital can hold one each: an electron spinning clockwise and an electron spinning counterclockwise.

To describe this, a fourth quantum number, the *spin quantum number*—designated m_s—is used. It can have values of $+\frac{1}{2}$ or $-\frac{1}{2}$, referred to as "up" and "down" spins, and it is not constrained by the other quantum numbers.

Wolfgang Pauli proposed the **Pauli exclusion principle**, which states that no two electrons in the same atom can have the same set of quantum numbers. Each orbital has the ability to hold two paired electrons: ↑↓.

Workbook Example 5.8

PROBLEM:
State the four quantum numbers for each of the electrons in a $2p$ subshell.

SOLUTION:
A $2p$ orbital has a principal quantum number n of 2, an angular momentum quantum number l of 1, and magnetic quantum numbers m_l of -1, 0, and 1. For each of these orbitals, two electrons with $+\frac{1}{2}$ and $-\frac{1}{2}$ are possible. So the quantum numbers possible in this subshell are

n	**l**	**m$_l$**	**m$_s$**
2	1	−1	$+\frac{1}{2}$
2	1	−1	$-\frac{1}{2}$
2	1	0	$+\frac{1}{2}$
2	1	0	$-\frac{1}{2}$
2	1	+1	$+\frac{1}{2}$
2	1	+1	$-\frac{1}{2}$

There are six unique sets of quantum numbers for a p subshell, so it can hold six electrons.

Section 5.10   Orbital Energy Levels in Multi-electron Atoms

In hydrogen atoms, only the distance from the nucleus determines the energy level of an electron. In multielectron atoms, electron–electron repulsions also play a part in the energy level of an electron.

Outer-shell electrons are farther from the nucleus and held less tightly. The electrons that are closer to the nucleus also shield the outer electrons from the nuclear charge. These outer electrons are not subject to the full charge of the nucleus, only to an **effective nuclear charge, Z_{eff}**.

$$Z_{eff} = Z_{actual} - \text{Electron shielding}$$

We can see how this relates to the shapes of the orbitals: a lower angular momentum quantum number will correspond to less electron shielding. For example, a $2s$ electron and a $2p$ electron are in the same shell, but because of the shape of the orbital, a $2s$ electron will spend more time close to the nucleus than a $2p$ electron. Therefore, the $2s$ electron is lower in energy than the $2p$ electron.

Section 5.11   Electron Configurations of Multielectron Atoms

Given the information above, we can now describe the **ground-state electron configuration**, or the lowest-energy configuration of any atom. Orbitals with the same energy level, such as the three orbitals in a p subshell, are referred to as **degenerate**.

Three rules are used, collectively known as the **aufbau principle** (*aufbau* means "building up"):

- Lower-energy orbitals fill before higher-energy orbitals.
- An orbital can hold only two electrons with opposite spins.
- If degenerate orbitals are available, they will first fill singly with electrons with parallel spins. Only after there is an electron in each orbital will electrons begin to double up in orbitals. This is **Hund's rule**.

There are two ways of representing these configurations. The first gives the principal quantum number, n, followed by the letter designation of the subshell, with the number of electrons in the subshell represented by a superscript:

H: $1s^1$ He: $1s^2$ Li: $1s^2 2s^1$ Be: $1s^2 2s^2$

One of the consequences of the rules about quantum numbers is that there is no $1p$ subshell: with a principal quantum number of 1, the only angular momentum quantum number allowed is 0, because the number must be less than n.

B: $1s^2 2s^2 2p^1$

The s subshells hold two electrons, p subshells hold six electrons, d subshells hold ten electrons, and f subshells hold 14. These numbers can all be pulled straight from the periodic table, as will be explained in a later section. After argon, the energy levels begin to cross one another, but these can still be taken from the periodic table, as will also be explained later.

The second way of representing these configurations uses *orbital filling diagrams* with the electrons represented by arrows. This method better shows the functioning of Hund's Rule:

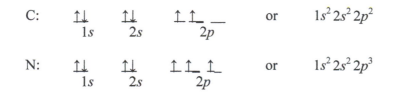

O: ⇅   ⇅   ⇅ ↑ ↑   or $1s^2 2s^2 2p^4$
    1s   2s   2p

Workbook Example 5.9

PROBLEM:
Give the electron configuration for neon.

SOLUTION:
The electron configuration for neon, with 10 electrons, is $1s^2 2s^2 2p^6$.

Noble gases, which are known to be unreactive, all have completely full energy levels: there are no more subshells available in the $n = 2$ energy level. This pattern continues with other noble gases as well.

Workbook Example 5.10

PROBLEM:
Give the orbital-filling diagram for magnesium.

SOLUTION:
The orbital-filling diagram for magnesium is

Mg: ⇅   ⇅   ⇅ ⇅ ⇅   ⇅   or $1s^2 2s^2 2p^6 3s^2$.
     1s   2s   2p   3s

For elements that have more electrons than a noble gas, the configuration of the noble gas can be inserted in brackets. So, the electron configuration of magnesium can also be written as

Mg: [Ne] $3s^2$.

Workbook Problem 5.6

Give the ground-state electronic configuration, both complete and shorthand, for calcium. Also draw the orbital-filling diagram.

Strategy: Using the periodic table, determine the number of electrons in calcium.

Step 1:   Determine the number of electrons in calcium.

Step 2:   Use the aufbau principle to determine the ground-state electronic configuration.

Step 3:   Determine the noble gas in the previous row, and remove the electrons associated with it to a shorthand notation. Specifically name the electrons in unfilled subshells.

Step 4:   Draw the orbital-filling diagram.

Section 5.12   Anomalous Electron Configurations

The larger the atom, the closer together the energy levels are found: in the transition metals, a number of electron configurations are seen that are not accounted for by the previous rules.

Completely-filled and exactly half-filled subshells confer stability. As a result, when energy levels are very close together, electrons will move from lower-energy shells into higher energy shells to gain these configurations and become more stable.

Workbook Example 5.11

PROBLEM:
Give the orbital-filling diagram of chromium.

SOLUTION:
Chromium is the smallest element that will transfer electrons to gain stability. Its electron configuration is $[Ar]4s^1 3d^5$. The reason for the increased stability can be seen in the orbital-filling diagram:

Cr: ↑↓ ↑↓   ↑↓ ↑↓ ↑↓   ↑↓   ↑↓ ↑↓ ↑↓   ↑   ↑ ↑ ↑ ↑ ↑
        1s 2s   2p   3s   3p   4s   3d

Section 5.13   Electron Configurations and the Periodic Table

The electrons that cannot be put into a noble gas shorthand are **valence shell** electrons. They are the most loosely held electrons, and they determine the chemical behavior of an element. All of the elements in a given group will have the same valence shell configuration: only the principal quantum number will change.

All alkali metals (group 1A) have a valence configuration of s^1. They are all reactive and tend to lose one electron to form +1 ions. The electron configuration supports this observation: there is one loosely held electron that is the only electron in its energy level. The same pattern can be seen across the groups: metals have loosely held electrons that are easily lost, and non-metals have gaps in their valence shells that they fill by taking electrons to form negative ions.

The periodic table is divided into "blocks" based on what type of subshell the outermost valence electrons are in:

- *s*-block elements: Groups 1A and 2A (filling of an *s* orbital); *n* = row number.
- *p*-block elements: Groups 3A through 8A (filling of *p* orbitals; *ns* orbitals are filled); *n* = row number.
- *d*-block elements — transition metals; *n* = row number −1.
- *f*-block elements — lanthanide and actinide elements; *n* = row number −2.

Workbook Example 5.12

PROBLEM:
Using only the periodic table, write the shorthand electron configuration for technetium (Tc).

SOLUTION:
Technetium is in the *d*-block, in the fifth row of the periodic table. The noble gas that is below it in electron configuration is krypton (the end of row 4). The electron configuration will begin with krypton. Because technetium is in the fifth row of the periodic table, its *s* electrons that will immediately follow will be 5*s*, and this subshell is filled. The *d* electrons will be in the *n* − 1 energy level, so they will be 4*d* electrons. There will be five of them.

$$\text{Tc} = [\text{Kr}] \, 5s^2 \, 4d^5$$

Section 5.14   Electron Configurations and Periodic Properties: Atomic Radii

Although atoms cannot be said to have a definite size, atomic radii can be assigned to atoms by measuring covalent bonds. When two identical atoms are covalently bonded together, half of the bond length is known as the atomic radius.

As *n* increases down through a group, atomic radii increase as larger valence-shell orbitals are being occupied.

As subshells fill across the periodic table from right to left, the effective nuclear charge felt by each electron increases: electrons within the same subshell only shield each other weakly, so the atomic radius decreases, even as the atomic number increases.

Workbook Example 5.13

PROBLEM:
Choose the smaller of the following pairs:

    K and Fe   N and Be   S and Al   Br and Ca

SOLUTION:
These element pairs are all in the same row as one another, so in each case the element that is rightmost in the periodic table will be smaller.

    K > Fe   N < Be   S < Al   Br < Ca

Self–Test

This section is intended to test your knowledge of the material covered in this chapter. Think through these problems, and make certain you understand them. Ask yourself if your answer makes sense. Many of these questions are linked to the chapter learning goals. Therefore, successful completion of these problems indicates you have mastered the learning goals for this chapter. You will receive the greatest benefit from this section if you use it as a mock exam, as this will clarify which topics you have mastered and which topics you need to study in more detail. Some of the problems may require you to think slightly beyond the material presented in the text. Remember to use your units!

True–False
1. The amplitude of a wave is the number of crests that pass a certain point per second.

2. Each element has a unique spectral "signature."

3. The photoelectric effect provides evidence for the particle-like nature of light.

4. Even at the size of everyday objects, de Broglie wavelengths are relevant.

5. The Bohr model of the atom holds true for all elements.

6. The energy of a photon can be calculated from its frequency.

7. The set of numbers {4, 4, 3, +½} is an allowed set of quantum numbers.

8. Electrons reside in *f* orbitals in common nonmetals.

9. Electrons in the same subshell do not effectively shield one another from nuclear charge.

10. Electrons that share an orbital must have parallel spins.

Multiple Choice
11. Electromagnetic energy is characterized by
    a. wavelength   b. amplitude   c. frequency   d. all of the above

12. The Balmer–Rydberg equation accounts for the spectral lines of
    a. helium.   b. hydrogen   c. halfnium   d. any element

13. For photons to knock electrons out of a metal, they must
    a. have sufficient intensity   b. have values close to Planck's constant
    c. be above the threshold frequency   d. occur in large numbers

14. de Broglie determined that at the atomic level
    a. matter has a wavelike nature   b. electrons are very slow-moving
    c. elements are sometimes photons   d. noble gases are mostly energy

15. Which of the following are permitted sets of quantum numbers?
    a. 3, 4, −2, +½   b. 4, 3, −2, −½   c. 4, 2, 2, 1   d. 4, 2, 2, +⅔

16. Frequency, wavelength, and energy are related such that:
    a. low-frequency electromagnetic energy has a short wavelength and high-energy photons
    b. high-frequency electromagnetic energy has a long wavelength and high-energy photons
    c. low-frequency electromagnetic energy has a long wavelength and low-energy photons
    d. high-frequency electromagnetic energy has a short wavelength and low-energy photons

17. The Heisenberg uncertainty principle states that
    a. matter and energy are the same thing
    b. the position of an electron or its velocity can be known, but not both at the same time
    c. electrons need to fill degenerate orbitals with perpendicular spins first
    d. everything with momentum has a wavelength

18. The alkali metals are
    a. s block elements   b. p block elements   c. d block elements   d. f block elements

19. The electron configuration for F is
    a. $1s^2 1p^6 1d^7$ b. $1s^2 2s^2 2p^6 3s^2 3p^5$ c. $[Ar] 3s^2 3p^5$ d. $1s^2 2s^2 2p^5$

20. The size of elements *increases*
    a. down the groups and left to right along the periods
    b. up the groups and left to right along the periods
    c. down the groups and right to left along the periods
    d. up the groups and right to left along the periods

Matching

21. Degenerate   a. can hold two electrons with opposite spins

22. Wavelength   b. equal in energy

23. Frequency   c. unit of s^{-1} or 1/s

24. Amplitude   d. height of a wave from the midline

25. Hertz   e. number of crests that pass a certain point per second

26. Photon   f. set of rules that guide the filling of atomic orbitals

27. Orbital   g. distance from peak to peak of a wave

28. Aufbau principle   h. ejection of electrons from a clean metal surface by certain wavelengths of light

29. Photoelectric effect   i. packet of light energy

Fill-in-the-Blank

30. When discussing electromagnetic energy, the Greek letter λ indicates _____,

    while the Greek letter ν indicates _____.

31. When excited with electricity or heat, elements give off a characteristic_____.

32. The photoelectric effect was explained by Einstein and Planck as demonstrating that light contained _____, or particles of light.

33. In the Bohr model of the atom, a transition from a higher energy level to a lower energy level corresponds with the _____ of a photon with specific energy.

34. An electron can be specifically described using four _____ _____.

35. _____ are spherical. Their energy depends only on their distance from the nucleus.

36. _____ subshells are dumbbell shaped, with nodes at the nucleus. Each of the orbitals can hold _____ electrons, for a total of _____ electrons in the subshell.

37. Hund's rule states that when _____ orbitals are being filled, they must first fill with electrons having _____.

38. The outermost electrons in an atom are called the _____. These determine the chemical reactivity of the element.

39. One-half of the bond length when two identical atoms are covalently bonded is known as the

_____ _____.

Problems
40. Cellular telephones commonly transmit at 835.6 MHz. What is the wavelength of this energy? How much energy is contained in 1 mole of cell phone photons?

41. Gamma rays have a wavelength of 2.15×10^{-5} nm. What is the frequency of this energy? How much energy is contained in one mole of gamma-ray photons?

42. A 432 nm laser emits a quick pulse that contains 2.5 mJ of energy. How many photons were in the pulse?

43. The ionization energy of sodium metal is 495.8 kJ/mol. If this much energy is required to eject 1 mol of electrons from the surface, what frequency and wavelength of light corresponds to the energy of photons required? Will light with wavelength 540 nm dislodge electrons from sodium? Will light with wavelength 195 nm?

44. An electron has an uncertainty in its position of 327 pm. What is the uncertainty in its velocity?

45. If a hydrogen electron in the $n = 6$ energy level returns to ground state, what is the wavelength of the light emitted? What is the energy of the photon emitted?

46. Without using shorthand notation, write the electron structure for ytterbium.

47. Write the four quantum numbers for every electron in the ground state of neon.

48. Write the first two quantum numbers for the highest-energy electron in bromine.

49. When electrons are unpaired, the weak magnetic field they generate is reinforced by any other unpaired electrons in the same atom. This leads to a phenomenon called *paramagnetism*: the more unpaired electrons, the more the atom is attracted to a magnetic field. Draw the orbital-filling diagrams for Si, P, and Cl and rank them from most paramagnetic to least paramagnetic.

50. If an excited hydrogen atom falls to $n = 2$ and emits a photon of light with 4.58×10^{-19} J, at what energy level did it begin?

51. If an electron is accelerated to 2.57×10^6 m/s, what is its de Broglie wavelength? The mass of an electron is 9.109×10^{-31} kg.

52. Identify the atoms with the following electron configurations:
    a. $[Ar]4s^1 3d^{10}$
    b. $[Xe]6s^2 4f^{14} 5d^{10} 6p^5$
    c. $[Kr] 4d^{10}$
    d. $[Ar]4s^1 3d^5$

53. Group the following atoms in order of increasing atomic radius: Os, Rn, Fe, Ru

54. Group the following atoms in order of increasing atomic radius: S, Mg, Cl, Ne

Challenge Problems

55. Watts are units of power defined as J/s. If an incandescent bulb consumes 40 watts at 550 nm for 4 hours, how many photons would it have emitted if it were 100% efficient? If only 5% of the electricity consumed by the bulb was converted to light and the rest was converted to heat, how long would the bulb need to be lit to heat 200 g of water from 20 °C to 95 °C? The specific heat of water is 4.184 J/g•°C. Use the units of specific heat to figure out how to determine the temperature change.

56. If a laser consumes 1500 W of power, and produces 2.73×10^{20} photons per second, what would the wavelength be if the efficiency of the laser were 100%? 80%? 20%? If the wavelength at which the laser operates is 1542 nm, how efficient is it?

CHAPTER SIX

Ionic Compounds: Periodic Trends and Bonding Theory

Learning Objectives

As a result of reading and studying this chapter, you should be able to

Section 6.1	**Electron Configurations of Ions**
1.	Write ground-state configurations for main-group and transition metal ions.
2.	Determine the number of unpaired electrons in a transition metal ion.

Section 6.2	**Ionic Radii**
3.	Predict the relative size of anions, cations, and atoms.
4.	Predict the relative size of isoelectronic ions.

Section 6.3	**Ionization Energy**
5.	Order elements from lowest to highest ionization energy.
6.	Explain the periodic trend in ionization energy.

Section 6.4	**Higher Ionization Energies**
7.	Compare successive ionization energies for different elements.
8.	Identify elements based on values of successive ionization energies.

Section 6.5	**Electron Affinity**
9.	Compare the value of electron affinity for different elements.
10.	Explain the periodic trend in electron affinity.

Section 6.6	**The Octet Rule**
11.	Use the octet rule to predict charges on main group ions, electron configurations of main group ions, and formulas for ionic compounds.

Section 6.7	**Ionic Bonds and the Formation of Ionic Solids**
12.	Visualize ionic compounds on the molecular level.
13.	Draw a Born–Haber cycle and calculate the energy change that occurs when an ionic compound is formed from its elements.
14.	Use the Born–Haber cycle to solve for the energy change associated with one of the steps.

Section 6.8	**Lattice Energies in Ionic Solids**
15.	Predict the relative magnitude of lattice energy given the formula or molecular representation of an ionic compound.

Chapter Summary

Chemical bonds are the forces that hold atoms together. There are two types: ionic bonds and covalent bonds. This chapter concentrates on the reasons ionic bonds are formed and the energies involved in the formation of these bonds. You will learn that the underlying reason for the formation of ions and ionic compounds is the valence-shell configuration of the elements, which determines the amount of energy needed for losing or gaining electrons. From the periodic trends in the energies for ion formation, you will learn how to predict which elements form cations or anions and if the formation of ionic compounds is energetically favorable. You will also learn about the Born–Haber cycle, a series of hypothetical steps used to calculate the overall energy involved in the formation of ionic compounds.

The Chapter in Detail

Section 6.1   Electron Configurations of Ions

Metal elements reside on the left-hand side of the periodic table, and tend to give up electrons to form cations. When they do so, they adopt the electron configurations of noble gases: alkali metals lose one electron to ionize and adopt the electron configurations of the noble gases one down from them on the periodic table, alkaline earth metals lose two electrons and adopt noble-gas configurations. In groups IA and 2A, the topmost energy level is emptied to leave behind a noble-gas configuration.

Nonmetals, on the right side of the periodic table, gain electrons and move up to the electron configurations of the noble gases above them. Halogens all gain one electron to ionize; the group 6A elements gain two electrons and adopt the same electron configurations as the noble gas elements in family 8A. In these elements, the topmost energy level is filled to gain a noble-gas configuration.

Transition metals ionize so that they lose their s electrons first, then they begin losing d electrons. This sometimes gives multiple stable ions, which accounts for the multiple ionization states of some transition metals.

Workbook Example 6.1

PROBLEM:
Write shorthand electron configurations for the following elements and their ions:

$$Mg^{2+} \quad F^- \quad S^{2-} \quad Fe^{3+}$$

SOLUTION:

Mg =	$[Ne]3s^2$	Mg^{2+} – [Ne]	Noble-gas configuration	
F =	$[He]2s^22p^5$	F^- – [Ne]	Noble-gas configuration	
S =	$[Ne]3s^23p^4$	S^{2-} – [Ar]	Noble-gas configuration	
Fe =	$[Ar]4s^23d^6$	Fe^{3+} – $[Ar]3d^5$	Half-filled d orbital provides stability	

Section 6.2   Ionic Radii

Cations are smaller than neutral atoms. When cations are formed, electrons are removed from valence-shell orbitals. In the case of groups 1A and 2A, the outermost energy levels are removed entirely. In addition, the effective nuclear charge (Z_{eff}) on all the remaining electrons increases, and those electrons are held more closely.

Anions are larger than neutral atoms. When anions are formed, valence-shell orbitals are completed. There are more electrons than protons, so the effective nuclear charge felt by the electrons will decrease. In addition, the additional electrons will add to the electron–electron repulsions felt by the outer electrons. The changes can be dramatic: a chlorine ion is almost twice the size of a neutral chlorine atom.

Workbook Example 6.2

PROBLEM:
Which atom or ion in the following pairs has the larger radius:

    a. Cr^{2+} or Cr^{4+} b. P or P^{3-} c. S or S^{2-} d. Na or Na^{+}

SOLUTION:
    a. Cr^{2+}: The radius of Cr^{4+} is smaller, because both the s and d orbitals are empty, so the pull on the remaining electrons is correspondingly stronger.
    b. P^{3-}: The addition of three electrons to a neutral phosphorus atom will greatly decrease the Z_{eff} felt by the electrons.
    c. S^{2-}: The addition of electrons causes a decrease in Z_{eff} so that the outer electrons are more loosely held, and the radius is greater.
    d. Na: The removal of electrons causes an increase in Z_{eff}, creating a decrease in the radius.

Section 6.3   Ionization Energy

Ionization energy (abbreviated E_i) is the amount of energy required to create a +1 cation from a gas-phase atom. It is always positive: energy must always be added to an atom to remove an electron, no matter how stable the resulting ion will be.

Ionization energy shows periodic trends:

- The lowest ionization energies are those for the alkali metals, because the loss of one electron empties the highest energy level.
- The highest ionization energies are those of the noble gases, because the loss of an electron at this level means that you would have to break into a filled energy level.
- Ionization energy decreases down a group: it takes less energy to remove the outermost electron from a large atom than from a small atom, because it is farther from the nucleus.

The periodic trends are not perfectly smooth:

- group 3A ionization energies are lower than those of group 2A. When group 3A ionizes, a shielded p subshell is emptied, while when group 2A ionizes, a less-stable, half-filled s subshell is created.
- group 6A ionization energies are lower than those of group 5A, because the loss of one electron in this group creates a more-stable, half-filled p subshell.

Workbook Example 6.3

PROBLEM:
Draw the electron-filling diagrams for Be, B, N, and O. Explain how these demonstrate the reasons behind the lower E_i for group 3A compared to group 2A and group 6A compared to group 5A.

SOLUTION:

Be: ⇅ ⇅   Ionization requires disruption of filled subshell
     $1s$ $2s$

B: ⇅ ⇅ ↑ __ __   Ionization removes single, well-shielded electron
     $1s$ $2s$ $2p$

N: ⇅ ⇅ ↑ ↑ ↑   Ionization requires disruption of half-filled subshell
     $1s$ $2s$ $2p$

O: ⇅ ⇅ ⇅ ↑ ↑   Ionization creates stable half-filled subshell
     $1s$ $2s$ $2p$

Section 6.4   Higher Ionization Energies

It is possible to remove more electrons sequentially from an atom, and the energy required for each step can be measured. Each step will take more energy, because it is the energy required to remove an electron from an already positively charged ion. In other words, it will take more energy to remove an electron because it really doesn't want to remove another electron. When all of the valence electrons have been removed, there will be a large jump (4x or more) in ionization energy.

Workbook Example 6.4

PROBLEM:
Identify the following fourth-period element from these ionization energies:

$E_i = 590$ kJ/mol $E_{i2} = 1145$ kJ/mol $E_{i3} = 4912$ kJ/mol
$E_{i4} = 6491$ kJ/mol $E_{i5} = 8153$ kJ/mol

SOLUTION:
To determine which group this element is in, it is necessary to look for a jump of 4x or more in the ionization energies. We find this between E_{i2} and E_{i3}, indicating that two electrons can be lost fairly easily, but that once two electrons are lost, the ion has a noble-gas configuration and is difficult to ionize further. This describes the alkaline earth metals; the period four alkaline earth metal is calcium.

Workbook Problem 6.1

From the following list of ionization energies, identify the group of the following element:

E_{i1} = 1012 kJ/mol	E_{i2} = 1907 kJ/mol	E_{i3} = 2914 kJ/mol
E_{i4} = 4964 kJ/mol	E_{i5} = 21 267 kJ/mol	E_{i6} = 25 431 kJ/mol

Strategy: Look for a jump of 4x or more between ionization energies: this will show the number of valence electrons and thus the distance from the nearest noble gas.

Step 1:   Find the large gap in ionization energies, and identify the group number.

Section 6.5   Electron Affinity

Electron affinity (abbreviated E_{ea}) is a measure of the energy released when an electron is added to a neutral gas-phase atom. For an atom that will accept an electron, the value of electron affinity is negative. The more stable the anion, the more energy is released. Many atoms—noble gases, for example—will not accept additional electrons, so their electron affinity is only theoretical. If it were possible, the electron affinity would be positive, because energy would be required to hold an electron where it does not want to be, but, of course, it is impossible to measure a process that can't be made to occur.

Remember!
Positive energy change = energy required for a process (ionization)
Negative energy change = energy released as a result of a process (electron affinity)

These definitions will recur throughout chemistry, and they are not particularly intuitive. Learn them *now*!

Periodic trends in electron affinity are related to electron configurations. In the nonmetals, electron affinities generally become more favorable in moving toward the halogens, with the exception that adding an electron to a somewhat stable half-filled p subshell will be less favorable. For example, nitrogen, with a half-filled p subshell, is more reluctant to accept an electron than is either of its neighbors, because its stability is disrupted by adding another electron.

In the metals, those that can accept an electron into a partially filled s orbital (alkali metals, for example), will have a slightly negative electron affinity. Atoms that must accept electrons into an otherwise unoccupied subshell will be very reluctant to do so: alkaline earth metals need to add the electron into a p subshell that is empty, but at an energy level that already contains electrons.

Noble gases, which would need to add an electron not only to a new subshell but also to a new energy level, simply will not do so.

Workbook Example 6.5

PROBLEM:
Using electron filling diagrams, rank the E_{ea} of Si, P, S, and Cl.

SOLUTION:

	$1s$	$2s$	$2p$	$3s$	$3p$	
Si:	⇅	⇅	⇅ ⇅ ⇅	⇅	↑ ↑ __	Ionization adds to stability by half-filling the $3p$ subshell.
P:	⇅	⇅	⇅ ⇅ ⇅	⇅	↑ ↑ ↑	Ionization destabilizes the half-filled subshell.
S:	⇅	⇅	⇅ ⇅ ⇅	⇅	⇅ ↑ ↑	Ionization moves toward the noble-gas configuration.
Cl:	⇅	⇅	⇅ ⇅ ⇅	⇅	⇅ ⇅ ↑	Ionization completes noble-gas configuration.

Based on the electron-filling diagrams, and the general trend that the electron affinities increase in moving toward the halogens, one would expect that Cl would have the largest negative E_{ea}, followed by S, then Si, with P having the smallest negative E_{ea} due to the destabilizing effect of adding to a half-filled p subshell. This is in fact the case.

Section 6.6   The Octet Rule

From the past few sections it can be seen that:

- Alkali metals (group 1A) have low first ionization energies and so tend to lose their ns_1 electrons in reactions to gain a noble-gas configuration.
- Alkaline earth metals (group 2A) have low first and second ionization energies and so tend to lose all of their ns electrons in reactions so as to gain a noble-gas configuration.
- Halogens (group 7A) have a large negative electron affinity and so tend to ionize readily to obtain a noble-gas configuration.

All of these observations can be summarized in the **octet rule**: main-group elements tend to undergo reactions that leave them with eight outer-shell electrons.

Why does it work this way? The group 1A and 2A metals lose valence electrons easily, because the electrons they lose have a low Z_{eff}: they are the only electrons in the outermost s orbital. The group 7A elements gain electrons easily, because the electrons have a high Z_{eff}: they are in p orbitals that do not shield one another very well. In both cases, the elements have gotten to the point where they have eight electrons in their outermost shell.

Workbook Example 6.6

PROBLEM:
Which noble-gas configurations are adopted by the atoms in the following ionic compounds:

$$Li_2O \qquad AlN \qquad CaBr_2$$

SOLUTION:
In lithium oxide, the two lithium atoms each lose an electron and adopt the electron configuration of helium. The two lost electrons are gained by oxygen, which then has the electron configuration of neon.

In aluminum nitride, the aluminum metal gives up three valence electrons to adopt the electron configuration of neon, while the nitrogen takes on the three electrons and also has the electron structure of neon.

When calcium loses two electrons, it gains the electron structure of argon. The bromine atoms each take one of the lost electrons to take on the electron structure of krypton.

Workbook Problem 6.2

Will the following elements be more likely to gain or lose electrons when they form ions? How many electrons will they gain or lose?

$$Rb \qquad Sc \qquad Se \qquad As$$

Strategy: Find these elements on the periodic table, and locate their nearest noble gas. Remembering the general rule that metals lose electrons and nonmetals gain them, and that main-group elements ionize in predictable ways, determine the most likely ionized form of the element.

Step 1: Identify the group number of the elements, and identify them as metals or nonmetals.

Step 2: Find the nearest noble gas.

Step 3: Identify the number of electrons that must be gained/lost to achieve the electron configuration of that noble gas.

Step 4: Write the ionized form of the element.

Section 6.7 Ionic Bonds and the Formation of Ionic Solids

When a metal with a low ionization energy comes into contact with a nonmetal with a favorable electron affinity, a reaction can occur in which an electron is transferred from the metal to the nonmetal, forming a cation and an anion.

Energetically, the electron affinity of the nonmetal is unlikely to completely offset the energy required to remove an electron from a metal. But the reaction between sodium and chloride described in the text proceeds explosively, with a large release of energy. What accounts for this?

These reactions not only form two ions with very stable noble-gas configurations but they also neutralize the charges that have been created in forming an ionic bond and, by extension, an **ionic solid**.

Ionic solids are networks of ions, in which the charge of one ion is neutralized by surrounding oppositely charged ions. These do not form molecules: we cannot assign one cation and one anion to one another. It is only possible to give the ratio required for the charges to be neutralized.

Na Cl Na Cl Na
Cl Na Cl Na Cl
Na Cl Na Cl Na
Cl Na Cl Na Cl
Na Cl Na Cl Na

If you look at this "salt crystal," you can see that there are four chlorine ions around each sodium ion and four sodium ions around each chlorine ion. If this were extended into three dimensions, there would be even more ions surrounding and neutralizing each other. This provides exceptional stability.

These reactions proceed rapidly and cannot be observed to occur in stages. However, when dealing with reaction energetics, it is possible to break apart the reaction mathematically. In the case of ionic solid formation, this is done by what is known as the **Born–Haber cycle**. There are five common steps in this cycle:

Sublimation of the metal atoms: conversion of the metal to gas	Unfavorable
Nonmetal molecules broken into individual atoms	Unfavorable
Ionization of the gaseous metal atoms	Unfavorable
Ionization of the gaseous nonmetal atoms	Favorable
Formation of ionic bond	Very favorable

Each of these steps can be assigned numerical values. The energetically favorable steps will have negative values, because they are releasing energy, and the unfavorable steps will have positive values, because they are consuming energy.

Workbook Example 6.7

PROBLEM:
Write the Born–Haber cycle for the formation of lithium fluoride from lithium metal and gaseous fluorine.

SOLUTION:

Sublimation of lithium	$Li\ (s) \rightarrow Li\ (g)$	+159.4 kJ/mol
Splitting of fluorine molecules	$\frac{1}{2}\ F_2\ (g) \rightarrow F\ (g)$	+158 kJ/mol
Ionization of lithium	$Li\ (g) \rightarrow Li^+\ (g) + e^-$	+520 kJ/mol
Ionization of fluorine	$F\ (g) + e^- \rightarrow F^-\ (g)$	−328 kJ/mol
Formation of ionic solid	$Li^+\ (g) + F^-\ (g) \rightarrow LiF\ (s)$	−1036 kJ/mol
	Total:	−527 kJ/mol

Workbook Problem 6.3

Write the Born–Haber cycle for the formation of sodium bromide using the following information:

Sublimation of Na:	+107.3 kJ/mol	Vaporization of $Br_2(l)$:	+ 30.9 kJ/mol
Breaking of Br_2 bond:	+224 kJ/mol	Ionization of Na:	+495.8 kJ/mol
Ionization of Br:	−325 kJ/mol	Formation of NaBr:	−747 kJ/mol

Strategy: Write and sum the energies of each of the steps in the reaction to calculate the overall energy change of the formation of sodium bromide from its elements.

Step 1: Write each step of the reaction with its energy required or produced per mole.

Step 2: Sum the energy terms to find the overall energy change of the reaction.

Section 6.8   Lattice Energies in Ionic Solids

Ionic solids have a structure that is referred to as a *crystal lattice*. The energy required to break this crystal lattice back into gaseous ions is called **lattice energy**, abbreviated U. It has the same magnitude but the opposite sign of the final step of the Born–Haber cycle: just as that last step—the formation of the ionic solid from gaseous ions—is always negative, the reverse will always be positive.

$$Li^+ (g) + F^- (g) \rightarrow LiF (s) \qquad -U = -1007 \text{ kJ/mol}$$
$$LiF (s) \rightarrow Li^+ (g) + F^- (g) \qquad U = +1007 \text{ kJ/mol}$$

The strength of ionic interactions is described by **Coulomb's law**, which has the mathematical form:

$$F = k \frac{z_1 z_2}{d^2}$$

where k is a mathematical constant, z_1 and z_2 are the charges on the ions, and d is the distance between the centers of the ions.

The practical result of this solid structure is that the lattice energy is largest—meaning the ions are most difficult to separate—when the charges are large and the distance is small. Small, highly charged ions will have the strongest crystal lattice.

If one ion is changed and the other held constant, the lattice energy will drop as the ion grows (assuming no change in the charge of the ion). This allows for the prediction of relative lattice energies.

Workbook Example 6.8

PROBLEM:
Rank the following compounds from strongest to weakest lattice energy:

$$LiCl \quad CsCl \quad KCl \quad NaCl \quad RbCl$$

SOLUTION:
The anion in each of these cases is chlorine, so only the size of the cation is varying. According to Coulomb's law, the smaller the distance between the ions, the stronger the lattice energy; so the strongest lattice energy will occur with the smallest cation and the weakest lattice energy with the largest cation.

$$LiCl > NaCl > KCl > RbCl > CsCl$$

Workbook Example 6.9

PROBLEM:
Rank the following compounds from strongest to weakest lattice energy:

$$MgCl_2 \qquad TiCl_4 \qquad CaCl_2 \qquad KCl \qquad AlCl_3$$

SOLUTION:
This is a slightly more complex situation than the previous example, but changes in charge have a huge effect on the lattice energy. Ti, with a charge of +4, will have a stronger lattice energy than Al, with a charge of +3. Only when the charges are the same does the size of ions play a role. Mg is smaller than Ca, so it will form a stronger lattice.

$$TiCl_4 > AlCl_3 > MgCl_2 > CaCl_2 > KCl$$

Workbook Problem 6.4

Which of each of the following pairs of ions has the greatest lattice energy?

$$NaCl \text{ or } KBr \quad AlF_3 \text{ or } MgCl_2 \quad CuBr_2 \text{ or } CuCl$$

Strategy: Determine the relative sizes and charges of the ions and assign relative strengths based on charge and size.

Step 1:   Identify the relative charges in the two sets of ions and how they change from one set to the other.

Step 2:   If necessary, identify the relative ion sizes of the two sets of ions and show how they change from one to the other.

Step 3:   Determine the relative lattice energies of the ion pairs.

Self–Test

This section is intended to test your knowledge of the material covered in this chapter. Think through these problems, and make certain you understand what is being asked and that your answer makes sense. By using this section as a mock exam, you can discover which topics you have mastered and which topics you need to study in more detail.

True–False

1. Main-group metals tend to form cations that give them noble-gas electron configurations.

2. Nonmetals lose electrons to adopt lower energy levels.

3. Transition metals always ionize the same way.

4. When ions form, their radii increase if they are forming anions and decrease if they are forming cations.

5. Ionization energy is always positive.

6. Ionization energies show no periodic properties.

7. Electron affinity applies primarily to metals, and it is difficult to determine for nonmetals.

8. The octet rule can be used to predict how ions will form in reactions.

9. The Born–Haber cycle provides information about the strength of covalent bonds.

10. Lattice energy is related only to ionic size.

Multiple Choice

11. Main-group elements form ions by
    a. losing electrons
    b. gaining electrons
    c. working toward a half-filled *d* orbital
    d. losing or gaining electrons to obtain a noble-gas configuration

12. What +3 ion has the electron configuration $1s^2 2s^2 2p^6$?
        a. B^{3+} b. Al^{3+} c. P^{3+} d. Fe^{3+}

13. Rank the following elements in order of greatest to least first ionization energy (IE_1):

Rb   Li   Cs   K   Na

a. Na > Li > K > Cs > Rb   b. Cs > Rb > K > Na > Li
c. Li > Na > K > Rb > Cs   d. Li > K > Na > Cs > Rb

14. Given the following ionization energies, which element is this most likely to be?

IE_1: 738 IE_2: 1451 IE_3: 7733 IE_4: 10 540
a. Na   b. Mg   c. Al   d. Si

15. Ionization energies are always positive, but electron affinities are generally negative because
    a. energy is released when an atom achieves a more stable configuration
    b. an electron can only be added to a cation
    c. energy is required to hold an extra electron onto an atom
    d. atoms will not accept extra electrons

16. Born–Haber cycles do NOT require which of the following?
    a. ionization energy of the metal   b. formation of the ionic solid
    c. liquefying the nonmetal   d. sublimation of the metal

17. Lattice energies vary with
    a. ion charge   b. electron affinity   c. ionic radius   d. A and C

18. Which of the following will have the highest lattice energy?
    a. AlN   b. $MgCl_2$ c. NCl_3 d. LiCl

Matching

19. Ionization energy   a. the energy change that occurs when an electron is added to an isolated atom in the gaseous state.

20. Coulomb's law   b. sum of interaction energies of ions in a crystal.

21. Electron affinity   c. main-group elements tend to undergo reactions that leave them with eight valence electrons.

22. Lattice energy   d. mathematical relationship between charge interaction and energy.

23. Born–Haber cycle   e. the amount of energy required to remove the outermost electron from an isolated atom in the gaseous state.

24. Core electron   f. a bond maintained by electrostatic interactions.

25. Ionic bond   g. electrons that form the inner shell of an element and do not participate in reactions.

26. Octet rule   h. a series of hypothetical steps, each of which contributes to the overall energy change during the formation of an ionic compound.

Fill-in-the-Blank

27. Chlorine and potassium will both ionize so as to have the electron configuration of
_____.

28. When metals ionize to form cations, the atomic radius _____.

29. Valence electrons are strongly shielded by inner _____ electrons, making them relatively _____ to remove.

30. Multiple ionization energies are possible with multielectron atoms. A very large _____ in ionization energies indicates that all of the _____ electrons have been removed.

31. Electron affinities cannot be determined for elements that do not form _____.

32. The higher the ionic _____, the stronger the lattice energy is a use of _____ law.

33. Main-group elements tend to undergo reactions that leave them with _____ outer shell electrons. This is the _____.

Problems

34. Write the electron configurations of the following elements first as neutral atoms and then as their most stable ion.

    P   Ti   Mn   K   Se

35. Rank these ions and atoms in order from smallest to largest.

    Cl^- Al Li^+ Cl Al^{3+} Li

36. Identify the following third-period element from these ionization energies:
    E_{i1} = 578 kJ/mol E_{i2} = 1817 kJ/mol E_{i3} = 2745 kJ/mol
    E_{i4} = 11 575 kJ/mol E_{i5} = 14 830 kJ/mol

37. Draw electron configurations for the following, and use them to rank these elements by electron affinity, lowest to highest. Justify your answer.

    Ga   Br   Ge

38. Calculate the heat of sublimation for aluminum given that the net energy change for formation of $AlCl_3$ (s) from the elements, Al (s) + 3/2 Cl_2 (g) → $AlCl_3$ (s), is −704.2 kJ/mol. Use the following information:

    Al: E_{i1} = 578 kJ/mol, E_{i2} = 1817 kJ/mol, E_{i3} = 2745 kJ/mol

    Cl: E_{ea} = −348.6 kJ/mol

    Bond-dissociation energy for Cl_2 = 244 kJ/mol
    Lattice energy for $AlCl_3$ (s) = 5492 kJ/mol

39. Using the octet rule, write and balance equations for the reactions of Mg, Li, and Al with Cl, O, and P.

40. Write the Born–Haber cycle for the formation of KF from its elements given the following information:

Sublimation of K:	+89.2 kJ/mol	$F_2 \rightarrow 2\ F\ (g)$:	+158 kJ/mol
Ionization of K:	+418.8 kJ/mol	Ionization of F:	−328 kJ/mol
Formation of KF:	−821 kJ/mol		

41. Rank the following from strongest to weakest lattice energy, and justify your answers:

LiCl   MgCl₂   KCl   AlCl₃   CsCl

Challenge Problem

42. A 1.53 g sample of an alkaline earth metal was reacted with 4.75 L of chlorine gas (3.17 g/L). The resulting metal chloride was analyzed by dissolving a 5.25 g sample in water and adding an excess of silver nitrate. The analysis yielded 18.8 g of silver chloride.
    a. What is the percent chlorine in the alkaline earth chloride?
    b. What is the identity of the alkaline earth metal?
    c. Write the balanced chemical equations for all reactions.
    d. What was the limiting reactant in the reaction between the alkaline earth metal and the chlorine? If the metal is in excess, how many grams did not react? If chlorine, how many liters remained?

CHAPTER SEVEN

Covalent Bonding and Electron-Dot Structures

Learning Objectives

As a result of reading and studying this chapter, you should be able to

Section 7.1	**Covalent Bonding in Molecules**	
	1.	Describe the difference between an ionic and covalent bond.
	2.	Describe changes in energy that occur as two nuclei approach to form a covalent bond.
	3.	Name a covalent compound given the chemical formula.
Section 7.2	**Strengths of Covalent Bonds**	
	4.	Predict trends in bond length and bond dissociation energy based on bond order and atomic size.
Section 7.3	**Polar Covalent Bonds: Electronegativity**	
	5.	Rank elements by increasing value of electronegativity.
	6.	Classify bonds as nonpolar covalent, polar covalent, or ionic.
	7.	Visualize regions of high and low electron density in a polar covalent bond.
	8.	Predict trends in bond dissociation energy based upon both atomic size and polarity of the bond.
Section 7.4	**A Comparison of Ionic and Covalent Compounds**	
	9.	Explain the different physical properties of NaCl and HCl (Table 7.3) based upon the ionic and covalent bonding models.
Section 7.5	**Electron-Dot Structures: The Octet Rule**	
	10.	Draw an electron-dot structure by using valence electrons to give all atoms (except H) an octet.
Section 7.6	**Procedure for Drawing Electron-Dot Structures**	
	11.	Use the five-step procedure for drawing electron-dot structures for all molecules, including those with expanded octets and those containing multiple bonds.
Section 7.7	**Drawing Electron-Dot Structures for Free Radicals**	
	12.	Draw electron-dot structures for free radicals.
Section 7.8	**Electron-Dot Structures of Compounds Containing Only Hydrogen and Second Row Elements**	
	13.	Draw electron-dot structures for molecules with more than one central atom.
	14.	Draw a complete electron-dot structure given only the connections of atoms.

Section 7.9   **Electron-Dot Structures and Resonance**
15.   Draw resonance structures and use curved arrows to depict how one structure can be converted to another.

Section 7.10   **Formal Charge**
16.   Calculate the formal charge on atoms in an electron-dot structure.
17.   Use the formal charge to evaluate the contribution of different resonance structures to the resonance hybrid.

Chapter Summary

Covalent bond formation is the result of electron sharing between atoms. Covalent bonds have specific lengths and energies; energy is released when a bond is formed and absorbed when a bond is broken. Covalent compounds generally obey the octet rule, but atoms in the third period and below can form expanded octets as well. These structures can be represented by electron-dot structures, which can also be used to predict the existence of resonance hybrids. When dissimilar atoms participate in covalent bonding, differences in electronegativity can lead to a polar bond. These polarity differences can be predicted with the periodic table.

The Chapter in Detail

Section 7.1   Covalent Bonding in Molecules

Hydrogen forms the simplest covalent molecules and provides a good model for covalent bonding: as the hydrogen atoms approach each other, the positive nuclei repel each other and the electrons repel each other, but the nuclei pull on the electrons of both atoms. If the attractive forces are greater than the repulsive forces, a covalent bond is formed, with the electrons shared between the nuclei.

There is an optimum distance for atoms: too far away and no bond can form; too close and the nuclei repel. Every covalent bond has a characteristic bond length that leads to maximum stability. These can be roughly predicted from atomic radii.

Section 7.2   Strengths of Covalent Bonds

The energy of a covalently bonded molecule is lower than that of the individual molecules: energy is released when they come together. This same amount of energy has to be put in to the bond to break it. This is the **bond dissociation energy (D).**

Bond dissociation energies are defined as the amount of energy needed to break a bond in an isolated, gaseous molecule. Bond dissociation energies are always positive and equal in magnitude, but are opposite in sign to the energy released when the bond forms.

Although atoms in different molecules are in differing environments, bond dissociation energies tend to be the same: A C–C bond always requires about the same energy to break no matter what molecule it is in.

Section 7.3   Polar Covalent Bonds: Electronegativity

Chemical bonding exists on a continuum from purely covalent bonds, such as those found in diatomic molecules, to purely ionic molecules. In between are molecules in which electrons are unequally shared: these are known as **polar covalent** molecules.

In polar covalent molecules, one atom will pull harder on the electrons and will therefore carry a slight negative charge. The other atom, deprived of its share of the electrons, will have a slight positive charge. These partial charges are represented with the lowercase Greek delta: $\delta-$ for the electron-rich atom, $\delta+$ for the electron-poor atom.

Bond polarity is the result of differences in **electronegativity**—the ability of an atom in a molecule to attract the shared electrons in a covalent bond. Fluorine is the most electronegative element, and it is given a unitless value of 4. Electronegativity decreases down a group and left across the periodic table. Metals have very low electronegativities: cesium has such a low electronegativity that it is sometimes referred to as "electropositive"! Electronegativity is related to electron affinity and ionization energy.

Bond polarity is generally predicted using the following scale:

- Atoms with similar electronegativities ($\Delta EN \leq 0.4$) form nonpolar bonds.
- Atoms whose electronegativities differ by more than 1.8 ($\Delta EN > 1.8$) form ionic bonds.
- Atoms whose electronegativities between ($0.4 \leq \Delta EN < 1.8$) form polar covalent bonds.

Workbook Example 7.1

PROBLEM:
Using the electronegativity values above, predict whether the following bonds are polar, nonpolar, or ionic:

Cl–Cl	O–F	Mg–Se	C–H	Li–F

SOLUTION:

Cl–Cl:	nonpolar	$3.0 - 3.0 = 0$
O–F:	polar	$4.0 - 3.5 = 0.5$
Mg–Se:	polar	$2.4 - 1.2 = 1.2$
C–H:	nonpolar	$2.5 - 2.1 = 0.4$
Li–F:	ionic	$4.0 - 1.0 = 3$

Workbook Problem 7.1

Predict whether the following bonds are polar, nonpolar, or ionic:

H–F	F–Cl	C–O	K–Cl	Ge–Br

Strategy: First, make your predictions based on periodic properties; then confirm your answer with the numbers in Table 7.4 in the textbook.

Step 1:   Predict the polarity of the bonds based on their locations, then check with electronegativity values from Table 7.4 in the textbook:

Step 2:   Plug in the numbers to check.

Section 7.4   A Comparison of Ionic and Covalent Compounds

Ionic compounds are solids with very high melting points, because the entire crystal lattice must be disrupted for the compound to melt.

Covalent compounds are low-melting-point solids, liquids, or gases. The atomic interactions within the molecules are strong, but the interactions between molecules are fairly weak.

Section 7.5   Electron-Dot Structures: The Octet Rule

Electron-dot structures, or Lewis structures, represent a molecule's valence electrons using dots and make it possible to simply model the formation of stable covalent compounds by combining atoms to fill valence shells.

The octet rule is the guiding principle for these structures, just as it was for ionic structures. The number of valence electrons an element has is the same as its group number: Group 7A elements (halogens) have seven valence electrons, Group 5A elements have five, and so on. Electrons are first placed around the element singly, then paired, until there are no more electrons to distribute. The single electrons are bonding electrons, and the paired electrons are nonbonding pairs or lone pairs.

Workbook Example 7.2

PROBLEM:
Draw electron-dot structures for the following neutral atoms.

| N | Si | F | S |

SOLUTION:

$\cdot \ddot{N} \cdot$ $\cdot \dot{Si} \cdot$ $\cdot \ddot{F} :$ $\cdot \ddot{S} \cdot$

The octet rule can be used to determine how bonds will form: only unpaired electrons will form bonds, and an atom will share electrons until it is surrounded by four pairs of electrons (an octet) or has no more electrons to share. Bonding pairs share electrons, and these shared electrons can be represented with a line.

Workbook Example 7.3

PROBLEM:
Draw the electron-dot structure for CH_4.

SOLUTION:
First draw the electron-dot structure for carbon. Carbon has four valence-shell electrons.

Each hydrogen has an electron that it can share with an unpaired electron on carbon.

H•

The hydrogens can be bonded to the carbon to complete the octet.

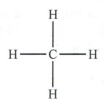

The paired electrons can also be represented as bonds:

H
|
H——C——H
|
H

Workbook Problem 7.2

Draw electron-dot structures for the following compounds.

HF CH_2Br_2 Cl_2

Strategy: Determine the valences of compounds, then combine them to form complete octets.

Step 1:   Determine the number of electrons on each atom based on its position on the periodic table.

Step 2: Combine the atoms to complete octets, substituting bond lines for paired electrons.

It is possible for atoms to share more than one pair of electrons: when this happens, multiple bonds are formed. These are shorter and stronger than single bonds. Double bonds are formed when two pairs of electrons are shared, and triple bonds are formed when three pairs of electrons are shared. The **bond order** is the number of electron pairs: a double bond has a bond order of 2, a triple bond has a bond order of 3.

Sometimes an atom donates a lone pair of electrons to a bond, so that both of the electrons in that bond come from one atom. These are called **coordinate covalent bonds**: they are frequently formed by N, O, P, and S, which have lone pairs to donate.

Section 7.6   Procedure for Drawing Electron-Dot Structures

Many compounds contain only hydrogen and second-row elements. Structures for these are straightforward to draw because the octet rule almost always applies and the number of bonds formed by each element is easy to predict. The second-row atoms are bonded to one another to form the core of the molecule, and the hydrogens are found around the periphery.

The general method that chemists use to draw Lewis structures of molecules are:

Step 1.   Find the total number of valence electrons in the molecule or ion.

Step 2.   Decide what the connections are between atoms, and draw lines to represent the bonds.

Step 3.   Subtract the number of valence electrons used for bonding from the total number calculated in Step 1 to find the number that remains.

Step 4.   If unassigned electrons remain after Step 3, place them on the central atom.

Step 5.   Form multiple bonds to fill an incomplete octet on the central atom.

Workbook Example 7.4

PROBLEM:
Draw the electron-dot structure for CH_3NH_2.

SOLUTION:
To draw the electron-dot structure of CH_3NH_2:

1. First draw the electron-dot structure of the core atoms.

$$\cdot\overset{\displaystyle\cdot}{\underset{\displaystyle\cdot}{C}}\cdot \qquad \cdot\overset{\displaystyle\cdot\cdot}{N}\cdot$$

2. Form a bond between the two core elements by forming an electron pair.

$$\cdot\overset{\displaystyle\cdot}{\underset{\displaystyle\cdot}{C}}-\overset{\displaystyle\cdot\cdot}{N}\cdot$$

3. Five hydrogens need to be placed; pair each one with a single electron to complete the octets.

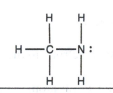

Workbook Problem 7.3

Draw electron-dot structures for the following compounds.

$$CH_3CH_2CH_3 \qquad CH_3OH \qquad C_2H_4$$

Strategy: Determine valences of compounds, then combine them to form complete octets.

Step 1:   Determine the number of electrons on each atom based on its position on the periodic table.

Step 2:   Combine the atoms to complete octets, substituting bond lines for paired electrons.

Compounds that contain elements beyond the second row can form expanded octets, and they do not always abide by the octet rule: their larger size makes it possible for them to accommodate more electrons. All electron-dot structures can be drawn using the same basic rules:

- Find the total number of valence electrons in the molecule. Add electrons for any negative charges, and subtract them for positive charges.
- Determine how the atoms are connected; the most electronegative atom is generally central, halogens and hydrogens are generally peripheral.
- Place two electrons in between each pair of atoms.
- Distribute nonbonding electrons to terminal atoms, and assign any leftover atoms to the central atom, expanding its octet.
- When in doubt, choose the most symmetrical arrangement.

Workbook Example 7.5

PROBLEM:
Draw the electron-dot structure for SO_4^{2-}.

SOLUTION:
To draw the electron-dot structure of SO_4^{2-}:

1. Find the total number of valence electrons in the molecule.

    Each oxygen has 6 electrons, the sulfur atom has 6 valence electrons, and 2 more need to be added for the 2 negative charges: the total number of electrons is 32.

2. Determine how the atoms are connected:

    The sulfur can form an expanded octet, and placing it in the center gives the most symmetrical arrangement.

    O

    O S O

    O

3. Place electrons between each atom to form single bonds, and putting the correct number of valence electrons around the oxygen.

    This accounts for 28 electrons, leaving 4 to be distributed. Two of the remaining electrons will form the charges on the ion, but even after that, 2 more remain. Also, the octets of the oxygens are not satisfied.

4.   Satisfy the octets of the oxygens with the four remaining electrons.

This accounts for all of the electrons, but oxygens typically form two bonds, not one. Two of the oxygens are ionized, making them stable with only one bond, but the other two will be happier double-bonded to the sulfur, giving it an expanded octet of 10 electrons.

Workbook Problem 7.4

Draw the electron-dot structure for PF_5.

Strategy:

Step 1:   Determine the total number of valence-shell electrons.

Step 2:   Determine the connections.

Step 3:   Draw the bonds and subtract the number of electrons used from the total number of valence electrons available.

Step 4:   Complete the octets of the outer atoms.

Step 5:   Place any remaining electrons on the central atom.

Section 7.7   Drawing Electron-Dot Structures for Free Radicals

Most molecules or ions that we deal with in chemistry have an even number of electrons that are paired up in orbitals. **Free radicals** are ions or molecules that have an odd number of electrons, which means that one orbital will be half-filled. Because of this, free radicals are very reactive and will seek to undergo some sort of reaction in order to form a more stable product.

The rules that we use to draw Lewis structures apply to free radicals.

Section 7.8   Compounds Containing Only Hydrogen and Second-Row Elements

We use Lewis structures to help us visualize **organic molecules**, or compounds that contain carbon. Organic compounds typically contain carbon, hydrogen, nitrogen, and oxygen and are essential to the study of the chemistry of biomolecules such as carbohydrates, lipids, nucleic acids, proteins, and others. The rules that we use to draw the Lewis structures for organic molecules are the same as those stated above.

Section 7.9   Electron-Dot Structures and Resonance

It is often the case, as in the sulfate ion (from Example 7.5) that there is more than one possibility for an electron structure. When deciding which of the sulfur atoms was double-bonded to oxygen, there was no reason that up and down double bonds were any more or less valid than side-to-side double bonds. When this occurs, the "real" structure is an average of all possible structures. This going back and forth between equivalent structures is known as **resonance**. It is indicated by a single, straight, double-headed arrow, ↔. Resonance structures differ *only* in placement of valence electrons, and not the movement of atoms!

Workbook Example 7.6

PROBLEM:
Draw the resonance structures for NO_3^-.

SOLUTION:
1.   Total number of valence-shell electrons = 5 (from N) + 18 (from 3 O) + 1 (from the 1– charge) = 24

2.   Determine the connections. In this anion, N is the central atom.

3.   Six of the 24 electrons are used for forming the three N–O bonds, leaving 18 electrons. All of these electrons are used to form the octet around the three oxygen atoms.

4.   There are no more electrons to distribute; however, nitrogen does not have a completed octet. To form a completed octet, we can borrow a pair of lone electrons from one of the oxygen atoms. However, since all three N–O bonds are equal, we must draw three

---

different resonance structures, each of which has a N=O bond between nitrogen and each of the different oxygens.

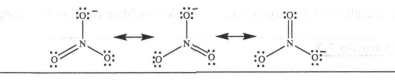

Workbook Problem 7.5

Draw as many resonance structures as possible for the carbonate (CO_3^{2-}) ion.

Strategy:

Step 1:   Determine the total number of valence-shell electrons.

Step 2:   Determine the connections.

Step 3:   Draw the bonds and subtract the number of electrons used from the total number of valence electrons available.

Step 4:   Complete the octets of the outer atoms.

Step 5:   Place any remaining electrons on the central atom, looking for multiple placement possibilities.

Section 7.10   Formal Charge

Sometimes it is difficult to distinguish which of a number of resonance possibilities is the "best," meaning the most stable configuration. Formal charges make it possible to distinguish between structures and find the one that is the most stable.

To determine the formal charge on an atom, take the number of valence electrons in a neutral, unbonded atom and subtract from it one-half the number of bonded electrons and all of the nonbonded electrons:

Formal charge = valence electrons – ½(bonding electrons) – nonbonding electrons

The most stable structures are those with the lowest formal charges overall and those that isolate negative charges on electronegative atoms.

For an ion, the overall formal charge of the molecule should be equal to the charge on the ion.

Workbook Example 7.7

PROBLEM:
Distinguish the best of these two structures for carbon dioxide by checking their formal charges.

$$:O\!\!-\!\!C\!\!\equiv\!\!O: \quad \longleftrightarrow \quad O\!\!=\!\!C\!\!=\!\!O$$

SOLUTION:
Symmetry makes the structure on the right appear more favorable, but both structures satisfy the octets of all the atoms. Their formal charges will make it possible to distinguish between them.

For the structure on the left:

Left oxygen	$= 6 - 1 - 6 = -1$
Carbon	$= 4 - 4 - 0 = 0$
Right oxygen	$= 6 - 3 - 2 = 1$

For the structure on the right:

Oxygen	$= 6 - 2 - 4 = 0$
Carbon	$= 4 - 4 - 0 = 0$

The formal charges confirm that the intuitive choice is correct: although both structures have an overall charge of 0, the asymmetrical structure has two of the same atom carrying opposite charges, which is an unstable situation.

Workbook Problem 7.6

Draw three resonance structures for the cyanate ion (CNO⁻) and use formal charges to determine which is the most stable. Carbon is the central atom.

Strategy: Draw electron-dot structures, then check the formal charges on the atoms to determine which is the most stable.

Step 1: Determine the number of valence electrons in the ion.

Step 2: Determine three reasonable electron-dot structures with carbon central and the number of valence electrons from Step 1.

Step 3:   Calculate the formal charge on each of the atoms in each of the structures.

Step 4:   Choose the most stable structure based on your comparison of the formal charges.

Self–Test

This section is intended to test your knowledge of the material covered in this chapter. Think through these problems, and make certain you understand what they are asking. Make sure your answers make sense. Successful completion of these problems indicates that you have mastered the material in this chapter. You will receive the greatest benefit from this section if you use it as a mock exam, as this will allow you to determine which topics you need to study in more detail.

True–False
1. Covalent bonds involve transfer of electrons from one atom to another.

2. Bond dissociation energies are always positive.

3. Covalent compounds have lower melting points and weaker intermolecular interactions than ionic compounds.

4. Bond polarity is usually the result of differences in electronegativity.

5. Electron-dot modeling can predict stable covalent compounds.

6. Valences of main-group elements cannot be predicted based on their position on the periodic table alone.

7. Only the structures of binary compounds can be predicted with electron-dot modeling.

8. It is not possible to model charged compounds using electron-dot structures.

9. The more symmetrical an atomic structure is, the more likely it is to be stable.

10. Free radicals contain an unpaired electron in an orbital.

Fill-in-the-Blank
11. Covalent bonds have characteristic _____. If the atoms are too far away, no _____ can form. Too close, and the _____ repel each other.

12. The energy required to break a bond in an isolated gaseous atom is the _____

    _____ _____. It is always _____, and equal in magnitude to

    the energy _____ when the bond forms.

13. Chemical bonds exist on a continuum from pure covalent to _____ _____ ,

    to pure _____.

14. In electron-dot structures, single electrons are _____ electrons, and paired

    electrons are _____.

15. Second-row elements are sometimes exceptions to the _____ _____, because

    they can form _____ _____.

16. Sometimes multiple structures can be drawn for a compound. In this situation, _____

    _____ can be used to distinguish the structure that is the most stable.

Matching

17. Bond length

a. the energy required to break a chemical bond in an isolated molecule in the gaseous state

18. Bond dissociation energy

b. a bond in which the electrons are more strongly attracted to one atom

19. Coordinate covalent bond

c. an average of several valid electron-dot structures for a molecule

20. Polar covalent bond

d. a pair of electrons not participating in a chemical bond.

21. Lone pair

e. "bookkeeping" charge assigned to an atom

22. Electronegativity

f. pair of electrons participating in a chemical bond

23. Bonding pair

g. the strength with which an atom pulls on the electrons associated with it

24. Formal charge

h. the distance at which attractive forces and nuclear repulsions are in balance

25. Resonance hybrid

i. a bond in which one atom donates a pair of electrons to an atom with a vacant orbital

Problems

26. Identify each of the following compounds as primarily ionic or covalent:

$LiBr$ HCl MgO NF_3

27. Draw electron-dot structures for H_2O_2, NH_4^+, Cl_2CO, SF_4, and XeF_4. Use resonance structures if necessary.

28. Two white crystalline compounds are tested. The first has a melting point of 800 °C, a solubility of 36 g/100 mL of water, and conducts electricity in solution. The second has a melting point of 180 °C, a solubility of 200 g/100 mL of water, and does not conduct electricity in solution. What conclusions can you make about the chemical bonds in these compounds?

29. Draw all resonance structures for CO_3^{2-}, and assign formal charges to each atom.

30. Rank the following bonds from most polar to least polar, using only the periodic table.

  C—H   Al—F   Li—Cl   Cl—Cl   H—Cl

  Justify your answers.

31. Rank the following atoms in order of increasing electronegativity.

  Cs   F   Fe   O

32. Given the following three structures for nitrous oxide, which is likely to contribute most to the structure? Use formal charges to justify your answer.

  :N̈—N≡O: :N̈═N═Ö: :N≡N—Ö̈:

CHAPTER EIGHT

Covalent Compounds: Bonding Theories and Molecular Structure

Learning Objectives

As a result of reading and studying this chapter, you should be able to

Section 8.1   **Molecular Shapes: The VSEPR Theory**
1. Use the VSEPR model to predict geometry from the total number of charge clouds and lone pairs of electrons around an atom.
2. Use the VSEPR model to predict bond angles and overall shape of a molecule or ion with one central atom.
3. Use the VSEPR model to predict bond angles and overall shape of a molecule with more than one central atom.

Section 8.2   **Valence Bond Theory**
4. Describe the difference between a sigma bond and a pi bond.

Sections 8.3–8.4   **Hybrid Orbitals**
5. Determine the type of hybrid orbitals based upon the number of charge clouds around an atom.
6. Write an electron-dot structure for a molecule and determine the hybridization and bond angles on nonterminal atoms.
7. Identify which orbitals overlap to form sigma and pi bonds in molecules.

Section 8.5   **Polar Covalent Bonds and Dipole Moments**
8. Predict whether a given molecule has a dipole moment, and if so draw its direction.
9. Interpret electrostatic potential maps of molecules.
10. Calculate the percent ionic character in a bond.

Section 8.6   **Intermolecular Forces**
11. Identify the types of intermolecular forces affecting a molecule.
12. Relate the strength of intermolecular forces to physical properties such as melting point and boiling point.
13. Sketch the hydrogen bonding that occurs between two molecules.

Sections 8.7–8.8   **Molecular Orbital Theory**
14. Interpret the molecular orbital diagram for a first-row diatomic molecule or ion.
15. Interpret the molecular orbital diagram for a second-row diatomic molecule or ion. Calculate the bond order and predict magnetic properties.

Section 8.9   **Combining Valence Bond Theory and Molecular Orbital Theory**
16. Draw orbital overlap diagrams for molecules and describe the use of both valence bond theory and molecular orbital theory.

Chapter Summary

In Chapter 7, you learned about formation of a covalent bond and how to draw Lewis structures to represent molecules; Chapter 8 continues this by expanding upon the two-dimensional Lewis structures that you drew in the last chapter. You will also learn in this chapter how to transfer this information to a three-dimensional image by using the valence-shell electron-pair repulsion model, or VSEPR, which takes both atoms and unbonded electrons into account. You will then be able enhance your picture of the three-dimensional model by using valence bond theory, which describes how electrons are spread throughout hybrid orbitals that permit the electrons in pairs and bonds to maximize their distances from each other. We end our overview of molecular modeling by looking at molecular orbital theory, which provides mathematical precision to the visualization of molecular structure and solves some problems that the previous models cannot. In addition to the above, we will talk about how three-dimensional molecules interact through intermolecular forces.

The Chapter in Detail

Section 8.1   Molecular Shapes: The VSEPR Model

Molecules exist not only in two dimensions; their three-dimensional shapes are also very important for their biological functions, and these shapes can often be predicted using the **valence-shell electron-pair repulsion (VSEPR) model**. Electrons both in lone pairs and in bonds try to repel each other because of their negative charges.

The result of these repulsions between electrons is molecular shape, which can be predicted by counting the number of electron clouds (both bonds and lone pairs), as follows. Note that a double or triple bond counts as only one electron cloud.

- Two electron clouds (e.g., CO_2) – linear.
- Three electrons clouds (e.g., BH_3) – trigonal planar. Molecular shapes (as opposed to electronic shapes) can vary; for example, two bonds and one nonbonding pair is bent.
- Four electron clouds (e.g., NH_3) – tetrahedral. Molecular shapes can vary: three bonds, one nonbonding pair is trigonal pyramidal; two bonds, two nonbonding pairs is bent.
- Five charge clouds (e.g., PCl_5) – trigonal bipyramidal: three equatorial, two axial groups. Molecular shapes can vary: four bonds, one nonbonding pair is seesaw-shaped; three bonds, two nonbonding pairs is T-shaped; two bonds, three nonbonding pairs is linear.
- Six charge clouds (e.g., SF_6) – octahedral: four equatorial, two axial groups. Molecular shapes can vary: five bonds, one nonbonding pair is square pyramidal; four bonds, two nonbonding pairs is square planar.

The shapes of larger molecules can be predicted by looking at the geometries around individual atoms.

Workbook Example 8.1

PROBLEM:
Predict the shape of XeF_2.

SOLUTION:
First, draw the electron-dot structure of the molecule. There is a total of 8 valence electrons from xenon and 14 from fluorine, for a total of 22.

$$:\ddot{F}-\ddot{X}e-\ddot{F}:$$

There are five charge clouds around xenon, two of which are attached to atoms. Referring back to the list above, five electron groups, two in bonds and three nonbonding pairs, is linear.

Workbook Example 8.2

PROBLEM:
Ethanol has the formula CH_3CH_2OH. It is the component of alcoholic beverages that gives them their vivifying effects. The structural formula for this compound is

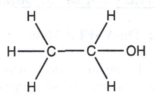

Describe the geometry around the carbon and oxygen atoms.

SOLUTION:
The two carbons each have four bonds and no nonbonding pairs: the geometry around these atoms is tetrahedral. The oxygen atom has two bonds and two nonbonding pairs: the geometry around this atom is bent.

Workbook Problem 8.1

Predict the molecular shape of CS_2 and SF_4.

Strategy: Draw the electron dot structure for each of the molecules and use the list above to determine the molecular shape.

Step 1:   Draw the electron-dot structure for the molecules, using the number of valence electrons and the most symmetrical structure.

Step 2:   Determine the number of charge clouds around the central atom. How many are used for bonding and how many are nonbonding?

Step 3:   Based on the chart, determine the shape.

Section 8.2   Valence Bond Theory

VSEPR theory provides information about molecular shape, but says nothing about the nature of covalent bonds.

Valence bond theory states that when atoms approach one another, singly occupied electronic orbitals will overlap, allowing the now-paired electrons to be influenced by both nuclei. The orbitals must be in the same phase, and they will contain electrons of opposite spin.

If anything other than s orbitals overlaps, there will be a directionality to the bond.

Bonds resulting from head-on overlap of orbitals are called **sigma (σ) bonds**.

Section 8.3   Hybridization and sp^3 Hybrid Orbitals

For carbon in particular, the electron structure of the atom does not mesh easily with the VSEPR structure of molecules. Carbon has two unpaired p electrons and no unpaired s electrons, but it forms four bonds.

Linus Pauling provided a mathematical solution, showing that the Schrödinger wave equations for s and p orbitals can be combined to form **hybrid orbitals**. In the case of carbon (and others), one s combines with three p to form four sp^3 orbitals.

Each sp^3 orbital has two lobes, a large and a small. The large lobes point toward the corners of a tetrahedron. Bonds formed by overlapping sp^3 orbitals are very strong: a tetrahedral arrangement of charge clouds always implies sp^3 hybridization.

Section 8.4   Other Kinds of Hybrid Orbitals

Atoms with three electron clouds are formed by sp^2 hybridization: the combination of one s orbital and two p orbitals. The hybridized orbitals interact head-on to form σ bonds between the nuclei, but this leaves an empty p orbital above and below the plane of the hybrid orbitals. The unhybridized p orbitals can interact sideways, forming a π bond, which shares electrons not between the nuclei, but above and below a line that connects the nuclei. The combination of σ and π bonds results in four electrons being shared: a double bond.

Atoms with two electron clouds are formed by sp hybridization: the combination of one s orbital and one p orbital. The two unhybridized p orbitals can interact sideways, forming two π bonds that share electrons above and below the nucleus. The combination of one σ and two π bonds results in six electrons being shared: a triple bond.

Atoms with five and six charge clouds have recently been found to be more complex than can be explained by valence bond theory.

Workbook Example 8.3

PROBLEM:
Determine the hybridization for the central atom in H_2CO (formaldehyde) and its bond types.

SOLUTION:
This molecule has the following electron-dot structure.

With three charge clouds, the central carbon atom is sp^2 hybridized. It has a σ bond to each hydrogen, and one σ and one π bond with the oxygen atom.

Workbook Problem 8.2

Describe the hybridization and bonding of each of the carbon atoms in the structure below.

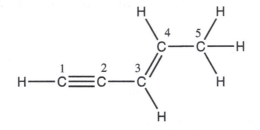

Strategy: Determine the number of electron clouds around each carbon to determine hybridization, then identify the number of σ and π bonds.

Step 1:   Determine the number of electron clouds around each carbon and its hybridization.

Step 2:   Determine the bond types present.

Section 8.5   Polar Covalent Bonds and Dipole Moments

Polar covalent bonds result from differences in electronegativity between atoms: these differences lead to electrons being displaced toward one part of a molecule, making it electron-rich and giving it a slight negative charge, while the other end of the molecule becomes electron-poor with a slight positive charge.

Graphically, the dipole, or charge separation, is indicated by a crossed arrow: +--> The crossed end is the positive end of the molecule. The point of the arrow indicates where the electrons are concentrated: the negative end of the molecule.

Molecules larger than two atoms can also have dipoles: the contributions of lone pairs and individual bond polarities can be summed to calculate a net dipole for a shaped molecule. In symmetrical molecules, the dipoles will sum to zero:

$$<\text{----}+$$
$$O = C = O$$
$$+\text{----->}$$

To determine the percent ionic character of a bond, the calculated dipole (full charges separated by the bond length) can be compared to the measured bond dipole. The following formula is used to calculate dipoles:

$$\mu = Q \times r$$

where Q is the charge (1.160×10^{-19} C) and r is the bond length. The result is then converted to the unit *debye* (D), where 1 debye is 3.336×10^{-30} C·m.

We can then calculate the percentage of ionic character in a type of bond if we know both the measured dipole (the experimental value) and the calculated dipole (the theoretical value) by using the equation

$$\text{Percent ionic character} = \frac{\text{measured dipole}}{\text{calculated dipole}} \times 100\%$$

Workbook Example 8.4

PROBLEM:
Determine the % ionic character of carbon monoxide, given the bond length of 112.8 pm and the measured dipole of 0.112 D.

SOLUTION:
Calculated dipole:

$$\mu = Q \times r$$

$$\mu = \left(1.160 \times 10^{-19} \, C\right)\left(112.8 \times 10^{-12} \, m\right)\left(\frac{1D}{3.336 \times 10^{-30} \, C \cdot m}\right)$$

$$\mu = 3.92 \, D$$

$$\text{Percent ionic character} = \frac{0.112 \, D}{3.92 \, D} \times 100\%$$

$$\text{Percent ionic character} = 2.86\% \text{ ionic}$$

Not surprisingly, carbon monoxide has very little ionic character.

Workbook Example 8.5

PROBLEM:
Which of the following would be expected to be polar and which nonpolar:

$$CH_4 \qquad CH_3Cl \qquad HCl \qquad Cl_2$$

SOLUTION:
CH_4 and Cl_2 are both symmetrical molecules, which makes them nonpolar. CH_3Cl and HCl are asymmetrical and contain polar bonds. Drawing the electron-dot structures can help to confirm this initial determination:

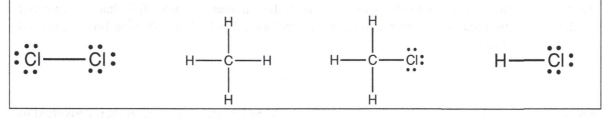

Workbook Problem 8.3

Indicate which of the following are likely to have dipole moments, and indicate the element with the highest electron density:

H_2CO – formaldehyde $N(CH_3)_3$ – trimethylamine   HBr – hydrogen bromide

Strategy: Look for asymmetrical molecules with electronegativity differences.

Step 1:   Determine which molecules are asymmetrical.

Step 2:   Determine which element is the most electronegative and thus the area of highest electron concentration.

Workbook Problem 8.4

BrF has a bond length of 176 pm and a measured dipole moment of 1.42 D. What is its percent ionic character?

Strategy: Calculate the dipole as if it were purely ionic, then calculate the percent ionic character.

Step 1:   Calculate ionic dipole.

Step 2:   Determine percent ionic character.

Section 8.6   Intermolecular Forces

Covalent and ionic bonds are **intramolecular forces**: they occur *within* molecules. **Intermolecular** forces occur *between* molecules. Collectively, intermolecular forces are often called **van der Waals** forces. They are divided into four general categories:

- **Ion–dipole forces**: these occur between ions and polar molecules. The charge on the ion will attract the portion of a polar molecule that has an opposite partial charge. These are particularly important in aqueous solutions of ionic substances. Ion–dipole forces have moderate strength: 10–50 kJ/mol.
- **Dipole–dipole forces**: dipoles will interact in liquids; the stronger the dipole, the stronger the interaction. Increasing strength of dipole moments correlates roughly with an increase in boiling points: stronger dipole–dipole interactions require more kinetic energy to disrupt. Dipole–dipole forces are weak: 3–4 kJ/mol.
- **London dispersion forces**: these forces are found in all types of molecules, but are the only ones that occur between nonpolar molecules. They are the result of varying electron densities; average electron densities are uniform, but a small variation in electron density can create a small dipole that can then induce an opposite dipole in another nearby molecule. Such *polarizability* increases with molecular size and varies with molecular shape: an extended molecule has more opportunity to create dipoles than one that is compact. Polarizability is related to boiling point: more dispersion forces mean a higher boiling point. London dispersion forces are weak: 1–10 kJ/mol.
- **Hydrogen bonds**: there are very strong dipole–dipole forces between a hydrogen atom bonded to oxygen, nitrogen, or fluorine and a hydrogen atom bonded with an unshared electron pair on another electronegative atom. Molecules that can hydrogen bond have boiling points that are unusually high for their small size, because these strong interactions must be broken for the molecules to enter the gas phase. Hydrogen bonds have moderate strength as compared to ionic and covalent bonds: 10–40 kJ/mol.

Workbook Example 8.6

PROBLEM:
Identify the intermolecular forces experienced by the following molecules, and rank them in order of increasing intermolecular forces:

$$NH_3 \qquad CH_4 \qquad HBr \qquad CH_3CH_2CH_2CH_3$$

SOLUTION:
NH_3 can hydrogen-bond, which is the strongest intermolecular force. It will also experience dipole–dipole and dispersion forces.

CH_4 cannot hydrogen-bond and has no dipole. It will experience only dispersion forces.

HBr cannot hydrogen-bond, but is polar and will have a dipole. This molecule will have both dipole–dipole and dispersion forces.

$CH_3CH_2CH_2CH_3$ cannot hydrogen-bond and has no dipole. It will experience only dispersion forces.

CH_4 will experience the weakest intermolecular forces. $CH_3CH_2CH_2CH_3$ will have stronger dispersion forces, because it is larger. HBr will have the next strongest, because it has both dipole–dipole and dispersion forces. NH_3, as a hydrogen-bonder, will have the strongest intermolecular interactions. This will also be the ranking in order of increasing boiling point.

$$CH_4 \quad < \quad CH_3CH_2CH_2CH_3 \quad < \quad HBr \quad < \quad NH_3$$

Workbook Problem 8.5

Determine the intermolecular forces that are present in the following molecules and order by increasing strength.

$$CH_3OH \qquad Xe \qquad CH_3Cl$$

Strategy: Examine the structures of the molecules and determine what interactions are present, then rank them in order of strength.

Step 1:   List the interactions experienced by each molecule.

Step 2:   Rank the molecules in order of increasing strength of interactions.

Section 8.7   Molecular Orbital Theory: The Hydrogen Molecule

The valence bond model does not always provide agreement with experimental observations. A more complex model known as **molecular orbital theory** provides results that come closer to observed molecular behavior. In contrast to the previous models described, which focused on atomic orbitals, molecular orbital theory looks at the molecule as a whole.

When two hydrogen atoms interact, the wave portions of the electrons will interfere with each other. They can interfere constructively and add together, or they can interfere destructively and subtract from one another.

If the orbitals of hydrogen atoms add together, they will form a single, egg-shaped *bonding* orbital. If they subtract from each other, they will form two small orbitals with a node in between the nuclei, which is an *antibonding* orbital, denoted σ*. The bonding orbital is stable and low in energy, and the antibonding orbital is high-energy and unstable. If more electrons are added to an H_2 molecule, they will have to be added to the antibonding orbital, thus destabilizing the molecule.

Bond order is defined as

$$\text{Bond order} = \frac{\text{number of bonding electrons} - \text{number of antibonding electrons}}{2}$$

Workbook Example 8.7

PROBLEM:
What is the bond order of an He_2^{2+} ion? Is it stable or unstable?

SOLUTION:
This molecule has two electrons, just as hydrogen does. The electrons will go first into the bonding orbital, leaving no electrons to go into the antibonding orbital. The bond order will be

$$\text{Bond order} = \frac{2 \text{ bonding electrons} - 0 \text{ antibonding electrons}}{2} = 1$$

Contrary to expectations, this will be a stable ion because it will form one bond.

Section 8.8 Molecular Orbital Theory: Other Diatomic Molecules

Experimental evidence shows that electron-dot structures fail to predict the properties of some molecules, particularly the diatomic gases in period 2, nitrogen, oxygen, and fluorine. Of these three gases, nitrogen and fluorine are **diamagnetic**, meaning that they are unaffected by magnetic fields, but oxygen is **paramagnetic**, meaning that it is attracted to a magnetic field. This phenomenon is only seen in atoms or molecules with unpaired electrons.

The electron-dot structures of these do not indicate that they have any unpaired electrons, so it is necessary to turn to molecular orbital theory to demonstrate why oxygen is paramagnetic: when a molecular orbital diagram is drawn, we see that two electrons are found in antibonding π* *2p* orbitals.

Section 8.9 Combining Valence Bond Theory and Molecular Orbital Theory

Both of these theories are useful to chemists. The simplicity of valence bond theory makes it a desirable tool whenever it can be useful; but where it fails to provide answers, the mathematical precision of molecular orbital theory can save the day.

For example, sigma (σ) bonds are "local" bonds—they occur between the nuclei of two atoms and go no further. Valence bond theory describes them well. On the other hand, Pi (π) bonds are not local; they occur outside of the internuclear space, and as a result, they can spread out over more than two atoms in resonance structures. When resonance structures occur, electrons are said to be "delocalized." Valence bond theory deals with this awkwardly at best: molecular orbital theory explains it perfectly.

Putting It Together

Oxalic acid can be used in low concentrations to prevent mites in beehives. Its molar mass is 93.03 g/mol and its percent composition is 26.68% carbon, 71.09% oxygen, and 2.224% hydrogen. The structure for this molecule includes a C–C single bond, with two carbons that are sp^2-hybridized. The hydrogen atoms are bonded to oxygen. Determine the empirical formula, molecular formula, and structure.

Self–Test

This section is intended to test your knowledge of the material covered in this chapter. Think through these problems, and make certain you understand what they are asking. Make sure your answers make sense. Successful completion of these problems indicates that you have mastered the material in this chapter. You will receive the greatest benefit from this section if you use it as a mock exam, as this will allow you to determine which topics you need to study in more detail.

True–False
1. Bond polarity is usually the result of differences in electronegativity.

2. Electron-dot modeling can predict stable covalent compounds.

3. The number of electron clouds in a molecule can be used to determine its three-dimensional shape.

4. Methane (CH_4) molecules have a pyramidal shape.

5. Molecular shape can be used to determine orbital hybridization.

6. When orbitals overlap head-on, π bonds are formed.

7. Molecular orbital theory can be used to explain molecular properties that cannot be explained by valence bond theory.

8. Multiple bonds can be explained by electron-dot and valence bond theories.

9. If a molecule has polar bonds, it will always have a dipole moment.

10. Intermolecular forces all are equal in strength.

11. Intermolecular forces can be related to melting and boiling points.

Fill-in-the-Blank
12. A molecule with four electron clouds will have a _____ electronic shape, but the molecular shape may also be _____, or _____ _____.

13. A trigonal planar arrangement of charge clouds always implies _____ hybridization.

14. _____ bonds are always local, but _____ bonds can be delocalized over more than two atoms.

15. _____ is a measure of the strength with which an element pulls on its electrons.

16. There are several types of intermolecular forces. _____ result from the interactions of two polar molecules. _____ occur in all molecules and are the result of temporary charge imbalances. _____ are the strongest intermolecular forces, and they result from interactions between a H atom bonded to an N, O, or F, and a nonbonding pair on another atom.

Matching

17. VSEPR model

    a.   intermolecular force characterized by attraction between a hydrogen atom with a partial positive charge and an unbonded electron pair

18. hybrid orbitals

    b.   attractive force caused by temporary dipoles

19. σ bond

    c.   substance that is weakly repelled by a magnetic field because all of its electrons are spin-paired

20. π bond

    d.   attracted to a magnetic field as a result of unpaired electrons

21. molecular orbital

    e.   an area in which the electrons involved in an atom or bond are most likely to be found

22. paramagnetic

    f.   a bond formed by sideways interactions with parallel orbitals in which the electrons do not reside between the nuclei

23. diamagnetic

    g.   a bond formed by head-on overlap of orbitals in which the electrons reside primarily between the two nuclei

24. London dispersion force

    h.   a combined solution for the Schrödinger wave equations that gives a new set of values

25. hydrogen bonding

    i.   a method for determining the three-dimensional structures of molecules using electron repulsions

Problems

26. Determine the electron geometry and molecular geometry for the following molecules:

$$NF_3 \quad H_2S \quad IF_4^+ \quad PF_5$$

27. Predict the geometry and hybridization of the carbon atoms in acetaldehyde:

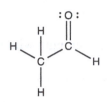

What types of bonds are found in this molecule? What atoms are linked by these bond types?

28. Sketch bonding and antibonding orbitals that come from the combination of two *s* orbitals. Indicate *where* they are interfering and *how* they are interfering (constructively or destructively).

Challenge Problem
29. 360 mg of an unknown acid require 78.24 mL of 0.10 M NaOH to titrate to neutrality. Elemental   analysis shows that the acid is 71.1% oxygen, 26.7% carbon, and 2.22% hydrogen. Determine the formula of the acid and draw its electron-dot structure. Using VSEPR theory, provide the molecular shape around each of the central atoms. Using valence bond theory, explain how the oxygen and carbon atoms are hybridized and what bond types—σ and π—each are participating in.

CHAPTER NINE

Thermochemistry: Chemical Energy

Learning Objectives

As a result of reading and studying this chapter, you should be able to

Section 9.1	**Energy and Its Conservation**
1.	Calculate the kinetic energy of an object in motion.
2.	Convert between common units for energy.

Section 9.2	**Internal Energy and State Functions**
3.	Identify state functions.
4.	Identify the signs of heat and work.

| **Section 9.3** | **Expansion Work** |
| 5. | Calculate PV work. |

| **Section 9.4** | **Energy and Enthalpy** |
| 6. | Calculate the internal energy change (ΔE) for a reaction. |

Sections 9.5 – 9.6	**Thermochemical Equations and Enthalpies of Chemical and Physical Change**
7.	Given a thermochemical equation and the amount of reactant or product, calculate the amount of heat transferred.
8.	Classify endo- or exothermic reactions.

Section 9.7	**Calorimetry and Heat Capacity**
9.	Calculate heat capacities, temperature changes, and heat transfer using equation for heat capacity (C), specific heat (c), or molar heat capacity (C_m).
10.	Calculate enthalpy changes in a calorimetry experiment.

| **Section 9.8** | **Hess's Law** |
| 11. | Use Hess's law to find ΔH for an overall reaction, given reaction steps and their ΔH values. |

Section 9.9	**Standard Heats of Formation**
12.	Identify standard states of elements.
13.	Write the standard enthalpy of formation reactions (ΔH^{o}_{f}) for compounds from their elements.
14.	Use values of ΔH^{o}_{f} for elements and compounds to calculate ΔH^{o} for a reaction.

| **Section 9.10** | **Bond Dissociation Energies** |
| 15. | Use bond dissociation energies to estimate ΔH^{o} for a reaction. |

Section 9.11   **Fossil Fuels, Fuel Efficiency, and Heats of Combustion**

16.   Calculate $\Delta H°_c$ for various fuels using thermochemical principles such as Hess's law, calorimetry, and/or bond dissociation enthalpies.

Sections 9.12–9.13 An Introduction to Entropy and Free Energy

17.   Predict the sign of the entropy change (ΔS) in a reaction given the chemical equation or a molecular diagram.

18.   Using the relationship between Gibbs free energy (ΔG) and spontaneity, predict the sign of ΔG, ΔH, and ΔS in a reaction.

19.   Use the Gibbs free energy equation to calculate an equilibrium temperature.

Chapter Summary

In this chapter, we introduce you to the concept of thermochemistry: the heat changes that take place during reactions. You'll begin the study of this topic by learning the difference between heat and energy, and the types of energy changes that can take place. You are then introduced to the law of conservation and energy, which is the first law of thermodynamics, and the concept of state functions. With this background knowledge, you will then learn how to calculate the internal energy of a system using $P\Delta V$ work and how the internal energy of a system is related to the enthalpy (ΔH) of the system. You will spend much of the rest of the chapter exploring how to use specific heat calculations in the laboratory and how to use Hess's law, standard heats of formation, and bond dissociation energies to calculate heats of reaction. Finally, you are introduced to the topics of entropy and free energy, topics that will be explored in more detail in later chapters.

The Chapter in Detail

Section 9.1   **Energy and Its Conservation**

Energy is the capacity to do work or supply heat. Energy can be changed from one form to another, but it cannot be created or destroyed. This is the conservation of energy law, also known as the first law of thermodynamics.

Kinetic energy (E_K) is the energy of motion, and it is described by the following formula:

$$E_K = \frac{1}{2}mv^2$$

where m = mass (measured in kg) and v = velocity (measured in m/s). Thermal energy is the energy of molecular motion: fast-moving molecules have a higher temperature than slow-moving molecules; in other words: more collisions, higher temperature. When thermal energy is transferred from one object to another, it is transferred as heat.

Potential energy (E_P) is stored energy. This can be the energy of position, as in a marble sitting at the top of a hill, or chemical potential energy, which is the capacity to react and thereby release energy. A bucket of gasoline has a great deal of chemical potential energy: a small spark will release a tremendous amount as the hydrocarbons react with the oxygen in the air.

Workbook Example 9.1

PROBLEM:
Describe the energy changes as a rubber band is shot.

SOLUTION:
As the rubber band is stretched, potential energy is added to it. When the rubber band is released, the potential energy is converted to kinetic energy as it flies across the room.

Section 9.2   Internal Energy and State Functions

When accounting for the energy changes in a reaction, the reaction is thought of as being isolated. The reactants and products are referred to as the *system*, while everything else is the *surroundings*.

In a completely isolated system—a theoretical situation—no energy is transferred to the surroundings, so the internal energy of the system is constant. (This is the first law of thermodynamics stated in a different way.)

It is not really possible to completely isolate a system, so energy will flow into and out of any system. When this is measured, it is expressed as $\Delta E = E_{final} - E_{initial}$.

Changes in energy are expressed in terms of the system. So if energy flows into the surroundings, the energy change is negative: the system has lost energy. If energy flows into the system, the energy change is positive: the system has gained energy. What makes this difficult to get a handle on is that in general, *you* are a part of the surroundings. When you hold a glass of cold water in your hand, the system of the water is gaining energy, while you are losing it. So a process that generates heat has a negative ΔE, while a process that requires heat has a positive ΔE.

Internal energy is a state function: it depends only on current conditions and not at all on how the system came to be in that condition. State functions are reversible: if you start at sea level and climb 2000 feet to the top of a mountain your current altitude will change, but if you then return to sea level, your overall change in altitude is zero; altitude is a state function.

Workbook Example 9.2

PROBLEM:
Which of the following are state functions:

                Temperature   Volume   Mass   Work

SOLUTION:
Temperature is a state function: if you warm a beaker in your hands, then put it back down on the lab bench, it will return to its previous temperature. The process is reversible, and you have a $\Delta T = 0$

A balloon carried outside on a cold day will shrink and look miserable, but when brought back indoors, it will resume its previous cheerful volume. It is a reversible process, $\Delta V = 0$, so volume is a state function.

Mass is a state function: how an object got its mass is irrelevant.
Work is *not* a state function. If an object has been moved from one side of a room to the other and

---

back again, its overall position has not changed, but work was still expended moving it from one place to the other. The sofa is in the same place it was, but you are still tired!

Section 9.3   Expansion Work

In physics (and by extension, chemistry), work is defined as force × distance ($w = F \times d$). In a system in which a cylinder is fitted with a piston, changes in the pressure of the gas can force the piston upwards, doing work. In this case, the equation used is

$$w = -P\Delta V$$

This is referred to as "PV work" and has the units of $L \cdot atm$ (we use J as the units of energy in chemistry; to convert to J, 1 L·atm = 101 J). An expanding system is doing work, so it is losing energy to the surroundings. This will have a negative value. A contracting system is having work done on it by the surroundings and so is gaining energy and will have a positive value. If the volume of the system does not change, $\Delta V = 0$, and no work is done.

Workbook Example 9.3

PROBLEM:
Calculate the work done when the volume of a reaction changes from 3.0 L to 12 L against a pressure of 15 atm.

SOLUTION:
Using the equation $w = -P\Delta V$:

$$w = -(15\,atm)(12\,L - 3\,L) = 135\,L \cdot atm$$

$$135\,L \cdot atm \times \frac{101\,J}{1\,L \cdot atm} = 13,635\,J \text{ or } 14\,kJ$$

Workbook Problem 9.1

If 7.4 L of hydrogen gas are reacted with 3.7 L of oxygen gas at 13.0 atm, 7.4 L of water vapor are produced. How many kilojoules of work have been done? Have they been done *by* the system or *to* the system?

Strategy: Use the equation $w = -P\Delta V$, then convert to kilojoules.

Step 1:   Use $w = -P\Delta V$:

Step 2:   Convert from $L \cdot atm$ to joules to kilojoules.

Step 3:   What does the sign tell you? Was work done by or to the system?

Section 9.4   Energy and Enthalpy

Reactions can transfer energy through heat (q) or PV work. These terms can be combined to give the overall energy change of a system:

$$\Delta E = q - P\Delta V,$$

which can also be rearranged so that we can solve for q:

$$q = \Delta E + P\Delta V$$

If a reaction is run in a closed container, the volume is constant and no PV work can be done. In this case, $\Delta E = q$.

If instead the reaction is carried out at constant pressure (such as the atmospheric pressure in the lab), PV work can be done, so $q = \Delta E + P\Delta V$.

Because reactions at constant pressure are so common, the change of heat in these systems is called the **heat of reaction**, or **enthalpy**, and abbreviated ΔH. Enthalpy is also a state function: it depends only on the initial and final states of the reaction, not the path the reaction takes. As with all state changes, $\Delta H = H_{products} - H_{reactants}$.

When the PV work done by a reaction at constant pressure is calculated, we find that it generally makes only a very small contribution to the overall energy change of a reaction. As a result, $\Delta H \approx \Delta E$. If there is no volume change as a result of the reaction, then $\Delta H = \Delta E$.

Workbook Example 9.4

PROBLEM:
The equation for the reaction of nitrogen and hydrogen to form ammonia is

$$N_2\ (g) + 3\ H_2\ (g) \rightarrow 2\ NH_3\ (g)$$

If 2 L of nitrogen and 6 L of hydrogen are reacted together, 4 L of ammonia are formed against 1 atm of pressure. At the same time, 8.23 kJ of heat are released by the reaction. What is the overall energy change of the reaction, and what is the percent contribution of PV work?

SOLUTION:
Using the equation $w = -P\Delta V$:

$$w = -(1\,atm)(4\,L - 8\,L) = 4\,L \cdot atm$$

$$4\,L \cdot atm \times \frac{101\,J}{1\,L \cdot atm} = 404\,J \; or \; 0.4\,kJ$$

Using the equation $\Delta E = q - P\Delta V$

$$\Delta E = -8.23\,kJ - 0.4\,kJ = -8.63\,kJ$$

The percent contribution of the PV work is

$$\frac{0.4\,kJ}{8.63\,kJ} \times 100\% = 4.6\%$$

Workbook Problem 9.2

When 7.4 L of hydrogen gas are reacted with 3.7 L of oxygen gas at 13.0 atm, 7.4 L of water vapor are produced, and 79.9 kJ of heat are released. What is the total energy change of the system?

Strategy: Use the answer to the previous workbook problem, substituting into
$\Delta E = q - P\Delta V$

Step 1: Use $\Delta E = q - P\Delta V$

Section 9.5 Thermochemical Equations and the Thermodynamic Standard State

The value of the enthalpy change of a reaction is stated as per mole of the reactant of interest. So if 200 kJ are released for one mole of the reactant, 400 kJ would be released if two moles reacted.

In calculating heat and energy change, it is essential to know the phases of all the reactants and products: for instance, the difference in energy change for producing liquid water is more than for producing gaseous water vapor.

To ensure a common understanding regarding the amounts and phases of products and reactants, a thermodynamic standard state has been defined as a substance in its most stable form at 1 atm pressure and 25 °C. For solutions, the standard state is a solution under the above conditions and at 1 M concentration.

Measurements made at standard state are identified with a superscript (°) and can be used to calculate other quantities at standard state.

Workbook Example 9.5

PROBLEM:
Identify the standard state for each of the following:

bromine   oxygen   iron   mercury   sodium

SOLUTION:
At 25 °C and 1 atm, bromine is a diatomic liquid, Br_2 (l); oxygen is a diatomic gas, O_2 (g); iron is a metallic solid, Fe (s); mercury is a metallic liquid, Hg (l); and sodium is a metallic solid, Na (s).

Workbook Example 9.6

PROBLEM:
What is the value of $\Delta E°$ if a reaction having $\Delta H° = 157.3$ kJ is carried out at a constant pressure of 12.3 atm and the volume change is 453 L?

SOLUTION:
$$\Delta H° = \Delta E° + P\Delta V$$

Replacing all of the known terms with the numbers from the problem, we have

$$157.3\,kJ = \Delta E° + \left(12.3\,atm \times 453\,L\right)$$
$$157.3\,kJ = \Delta E° + \left(5571.9\,L \cdot atm\right)$$

Converting 5571.9 $L \cdot atm$ to J (remember that 1 $L \cdot atm$ = 101 J), we have

$$157.3\,kJ = \Delta E° + 562.8\,kJ$$

Solving for $\Delta E°$:

$$\Delta E° = 157.3\,kJ - 562.8\,kJ = -405\,kJ$$

Workbook Problem 9.3

When 0.5 mol of aqueous carbonic acid dissociates into water and carbon dioxide, the volume change at 1 atm is 11.2 L. If $\Delta H°$ for this reaction is –20.7 kJ/mol, what is ΔE?

Strategy: Use the equation $\Delta H = \Delta E + P\Delta V$.

Step 1:   Determine the ΔH for the reaction given the amount of carbonic acid that is dissociated.

Step 2:   Determine $P\Delta V$ and convert from $L \cdot atm$ to kJ.

Step 3:   Determine ΔE from the equation $\Delta H = \Delta E + P\Delta V$.

Section 9.6   Enthalpies of Physical and Chemical Change

There are enthalpies associated with all physical changes or phase changes. It is important to realize that because all of these changes are reversible, the energies associated with the changes are also reversible: the heat required to melt a certain amount of a substance is the same amount of heat that will be released when it freezes again. The two heats will be equal in magnitude, but opposite in sign.

Types of heat change:

- Heat of fusion ($\Delta H°_{fus}$) is the energy required to melt a substance at its normal melting/freezing point (for water, this is 32 °F or 0 °C). The name is a reference to the fact that when ice chips, for example, are melted, they will *fuse* together to form liquid with no air spaces. The heat of fusion is also the heat a substance releases when it freezes.
- Heat of vaporization ($\Delta H°_{vap}$) is the energy required to convert a substance from liquid to gas at its normal boiling point. This energy tends to be high, because all intermolecular interactions must be broken to vaporize a substance. Heat of vaporization is also the energy released by a substance when it condenses from gas to liquid, which is why a steam burn can be so severe.
- Heat of sublimation/deposition ($\Delta H°_{sub}$): when a substance is converted straight from a solid to a gas, it is said to have sublimed. The heat of sublimation is the sum of the heats of fusion and vaporization for a substance, because both of these phase changes are contained in sublimation.

Chemical changes also have associated enthalpies. At constant pressure, heat can flow into or out of a reaction system.

Reactions that emit heat into the surroundings are exothermic: the enthalpy of the products is lower than that of the reactants, so $H_{products} - H_{reactants}$ is negative. $\Delta H < 0$ is **exothermic**.

A reaction that absorbs heat from the surroundings is endothermic: in this case, the enthalpy of the products is higher than that of the reactants, so $H_{products} - H_{reactants}$ is positive. $\Delta H > 0$ is **endothermic**.

If an equation is balanced, and all substances are in their standard states, it is possible to calculate the enthalpy change of the reaction in either direction using the moles of a reactant. All reactions can be reversed, so $\Delta H°$ for the forward reaction = $-\Delta H°$ for the reverse reaction. It is also important to use balanced reactions: the coefficients are very important!

Workbook Example 9.7

PROBLEM:
The standard enthalpy for the combustion of propane (C_3H_8) is –2044 kJ. If a propane tank contains 3.5 kg of propane, how much energy will be released or absorbed when the propane is burned?

SOLUTION:
The $\Delta H°$ reported is for the combustion of 1 mol of propane. First, it is necessary to determine the number of moles of propane present. From the formula, the molecular weight of propane is 44 g/mol. 3.5 kg is 3500 g.

$$3500\,g\,propane \times \frac{1\,mol\,propane}{44\,g\,propane} = 79.5\,mol\,propane$$

The amount of heat involved in the combustion of this amount of propane is

$$79.5\,mol\,propane \times \frac{-2044\,kJ}{1\,mol\,propane} = -162,500\,kJ$$

This is an exothermic reaction. Two things make this clear: the first is that it is a combustion reaction; common sense says that heat is released when things are burned, and this is no exception. But the primary reason that it is obviously an exothermic reaction is the sign on the enthalpy change. When the enthalpy change is negative, heat is being lost by the reactants and gained by the surroundings—an exothermic reaction.

Workbook Example 9.8

PROBLEM:
The reaction of boron and hydrogen proceeds by the following reaction:

$$2\,B\,(s) + 3\,H_2\,(g) \rightarrow B_2H_6\,(g) \qquad \Delta H° = 36.4\,kJ/mol$$

If 25.4 g of boron react, what is the energy change? Is this reaction endothermic or exothermic?

SOLUTION:
The stoichiometric ratios here must be considered very carefully. The reported $\Delta H°$ is per mole of boron hydride, but two moles of boron are required per mole of boron hydride.

$$25.4\,g\,B \times \frac{1\,mol\,B}{10.81\,g\,B} \times \frac{1\,mol\,B_2H_6}{2\,mol\,B} \times \frac{36.4\,kJ}{1\,mol\,B_2H_6} = 42.8\,kJ$$

No calculations are required to determine that this is an endothermic reaction—this can be seen immediately from the positive value of ΔH.

Workbook Problem 9.4

How much heat is released or required when 50 g of water are converted from liquid to gas at 100 °C? ΔH_{vap} for water is 40.7 kJ/mol.

Strategy: Use the heat of vaporization to determine the amount of heat involved.

Step 1: Convert the grams of water to moles.

Step 2: Multiply by the heat of vaporization.

Step 3: Make sure the sign of the heat transfer makes sense: the water is the system.

Workbook Problem 9.5

For the reaction

$$2\ Ca\ (s)\ +\ O_2\ (g)\ \rightarrow\ 2\ CaO\ (s)\qquad \Delta H = -634.9\ kJ/mol,$$

how much heat is involved if 3.24 g of calcium metal react with an excess of oxygen gas? Is this an exothermic or an endothermic reaction?

Strategy: Convert to moles and use the stoichiometry of the reaction.

Step 1: Determine the number of moles of calcium.

Step 2: Using the stoichiometric ratio, determine the heat involved.

Step 3: Based on the sign of the heat transfer, determine whether the reaction is exothermic or endothermic.

Section 9.7   Calorimetry and Heat Capacity

Calorimetry is an experimental method for measuring ΔH for a reaction. It can be carried out at either constant pressure or constant volume.

A constant-pressure calorimeter can be as simple as an insulated container with a loosely fitting lid to maintain the contents at atmospheric pressure, a stirrer, and a thermometer. This setup measures ΔH.

A constant-volume calorimeter—also called a bomb calorimeter—is more complex, and it measures the ΔE of combustion reactions. A sample of a substance is placed in an oxygen environment, and the contents are electrically ignited. The heat change is calculated from the increase in temperature of the surrounding water.

To calculate a heat change from a temperature change, it is necessary to know the relationship between the two, something that is distinctive for different materials.

The heat capacity (C) is calculated using the following formula:

$$C = \frac{q}{\Delta T}$$

where q is heat (measured in joules), and ΔT is the temperature change (remember to calculate it as the final temperature – the initial temperature). This is an extensive property: the amount of heat required will change based on the amount of the material present.

The **specific heat** is the amount of heat required to increase one gram of a material by 1 °C. For water, 4.184 J will raise the temperature of 1 g of water by 1 °C. The units of this heat capacity are therefore J/g °C. Molar heat capacities are also sometimes used. The units for these are J/mol °C. This is an intrinsic property, and can be used to identify unknown samples.

We can calculate specific heat by using the following equation:

$$q = mc\Delta T$$

(If you look back, you will see that heat capacity (C) is simply mass × specific heat.) In calorimetry, the heat gained or lost by the water is the heat lost or gained by whatever is in it. In other words, the water in the calorimeter is the *surroundings*.

Workbook Example 9.9

PROBLEM:
Calculate the amount of heat lost when the temperature of 14.3 g of iron drops 17 °C. The specific heat of iron is 0.449 J/g·°C.

SOLUTION:
Use the formula $q = mc\Delta T$ to determine the amount of heat involved.

$$q = (14.3\,g)\left(0.449\,\frac{J}{g\cdot °C}\right)(17.0\,°C) = 109\,J$$

Workbook Example 9.10

PROBLEM:
If a 7.3 g sample of an unknown metal at 100 °C is dropped into 25 g of water at 20.0 °C, and the final temperature of the water is 22.1 °C, calculate the specific heat of the metal. The specific heat of water is 4.184 J/g·°C.

SOLUTION:
Use the formula $q = mc\Delta T$ to determine the amount of heat absorbed by the water. This is also the heat given off by the metal, which will allow the specific heat to be calculated.

$$q(water) = (25\,g)\left(4.184\frac{J}{g\cdot\text{℃}}\right)(22.1\text{℃} - 20.0\text{℃}) = 220\,J$$

220 J are absorbed by the water, so 220 J were given off by the metal:

$$-220\,J = (7.3\,g)c(22.1\text{℃} - 100\text{℃})$$

$$c = \frac{-220\,J}{(7.3\,g)(22.1\text{℃} - 100\text{℃})} = 0.386\frac{J}{g\cdot\text{℃}}$$

If you were to consult a data table, you would see that the specific heat indicates that the metal is copper.

Workbook Problem 9.6

A student mixes 1.50 g of NaOH with 50.0 mL of water at 22.73 °C in a coffee-cup calorimeter. The final temperature of the reaction is 30.70 °C. Assuming that the calorimeter absorbs a negligible amount of heat, and that the density of the solution is the same as that of water, calculate the amount of heat evolved in this dissociation reaction. What is the heat of dissociation of NaOH in kJ/mol?

Strategy: Determine the heat absorbed by the water, then reverse the sign to determine the number of joules evolved by the dissociation. Convert the grams of NaOH to moles, and divide the heat evolved by the number of moles of NaOH to determine the molar heat of dissociation.

Step 1:   Determine the amount of heat gained by the water using the equation
$q = mc\cdot T$

Step 2:   Determine the amount of heat released by the reaction.

Step 3:   Determine the number of moles of NaOH involved.

Step 4:   Determine the heat of dissociation per mole of NaOH.

Section 9.8   Hess's Law

Reactions do not always go cleanly and often produce a mixture of products. This makes direct measurement of the energetics of these reactions problematic. With enthalpy being a state function, the reactions can be broken down into steps and added together to determine the energetics of a difficult-to-measure step. This is Hess's law: the overall enthalpy change for a reaction is equal to the sum of the enthalpy changes for the individual steps in the reaction.

Here are the rules for calculating the overall enthalpy change:

- Reactants must all appear on the left. It is acceptable to reverse reactions to make this happen.
- Products must all appear on the right. Again, reverse reactions if necessary.
- Intermediates in the reaction must appear on both sides so that they cancel out.
- When a reaction is reversed, the sign on that reaction's ΔH must also be reversed.
- Reactions and their energies can be multiplied through by any necessary factor.
- All the reactions and their energies are added to get the desired reaction and its energy.

Workbook Example 9.11

PROBLEM:
Calculate $\Delta H°$ for the reaction $3\,C\,(s) + 4\,H_2\,(g) \rightarrow C_3H_8\,(g)$ given the following information:

$$C_3H_8\,(g) + 5\,O_2\,(g) \rightarrow 3\,CO_2\,(g) + 4\,H_2O\,(g) \quad \Delta H° = -2043 \text{ kJ}$$
$$C\,(s) + O_2\,(g) \rightarrow CO_2\,(g) \quad \Delta H° = -393.5 \text{ kJ}$$
$$2\,H_2\,(g) + O_2\,(g) \rightarrow 2\,H_2O\,(g) \quad \Delta H° = -483.6 \text{ kJ}$$

SOLUTION:
Reactants
Place the reactants on the left and the products on the right, flipping reactions around if needed, and multiplying them through if necessary to get the equation as written.

$$C\,(s) + O_2\,(g) \rightarrow CO_2\,(g) \quad \Delta H° = -393.5 \text{ kJ}$$
$$2\,H_2\,(g) + O_2\,(g) \rightarrow 2\,H_2O\,(g) \quad \Delta H° = -483.6 \text{ Kj}$$

These reactions both need to be reacted through by coefficients, the first equation by 3, and the second equation by 2 to give them the coefficients needed:

$$3 \text{ C } (s) + 3 \text{ O}_2 (g) \rightarrow 3 \text{ CO}_2 (g) \qquad \Delta H° = -1180.5 \text{ kJ}$$
$$4 \text{ H}_2 (g) + 2 \text{ O}_2 (g) \rightarrow 4 \text{ H}_2\text{O} (g) \qquad \Delta H° = -967.2 \text{ kJ}$$

Products
Take the first reaction in the given list and turn it around, also reversing the sign on $\Delta H°$.

$$3 \text{ CO}_2 (g) + 4 \text{ H}_2\text{O} (g) \rightarrow \text{C}_3\text{H}_8 (g) + 5 \text{ O}_2 (g) \qquad \Delta H° = 2043 \text{ kJ}$$

Stack up the rewritten reactions, and add them together:

$$3 \text{ C } (s) + 3 \text{ O}_2 (g) \rightarrow 3 \text{ CO}_2 (g) \qquad \Delta H° = -1180.5 \text{ kJ}$$
$$4 \text{ H}_2 (g) + 2 \text{ O}_2 (g) \rightarrow 4 \text{ H}_2\text{O} (g) \qquad \Delta H° = -967.2 \text{ kJ}$$
$$3 \text{ CO}_2 (g) + 4 \text{ H}_2\text{O} (g) \rightarrow \text{C}_3\text{H}_8 (g) + 5 \text{O}_2 (g) \qquad \Delta H° = 2043 \text{ kJ}$$

$$3 \text{ C } (s) + 3 \text{ O}_2 (g) + 4 \text{ H}_2 (g) + 2 \text{ O}_2 (g) + 3 \text{ CO}_2 (g) + 4 \text{ H}_2\text{O} (g) \rightarrow$$
$$3 \text{ CO}_2 (g) + 4 \text{ H}_2\text{O} (g) + \text{C}_3\text{H}_8 (g) + 5\text{O}_2 (g)$$

Cancel-out like terms on the reactant and product sides of the equation:

$$3 \text{ C } (s) + 4 \text{ H}_2 (g) \rightarrow \text{C}_3\text{H}_8 (g) \qquad \Delta H° = -1180.5 \text{ kJ} + (-967.2 \text{ kJ}) + 2043 \text{ kJ}$$
$$3 \text{ C } (s) + 4 \text{ H}_2 (g) \rightarrow \text{C}_3\text{H}_8 (g) \qquad \Delta H° = -104.7 \text{ kJ}$$

Workbook Problem 9.7

Calculate ΔH for the reaction

$$\text{Fe}_2\text{O}_3 (s) + 3 \text{ CO} (g) \rightarrow 2 \text{ Fe } (s) + 3 \text{ CO}_2 (g)$$

given the following information:

$$4 \text{ Fe } (s) + 3 \text{ O}_2 (g) \rightarrow 2 \text{ Fe}_2\text{O}_3 (s) \qquad \Delta H° = -1648.4 \text{ kJ}$$
$$2 \text{ CO} (g) + \text{O}_2 (g) \rightarrow 2 \text{ CO}_2 (g) \qquad \Delta H° = -565.4 \text{ kJ}$$

Strategy: Place the reactants on the left and the products on the right, flipping reactions around if needed, and multiplying them through if necessary to get the equation as written.

Step 1:   Place reactants on the left, multiplying through if necessary.

Step 2:   Place products on the right, multiplying through if necessary.

Step 3:   Stack up the rewritten reactions, and add them together.

Step 4:   Cancel terms that appear on both sides.

Section 9.9   Standard Heats of Formation

Another method can be used to determine the energy changes for any reaction, whether actual or theoretical. The **standard heat of formation** for a compound is the enthalpy change for the hypothetical formation of 1 mole of that compound in its standard state from its constituent elements in their standard states. The standard enthalpy of formation of an element in its standard state is zero.

Workbook Example 9.12

PROBLEM:
Write the equation for the reaction corresponding to the standard heat of formation of acetone, C_3H_6O.

SOLUTION:
The standard state of carbon is graphite, while hydrogen and oxygen are diatomic gases:

$$3 \text{ C } (graphite) + 3 \text{ H}_2 \text{ } (g) + \tfrac{1}{2} \text{ O}_2 \text{ } (g) \rightarrow C_3H_6O \text{ } (l)$$

It's as simple as that. You will see fractional coefficients in thermochemical equations, because they are written to provide a coefficient of 1 on the product.

There are extensive tables of heats of formation (one of which is in the appendix of your textbook), and they can be used to calculate heats of reaction in a variation on final – initial: The sum of the heats of formation of the products, minus the sum of the heats of formation of the reactants will be the overall heat of reaction. This is summarized in the following formula:

$$\Delta H° = \sum \Delta H°_{products} - \sum \Delta H°_{reactants}$$

It is important to note that the standard heats of formation must be multiplied by the coefficients in the balanced equation.

Workbook Example 9.13

PROBLEM:
Calculate $\Delta H°$ for the reaction C_3H_6O (l) + 4 O_2 (g) → 3 CO_2 (g) + 3 H_2O (l) using the following information:

$$\Delta H_f^o\left(C_3H_6O(l)\right)=-248.4\,kJ\,/\,mol \qquad \Delta H_f^o\left(O_2(g)\right)=-0\,kJ\,/\,mol$$

$$\Delta H_f^o\left(CO_2(g)\right)=-393.5\,kJ\,/\,mol \qquad \Delta H_f^o\left(H_2O(l)\right)=-285.8\,kJ\,/\,mol$$

SOLUTION:
Subtract the total heats of formation of the reactants from the total heats of formation of the products.

$$\Delta H°=\left[3\left(-393.5\,kJ\,/\,mol\right)+3\left(-285.8\,kJ\,/\,mol\right)\right]-\left[1\left(-248.4\,kJ\,/\,mol\right)+4\left(0\,kJ\,/\,mol\right)\right]$$

$$\Delta H°=\left[-1180.5\,kJ+^-285.8\,kJ\right]-\left[-248.4\,kJ\right]=-1790\,kJ$$

Workbook Problem 9.8

During glycolysis, glucose ($C_6H_{12}O_6$) is taken through a series of reactions that combine it with oxygen and release carbon dioxide and water. How much energy is released through the slow combustion of 10.0 grams of glucose in this process? What would be the temperature change if this energy were used to heat 500 g of water?

You may use the following information:

$$\Delta H_f^o\left(C_6H_{12}O_6(s)\right)=-1273.3\,kJ\,/\,mol \qquad \Delta H_f^o\left(O_2(g)\right)=-0\,kJ\,/\,mol$$

$$\Delta H_f^o\left(CO_2(g)\right)=-393.5\,kJ\,/\,mol \qquad \Delta H_f^o\left(H_2O(l)\right)=-285.8\,kJ\,/\,mol$$

Strategy: Write a balanced equation for the reaction, and use the standard heats of formation to determine the overall heat of reaction per mole of glucose, then convert grams to moles and determine the heat released by 10 grams of glucose. Use this heat and the equation $q=mc\Delta T$ to determine the temperature change.

Step 1: Write and balance the equation.

Step 2: Determine the overall heat of reaction using

$$\Delta H°=\sum\Delta H°_{products}-\sum\Delta H°_{reactants}\,\cdot$$

Step 3:   Convert the grams of glucose to moles.

Step 4:   Determine the energy released by 10 g of glucose.

Step 5:   Using $q = mc\Delta T$, determine the temperature change of 500 g of water.

Section 9.10   Bond Dissociation Energies

Heats of formation are not available for every compound. When there is no data for the heat of formation of a compound, it can be estimated using the known bond dissociation energies for every bond in the compound. The bond energies are positive: energy must always be put in to break a bond. The bond formation energies will be equal in magnitude and opposite in sign.

To determine the approximate enthalpy of any reaction, subtract the sum of the bond dissociation energies for the products from the bond dissociation energies in the reactants.

Workbook Example 9.14

PROBLEM:
Using bond dissociation energies, estimate the heat of reaction of hydrogen and oxygen to form water.

Type of Bond	Bond Energy
H–H	436 kJ/mol
O=O	498 kJ/mol
O–H	464 kJ/mol

SOLUTION:

$$2 \text{ H}_2 \text{ (g)} + \text{O}_2 \text{ (g)} \rightarrow 2 \text{ H}_2\text{O (g)}$$

On the product side, four H–O bonds are formed:

$$4 \times 464 \text{ kJ/mol} = 1856 \text{ kJ/mol released}$$

On the reactant side, two H–H bonds are broken, and one O=O bond is broken:

$$(2 \times 436 \text{ kJ/mol}) + 498 \text{ kJ/mol} = 1370 \text{ kJ/mol absorbed}$$

$$\text{Reactants} - \text{products} = -486 \text{ kJ/mol}$$

This is the answer for the problem as written—for the production of two moles of water. For one mole of water, the answer will be –486 kJ/mol × 0.5 = –243 kJ/mol. This is very close to the heat of formation of water vapor.

Workbook Problem 9.9

Predict the energy change for combustion of methane (CH_4) using bond dissociation energies. Use the following information:

Type of Bond	Bond Energy
C–H	414 kJ/mol
O=O	498 kJ/mol
O–H	464 kJ/mol
C=O	799 kJ/mol

Strategy: Write a balanced equation for the reaction, and determine the bond energies for both products and reactants. Subtract products from reactants to determine the overall energy change of the reaction.

Step 1:   Write and balance the equation.

Step 2:   Determine the energy of the bonds on the reactant side.

Step 3:   Determine the energy of the bonds on the product side.

Step 4:   Determine the energy released by the reaction.

Section 9.11   Fossil Fuels, Fuel Efficiency, and Heats of Combustion

Fuels all have standard heats of combustion (ΔH_c^o), which is the energy released when they are burned.

Different applications have different fuel concerns. For example, mass is critical for a rocket, and a large volume of fuel is problematic in an automobile. These energies can be converted to a variety of units for comparison: kJ/g where mass is a concern, kJ/mL where volume is a concern. These quantities of energy per milliliter or per gram are referred to as fuel efficiencies.

Workbook Example 9.15

PROBLEM:
Toluene (C_7H_8) has a density of 0.8669 g/mL and a molar heat of combustion of –3910 kJ/mol. What is its heat of combustion in kJ/g and J/mL?

SOLUTION:
This is a pretty straightforward conversion problem:

$$\frac{-3910\,kJ}{1\,mol} \times \frac{1\,mol}{92\,g} = -42.5\,\frac{kJ}{g}$$

$$\frac{-42.5\,kJ}{g} \times \frac{0.8669\,g}{1\,mL} = -36.8\,\frac{kJ}{mL}$$

With the exception of hydrogen gas, all common fuels are hydrocarbons. Fossil fuels gain their energy from photosynthesis—coal is primarily vegetable in origin, petroleum is largely marine, and natural gas is mostly methane. Coal and natural gas can be used in their natural state, but petroleum requires fractionation.

Alternative fuels are a burgeoning new industry. Ethanol can be produced by fermenting corn or cane sugar, but processes to produce it from waste wood are being investigated.

Section 9.12   An Introduction to Entropy

Chemical reactions generally release energy in proceeding from higher-energy reactants to lower-energy products. But this is not always the case: there are endothermic processes that occur even though energy must be added to them. For an ice cube to melt, it must spontaneously absorb energy from the surroundings. Release of energy cannot be the only factor in reactions.

Spontaneous processes are those that occur on their own without external influence. Nonspontaneous processes require constant inputs of energy. For example to roll a marble down a hill simply requires that you get it started; rolling down the hill is a spontaneous process. Rolling it back up the hill requires a constant input of energy, and is nonspontaneous.

The unifying feature of spontaneous endothermic processes is an increase in the entropy (S) of the reaction mixtures. **Entropy** is an increase in molecular disorder or randomness. When an ice cube melts, the rigid crystalline structure of the solid water becomes the far-less-orderly liquid water. The units for entropy are J/K: larger values indicate a greater degree of disorder.

Freedom to move and the ability to adopt numerous conformations are all related to entropy. Phase also plays a major role: solids have less entropy than liquids, with gases having the most entropy of all. Solid iodine has an entropy of 116 J/K, while gaseous iodine has an entropy of 261 J/K.

As with all change measurements, $\Delta S = S_{final} - S_{initial}$. In contrast to ΔH, a positive ΔS is favorable for a reaction, indicating an increase in disorder.

Workbook Example 9.16

PROBLEM:
Predict whether $\Delta S°$ is likely to be positive or negative for the following processes:
    a. $H_2O\ (s) \rightarrow H_2O\ (g)$
    b. $Na\ (s) + \frac{1}{2}\ Cl_2\ (g) \rightarrow NaCl\ (s)$
    c. $CH_3CH_2OH\ (l) +\ 3\ O_2\ (g) \rightarrow\ 2\ CO_2\ (g) +\ 3\ H_2O\ (g)$

SOLUTION:
    a. Converting water from a crystalline solid to a gas will generate a large positive ΔS.
    b. Going from a solid and a gas to a crystal lattice will cause a large negative ΔS.
    c. Four moles of reactants (one a liquid) being converted to 5 moles of gas will increase the entropy, giving a positive ΔS.

Section 9.13   An Introduction to Free Energy

A spontaneous process is one where the overall energetics are favorable: ΔH and ΔS can both be favorable, with one negative and one positive, or one can be favorable and the other unfavorable, with the favorable outweighing the unfavorable. Because entropy is temperature dependent, a reaction can be spontaneous at one temperature and nonspontaneous at another.

All of this is bundled into the equation for the Gibbs free-energy change (ΔG):

$$\Delta G = \Delta H - T\Delta S$$

The sign of ΔG determines whether a process is spontaneous. A negative ΔG indicates a spontaneous process in which free energy is being released and a positive ΔG indicates a nonspontaneous process in which free energy is being absorbed. When $\Delta G = 0$, the reaction is at equilibrium, and no net change is occurring.

Entropy is temperature dependent. If the free-energy equation is set to zero (equilibrium) it is possible to solve for the temperature at which the process will become spontaneous (or nonspontaneous) using the following equation:

$$T = \frac{\Delta H}{\Delta S}$$

Workbook Example 9.17

PROBLEM:
At what temperature is the vaporization of methanol at equilibrium, given the following values:
    ΔH_f methanol (l) = –238.7 kJ/mol ΔH_f methanol (g) = –201.2 kJ/mol
    $\Delta S°$ methanol (l) = 127 J/K·mol $\Delta S°$ methanol (g) = 238 J/K·mol

SOLUTION:

$$T = \frac{\Delta H}{\Delta S} = \frac{-201.2\,\frac{kJ}{mol} - \left(-238.7\,\frac{kJ}{mol}\right)}{0.238\,\frac{kJ}{K\Delta mol} - 0.127\,\frac{kJ}{K\Delta mol}} = 338\,K = 65\,℃$$

This is the boiling point of methanol.

Workbook Problem 9.10

Given the following data, is the reaction of hydrogen and chlorine gas spontaneous at room temperature?

ΔH_f HCl (g) = –92.3 kJ/mol $\qquad$ $\Delta S°$ H$_2$ (g) = 130.6 J/K·mol

$\Delta S°$ Cl$_2$ (g) = 223 J/K·mol $\qquad$ $\Delta S°$ HCl (g) = 186.8 J/K·mol

Strategy: Write a balanced equation for the reaction, and determine ΔH and ΔS for the reaction. Using the Gibbs free-energy equation, determine ΔG for the reaction at 25 °C, remembering to convert to kelvins first.

Step 1: Write and balance the equation.

Step 2: Determine ΔH.

Step 3: Determine ΔS.

Step 4: Determine the free energy of the reaction at 25 °C, first converting the temperature to K.

Step 5: Determine the spontaneity of the reaction.

Putting It Together

Liquid hydrazine (N$_2$H$_4$) is used as the propellant in the maneuvering thrusters of spacecraft. When it is exposed to a catalyst, it undergoes a multistep decomposition that eventually results in hydrogen and nitrogen gas. If 3 g of hydrazine decompose, what is the free-energy change of the system at 298 K? Use the following data:

ΔH_f hydrazine = + 95.4 kJ/mol $\qquad$ $S°$ hydrazine = + 121.2 J/mol·K

$S°$ N$_2$ = + 191.5 J/mol·K $\qquad$ $S°$ H$_2$ = + 130.6 J/mol·K

Self–Test

This section is intended to test your knowledge of the material covered in this chapter. Think through these problems, and make certain you understand what they are asking. Make sure your answers make sense. Successful completion of these problems indicates that you have mastered the material in this chapter. You will receive the greatest benefit from this section if you use it as a mock exam, as this will allow you to determine which topics you need to study in more detail.

True–False
1. Energy can be converted from one form to another.

2. Changes in energies are always calculated *initial – final*.

3. The energy changes in a state function depend on path.

4. When a system expands against pressure, it is doing work.

5. The thermodynamic standard state of an element is its most stable form at 0 °C.

6. A reaction can do work or transfer heat, but not both.

7. Melting is an exothermic process.

8. Specific heat is an extensive property, and it cannot be used to identify a material.

9. Fractional coefficients can be used in thermochemical equations.

10. Standard heats of formation can be used to calculate reaction enthalpies.

11. Depending on the needs of the end user, different units can be used for heats of combustion of fuels.

12. A negative ΔS is favorable for a reaction.

13. Gases have lower entropy than liquids.

14. Free-energy changes can be used to predict spontaneity of reactions.

15. Entropy is a temperature-dependent property.

Matching

16. Conservation of energy	a. heat released on reaction with oxygen
17. Energy	b. a process that proceeds without external influence
18. Temperature	c. going directly from solid to gas
19. System	d. a measure of molecular kinetic energy

20. Entropy

    e.   energy can be transferred between forms, but cannot be created or destroyed

21. State function

    f.   absorbing energy

22. Heat

    g.   a measure of randomness or disorder

23. Work

    h.   capacity to do work

24. Enthalpy

    i.   heat required to convert a substance from solid to liquid

25. Heat of fusion

    j.   releasing energy

26. Sublimation

    k.   the amount of energy required to raise the temperature of one gram of a substance 1 °C

27. Specific heat

    l.   a measurement that is independent of path

28. Exothermic

    m.   the focus of a thermochemical process

29. Hess's law

    n.   exerting a force over a distance

30. Heat of combustion

    o.   the overall enthalpy of a reaction can be determined by combining any number of theoretical steps

31. Spontaneous

    p. $E + P\Delta V$

32. Endothermic

    q.   energy transferred as the result of a temperature difference

Fil-in-the-Blank

33. Energy can be either _____ , the energy of motion, or _____ , stored energy.

34. Molecular kinetic energy is measured as _____. When this property differs between objects, the energy transferred is called _____.

35. Energy that flows out of a system is given a _____ sign. Reactions in which this occurs are called _____.

36. It is possible for state functions to return to their original condition. As a result, they are said to be completely _____.

37. If there is no change in the _____ of a system, no _____ is done.

38. The most stable form of a substance at 25 °C and 1 atm is referred to as its

_____.

39. When calorimetry is used to measure heat flows, changes in the temperature of the water can

be converted to changes in heat by using the mass of the water and its

_____.

40. During an endothermic process, heat flows from the _____ to the

_____ and ΔH is _____.

41. The energy required to break a chemical bond is its _____.

It is always _____.

42. The greater the disorder in a system, the higher the _____. If the

temperature of a system is increased, this property will _____.

43. The _____ can determine the spontaneity of a process. When it

is zero, the system is at _____. When it is _____ than

zero, the reaction will proceed spontaneously. When it is _____ than

zero, the reaction will be nonspontaneous.

Problems

44. Kinetic energy is defined as $1/2\ mv^2$. How fast would a 2-gram marble need to be moving to
have the same kinetic energy as a 1.3 kg bowling ball moving at 5 m/s?

45. If 950 mL of a gas is compressed to 550 mL under a constant external pressure of 8.00 atm
and the gas absorbs 12 kJ, what are the values of q, w, and ΔE for the gas? Is work being
done on the system or by the system? What is the value of ΔE for the surroundings?

46. If the specific heat of mercury is 0.140 J/g·°C, how much will the temperature of 15 grams of
mercury change if 150 joules of heat are added to it?

47. If 2.00 L of chlorine gas react stoichiometrically with 1.00 L of oxygen at 1 atm to form
1.00 L of Cl_2O gas according to the following reaction, what is the overall energy change of
the system (the density of chlorine gas is 3.17 g/L)? Is this an endothermic reaction or an
exothermic reaction?

$$Cl_2\ (g) + \tfrac{1}{2}\ O_2\ (g) \rightarrow Cl_2O\ (g) \qquad \Delta H = 80.3 \text{ kJ/mol}$$

48. Given the reaction below, and the following entropy values, is this reaction spontaneous at
25 °C? Will it be spontaneous at any temperature?

$$\Delta S°\ Cl_2 = 223.0 \text{ J/mol·K} \qquad \Delta S°\ O_2 = 205 \text{ J/mol·K} \quad \Delta S°\ Cl_2O = 266.1 \text{ J/mol·K}$$

49. The ΔH of solution for $MgSO_4 \cdot 7\ H_2O$ (Epsom salts) is $+16.11$ kJ/mol. If 15 grams of Epsom salts are dissolved in 100 grams of water at $24.36\ °C$, what is the final temperature of the solution? (Assume that the specific heat and density of the solution are the same as that of pure water.)

50. How much energy, in kJ, is required to take 50 grams of water from $-10\ °C$ to steam at $100\ °C$? Use the following values:

$$c_{ice} = 2.108\ \text{J/g·°C} \qquad c_{water} = 4.184\ \text{J/g·°C}$$
$$\Delta H_{fus} = 6.01\ \text{kJ/mol} \qquad \Delta H_{vap} = 40.7\ \text{kJ/mol}$$

    What is the largest contributor to the energy consumption? Why is this?

51. A 6.73 gram chunk of metal known to be a mixture of copper ($c = 0.385$ J/g·°C) and gold ($c = 0.129$ J/g·°C) is taken from boiling water and dropped into 25 grams of water at $25.00\ °C$. The final temperature is $26.65\ °C$. How many grams of copper and how many grams of gold are present in the sample?

52. Use Hess's law to determine the value of ΔH for the following reaction:

$$2\ Fe\ (s) + 3/2\ O_2\ (g) \rightarrow Fe_2O_3\ (s)$$

    from the following reactions:

$$Fe_2O_3\ (s) + 3\ CO\ (g) \rightarrow 2\ Fe\ (s) + 3\ CO_2(g) \qquad \Delta H° = -26.7\ \text{kJ}$$
$$CO + ½\ O_2 \rightarrow 3\ CO_2\ (g) \qquad \Delta H° = -283.0\ \text{kJ}$$

53. Spraying water on a kitchen grease fire can spread it, so in commercial kitchens a large box of baking soda is often kept in case of fire. When sodium bicarbonate is thrown on the flames, it physically puts the fire out, but also decomposes to form carbon dioxide and water by the following reaction:

$$2\ NaHCO_3\ (s) \rightarrow Na_2CO_3\ (s) + H_2O\ (l) + CO_2\ (g)$$

    Given the following heats of formation, what is the approximate ΔH of this decomposition reaction?

$$\Delta H_f° NaHCO_3 = -947.7\ \text{kJ/mol}$$
$$\Delta H_f° Na_2CO_3 = -1131\ \text{kJ/mol}$$
$$\Delta H_f° H_2O = -285.9\ \text{kJ/mol}$$
$$\Delta H_f° CO_2 = -393.5\ \text{kJ/mol}$$

54. Using the bond dissociation energies given, calculate the enthalpy of the following reaction:

$$2\ Cl_2\ (g) + CH_4\ (g) \rightarrow 2\ H_2\ (g) + CCl_4\ (g)$$

$$Cl–Cl = 243\ \text{kJ/mol} \qquad H–H = 436\ \text{kJ/mol}$$
$$C–H = 410\ \text{kJ/mol} \qquad C–Cl = 330\ \text{kJ/mol}$$

55. Given the entropy values below, at approximately what temperature will the reverse of the above reaction become spontaneous?

$$\Delta S° \; Cl_2 = 223 \; J/mol·K \qquad \Delta S° \; H_2 = 130.6 \; J/mol·K$$
$$\Delta S° \; CH_4 = 188 \; J/mol·K \qquad \Delta S° \; CCl_4 = 214 \; J/mol·K$$

56. Using the following standard heats of formation, calculate the heat of combustion of pentane (C_5H_{12}) in kJ/mol and kJ/g. The density of pentane is $0.626 \; g/cm^3$. What is the heat of combustion per mL?

$$\Delta H_f^o \, pentane \;\; = -146.3 \; kJ/mol$$

$$\Delta H_f^o \, H_2O \quad\;\; = -285.9 \; kJ/mol$$

$$\Delta H_f^o \, CO_2 \quad\;\; = -393.5 \; kJ/mol$$

CHAPTER TEN

Gases: Their Properties and Behavior

Learning Objectives

As a result of reading and studying this chapter, you should be able to

Section 10.1	**Gases and Gas Pressure**
1.	Convert between different units of pressure.
2.	Describe how a barometer and manometer measure pressure.

Section 10.2	**The Gas Laws**
3.	Use the individual gas laws to calculate pressure, volume, molar amount, or temperature for a gas sample when conditions change.

Section 10.3	**The Ideal Gas Law**
4.	Use the ideal gas law to calculate pressure, volume, molar amount, or temperature for a gas sample.

Section 10.4	**Stoichiometric Relationships with Gases**
5.	Calculate volumes of gases in chemical reactions.
6.	Calculate the density or molar mass of a gas using the formula for gas density.

Section 10.5	**Mixture of Gases: Partial Pressure and Dalton's Law**
7.	Calculate the partial pressure, mole fraction, or amount of each gas in a mixture.

Section 10.6	**The Kinetic–Molecular Theory of Gases**
8.	Use the assumptions of the kinetic–molecular theory to predict gas behavior.
9.	Calculate the average molecular speed of a gas particle at a given temperature.

Section 10.7	**Gas Diffusion and Effusion: Graham's Law**
10.	Visualize the process of effusion and diffusion.
11.	Use Graham's law to estimate relative rates of diffusion for two gases.

Section 10.8	**The Behavior of Real Gases**
12.	Understand the conditions under which gases deviate the most from ideal behavior.
13.	Use the van der Waals equation to calculate the properties of real gases.

Section 10.9	**The Earth's Atmosphere and Air Pollution**
14.	Convert between different units to express the concentration of pollutants.
15.	Use the gas laws, Dalton's law, and stoichiometry to calculate amounts of pollutant gases in the atmosphere.
16.	Identify the components and causes of photochemical smog.

Sections 10.10 – 10.11 The Greenhouse Effect and Climate Change

17.   Explain the principle of the greenhouse effect.
18.   Describe the trends in greenhouse gas concentrations over time and measured and predicted effects of climate change.

Chapter Summary

Gases behave very differently than solids or liquids. They are compressible, because they are mostly empty space, and they exert pressure as a result of collisions between their rapidly moving particles and the walls of any container in which they are placed. Extensive experimental work in the early days of the science of chemistry worked out the gas laws: simple equations that can be used to predict how gases will behave when conditions are varied. The four variables that are relevant to gases are temperature, pressure, volume, and amount. Knowing the mathematical relationships between these variables makes it possible to perform stoichiometry with gas-generating equations. Mixtures of gases can also be dealt with using the gas laws, in particular Dalton's law of partial pressures. Kinetic–molecular theory provides a model for gas behavior that explains the gas laws, and is a basis for understanding effusion and diffusion. Generally, it is assumed that gases will behave in an "ideal" way, that is, they will adhere closely to the tenets of kinetic–molecular theory. At high pressures, this ceases to be true, so additional strategies are needed to determine the behavior of gases under these conditions. The chapter ends with a brief look at the Earth's atmosphere.

The Chapter in Detail

Section 10.1   Gases and Gas Pressure

The air that we breathe is primarily nitrogen—78% by volume—with the rest being made up primarily of oxygen and about 1% argon. All other gases, including carbon dioxide, exist only in very small amounts in the atmosphere.

Gases have very consistent properties among the different species. All gases will mix with all other gases to form homogeneous mixtures. Gas molecules exist very far apart and interact very little, so the properties of one gas molecule compared to another do not really matter in terms of the gas's overall behavior.

Gases are highly compressible, also as a result of the distance between molecules—the volume of a gas is 99.0% empty space.

Gases exert a measurable pressure on the walls of their container. Pressure is measured as force per unit area, and force is measured as mass multiplied by acceleration.

The SI unit for force is the newton (N), and the unit for pressure is the pascal (Pa).

$$P = \frac{force}{area} = \frac{kg \cdot \frac{m}{s^2}}{m^2} = \frac{N}{m^2} = Pa$$

The pascal is very small and not convenient to use in chemistry measurements. There is a number of alternative pressure units, that are more practical. The millimeter of mercury (mmHg), also called a torr after the inventor of the barometer, Evangelista Torricelli, is the most well known. In a barometer, an open dish of mercury has in it a long, thin, glass tube full of mercury, closed at the top end and open at the bottom. An increase in atmospheric pressure forces mercury higher into the tube, and a drop in atmospheric pressure allows the mercury to drop. Atmospheric pressure is defined as 760 mmHg, which is about 101,000 Pa, or 101 kPa. In addition to millimeters of mercury, the atmosphere (atm) and the bar (1 bar = 100,000 Pa) are commonly used in chemistry.

To measure the pressure of a gas, an open-ended manometer can be used, which compares the pressure of the gas in the container to atmospheric pressure. If the liquid level in both is equal, the pressure of the contained gas is equal to atmospheric pressure. If the liquid is lower on the container side, the pressure in the container is higher than atmospheric pressure, and if the liquid is higher on the container side, the pressure is lower in the container. The differences can be measured to quantify the pressure in the container.

A barometer or manometer can be filled with a liquid other than mercury; in that case, the density of the liquid can be used as a conversion factor to mmHg.

With problems involving a manometer, when in doubt, draw a picture!

Workbook Example 10.1

PROBLEM:
A barometer filled with water has a height of 9.563 m. What is the atmospheric pressure in atmospheres? The density of Hg is 13.6 g/mL, and the density of water is 1.00 g/mL.

SOLUTION:
The height of the barometer can be converted to mmHg by using the relative densities of the two liquids as a conversion factor. The pressure in mmHg can then be converted to atm.

$$9.563\,m \times \frac{1000\,mm}{1\,m} \times \frac{1.00\,g/_{mL}}{13.6\,g/_{mL}} = 703\,mm\,Hg$$

$$703\,mm\,Hg = \frac{1\,atm}{760\,mm\,Hg} = 0.925\,atm$$

Workbook Example 10.2

PROBLEM:
If a gas is contained in an open-ended manometer, and the mercury on the side of the gas is 37 mm higher than the mercury in the side open to the air, what is the pressure of the contained gas?

SOLUTION:
The gas is at less than atmospheric pressure, because the pressure of the air on the open end of the tube pushes the mercury toward the contained gas. Atmospheric pressure is 760 mmHg, so the pressure of the gas must be 760 mmHg – 37 mmHg = 723 mmHg.

Workbook Problem 10.1

If the difference in height between the two sides of a manometer is 73 mmHg, with the mercury higher on the open side of the manometer, what is the pressure of the contained gas in atmospheres?

Strategy: Determine the pressure of the gas in mmHg; convert to atmospheres.

Step 1:   Determine the pressure of the gas in mmHg.

Step 2:   Convert from mmHg to atmospheres.

Section 10.2   The Gas Laws

Although they have very different chemical properties, gases have very similar physical properties that can be defined by four variables: pressure, temperature, volume, and number of moles (amount). The relationships among these variables form the gas laws, and a substance that follows them exactly is known as an **ideal gas**.

The gas laws in this section may appear to be almost absurdly simple, but they were worked out in the seventeenth and eighteenth centuries through extensive experimentation, and these laws have contributed tremendously to the understanding of matter that students today take for granted.

Boyle's law – When only volume and pressure are varying, they are inversely related: when volume increases, pressure decreases. Mathematically, Boyle's law is stated in a few ways:

$$V \propto \frac{1}{P} \qquad PV = k \text{ (where } k \text{ is a constant)} \qquad P_1V_1 = P_2V_2$$

Charles's law – When only volume and temperature are varying, they are directly related: when temperature drops, the volume drops as well. Temperatures *must* be expressed in kelvins for the proportionality to be seen. Mathematically, Charles's law is stated as

$$V \propto T \qquad \frac{V}{T} = k \text{(where } k \text{ is a constant)} \qquad \frac{V_1}{T_1} = \frac{V_2}{T_2}$$

Charles's law is also the basis of the Kelvin scale: if volume is plotted against temperature, there is a point at which the volume drops (theoretically) to zero. This occurs at –273.15 °C, a temperature which is known as *absolute zero*. This is the lowest possible temperature, and therefore it is 0 K.

Avogadro's law – If only amount and volume are varying, they are directly related: more moles of gas will always mean more volume, with all else being equal. Mathematically, Avogadro's law is stated as

$$V \propto n \qquad\qquad \frac{V}{n} = k \text{(where } k \text{ is a constant)} \qquad\qquad \frac{V_1}{n_1} = \frac{V_2}{n_2}$$

It can also be seen from Avogadro's Law that one mole of any gas will have the same volume at the same temperature and pressure. In fact, 1 mol of any gas at 0 °C and 1 atm will have a volume of 22.4 L. These conditions (0 °C and 1 atm) are referred to as STP—standard temperature and pressure—for gas law calculations.

Workbook Example 10.3

PROBLEM:
If 2 M of a gas at 150 K and 1 atm occupy 24.6 L, how many liters will 4.3 M of gas occupy under the same conditions?

SOLUTION:
There are a lot of numbers in this problem, making the primary challenge figuring out what is important and what is not. The temperature and pressure of the gas are given, but they do not vary and so need not be considered. The important numbers are the moles of gas and the one volume given. This is enough information to solve for the other volume using Avogadro's law:

$$\frac{24.6\,L}{2\,mol} = \frac{V_2}{4.3\,mol}$$

$$V_2 = \frac{(24.6\,L)(4.3\,mol)}{2\,mol} = 52.9\,L$$

Workbook Problem 10.2

If a quantity of oxygen occupies 27 L at 37 °C, how many liters will it occupy at 350 °C?

Strategy: Determine which gas law is involved, convert temperatures, solve for the final volume.

Step 1: Determine which gas law to use.

Step 2: Convert the temperatures to K.

Step 3: Solve for the final volume.

Section 10.3   The Ideal Gas Law

All three gas laws can be combined to form the **ideal gas law**:

$$PV = nRT$$

where R is the ideal gas law constant $\left(R = 0.08206 \dfrac{L \cdot atm}{K \cdot mol} \right)$. The ideal gas law constant can be calculated from the molar volume of a gas:

$$R = \frac{PV}{nT} = \frac{(1\,atm)(22.4\,L)}{(1\,mol)(273.15\,K)} = 0.08206 \frac{L \cdot atm}{K \cdot mol}$$

The ideal gas law can be used to calculate any one of the four variables that describe gases if the other three are known, and it can also be rearranged to solve for any of the four variables:

$$PV = nRT \qquad P = \frac{nRT}{V} \qquad V = \frac{nRT}{P} \qquad n = \frac{PV}{RT} \qquad T = \frac{PV}{nR}$$

If the problem involves a change, and you are unsure of which gas law to use, try the following trick: set your two conditions equal to each other like this:

$$\frac{P_1 V_1}{n_1 R T_1} = \frac{P_2 V_2}{n_2 R T_2}$$

Then cancel out anything that is not changing. For instance, if the experiment is being done at constant pressure, $P_1 = P_2$ and they can be cancelled out. R is always the same on both sides, so can always be cancelled out.

Note: the standard state for gas laws is not the same as the thermodynamic standard state (apparently gas-law chemists like to be cold).

It's important to note that there are no really "ideal" gases – all gases deviate from ideal behavior to a certain extent. However, the deviations are usually so small as to be negligible.

Workbook Example 10.4

PROBLEM:
What is the volume of 7.62 g of Cl_2 at 45.7 °C and 0.257 atm?

SOLUTION:
First, solve for the moles of chlorine and the temperature in K:

$$n = 7.62\,g\,Cl_2 \times \frac{1\,mol\,Cl_2}{71.0\,g} = 0.107\,mol\,Cl_2$$

$$45.7°C + 273.15 = 318.85\,K$$

Then solve for the volume:

$$V = \frac{nRT}{P} = \frac{(0.107\,mol)\left(0.08206 \dfrac{L \cdot atm}{K \cdot mol} \right)(318.85\,K)}{0.257\,atm} = 10.89\,L$$

Workbook Problem 10.3

If 6.7 g of fluorine were placed in a 3.0 L container at 25.0 °C, what pressure would there be in the container?

Strategy: Rearrange the ideal gas law to solve for pressure, convert grams to moles, and degrees Celsius to kelvins; solve.

Step 1:   Rearrange the ideal gas law.

Step 2:   Convert grams to moles and degrees Celsius to kelvins.

Step 3:   Solve for the final pressure.

Workbook Problem 10.4

If 12.3 g of neon were taken from 3.2 L at 1 atm and 25 °C to 18 atm and 750° C, what would the final volume of the gas be?

Strategy: Determine what form of the gas laws is needed, convert temperatures and moles if necessary, and solve for final pressure.

Step 1:   Determine the form of gas laws needed.

Step 2:   Convert grams to moles and degrees Celsius to kelvins if needed.

Step 3:   Solve for the final pressure.

Section 10.4 Stoichiometric Relationships with Gases

Many chemical reactions involve gases as either reactants or products. The gas laws make it possible to calculate the volumes of these gases as well.

A corollary of Avogadro's law is that the molar volumes of gases are equivalent. In other words, if you have 35 mL of one gas, and 35 mL of another gas, you have the same number of particles in each volume, and thus, the same number of moles.

The density of a gas can also be calculated using the ideal gas law: density is simply mass per volume. If either mass or number of moles is known, the density can be calculated.

If mass of a gas is known, the molar mass can also be calculated – solve for n in the ideal gas law, and then divide grams per mole to calculate the molar mass.

Workbook Example 10.5

PROBLEM:
If 2.52 g of zinc are dropped into concentrated HCl, how many liters of hydrogen will be produced at 33 °C and 1 atm? The chemical equation is

$$Zn\ (s) + 2HCl\ (aq) \rightarrow ZnCl_2\ (aq) + H_2\ (g)$$

SOLUTION:
First, solve for the moles of hydrogen gas produced, then use the ideal gas law to calculate the volume of the gas:

$$2.52\,g\,Zn \times \frac{1\,mol\,Zn}{65.39\,g} \times \frac{1\,mol\,H_2}{1\,mol\,Zn} = 0.0385\,mol\,H_2$$

$$V = \frac{nRT}{P} = \frac{(0.0385\,mol)\left(0.08206\,\frac{L \cdot atm}{K \cdot mol}\right)(33 + 273.15\,K)}{1\,atm} = 0.967\,L$$

Workbook Example 10.6

PROBLEM:
What is the density of chlorine gas at 85 °C and 1.5 atm?

SOLUTION:
It does not look like there is enough information here, but there is! Determine the volume of 1 mol of gas under the given conditions. The mass of 1 mol of Cl_2 is known, so the density can be calculated.

$$V = \frac{nRT}{P} = \frac{(1\,mol)\left(0.08206\,\frac{L \cdot atm}{K \cdot mol}\right)(85 + 273.15\,K)}{1.5\,atm} = 19.59\,L$$

1 mol of Cl_2 = 71 g, so the density of chlorine under these conditions is

$$\frac{71\,g}{19.59\,L} = 3.6\,\frac{g}{L}$$

Workbook Example 10.7

PROBLEM:
If a sample of gas weighing 11.2 g has a volume of 3.13 L at 300 K and 2.0 atm, what is the molar mass of the gas? What is a possible identity for the gas?

SOLUTION:
To figure this out, first solve for the number of moles of gas present:

$$n = \frac{PV}{RT} = \frac{(2.0\,atm)(3.13\,L)}{\left(0.08206\dfrac{L \cdot atm}{K \cdot mol}\right)(300\,K)} = 0.254\,mol$$

So 0.254 mol of the gas weighs 11.2 g. This makes it possible to solve for the molar mass:

$$\frac{11.2\,g}{0.254\,mol} = 44.0\,{}^{g}\!/_{mol}$$

It is likely that the gas is carbon dioxide.

Workbook Problem 10.5

In the following reaction, how many liters of oxygen will be needed to react with 7.45 L of NO? If the temperature of the reaction does not change, how many liters of NO_2 will be produced?

$$2\ NO\ (g) + O_2\ (g) \rightarrow 2\ NO_2\ (g)$$

Strategy: Determine which gas law is involved; solve for the volumes of oxygen and nitrogen dioxide.

Step 1:   Determine the gas law to use.

Step 2:   Solve for the volume of oxygen.

Step 3:   Solve for the volume of nitrogen dioxide.

Workbook Problem 10.6

What is the density of xenon at 300 mm Hg and 100 °C?

Strategy: Convert temperatures and pressures to make it possible to solve for the volume of 1 mol under these conditions. Divide the mass of 1 mol of xenon by the volume to determine the density.

Step 1:   Convert the temperature and pressure.

Step 2:   Use the ideal gas law to calculate the volume of the gas.

Step 3:   Determine the density of the gas.

Workbook Problem 10.7

The density of a gas was found to be 0.798 g/L at 700 °C and 3.75 atm. What is the molar mass of the gas? Can you suggest a possible identity for the gas?

Strategy: Decide on a volume of gas, then solve for *n*. Using the density makes it possible to solve for the mass for that volume, and then for the molar mass.

Step 1:   Determine the number of moles in 1 L of gas.

Step 2:   Solve for the mass in 1 L of gas using the density.

Step 3:   Determine the molar mass by dividing grams by moles.

Section 10.5   Partial Pressure and Dalton's Law

Because gas particles do not interact much with one another, each gas in a mixture remains independent. This makes dealing with mixtures less complex than might be expected.

Dalton's law of partial pressures states that the total pressure of a mixture of gases is the sum of the pressures of the individual gases:

$$P_{total} = P_1 + P_2 + P_3 + \cdots + P_n$$

Since all the gases in a mixture are by definition at the same temperature and pressure, the pressure exerted by a mixture of gases is purely dependent on the number of moles in the mixture.

To determine the pressures of individual components of the mixture, the mole fraction is used. This is calculated by dividing the number of moles of the component of interest by the total number of moles in the mixture:

$$mole\ fraction(\chi) = \frac{moles\ of\ component}{total\ number\ of\ moles\ in\ mixture}$$

The pressure of the individual component is then calculated as the total pressure of the gas times the mole fraction of the individual component:

$$P_1 = \chi_1 P_{total}$$

This equation also makes it possible to determine the mole fraction of a gas based on its partial pressure.

Workbook Example 10.8

PROBLEM:
A 3.0 L flask at 25 °C contains Ar at a partial pressure of 0.76 atm, He at a partial pressure of 0.32 atm, and Ne at a partial pressure of 0.42 atm. What is the total pressure of the mixture? What is the mole fraction of each gas?

SOLUTION:
$$P_{total} = 0.76\,atm + 0.32\,atm + 0.42\,atm = 1.5\,atm$$

The mole fraction of each gas can be calculated from the partial pressure:

$$\chi_{Ar} = \frac{0.76\,atm}{1.5\,atm} = 0.51 \qquad \chi_{He} = \frac{0.32\,atm}{1.5\,atm} = 0.21 \qquad \chi_{Ne} = \frac{0.42\,atm}{1.5\,atm} = 0.28$$

Note: Mole fractions are unitless and will always add up to 1, just as percentages always add up to 100.

Workbook Problem 10.8

A 30 L tank is filled with the following: 18.3 g CO_2, 23.1 g Ne, and 5.26 g H_2. If the container is at 25 °C, what is the total pressure in the container, and what are the partial pressures of the individual gases?

Strategy: Determine the number of moles of each gas, the total number of moles, and the mole fraction of each component. Using these, calculate the total pressure and the partial pressures using the ideal gas law and Dalton's law.

Step 1:   Determine the number of moles of each gas.

Step 2:   Use the total number of moles to calculate the total pressure.

Step 3:   Use the mole fraction of each component to determine the partial pressures.

Section 10.6   The Kinetic-Molecular Theory of Gases

The kinetic–molecular theory is a model developed over a century ago that describes the macroscopic behavior of gases based on a series of assumptions:

- Gas particles are tiny, and they move randomly.
- The volume of the gas particles themselves is negligible.
- The gas particles are independent of each other.
- Collisions of gas particles, with one another or with the walls of the container, are elastic. That is, no energy is lost in these collisions, so the total kinetic energy of a gas is constant at constant temperature.
- The average kinetic energy is proportional to the Kelvin temperature of the gas.

These assumptions can be used to explain the individual gas laws quite effectively:

- Boyle's law: Pressure increases as volume decreases because the smaller the space, the greater the number of collisions and thus the greater the pressure.
- Charles' law: Temperature is a measure of kinetic energy. If the pressure is constant, particles moving faster will require more space to avoid collisions.
- Avogadro's law: Volume increases as amount increases at constant pressure because adding more particles to the mixture will increase the number of collisions unless the space is also increased.
- Dalton's law: Because the identity of the particles is irrelevant and they do not interact, the pressures of individual component will depend only on the mole fraction of that component.

It is possible to derive from this the average speed of the particles of a gas:

$$u = \sqrt{\frac{3RT}{M}}$$

In this equation, u is the speed (m/s), M is the molecular mass (kg/mol), and a different form of R must be used: $8.314 \dfrac{J}{K \cdot mol}$. All of these changes are required to make the units work out to m/s—remember from the last chapter that joules are derived from the definition of kinetic energy $\left(kg \dfrac{m^2}{s^2} \right)$.

Gas particles move amazingly quickly! Larger molecules move more slowly than smaller molecules, but it important to remember that this is an average: in any gas there will be a distribution of actual speeds. However, the total kinetic energy in the gas will remain the same. Also, gas particles do not move very far before colliding either with another gas particle or with something else, so they proceed in a zigzag fashion. The *mean free path*, the average distance between collisions, is longer for a smaller, faster particle and shorter for a larger, slower-moving particle.

Workbook Example 10.9

PROBLEM:
Calculate the average speed of a xenon atom at 398 K.

SOLUTION:

$$u = \sqrt{\frac{3RT}{M}} = \sqrt{\frac{(3)\left(8.314 \frac{kg \cdot m^2}{s^2}\frac{}{K \cdot mol} \right)(398K)}{0.131 \frac{kg}{mol}}} = \sqrt{76000 \frac{m^2}{s^2}} = 275 \frac{m}{s}$$

Workbook Problem 10.9

What temperature is required to get gaseous iodine molecules moving at the same average speed achieved by hydrogen molecules at room temperature?

Strategy: Determine the speed of H_2 molecules at 25 °C, then solve the equation again for the temperature needed for the much larger I_2 molecules to move at that same speed.

Step 1:   Determine the speed of H_2 at 25 °C

Step 2:   Determine the temperature required for the iodine molecules to achieve the same speed.

Section 10.7   Gas Diffusion and Effusion: Graham's Law

Because of the high speeds at which gas molecules move, and their frequent collisions, gases mix rapidly with one another (this is why you can smell baking cookies from quite a distance). This rapid mixing of gases with frequent collisions is called **diffusion**. It is difficult to mathematically model diffusion because of the random nature of the collisions.

More readily characterized mathematically is **effusion**: the progress of gas particles through a tiny hole into a vacuum. Not surprisingly, speed is inversely proportional to molar mass: a bigger, and thus slower-moving gas will effuse more slowly than a small, faster-moving gas.

Graham's law describes this property mathematically:

$$Rate \propto \frac{1}{\sqrt{M}}$$

This equation can be used most simply to calculate the relative rates of effusion in a mixture of gases:

$$\frac{Rate_1}{Rate_2} = \sqrt{\frac{M_2}{M_1}}$$

Workbook Example 10.10

PROBLEM:
Calculate the ratio of effusion rates of Ne and Xe from the same container at the same temperature and pressure.

SOLUTION:

$$\frac{Rate\,of\,effusion\,of\,Ne}{Rate\,of\,effusion\,of\,Xe} = \sqrt{\frac{131.3\frac{Xe}{mol}g}{20.18\frac{Ne}{mol}g}} = 2.55$$

Workbook Problem 10.10

Oxygen-16 has an atomic mass of 15.995, and oxygen-18 an atomic mass of 17.999. What will be their relative rates of effusion?

Strategy: Solve using Graham's law.

Step 1:   Solve using Graham's law:

Section 10.8   The Behavior of Real Gases

The ideal gas law is a good approximation of the behavior of gases under normal conditions, but begins to fail when pressure is very high.

The actual volume of gas particles can no longer be neglected when they are very close together. As a result, the volume of a real gas is larger than that of an ideal gas at high pressure.

Attractive forces between gas particles will also become a factor at high pressures: these tend to reduce the volume of the gas.

Volume and attractive forces negate each other at moderate pressures, but above 350 atm, the volume of the gas particles becomes the primary factor.

These factors have been accommodated mathematically in the van der Waals equation:

$$\left(P + \frac{an^2}{V^2}\right)(V - nb) = nRT$$

where the *a* term is a correction for intermolecular attractions and the *b* term is a correction for molecular volume. Solving this equation for pressure can be done fairly simply, but solving for volume is not trivial.

Workbook Example 10.11

PROBLEM:
Calculate the pressure of 1 mol of N_2 gas in 50 mL at 300 K using the ideal gas law and the van der Waals equation. The van der Waals constants for nitrogen are $a = 1.35(L^2 \bullet atm)/mol^2$ and $b = 0.0387$ L/mol.

SOLUTION:
Ideal gas law:

$$P = \frac{nRT}{V} = \frac{(1\,mol)\left(0.08206\,\frac{L \cdot atm}{K \cdot mol}\right)(300\,K)}{0.050\,L} = 492\,atm$$

van der Waals equation:

$$P = \frac{nRT}{V - nb} = \frac{an^2}{V^2} = \frac{(1\,mol)\left(0.08206\,\frac{L \cdot atm}{K \cdot mol}\right)(300\,K)}{0.050\,L - \left[(0.0387\,\frac{L}{mol})\right]} = \frac{(1\,mol)^2\left(1.35\,\frac{L^2 \cdot atm}{mol^2}\right)}{(0.050\,L)^2}$$

$$P = \frac{24.6\,L \cdot atm}{0.0113\,L} - \frac{1.35\,L^2 \cdot atm}{0.00250\,L^2} = 2177\,atm - 540\,atm = 1637\,atm$$

At very high pressures, the ideal gas law fails rather dramatically.

Workbook Problem 10.11

Calculate the pressure of 1 mol of O_2 gas in 100 mL at 300 K using the ideal gas law and the van der Waals equation. The van der Waals constants for oxygen are $a = 1.38(L^2 \cdot atm)/mol^2$ and $b = 0.0318$ L/mol. Does the ideal gas law approximate the pressure acceptably under these conditions?

Strategy: Solve using the ideal gas law and the van der Waals equation.

Step 1: Solve using the ideal gas law.

Step 2: Solve using the van der Waals equation.

Step 3: Compare the two values.

Section 10.9   The Earth's Atmosphere and Air Pollution

The atmosphere has four major regions. The *troposphere* is closest to the Earth's surface and both has the greatest effect on the surface and is most affected by human activity. Above the troposphere is the *stratosphere*, which extends from approximately 12 to 50 km above the Earth's surface. The *mesosphere* extends another 35 km, and it is below the final layer of the atmosphere, the *thermosphere*, which extends to 120 km.

In the troposphere, there are three major effects of human activity:

- Air pollution: as a side product of the Industrial Revolution, there has been an increase in hydrocarbon molecules and NO in the air. NO reacts to form NO_2, which is split by sunlight into NO and free radical oxygen molecules. These can attack oxygen molecules, forming ozone (O_3), a highly reactive molecule that further reacts to form *smog*.
- Acid rain: when high-sulfur coal is burned, SO_2 is released into the air. This is further oxidized to form SO_3, which can combine with water to form sulfuric acid. In the form of "acid rain," this has damaged forests and lakes, causing the extinction of fish in some areas. It also causes marble and limestone—two materials common in buildings and monuments—to slowly dissolve.

- Global warming: Some scientists worry that the increase in the levels of CO_2 in the atmosphere as a result of industrialization may reduce the amount of radiation that can be reflected back out into space. This has the potential to cause the warming of the troposphere.

The ozone layer is an atmospheric band that stretches from 20 to 40 km above the Earth's surface. Ozone is an irritating pollutant at low levels in the atmosphere, but absorbs ultraviolet light in the upper atmosphere, acting like sunscreen for the Earth below. Changes in the ozone layer, particularly thinning around the poles, are attributed to chlorofluorocarbons, which also react with ultraviolet light to form chlorine radicals that destroy ozone. These molecules have been banned internationally, but the ban is largely ignored in China and Russia. It is expected that the levels of CFCs will return to pre-1980 levels by mid-century.

Section 10.10   The Greenhouse Effect

The greenhouse effect refers to the absorption of infrared, or IR, radiation by gases in the atmosphere, which causes an increase in planetary temperature. These gases, referred as greenhouse gases, include water, carbon dioxide, and methane: three compounds that are plentiful on Earth.

The way this works is this: the Sun emits radiation to the Earth (and all planets) in the form of UV and visible regions of the electromagnetic spectrum. Most UV radiation is absorbed by ozone and oxygen in the stratosphere, so most of it does not reach ground level. Once visible light reaches the ground level, it becomes absorbed by everything it touches, which causes the Earth's surface to heat up. The surface then gives off infrared radiation toward space. Once this radiation interacts with the greenhouse gases in the atmosphere, these greenhouse gases absorb the energy; the absorbed radiation makes the bonds in the molecules vibrate. However, the gas molecules will only absorb IR radiation if two conditions are met:

- The energy difference between the lower vibrational state and the excited vibrational state exactly matches the energy of the IR photon.
- The vibration results in a change in dipole moment.

Once these molecules absorb the IR energy, it "traps" the energy within our atmosphere and increases the temperature of our planet —very similar to how a greenhouse works!

Section 10.11   Climate Change

Two phrases are frequently used when we talk about pollution and our environment:

- **Global warming** refers to the theory that increasing concentrations of greenhouse gases will upset the delicate thermal balance of incoming and outgoing radiation on Earth (the greenhouse effect)
- **Climate change** is a newer hypothesis suggesting that there will not be a uniform increase in temperature throughout the planet: some areas will experience an increase in temperature, while some areas may stay the same or experience some cooling.

Climate science is complex because there numerous factors that influence climate such as cloud cover, particulate matter, solar energy, changing surface reflection due to melting polar ice caps, and deforestation.

Putting It Together

13.67 L of an unknown gas (kept at 3 atm and 500 K) weigh 197 g. What is the molecular formula of this gas if it is found to be 12.19% C, 0.51% H, 28.93% F, 18.02% Cl, and 40.56% Br?

Self-Test

This section is intended to test your knowledge of the material covered in this chapter. Think through these problems, and make certain you understand what they are asking. Make sure your answers make sense. Successful completion of these problems indicates that you have mastered the material in this chapter. You will receive the greatest benefit from this section if you use it as a mock exam, as this will allow you to determine which topics you need to study in more detail.

True–False
1. The atmosphere is about 70% oxygen.

2. Gases are more compressible than liquids.

3. Pascals, atmospheres, bars, and millimeters of mercury are all accepted units of pressure.

4. If the mercury in a manometer is lower on the side that is open to the air, then the pressure of the gas is lower than the pressure of the air.

5. Stoichiometric ratios can be determined by volume in gas reactions.

6. The smaller a gas particle, the slower it travels.

7. The ratio of effusion rates of two gases is directly proportional to the square roots of their masses.

8. The ideal gas law begins to fail at high temperatures.

9. When you inhale, you are breathing stratosphere.

10. Incomplete combustion of hydrocarbons is a major factor in air pollution.

Matching

11. Barometer	a.	force per unit area
12. Pressure	b.	the pressure of one component in a gas mixture is related to its mole fraction in the mixture
13. Atmospheric pressure	c.	the rate at which a gas effuses is inversely proportional to the square root of its molar mass
14. Boyle's law	d.	the escape of gas particles through a tiny hole in a membrane into a vacuum
15. Charles's law	e	the pressure exerted by one component in a gaseous mixture

16. Avogadro's law

    f. the chaotic mixing of gases through random collisions

17. Ideal gas law

    g. number of moles of a component divided by the total moles in a mixture

18. Dalton's law

    h. volume and temperature vary directly when pressure and amount are held constant

19. Partial pressure

    i. volume and pressure are inversely proportional when temperature and amount are constant

20. Kinetic–molecular theory

    j. a piece of equipment used to measure atmospheric pressure

21. Graham's law

    k. equipment for measuring gas pressures

22. Diffusion

    l. pressure generated by the atmosphere pressing down on the Earth's surface

23. Effusion

    m. allows any of the four variables to be calculated— T, P, V, or n—if the others are known

24. Ideal gas

    n. a model that accounts for the behavior of gases

25. STP

    o. moles and volume vary directly.

26. Standard molar volume

    p. gas that precisely follows the tenets of kinetic—molecular theory

27. Mole fraction

    q. 22.4 L at STP

28. Manometer

    r. 0 °C and 1 atmosphere pressure

Fill-in-the-Blank

29. Gases exert _____ on the walls of their container as a result of _____ with the walls.

30. The physical properties of a gas can be defined by four variables: _____, _____, _____, and _____.

31. The concentration of a component in a gas mixture is typically calculated as its _____. This can also be used to determine the _____ _____ of the component.

32. In kinetic–molecular theory, the _____ of individual gas particles is said to be _____. The assumption fails at high _____.

33. Effusion rates are proportional to the square root of molar mass. Gases with _____

    molar masses move significantly _____ than gases with high molar masses.

Problems

34. If atmospheric pressure is 0.983 atm, how high a column of silicone oil (density = 0.970 g/mL) will this pressure support (the density of mercury is 13.6 g/mL)?

35. If you are using an open-ended manometer filled with ethyl alcohol, rather than mercury, what is the gas pressure in mmHg if the level of ethyl alcohol in the arm connected to the bulb is
    a) 340 mm lower and P_{atm} = 760 mmHg; and b) 257 mm higher and P_{atm} = 760 mmHg?
    The density of the alcohol is 0.789 g/mL and the density of mercury is 13.6 g/mL.

36. How many moles of gas are present in a 50 mL flask at 5.78 atm and –20 °C? If the mass of the gas is 0.281 g, what is the gas?

37. What is the density of methane (CH_4) at 100 °C and 4 atmospheres?

38. A student carried out a reaction in the lab in which one of the products was a gas. She collected the gas for analysis and found that it contained 7.69% hydrogen and 92.3% carbon. She also observed that 250 mL of the gas at 35 °C and 763 mmHg had a mass of 0.258 g.
    a) What is the empirical formula of the gas? b) What is the molar mass of the gas?
    c) What is its molecular formula?

39. What is the molar mass of a gas with a density of 14.04 g/L at 15 atm and 300 °C?

40. When 25 mL of butane (C_4H_{10}, density 0.6014 g/mL) are burned, how many liters of oxygen are consumed, and how many liters of carbon dioxide and water vapor are produced at 30 °C and atmospheric pressure? What is the overall change in volume for the reaction? Is work being done as well as heat given off?

41. If 7.03 g Ne, 11.7 g N_2, and 15.7 g He are placed in a 5000 mL container at room temperature (25 °C), what is the total pressure, and what is the pressure of each component?

42. What are the relative effusion rates of H_2 and He?

43. If a 50 mL aerosol can at room temperature (25 °C) with an internal pressure of 7.8 atm is emptied into a 375 mL container at atmospheric pressure, what is the final temperature of the gas in Celsius?

CHAPTER ELEVEN

Liquids, Solids, and Phase Changes

Learning Objectives

As a result of reading and studying this chapter, you should be able to

Section 11.1   Properties of Liquids
1.   Predict which substance has higher viscosity or surface tension based on its molecular structure.

Section 11.2   Phase Changes between Solids, Liquids, and Gases
2.   Use the equation for Gibbs free energy to calculate the temperature or ΔS for a phase change.
3.   Calculate the amount of heat associated with phase changes and draw heating curves.

Section 11.3   Evaporation, Vapor Pressure, and Boiling Point
4.   Use the Clausius–Clapeyron equation to calculate vapor pressure at varying temperatures or ΔH_{vap}.

Section 11.4   Kinds of Solids
5.   Classify types of solids based on their chemical composition and properties.

Sections 11.5 – 11.8   Structures of Solids
6.   Use the Bragg equation to calculate spacing between atomic layers in a crystal.
7.   Identify a unit cell from a pattern.
8.   Identify the four kinds of spherical packing arrangements in crystalline solids and the three kinds of cubic unit cells.
9.   Calculate the density of a substance, atomic radii of its atoms, or molecular mass given its unit cell dimensions.
10.   Use the unit cell to determine the formula and geometry of ionic compounds.

Section 11.9   Phase Diagrams
11.   Use a phase diagram to interpret the behavior of a substance at a given temperature and pressure.
12.   Sketch a phase diagram, given the appropriate data.

Chapter Summary

Unlike in gases, the molecules in liquids and solids interact. This chapter enhances the previous discussion on intermolecular forces by discussing the effects they have on the properties of materials and their phase transitions. Free-energy changes accompany all physical and chemical changes: in this chapter we will examine the contributions of entropy and enthalpy to phase changes, as well as the contributions of changes in pressure and temperature. Solids take a number of forms. Those that are crystalline can be characterized by X-ray crystallography, which provides data about unit cells. This information can be used to gain insights into packing and allows for calculations of atomic radius and density.

The Chapter in Detail

Section 11.1   Properties of Liquids

Viscosity is the measure of a liquid's resistance to flow: the greater the intermolecular interactions, the more they hold on to one another and the greater the resistance to flow. Long-chain polar molecules like those found in honey will not flow as readily as those in a nonpolar solvent held together only by dispersion forces.

Surface tension is also caused by intermolecular forces: at a liquid/gas interface, molecules with strong intermolecular forces will arrange themselves to maximize those interactions. It is this phenomenon that makes it possible for insects to walk on water; it also causes water to bead up on a nonpolar surface: because a nonpolar surface provides nothing for the water to hydrogen bond with, the water molecules hold tightly together.

Both viscosity and surface tension are temperature dependent: increasing kinetic energy will reduce both properties.

Section 11.2   Phase Changes between Solids, Liquids, and Gases

There are six possible phase changes between solids, liquids, and gases, all associated with a change in free energy, ΔG.

- fusion (melting): solid $\rightarrow$ liquid
- freezing: liquid $\rightarrow$ solid
- evaporation: liquid $\rightarrow$ gas
- condensation: gas $\rightarrow$ liquid
- sublimation: solid $\rightarrow$ gas
- deposition: gas $\rightarrow$ solid

The free-energy change can be described using the equation $\Delta G = \Delta H - T\Delta S$, and it has two portions. The enthalpy change is related to the breaking or forming of intermolecular interactions, and the entropy change is associated with the increase or decrease in disorder associated with breaking or forming intermolecular interactions.

For melting, sublimation, and vaporization, heat must be added, but disorder increases: ΔH and ΔS are both positive.

For freezing, condensation, and deposition, heat is released, but order increases: ΔH and ΔS are both negative.

At equilibrium, $\Delta G = 0$. This makes it possible to calculate freezing and boiling points, or, more usefully, to calculate entropy changes from measured freezing or boiling points, using the following equations:

$$T = \frac{\Delta H}{\Delta S} \qquad\qquad \Delta S = \frac{\Delta H}{T}$$

Workbook Example 11.1

PROBLEM:
Methanol boils at 64.7 °C and has an enthalpy of vaporization of 35.3 kJ/mol. What is ΔS_{vap} for this phase transition?

SOLUTION:
Using the equation above, we calculate ΔS_{vap} to be

$$\Delta S = \frac{\Delta H}{T} = \frac{35300\ ^{J}/_{mol}}{(64.7 + 273.15)\ K} = 104\ ^{J}/_{K \cdot mol}$$

Workbook Example 11.2

PROBLEM:
A theoretical liquid has an enthalpy of fusion of −15.7 kJ/mol and an entropy of fusion of −142 J/mol•K. What is the freezing point of this liquid?

SOLUTION:
Using the above equation, we calculate T to be:

$$T = \frac{\Delta H}{\Delta S} = \frac{-15700\ \dfrac{J}{mol}}{-142\ \dfrac{J}{K \cdot mol}} = 111\ K = -163\ °C$$

Heating curves are graphical representations of phase changes. The vertical axis shows temperature, and the horizontal axis shows the heat added. In heating curves we see that the temperature increases up to the normal melting point then stops increasing as the phase change occurs. The same happens at the boiling point.

Heating curves have a number of notable features. The slopes when solids are being warmed to their melting point, liquids are being warmed to the boiling point, and gases are being heated are equivalent to the molar heat capacities of these phases. The amount of kJ/mol that is added in the times that the temperature does not change is the heat of fusion (melting) and the heat of vaporization.

The heat of vaporization will always be greater than the heat of fusion: the flat line that occurs while a liquid is being vaporized is always longer than the flat portion of the line that corresponds to melting. This is due to the fact that in melting, the molecules only need to overcome the intermolecular forces enough to be able to move around each other, while in vaporization they must completely overcome the intermolecular forces, so that the molecules can be fully separated from one another.

Workbook Problem 11.1

If the enthalpy of fusion of ammonia is 5.97 kJ/mol, and the normal freezing point of ammonia is −107 °C, what is the change in entropy associated with the freezing of ammonia?

Strategy: Rearrange the equation $\Delta G = \Delta H - T\Delta S$ to solve for the entropy change.

Step 1: Rearrange to solve for ΔS.

Step 2: Solve for ΔS, being careful to maintain the correct sign.

Section 11.3 Evaporation, Vapor Pressure, and Boiling Point

Liquid molecules at a certain temperature all have the same average kinetic energy, but some have more energy than others. If a molecule with a high kinetic energy reaches the surface, it can escape, or evaporate.

In an open container, the molecules will drift away, and the liquid will evaporate completely. In a closed container, a dynamic equilibrium will develop: the same number of molecules will be escaping into the air as are simultaneously returning to the liquid. As a result, at a specific temperature, a liquid will have a characteristic vapor pressure.

When the vapor pressure is equal to the external pressure, the liquid will boil. When the vapor pressure is 1 atm, the normal boiling point has been reached. If the external pressure is higher or lower than 1 atm, the boiling point will change correspondingly: the boiling point at lower pressures will be a lower temperature, the boiling point at a higher pressure will be a higher temperature.

When the natural log of the vapor pressure is plotted against the inverse of the temperature, the graph looks like a linear relationship (it fits the form y = mx + b) known as the Clausius–Clapeyron equation:

$$\ln P_{vap} = \left(\frac{-\Delta H_{vap}}{R}\right)\frac{1}{T} + C$$

This form of the Clausius–Clapeyron equation resembles the equation for a straight line (y = mx + b) where y is $\ln P_{vap}$, m (the slope) is equal to $\dfrac{-\Delta H_{vap}}{R}$, x is $\dfrac{1}{T}$, and b (the y-intercept) is C, where C is a constant that is characteristic of each substance.

This equation can also be rearranged so as to make it possible to solve for the vapor pressures and temperatures, provided that ΔH_{vap} is known:

$$\ln\left(\frac{P_1}{P_2}\right) = \frac{\Delta H_{vap}}{R}\left(\frac{1}{T_2} - \frac{1}{T_1}\right)$$

In all of these equations, R = 8.3145 J/K•mol.

Workbook Example 11.3

PROBLEM:
Methanol has a normal boiling point of 64.7 °C and a vapor pressure of 400 mm Hg at 49.9 °C. What is ΔH_{vap} for methanol?

SOLUTION:
There is a great deal of information here, but some of it is hidden. What you have is two vapor pressures and two temperatures with which to solve for ΔH. The vapor pressure of a liquid is always 760 mm Hg at its normal boiling point.

Using the two-point form of the Clausius–Clapeyron equation from above, we get:

$$\ln\left(\frac{400\ mm\ Hg}{760\ mm\ Hg}\right) = \frac{\Delta H_{vap}}{8.314\ ^J\!/_{K\cdot mol}}\left(\frac{1}{64.7 + 273.15} - \frac{1}{49.9 + 273.15}\right)$$

$$-0.6419 = \frac{\Delta H_{vap}}{8.314\ ^J\!/_{K\cdot mol}}(0.00296\ K - 0.00310\ K)$$

$$-0.6419 = \frac{\Delta H_{vap}}{8.314\ ^J\!/_{K\cdot mol}}(-0.00014\ K)$$

$$\frac{-0.6419}{-0.00015\ K} = \frac{\Delta H_{vap}}{8.314\ ^J\!/_{K\cdot mol}}$$

$$4585\ K = \frac{\Delta H_{vap}}{8.314\ ^J\!/_{K\cdot mol}}$$

$$\Delta H_{vap} = (4346\ K)(8.314\ ^J\!/_{K\cdot mol}) = 38119\ ^J\!/_{mol} = 38.1\ ^{kJ}\!/_{mol}$$

Workbook Example 11.4

PROBLEM:
Given the information in Workbook Example 11.3, what is the vapor pressure of methanol at 25 °C ?

SOLUTION:
With a variety of temperatures and vapor pressures to choose from, and having solved for ΔH_{vap} above, this is a fairly straightforward problem. Here, the equation above is used with 25 °C on one side and the normal boiling point on the other:

$$\ln\left(\frac{P_1}{760\ mm\ Hg}\right) = \frac{38100\ ^J\!/_{mol}}{8.314\ ^J\!/_{K \cdot mol}}\left(\frac{1}{273.15 + 64.7\ K} - \frac{1}{273.15 + 25\ K}\right)$$

$$\ln\left(\frac{P_1}{760\ mm\ Hg}\right) = \frac{38100\ ^J\!/_{mol}}{8.314\ ^J\!/_{K \cdot mol}}(0.00296\ K - 0.00335\ K)$$

$$\ln\left(\frac{P_1}{760\ mm\ Hg}\right) = \frac{38100\ ^J\!/_{mol}}{8.314\ ^J\!/_{K \cdot mol}}(-0.00039\ K)$$

$$\ln\left(\frac{P_1}{760\ mm\ Hg}\right) = -1.78723$$

$$e^{\ln\left(\frac{P_1}{760\ mm\ Hg}\right)} = e^{-1.69341}$$

$$\frac{P_1}{760\ mm\ Hg} = 0.16742$$

$$P_1 = 127\ mm\ Hg$$

Workbook Problem 11.2

Ethanol has a normal boiling point of 78.4 °C and a vapor pressure of 400 mm Hg at 63.5 °C. What is ΔH_{vap} for ethanol?

Strategy: Use the Clausius–Clapeyron equation to solve for ΔH.

Step 1:   Convert temperatures to kelvins.

Step 2: Solve for Δ*H*.

Workbook Problem 11.3

Given the information in the previous problem, what is the vapor pressure of ethanol at 25 °C ?

Strategy: Use the Clausius–Clapeyron equation to solve for the vapor pressure.

Step 1: Use the Clausius–Clapeyron equation to solve for the vapor pressure.

Section 11.4 Kinds of Solids

Crystalline solids have an ordered, long-range structure. This order is visible macroscopically as sharp edges and flat faces. The four types of crystalline solids are

- Ionic: the constituent particles are ions, held together by ionic bonds.
- Molecular: the constituent particles are molecules, held together by intermolecular forces.
- Covalent network: there are no individual particles, but vast covalent networks.
- Metallic: there are individual metal atoms, but they are all associated in an "electron sea."

Amorphous solids have randomly-arranged constituent particles.

Section 11.5 Probing the Structure of Solids: X-Ray Crystallography

Early in the twentieth century, techniques were developed that made it possible to look into structures at the atomic level. These were—and still are—based on the diffraction of X-rays. It is a principle of optics that the wavelength of light must be no more than twice the length of an object for it to be visible. As a result, the short wavelengths of X-rays are needed to visualize atoms.

When X-rays are directed at a crystal, a diffraction pattern develops as a result of the regular pattern of atoms: as the X-rays leave the crystal again, they interfere with one another both constructively and destructively. The diffraction pattern can then be interpreted to provide data about interatomic distances.

The basis for this interpretation is the Bragg equation, for which a father and son shared the Nobel Prize in physics in 1915:

$$d = \frac{n\lambda}{2\sin\theta}$$

The wavelength, λ, is known; the angle of reflection, θ, can be measured; and n is a small whole number, usually 1.

Computer-controlled diffractometers are now used for both data collection and analysis, making it possible to investigate the structures of even the largest macromolecules.

Section 11.6   The Packing of Spheres in Crystalline Solids: Unit Cells

In crystalline atomic packing, the coordination number is the number of other atoms touching the atom under discussion. Unit cells are the small repeating units of a crystal.

Four types of packing are possible:

- **Simple cubic packing**: each atom touches six others (coordination number of 6) and 52% of space is used. The unit cell is a primitive cubic: 1/8 of eight atoms are found on the corners of the unit cell: one atom per unit cell, atomic diameter = edge length.
- **Body-centered cubic packing**: a simple cubic packing arrangement with another central atom. The coordination number is 8 (four neighbors above, and four below), and 68% of space is used. There are two atoms per unit cell, atomic diameter $= \frac{2d}{\sqrt{3}}$, where d is the edge length of the unit cell.
- **Cubic closest packing**: in this arrangement, there are two alternating layers of simple cubic packing plus an additional atom halfway between the two unit cells—half in and half out, as it were. There are four atoms per unit cell, in face-centered cubic unit cells: the coordination number is 12, and 74% of the space is used. Atomic radius $= \sqrt{\frac{d^2}{8}}$, where d is the edge length of a unit cell.
- **Hexagonal closest packing**: this is a noncubic arrangement, and thus it does not have a cubic unit cell. The coordination number is 12, packing efficiency is 74%, just as in cubic closest packing.

Workbook Example 11.5

PROBLEM:
Aluminum has a face-centered cubic unit cell. If the edge of a unit cell is 404.5 pm, what is the density of aluminum in g/cm^3?

SOLUTION:
A face-centered cubic unit cell has a total of four atoms.

Based on this, it is possible to calculate the mass of a unit cell from the molar mass of aluminum and Avogadro's number:

$$4 \text{ atoms } Al \times \frac{1 \text{ mol } Al}{6.022 \times 10^{23} \text{ } Al \text{ atoms}} \times \frac{26.98 \text{ g } Al}{1 \text{ mole } Al} = 1.792 \times 10^{-22} \text{ g } Al$$

The volume of a unit cell can be calculated from the edge length, being sure to convert from picometers to centimeters:

$$\left(404.5 \times 10^{-12} \text{ } m \times \frac{100 \text{ cm}}{1 \text{ m}} \right)^2 = 6.618 \times 10^{-23} \text{ cm}^3$$

Since we know the mass and the volume, we can calculate the density:

$$d = \frac{m}{V} = \frac{1.792 \times 10^{-22} \text{ g}}{6.618 \times 10^{-23} \text{ cm}^3} = 2.71 \text{ }g/_{cm^3}$$

This agrees closely with the calculated density of aluminum.

Workbook Example 11.6

PROBLEM:
Nickel has a face-centered cubic unit cell. If the edge of a unit cell is 352.4 pm, what is the diameter of a nickel atom in pm?

SOLUTION:
The side of a face-centered cubic unit cell has one full atom and two radii as the diagonal, making the diagonal equal to 4r or 2d.

Using the Pythagorean theorem:

$$(352.4 \times 10^{-12} \text{ } m)^2 + (352.4 \times 10^{-12} \text{ } m)^2 = (2d)^2$$

$$2.484 \times 10^{-19} \text{ } m = 4d^2$$

$$2.492 \times 10^{-10} \text{ } m = d$$

The diameter of a nickel atom is approximately 249 pm.

Workbook Problem 11.4

Polonium has a density of 9.3 g/cm³ and a simple cubic structure. Estimate the radius of a polonium atom.

Strategy: Use the number of atoms per unit cell and the molecular mass to determine the mass of a unit cell, then use the density to solve for the volume of a cell. With the volume, determine the length of one side, then determine the atomic radius.

Step 1:   Determine the mass of a unit cell.

Step 2:   Solve for the volume of a unit cell.

Step 3:   Determine the edge length of a unit cell.

Step 4:   Determine the atomic radius.

Section 11.7   Structures of Some Ionic Solids

When ionic solids crystallize, they adopt a unit cell structure that allows for the accommodation of ions of different sizes: cations are smaller than their neutral atoms, and anions are larger.

The unit cell of an ionic solid is always electrically neutral.

Section 11.8   Structures of Some Covalent Network Solids

Carbon has more than 40 allotropes—different structural forms of the element with varying physical and chemical properties. Most are amorphous, but there are three famous exceptions:

- *Graphite*: the most common and most stable allotrope, graphite consists of two-dimensional sheets of fused, sp^2-hybridized, six-membered rings. Graphite is the "lead" in pencils, an electrode in batteries, and a lubricant for locks. All of these functions rely on the fact that the layers of graphite can slide over one another when air and water adsorb onto them. A closely related structure is carbon *nanotubes*, which are like sheets of graphite rolled into tubes. They are being investigated as structural composites, because they have 50 to 60 times the tensile strength of steel.
- *Diamond*: a covalent network solid in which sp^3-hybridized carbon atoms bond to one another in a vast network. Diamond is the hardest known substance, and because of this, it is used in industrial saw blades and drill bits, as well as in jewelry. It is an electrical insulator and has a melting point over 3550 °C.
- *Fullerene*: discovered in 1985 as a component of soot, fullerene is a soccer ball-shaped molecule of 60 sp^2-hybridized carbon atoms in alternating hexagons and pentagons.

When reacted with rubidium, a superconducting material called rubidium fulleride (Rb_3C_{60}) is formed.

Silicon has similar abilities to form network solids, but because it is larger than carbon, it is less able to double bond with oxygen. As a result, it forms four single bonds with oxygen that are used to bridge to other silicon atoms. Silica has the empirical formula SiO_2. Most minerals and rocks are formed by silicates: these account for 75% of the Earth's crust by mass.

When silica is heated above 1600 °C, many of the Si–O bonds break, converting the silica into a viscous liquid. When it cools, the bonds reform into a random arrangement and an amorphous solid—quartz glass—is formed. Addition of transition metal ions lends color, and the addition of B_2O_3 adds heat resistance. This *borosilicate* glass is sold as Pyrex.

Section 11.9   Phase Diagrams

Temperature can cause a spontaneous phase change, but pressure can as well. Graphs of temperature (horizontal axis) versus pressure (vertical axis) are called **phase diagrams**.

The lines on a phase diagram indicate interfaces between phases. It is fairly straightforward to remember which phase is which by thinking of water, a substance familiar in all three phases. As the temperature increases, the substance becomes first liquid, then gas.

A similar pattern is seen on the pressure axis: at the highest pressures, most substances are solids. The very few exceptions are substances, such as water and gallium, in which the liquid form is more dense than the solid. These are described with a backward-sloping line between the solid and liquid phases, which indicates that increasing pressure favors the liquid form.

There are a few features of phase diagrams that should be noted.

- *Triple point*: this is the temperature and pressure at which all three phases exist in equilibrium.
- *Critical point*: this is the temperature and pressure above which gases and liquids behave indistinguishably as *supercritical fluids*: gas and liquid phases have the same density and are miscible.
- *Critical pressure*: this is the pressure above which a liquid cannot be vaporized.
- *Critical temperature*: this is the temperature above which a gas cannot be liquefied.
- *Normal melting/boiling point*: this is the phase transition at 1 atm pressure.

Putting It Together

Sulfur dioxide is one of the compounds that causes acid rain. Acid rain is produced first by the oxidation of sulfur dioxide to sulfur trioxide in air. The sulfur trioxide then reacts with water, producing sulfuric acid. What is the molecular shape of sulfur dioxide? What intermolecular forces are present? Calculate the heat of reaction for the production of sulfur trioxide and sulfuric acid. (ΔH_f SO_2 = –296.8 kJ/mol, ΔH_f SO_3 = –395.7 kJ/mol, ΔH_f H_2O = –285.8 kJ/mol, and ΔH_f H_2SO_4 = –814.0 kJ/mol.)

Self–Test

This section is intended to test your knowledge of the material covered in this chapter. Think through these problems, and make certain you understand what they are asking. Make sure your answers make sense. Successful completion of these problems indicates that you have mastered the material in this chapter. You will receive the greatest benefit from this section if you use it as a mock exam, as this will allow you to determine which topics you need to study in more detail.

True–False
1. Freezing causes a positive entropy change and a negative enthalpy change.

2. Vapor pressure does not vary with temperature.

3. Covalent network solids are extremely large molecules.

4. X rays can be used to determine the strength of intermolecular forces in liquids.

5. The coordination number of a packing arrangement is the number of atoms in the unit cell.

6. Fullerene and graphite are both isotopes of carbon.

7. Increasing pressure increases the boiling point of a liquid.

Matching

8. Allotrope		a.	temperature and pressure at which the solid, liquid, and gas phases of the same substance are at equilibrium
9. Critical point		b.	heat required for phase change from solid to liquid
10. Heat of fusion		c.	point at which any increase in temperature or pressure produces a supercritical fluid
11. Molecular solid		d.	resistance to flow
12. Normal boiling point		e.	solid in which the constituent particles are randomly arranged
13. Viscosity		f.	different form of a pure element: can have different physical and chemical properties
14. Vapor pressure		g.	temperature at which the phase spontaneously changes from liquid to gas at 1 atm pressure
15. Triple point		h.	solid held together by intermolecular forces

16. Amorphous solid
        i.   solid in which the constituent particles have long-range order

17. Crystalline solid
        j.   the partial pressure of a gas in equilibrium with a liquid in a closed container.

Multiple Choice

18. Which of the following liquids has the highest boiling point?
    a. $CH_3CH_2CH_3$
    b. $CH_3(CH_2)_2CH_3$
    c. $CH_3(CH_2)_4CH_3$
    d. $CH_3(CH_2)_6CH_3$

19. Which of the following liquids has the lowest vapor pressure?
    (A) CH_3CH_3 (B) CH_3OCH_3 (C) CH_3CH_2OH
    a. A
    b. B
    c. C
    d. B and C will have comparable values

20. If the external pressure is more than 1 atm, a liquid boils at
    a. the normal boiling point
    b. a temperature below the normal boiling point
    c. a temperature above the normal boiling point
    d. there is not enough information to say

21. A solid whose constituent particles are held together by charge interactions is a(n)
    a. ionic solid
    b. molecular solid
    c. covalent network solid
    d. metallic solid

22. The type of packing present when the unit cell contains a complete central atom is
    a. simple cubic packing
    b. body-centered cubic
    c. hexagonal closest packed
    d. cubic closest packed

23. Most of the allotropes of carbon are amorphous, but diamond, fullerene, and graphite are all
    a. ionic solids
    b. covalent network solids
    c. forms of silica
    d. metallic

24. The point at which the phase transition from liquid to solid occurs at 1 atm is
    a. the normal freezing point
    b. the triple point
    c. the critical point
    d. the critical pressure

Fill-in-the-Blank

25. Phase changes have an associated change in free energy. The _____ portion of this change is involved in disrupting or forming intermolecular forces.

26. In a heating curve, the temperature increase stops during _____.

27. In a phase diagram, the phase change represented by the line between solid and gas is

    _____.

28. As temperature decreases, vapor pressure _____, and viscosity and

    surface tension _____.

29. A brittle, crystalline solid with sharp edges and a very high melting point is a(n)

_____. A relatively soft, low-melting-point crystalline solid is most

likely a _____, and an extremely hard, extremely high-melting-point

solid is probably a _____.

Problems

30. What is the most important intermolecular force in the following molecules?
    $CH_3CH_2CH_2OH$ $CH_3(CH_2)_4CH_3$ PCl_3

31. Which of the following would you expect to have the highest boiling point and why?
    a. Br_2 or Cl_2 b. CH_3OH or CH_3CH_2OH c. CH_4 or CH_3OH

32. For platinum, ΔH_{vap} = 565.3 kJ/mol and ΔS_{vap} = 150.8 J/K·mol. What is the boiling point of platinum?

33. If 0.237 g of water condensed on a 50.0 g block of iron at 25.0 °C, what would the final temperature of the iron be? ΔH_{vap} = 40.7 kJ/mol, c_{iron} = 0.449 J/g·°C.

34. If a prankster were to take a 45 g ice cube out of a household freezer at −20 °C and drop it down the back of an unsuspecting sibling, how much heat would be absorbed from the hapless sibling if she decided to pretend she did not notice? Normal body temperature is 37 °C, the molar heat capacity of ice is 36.57 J/mol·°C, ΔH_{fus} = 6.01 kJ/mol, and the specific heat of water is 4.184 J/g·°C.

35. The normal boiling point of benzene is 80.1 °C and ΔH_{vap} = 30.8 kJ/mol. What is the vapor pressure of benzene at 75.0 °C?

36. Gold has a density of 19.3 g/cm^3 and a face-centered cubic unit cell. What is the atomic radius of gold?

37. Palladium has a face-centered cubic unit cell. What is the density of palladium if the atomic radius is 137 pm?

38. For the phase diagram shown, identify the normal freezing point, the triple point, the critical point, and the lines that indicate boiling, sublimation, and melting. What does the slope of the central line indicate about the substance whose data is shown?

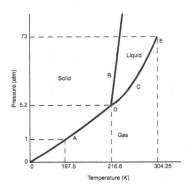

214

Challenge Problem

39. How many grams of water can be taken from –20 °C to 115 °C by burning 320 g of propane (C_3H_8)? Assume that all the energy of combustion goes into the water.

    Necessary information:

ΔH_{comb} for propane	= 2220 kJ/mol	Specific heat for ice	= 2.03 J/g•°C
Specific heat for water	= 4.184 J/g•°C	Specific heat of steam	= 2.08 J/g•°C
ΔH_{fus} for water	= 6.01 kJ/mol	ΔH_{vap} for water	= 40.67 kJ/mol

CHAPTER TWELVE

Solutions and Their Properties

Learning Objectives

As a result of reading and studying this chapter, you should be able to

Sections 12.1–12.2 Solutions and Energy Changes

1. Describe the types of intermolecular forces involved when a solute dissolves in a solvent.
2. Predict solubility based on the chemical structure of the solute and solvent.
3. Describe the effect of ion size and lattice energy on the solubility of ionic compounds.

Section 12.3 Concentration Units for Solutions

4. Calculate the concentration of a solution in units of mass percent.
5. Calculate the concentration of a solution in units of parts per million (ppm) or parts per billion (ppb).
6. Calculate the concentration of a solution in terms of molality.
7. Convert from one unit of concentration to another.

Section 12.4 Some Factors That Affect Solubility

8. Analyze a graph to determine the solubility of a solute at a given temperature.
9. Use Henry's law to calculate the solubility of a gas.

Sections 12.5–12.8 Physical Properties of Solutions: Colligative Properties

10. Qualitatively predict changes to the vapor pressure of a solution when a solute is added.
11. Calculate the vapor pressure of a solution containing a nonvolatile solute.
12. Calculate the vapor pressure of a solution containing a volatile solute.
13. Calculate the amount of boiling-point elevation and freezing-point depression for a solution.
14. Calculate osmotic pressure for a solution.
15. Use osmotic pressure or other colligative property to calculate the molecular weight of a solute.

Section 12.9 Fractional Distillation of Liquid Mixtures

16. Interpret the liquid/liquid phase diagram of a solution and apply the diagram to fractional distillation.

Chapter Summary

This chapter focuses on homogeneous mixtures, primarily solutions. The energetics of solution formation, the free-energy changes involved, and factors that affect solubility are covered, as are the great variety of mass- and mole-based concentration units, along with calculations involving these units and the ways that they are interconverted. We explain and demonstrate the effects of pressure and temperature on the solubility of gases and the four colligative properties: lowering of vapor pressure, raising of boiling point, lowering of freezing point, and production of osmotic pressure; methods of their calculation are also explained. Finally, the practical uses of these four properties are discussed.

The Chapter in Detail

Section 12.1   Solutions

Mixtures can be classified as *homogeneous* or *heterogeneous*. In heterogeneous mixtures, the components are visibly nonuniform; homogeneous mixtures appear the same throughout.

Homogeneous mixtures are classified according to the size and behavior of the constituent particles:

- *Solutions* contain particles the size of an ion or small molecule in the range of 0.1–2 nm. They are transparent and do not separate on standing. An example is saltwater.
- *Colloids* contain particles with diameters in the range of 2–500 nm. They scatter light and may be murky or opaque, but they do not separate on standing. An example is fog.
- *Suspensions* are only temporarily homogeneous mixtures: inputs of energy are required to keep them mixed. These contain the largest particles and separate when left standing. Paint is an example of a suspension.

Any state of matter can form a solution with another state. Seven possible solution types are known: gases can dissolve in solids, liquids, or other gases; liquids can dissolve in other liquids or in solids; and solids can dissolve in liquids or in one another.

When a gas or solid is dissolved in a liquid, the liquid is referred to as the **solvent** and the dissolved substance is the **solute**.

When two liquids are dissolved in one another, the main component is the solvent and the lesser component is the solute.

Workbook Example 12.1

PROBLEM:
Drugstore "rubbing alcohol" can be purchased as 90% isopropyl alcohol or as 70% isopropyl alcohol. For each of these, what is the solvent and what is the solute?

SOLUTION:
In both of these cases, the isopropyl alcohol is the solvent. The minor component, or solute, is water.

Workbook Problem 12.1

A grocery store bottle of hydrogen peroxide is labeled as 3% solution. What is the solute and what is the solvent?

Strategy: Determine which is the major component and which is the minor component.

Section 12.2   Energy Changes and the Solution Process

Except for mixtures of gases, intermolecular forces are at work in solutions as well as in pure substances, but the situation is made somewhat more complex by the fact that there are solvent–solvent interactions, solvent–solute interactions, and solute–solute interactions that all need to be considered.

When all three sets of interactions are similar, whether polar or nonpolar, dissolution will occur. This is often summarized simply in the expression "like dissolves like." Polar substances will dissolve in polar solvents, and nonpolar substances will dissolve in nonpolar solvents.

When ionic substances dissolve in water, water molecules surround the ions and neutralize their charges. When this occurs, the ions are said to be *solvated*, or, specific to water, *hydrated*.

Free-energy changes accompany solution formation: ΔS_{sol} will be positive, because disorder invariably increases when a solution is formed. ΔH_{sol} can be either positive or negative: some ionic compounds absorb heat when they dissolve and some release it.

ΔH_{sol} can be broken into three components, the sum of which will be the overall enthalpy change:

- Solvent–solvent interactions: these have a positive ΔH because solvent molecules must be separated and their intermolecular forces disrupted to make room for solute particles.
- Solute–solute interactions: these also have a positive ΔH, and for the same reason: the interactions between the solute particles have to be overcome. For ionic solids, this is proportional to the lattice energy.
- Solvent–solute interactions: these have a negative ΔH because the solvent molecules solvate the solute particles. This portion of the energy has a larger contribution in the case of smaller ions and larger charges.

Workbook Example 12.2

PROBLEM:
Which of the following would you expect to be soluble in hexane, a nonpolar solvent? Rank them in order from most soluble to least soluble.

$$KCl \qquad C_2H_6 \qquad CH_3OH$$

SOLUTION:
The polarity of the solutes will determine their solubility. Using "like dissolves like," the most nonpolar solute will be the most soluble. The least-polar solute here is C_2H_6, which has no dipole. Methanol (CH_3OH) is the next most soluble. It is able to hydrogen bond, but has a nonpolar component that may provide some solubility. The most-polar solute is KCl, which is unlikely to dissolve at all.

$$C_2H_6 \qquad > \qquad CH_3OH \qquad > \qquad KCl$$

Workbook Problem 12.2

Rank the following in order of increasing hydration energy (least negative to most negative):

$$Mg^{2+} \qquad Cl^- \qquad Li^+$$

> **Strategy:**   Charge and molecular size determine the hydration energy.

Section 12.3   Concentration Units for Solutions

There are four main methods for expressing the concentration of a solution, that is, the amount of solute per unit of solvent.

Method 1

$$Molarity = \frac{moles\ of\ solute}{Liter\ of\ solution}$$

The main advantage of this method is that stoichiometry calculations and titrations are greatly simplified by using moles. Disadvantages are temperature dependence (the volume of solvent will change slightly with increases and decreases in temperature) and the inability to determine solvent quantities without density.

Method 2

$$Mole\ fraction(\chi) = \frac{moles\ of\ component}{total\ moles\ in\ the\ solution}$$

The advantages of this method are temperature independence and convenience with calculations involving gases. This method is rarely used for liquid solutions because other methods are easier for calculations involving liquid solutions.

Method 3

$$mass\ percent(mass\ \%) = \frac{mass\ of\ component}{total\ mass\ of\ solution} \times 100\%$$

This method is useful for calculations involving small quantities, but density is needed to convert to molarity. Two variants on this method are often used for environmental calculations:

$$parts\ per\ million(ppm) = \frac{mass\ of\ component}{total\ mass\ of\ solution} \times 10^6$$

$$parts\ per\ billion(ppb) = \frac{mass\ of\ component}{total\ mass\ of\ solution} \times 10^9$$

For such dilute solutions, 1 kg solution = 1 L solution.

$$1\ ppm = \frac{1\ mg\ solute}{1\ L\ solution} \qquad\qquad 1\ ppb = \frac{1\ \mu g\ solute}{1\ L\ solution}$$

Method 4

$$molality = \frac{moles\ of\ solute}{kg\ of\ solution}$$

The temperature independence of this method makes it particularly well suited to measuring properties of solutions at temperature extremes, but it is difficult to use this method to measure the mass of liquids, and the density of the solution must be known in order to convert to molarity.

Workbook Example 12.3

PROBLEM:
A sample of water is found to have 65 ppm of arsenic. How many grams of arsenic will 1 L contain? What is the molarity of this solution?

SOLUTION:
A concentration of 1 ppm means that each liter of an aqueous solution contains 1 mg of solute. Therefore, the mass of arsenic in 1 L of this water is 65 mg. To calculate the molarity of this solution, we need to convert milligrams to moles.

$$65\,mg\,As \times \frac{1g}{1000\,mg} \times \frac{1\,mole\,As}{74.92\,g\,As} = 0.000868\,mol\,As = 868\,\mu mol\,As$$

Since we have 1 L of solution, the molarity is equal to 868 µM. Density is not needed to convert such dilute solutions from mass-based measures to molarity, because the solution will have the same density as water.

Workbook Example 12.4

PROBLEM:
A solution is made by dissolving 35.0 g of NaCl in 100 g of water. The resulting solution has a density of 1.16 g/mL. Determine the mass percent and molarity of the solution.

SOLUTION:
The mass percent of the solution is the mass of the solute divided by the total mass of the solution:

$$mass\,percent = \frac{35.0\,g}{135\,g} \times 100\% = 25.9\%\,NaCl$$

To convert to molarity, the mass of solute must be converted to moles and the solution converted to volume using the density:

$$35.0\,g \times \frac{1\,mol\,NaCl}{58.44\,g\,NaCl} = 0.599\,mol\,NaCl$$

$$\frac{135\,g}{1.16\,g/mL} = 116\,mL$$

$$M = \frac{0.599\,mol\,NaCl}{0.116\,L} = 5.16\,M\,NaCl$$

Workbook Problem 12.3

A solution is prepared by dissolving 15.43 g of glucose ($C_6H_{12}O_6$) in 300 g of water. The density of this solution is 1.09 g/mL. Calculate the mass percent, molality, and molarity of the solution.

Strategy:	Calculate mass percent of solution, volume of solution, and moles of solute to provide all units.
Step 1:	Determine mass percent.
Step 2:	Calculate the moles of solute and the molality of the solution.
Step 3:	Calculate the volume of the solution and the molarity of the solution.

Section 12.4   Some Factors That Affect Solubility

Solubility has natural limits: at some point, the solute molecules will be in equilibrium, meaning that they are going in and out of solution at an equal rate. When this occurs, the solution is said to be **saturated**.

Some solutes are more soluble at high temperatures. In this case, it is possible to produce a **supersaturated** solution with a greater-than-equilibrium amount of solute by creating a saturated solution at a high temperature and allowing it to cool undisturbed. These solutions are unstable.

The amount of solute that can be dissolved in a solvent to form a saturated solution at a given temperature is the **solubility** of that substance. A variation of solubility is **miscibility**: when the solution is a mixture of liquids, the liquids are sometimes soluble in one another regardless of proportions.

The temperature effect on solubility of solids is unpredictable, but gases become less soluble as the temperature of the solution increases. For example, this is why warm sodas quickly go flat; it is also why less oxygen dissolves in warmer lakes and rivers, making fish sensitive to thermal pollution.

Pressure does not affect the solubility of solids and liquids, but has a large effect on gases, which is described by **Henry's law**:

$$solubility = k \times P$$

where k is a constant related to the gas (with usual units of mol/L·atm), and P is the partial pressure of the gas. In contrast to other gas calculations, these measurements are reported at 25 °C.

Workbook Example 12.5

PROBLEM:
Determine the solubility of CO_2 (g) at 25 $^\circ$C and a partial pressure of 33.5 mmHg. Note: the k constant is equal to 0.0032 mol/L·atm.

SOLUTION:
Use Henry's law, being careful to convert units as needed.

$$solubility = 0.0032\frac{mol}{L \cdot atm} \times 33.5 \; mm \, Hg \times \frac{1 \, atm}{760 \; mm \, Hg} = 1.41 \times 10^{-4} \frac{mol}{L}$$

Workbook Problem 12.4

Given the Henry's law constant in the last problem (0.0032 mol/L·atm), how many liters of CO_2 will bubble out of a 500 mL soda after it is opened at 25 °C if it was previously under 2.3 atm of pressure with 100% CO_2 in the air, and it comes to equilibrium with the atmosphere where the partial pressure of CO_2 is only 0.3 mmHg?

Strategy: Calculate the initial and final amounts of CO_2 in a saturated solution; determine the number of moles and then the liters of CO_2.

Step 1:   Calculate the moles of dissolved CO_2 under pressure.

Step 2:   Calculate the moles of dissolved CO_2 after pressure is released and equilibrium with the atmosphere is reached.

Step 3:   Calculate the change in dissolved CO_2—this is the CO_2 that has been released.

Step 4:   Determine the volume of gas at 25 °C and 1 atm pressure.

Section 12.5   Physical Behavior of Solutions: Colligative Properties

Solutions behave in some ways differently than pure solvents. Properties that depend on the amount of dissolved solute but not its chemical identity are called **colligative properties**. They include

- Lowered vapor pressure
- Raised boiling point
- Lowered freezing point
- Osmosis: the migration of solvent and small molecules through a membrane.

These properties will be examined in detail in the following sections.

Section 12.6   Vapor-Pressure Lowering of Solutions: Raoult's Law

In a solution with a nonvolatile solute (one that has no vapor pressure of its own), the vapor pressure is lower than it would be in the pure solvent: the solution already has a higher entropy than the pure solvent, so the entropy gain in evaporating is reduced, and the vapor pressure is reduced.

Mathematically, this is described by Raoult's law:

$$P_{solution} = P_{solvent} \cdot X_{solvent}$$

where X is the mole fraction of the solvent in the mixture.

When calculating the mole fraction, the total number of moles of particles must be used for the solute. For a molecular solute, like glucose, this will be the same as the number of moles of molecules. But for an ionic solute, this will be the number of moles of ions generated.

Moles of ions generated are mathematically described by the **van't Hoff factor**, a multiplier that expresses the amount of dissociation of an ionic substance. Unfortunately, it is not as simple as 1 mol of NaCl breaking into 2 mol of ions: there is always some undissociated salt in a solution.

Just like the ideal gas law, Raoult's law applies to "ideal" solutions. It works best at low solute concentrations.

Workbook Example 12.6

PROBLEM:
Determine the vapor pressure of a solution made by dissolving 25.7 g of LiBr in 1.00 kg of water at 25 $^{\circ}$C. The vapor pressure of pure water at 25 °C is 23.76 mmHg. Assume complete dissociation.

SOLUTION:
Use Raoult's law. First, determine the mole fraction of water in the solution:

$$Moles\ of\ LiBr = 25.7\ g \times \frac{1\,mol\ LiBr}{86.85\,g\ LiBr} = 0.2959\,mol\ LiBr$$

$$\text{Moles } H_2O = 1000\,g \times \frac{1\,mol\,H_2O}{18.0\,g\,H_2O} = 55.56\,mol\,H_2O$$

$$\chi_{H_2O} = \frac{55.56\,mol\,H_2O}{(55.56\,mol\,H_2O + 0.2959\,mol\,LiBr)} = 0.995$$

Then multiply by the vapor pressure of pure solvent:

$$P_{solution} = 23.76\,mmHg \times 0.995 = 23.64\,mmHg$$

Workbook Example 12.7

PROBLEM:
How much LiBr would need to be dissolved in 1.00 kg water to lower the vapor pressure to 20 mmHg, assuming complete dissociation?

SOLUTION:
First, calculate the mole fraction of water:

$$20.0\,mmHg = 23.76\,mmHg \times \chi_{H_2O}$$

$$\chi_{H_2O} = 0.842$$

Next, determine the number of moles of solute needed for this mole fraction:

$$X_{H_2O} = \frac{55.56\,mol\,H_2O}{(55.56\,mol\,H_2O + x\,mol\,LiBr)} = 0.842$$

$$55.56\,mol = 0.842(55.56\,mol + x\,mol)$$

$$55.56 - 46.78 = 0.842x$$

$$x = 10.43\,mole\,ions = 5.21\,mol\,LiBr$$

$$5.21\,mol \times \frac{86.85\,g\,LiBr}{1\,mol\,LiBr} = 453\,g\,LiBr$$

As you can see, these effects are rather small: if 453 g of LiBr were dissolved in 1 kg of water, the solution would no longer be dilute enough for Raoult's law to apply.

Workbook Problem 12.5

Assuming complete dissociation for both, calculate the quantity of $CaCl_2$ and LiCl that are needed to lower the vapor pressure of 500 g of water by 3 mm Hg at 70 °C. The vapor pressure of water at this temperature is 233.7 mmHg.

Strategy: Determine the mole fraction of ions needed to lower the vapor pressure by this amount, then convert to grams of each ionic substance.

Step 1:   Calculate the mole fraction of water in a solution under these conditions.

Step 2:   Calculate the moles of dissolved ions that will lead to this mole fraction.

Step 3:   Convert to moles, then grams, of $CaCl_2$.

Step 4:   Convert to moles, then grams, of LiCl.

When two liquids with different vapor pressures are mixed, the total vapor pressure of the solution will be the combination of the vapor pressures of the two components. The vapor pressure of each component is calculated using Raoult's law, based on the mole fraction of each liquid:

$$P_{\text{total}} = P_A + P_B$$

The pressures of the individual components are calculated using Raoult's law:

$$P_A = \chi_A \left(P_A^{\circ} \right) \qquad P_B = \chi_B \left(P_B^{\circ} \right)$$
$$P_{\text{total}} = \chi_A \left(P_A^{\circ} \right) + \chi_B \left(P_B^{\circ} \right)$$

Workbook Example 12.8

PROBLEM:
The vapor pressure of water at 25 °C is 23.8 mmHg. The vapor pressure of ethyl alcohol (C_2H_5OH) at 25°C is 61.2 mmHg. What is the vapor pressure of a solution prepared by combining 50 g of each liquid?

SOLUTION:
Calculate the moles of each to determine the mole fractions:

$$50\,g\,C_2H_5OH \times \frac{1\,mol\,C_2H_5OH}{46.0\,g\,C_2H_5OH} = 1.09\,mol\,C_2H_5OH$$

$$50\,g\,H_2O \times \frac{1\,mol\,H_2O}{18.0\,g\,H_2O} = 2.78\,mol\,H_2O$$

$$\chi_{C_2H_5OH} = \frac{1.09\,mol\,C_2H_5OH}{\left(1.09\,mol\,C_2H_5OH + 2.78\,mol\,H_2O\right)} = 0.282$$

$$\chi_{H_2O} = \frac{2.78\,mol\,H_2O}{\left(1.09\,mol\,C_2H_5OH + 2.78\,mol\,H_2O\right)} = 0.718$$

$$P_{total} = \left(0.282 \times 61.2\,mmHg\right) + \left(0.718 \times 23.8\,mmHg\right) = 34.35\,mmHg$$

Workbook Problem 12.6

The label on a bottle of wine states that it is 13% ethanol. Using the vapor pressures listed above, what is the pressure above a 150 g glass of wine, ignoring the fact that other volatile components are present in the wine.

Strategy: Determine the mole fraction of each component, and calculate the partial pressure of each.

Step 1: Solve for the mass and moles of each component.

Step 2: Determine the mole fraction of each component.

Step 3:   Determine the partial pressure of each component and the total pressure.

Section 12.7   Boiling-Point Elevation and Freezing-Point Depression of Solutions

Another aspect of the reduced vapor pressure of a solution with a nonvolatile solvent is the raised boiling point. If the vapor pressure is being held down, and the solution boils when the vapor pressure is equal to the external pressure, the vapor pressure will reach the external pressure at a different temperature.

This is seen in the phase diagram of a solution, in which all of the lines are shifted down: the boiling point is raised and the freezing point lowered by the addition of solutes. As in the case of all colligative properties, it is the number of particles that causes the effect.

The fundamental reason for this is the same as for vapor-pressure depression: the increase in the entropy of the solution makes phase changes less favorable. More enthalpy needs to be added to get the liquid to boil, and more enthalpy needs to be removed to get the liquid to freeze.

Mathematically, this is described with the following equations:

$$\Delta T_b = K_b m \qquad\qquad \Delta T_f = K_f m$$

For these equations, K_b is the **molal boiling-point elevation constant** and K_f is the **molal freezing-point depression constant**. Both are dependent on the *solvent*, not the identity of the particles. m is the molal concentration of solute particles, whether they are ions or molecules. Molality rather than molarity is used for these calculations, because temperature extremes affect the volume of solutions.

The actual dissociation of ionic compounds can also be taken into account by adding in the van't Hoff factor for the compound:

$$\Delta T_b = K_b\, m\, i \qquad\qquad \Delta T_f = K_f\, m\, i$$

Workbook Example 12.9

PROBLEM:
What will be the freezing point and boiling point of an aqueous solution containing 55.0 g of KCl in 250 g of water? $K_b(H_2O) = 0.51\ ^{\circ}C/m$ and $K_f = 1.86\ °C/m$. Assume complete dissociation.

SOLUTION:
Determine the molality of the solution, then calculate the boiling-point elevation and the freezing-point depression:

$$55.0 \, g \, KCl \times \frac{1 \, mol \, KCl}{74.55 \, g \, KCl} \times \frac{2 \, mol \, ions}{1 \, mol \, KCl} = 1.48 \, mol \, ions$$

$$\frac{1.48 \, mol \, ions}{0.250 \, kg} = 5.92 \, m$$

$$\Delta T_b = \left(0.51 \, {}^{\circ}C/m\right)(5.92 \, m) = 3.02 \qquad\qquad 100 \, {}^{\circ}C + 3.02 \, {}^{\circ}C = 103 \, {}^{\circ}C$$

$$\Delta T_b = \left(1.86 \, {}^{\circ}C/m\right)(5.92 \, m) = 11.0 \qquad\qquad 0 \, {}^{\circ}C - 11.0 \, {}^{\circ}C = -11 \, {}^{\circ}C$$

Workbook Problem 12.7

How many grams of $(NH_4)_2SO_4$ need to be added to 300 g of H_2O so that the freezing point of the solution is lowered to $-15.0 \, {}^{\circ}C$? $K_f = 1.86 \, {}^{\circ}C/m$. Assume complete dissociation.

Strategy: Using the equation for freezing-point depression, solve for the molality of ions needed, then convert to moles and grams of ammonium sulfate.

Step 1: Determine the molality of ions needed to generate this freezing-point depression.

Step 2: Determine the moles of $(NH_4)_2SO_4$ needed to generate this quantity of ions.

Step 3: Convert to grams of ammonium sulfate needed.

Section 12.8   Osmosis and Osmotic Pressure

Cell membranes, along with some manufactured membranes, are *semipermeable*, meaning that solvent molecules and some very small molecules can pass freely through the membrane but large molecules and ions are blocked.

In the process of **osmosis**, solvent molecules pass preferentially from the side of low-solution concentration to the side of high-solution concentration. As a result, the level of the low concentration or pure solvent will drop, and the level of the high-concentration solution will rise.

This is an entropy-driven process: the dilution of a solution increases the disorder of the solution. The process will continue until the external pressure and the **osmotic pressure** are equal.

Osmotic pressure is calculated with the following equation:

$$\Pi = MRT$$

Where Π is the osmotic pressure, M is the molarity of solute particles in solution, and R and T are the gas law constant and the temperature in kelvins. Molarity is used for these calculations because the temperature is a variable as well.

Workbook Example 12.10

PROBLEM:
Determine the osmotic pressure of a 0.050 M solution of sodium chloride at 24.3 °C.

SOLUTION:

$$\Pi = \left(\frac{0.05\,mol}{L} \times \frac{2\,mol\,ions}{1\,mol\,NaCl} \right) \left(0.08206 \frac{L \cdot atm}{K \cdot mol} \right) (273.15 + 24.3\,K) = 2.44\,atm$$

Workbook Problem 12.8

A solution containing an unknown substance has an osmotic pressure of 7.34 atm at 25 °C. What is the molarity of the solution?

Strategy: Using the equation for osmotic pressure, solve for the molarity of the solution.

Colligative properties have some very practical uses. For example, spreading a salt such as calcium chloride, which dissociates into three particles in an exothermic reaction, thus lowering the freezing point of the snow, is a particularly efficient way to de-ice a sidewalk and keep it clear.

The coolant added to an automobile engine both raises the boiling point of the coolant and lowers its freezing point, maintaining it as a liquid over a greater range of temperatures.

The de-icer sprayed on airplane wings operates on the same principle: maintaining the liquid range of water to prevent ice build-up.

Desalination of seawater can be accomplished through **reverse osmosis**: putting more than 30 atm of pressure on seawater on one side of a semi-permeable membrane will force pure water through the membrane until the osmotic pressure of the increasingly concentrated seawater is equal to the external pressure.

Any of the colligative properties can be used to determine molecular weights, but osmotic pressure, as the effect is so large, can be used most accurately.

Workbook Problem 12.9

A solution is prepared from 23 mg of a compound in 100 mL of aqueous solution. The osmotic pressure of the solution is 8.428 mmHg at 25 °C. What is the approximate molecular weight of the compound?

Strategy: Using the equation for osmotic pressure, calculate the molarity of the solution, and from that the molecular weight of the compound.

Step 1:   Calculate the molarity of the solution.

Step 2:   Calculate the number of moles in the solution.

Step 3:   Determine the molecular weight.

Section 12.9 Fractional Distillation of Liquid Mixtures

Purification of hydrocarbons is accomplished using a method known as **fractional distillation**. When a group of volatile liquids is heated, the vapor component is enriched in the fraction with the lowest boiling point. Vapor is condensed by cooling, providing an enriched fraction of the lowest boiling-point component. Repeating the heating and cooling cycles allows for purification of the components of the volatile mixtures: this occurs naturally in a distillation column.

Putting It Together

Esters, organic compounds containing C, H, and O, often have pleasant odors. Butyl butanoate has the odor of pineapples. A 0.83 g sample of butyl butanoate was subjected to combustion analysis, which produced 2.026 g of CO_2 and 0.8288 g of H_2O. A 1.50 g sample was dissolved in enough solvent to make 250 mL of solution. The osmotic pressure of this solution at 25 °C was 1.02 atm. Determine the molecular formula of butyl butanoate.

Self–Test

This section is intended to test your knowledge of the material covered in this chapter. Think through these problems, and make certain you understand what they are asking. Make sure your answers make sense. Successful completion of these problems indicates that you have mastered the material in this chapter. You will receive the greatest benefit from this section if you use it as a mock exam, as this will allow you to determine which topics you need to study in more detail.

True–False
1. A solution of metals is called an *alloy*.

2. Entropy always increases when a solution is formed, but enthalpy can increase or decrease.

3. Molality is the most practical unit of concentration for environmental measurements.

4. Potassium nitrate will dissolve readily in oil.

5. The density of a solution is needed to convert from mass percent to molarity.

6. Supersaturated solutions are stable at low temperatures.

7. Increasing pressure causes ionic solids to become more soluble.

8. The more concentrated a solution, the lower the vapor pressure.

9. The more concentrated a solution, the higher the freezing point.

10. Osmotic pressure can be overcome with high external pressures.

Multiple Choice
11. When a gas or solid is dissolved in a liquid, the solute is
    a. the dissolved substance
    b. the liquid
    c. the major component
    d. the minor component

12. ΔH_{soln} will be exothermic if
    a. ΔH for solvent–solvent interactions is negative
    b. ΔH for solute–solute interactions is negative
    c. ΔH for solute–solvent interactions is negative
    d. the sum of the three types of interactions leads to a negative ΔH

13. Benzene, a nonpolar organic compound, is most likely to dissolve in
   a. water   b. CH_3CH_2OH c. CCl_4 d. NH_3

14. When the temperature of a solvent is lowered, the solubility of gases
   a. decreases   b. increases
   c. is not affected   d. cannot be generally predicted

15. A mixture that contains particles large enough to be visible with a low-power microscope is a
   a. heterogeneous mixture   b. solution
   c. suspension   d. colloid

Matching

16. Colligative property   a. a technique for purifying hydrocarbon mixtures

17. Colloid   b. passage of solvent from a less-concentrated to a more-concentrated solution

18. Fractional distillation   c. a solution holding the maximum quantity of solute

19. Heat of solution   d. able to mix together in any proportions.

20. Miscible   e. the liquid portion of a solution

21. Molality   f. a homogeneous mixture containing large particles that do not settle out

22. Osmosis   g. allows passage of small molecules only, not ions or large molecules

23. Semipermeable membrane   h. a solid or gas that is dissolved in a liquid

24. Saturated   i. homogeneous mixture containing very small molecules or ions

25. Solubility   j. a solution that has been heated to dissolve more solute, then cooled: unstable

26. Solute   k. property that depends on the amount of dissolved solute, not its identity

27. Solution   l. measure of concentration that is moles per kilogram of solvent

28. Solvent   m. enthalpy change, positive or negative, that occurs when a solution forms

29. Supersaturated   n. the amount of solute that a solution can contain at a specific temperature

Fill-in-the-Blank

30. Solution formation has two endothermic elements: disrupting _____

    interactions and disrupting _____ interactions both require energy.

31. ΔH_{soln} is _____ due to _____

    _____.

32. To convert between mass-based units and mole-based units requires that the

    _____ of a solution be known.

33. Gases become _____ soluble as the temperature decreases.

34. Adding a nonvolatile solute to a solution causes the vapor pressure to _____,

    the boiling point to _____, the freezing point to _____, and

    _____ _____ to develop across semipermeable membranes.

35. Colligative properties can be used to determine the _____ _____ of

    an unknown solute.

36. Dissolved ions that are surrounded and stabilized by a shell of solvent molecules are said to

    be _____. If water is the solvent, the term used is _____.

37. The spontaneous dissolution of ionic solids can be either _____ or

    _____, depending on the energetics of solution formation.

Problems

38. Rank the following in order of increasing solubility in benzene, a nonpolar solvent.

    I_2 NaCl $CH_3(CH_2)_2CH_2OH$

39. If 28.4 g of sucrose ($C_{12}H_{22}O_{11}$) are dissolved in 350 g of water, the final volume is 374 mL. Calculate the concentration of the solution in terms of: (a) mass percent, (b) mole fraction, (c) molality, and (d) molarity.

40. Household hydrogen peroxide is listed on the label as 3% H_2O_2. What is the molarity of this solution if the density of the solution is 1.01 g/mL?

41. A sample of industrial effluent is found to have an arsenic concentration of 15 ppb. What is the molarity of this solution? How many grams of arsenic would be found in 100,000 L of this effluent?

42. Goldfish prefer cold water to warm. If the mole fraction of oxygen in air is 0.21, what is the oxygen concentration in water at 25 °C? The Henry's law constant for oxygen is 1.32×10^{-3} mol/L · atm.

43. The Henry's law constant for nitrogen is 6.25×10^{-4} mol/L·atm. Calculate the solubility of nitrogen if its partial pressure above water is 976 mm Hg.

44. The vapor pressure of pure water is 42.175 mmHg at 35 °C. Calculate the vapor pressure of a solution made with 4.5 g of glucose ($C_6H_{12}O_6$) and 55 g of H_2O.

45. The vapor pressure of water at 25 °C is 23.8 mmHg. Calculate the vapor pressure of an aqueous solution that contains 5.8 g $FeCl_3$ and 72.1 g of water at this temperature. The van't Hoff factor for $FeCl_3$ is 3.4.

46. The vapor pressure of benzene, (C_6H_6) at 25 °C is 93.4 mmHg. The vapor pressure of toluene ($C_6H_5CH_3$) is 26.9 mmHg at 25 °C. Calculate the vapor pressure of a solution prepared from 10.0 g of benzene and 10.0 g of toluene.

47. The normal boiling point of carbon tetrachloride (CCl_4) is 76.7 °C. How many grams of naphthalene ($C_{10}H_8$) need to be added to 1.0 kg of carbon tetrachloride to raise the boiling point of the solution to 80.0 °C? K_b for CCl_4 is 5.03 °C/m.

48. Calculate the boiling point and freezing point of a solution prepared by mixing 10.0 g NaCl with 90.0 g of water. Assume complete dissociation. $K_b = 0.51$ °C/m, $K_f = 1.86$ °C/m

49. A solution was prepared by mixing 75.0 g of benzene (freezing point = 5.5 °C) and 2.50 g of an unknown substance. The freezing point of the solution was 3.5 °C. Calculate the molar mass of the unknown substance. $K_f = 5.12$ °C/m.

50. Calculate the osmotic pressure of a solution at 15 °C containing 18.0 g of glucose ($C_6H_{12}O_6$) in 1.2 L of solution.

51. Calculate the molar mass of 0.250 g of a peptide in 1 L of water at 15 °C that has an osmotic pressure of 16.4 mmHg.

Challenge Problem
52. The vapor above a room-temperature mixture of pentane and hexane is 43.2% pentane by mass. What is the mass percent composition of the solution if the vapor pressure of pure pentane is 425 mmHg and the vapor pressure of hexane is 151 mmHg at room temperature? What is the total pressure of the solution?

CHAPTER THIRTEEN

Chemical Kinetics

Learning Objectives

As a result of reading and studying this chapter, you should be able to

Section 13.1   **Reaction Rates**
1. Determine an average reaction rate over a specified period of time and estimate an instantaneous reaction rate from a graph of concentration versus time.
2. Relate the rate of consumption of any reactant to the rate of formation of any product using reaction stoichiometry.

Sections 13.2–13.3 Rate Laws, Reaction Order, and Method of Initial Rates
3. Find the order with respect to each reactant, the overall order of a reaction, and the units of the rate constant given the rate law.
4. Predict the change in reaction rate when the concentration of a reactant changes by a specified amount.
5. Determine the rate law and rate constant using initial rate and concentration data.

Sections 13.4–13.6 Integrated Rate Laws for Zeroth-, First-, and Second-Order Reactions
6. Use the integrated rate law to determine the half-life and the concentrations remaining at various times for a zeroth-order reaction.
7. Use the integrated rate law to determine the half-life and the concentrations remaining at various times for a first-order reaction.
8. Use the integrated rate law to determine the half-life and the concentrations remaining at various times for a second-order reaction.
9. Determine a reaction order and rate constant graphically.

Sections 13.7–13.8 Reaction Rates and Temperature: The Arrhenius Equation
10. Interpret potential energy diagrams and suggest geometries for successful collisions and transition states using collision theory.
11. Calculate rate constants, temperatures, and activation energy using the Arrhenius equation.

Sections 13.9–13.11 Reaction Mechanisms and Rate Laws for Elementary Reactions
12. Given a reaction mechanism, write the overall reaction, identify intermediates, and determine the molecularity for each elementary step.

Section 13.12   **Catalysis**
13. For mechanisms with an initial slow step (a) predict the rate law for the overall reaction or (b) propose a mechanism that agrees with the experimental rate law.

14.   For mechanisms with an initial fast equilibrium step, predict the rate law.

Section 13.13   **Homogeneous and Heterogeneous Catalysts**

15.   Use a molecular diagram for a catalyzed reaction to determine the rate law and propose a mechanism consistent with the rate law.

16.   Identify catalysts and intermediates in the mechanism of a catalyzed reaction.

17.   Interpret a potential energy diagram representing the mechanism in a catalyzed reaction.

Chapter Summary

In this chapter, you will learn about chemical kinetics: the study of reaction rates and mechanisms. The appearance of products can be related to the disappearance of reactants, and reaction orders can be determined from experimental data and plots of concentrations versus time in both logarithmic and non-logarithmic forms. Integrated rate laws can be used to calculate the concentration of a reactant at any given time point, the fraction of reactant remaining, or the time required for the reaction to progress to a certain point. Radioactive decay follows first-order kinetics with a decay constant in place of a reaction constant. This makes it possible to determine how long an isotope has been decaying, determine how long it will take for a specific amount or proportion of decay to occur, or solve for either half-life or decay constant given experimental data. Kinetics make it possible to identify a rate-determining step and propose a reaction mechanism. You will also learn how to use plots of concentration versus time to determine a reaction order. Kinetic molecular theory and the Arrhenius equation make it possible to derive equations to determine activation energy and rate constants at different temperatures. Catalysts, both heterogeneous and homogeneous, and their mechanisms are also discussed.

The Chapter in Detail

Section 13.1   Reaction Rates

Reaction rates are determined by how much the concentration of a reactant or product changes per unit of time, using the following equation.

$$rate = \frac{\Delta\, concentration}{\Delta\, time}$$

The concentration of reactants decreases and the concentration of products increases as time passes. As always, changes are calculated as final – initial, so for a disappearing reactant, the sign on the rate will be negative and for an appearing product, the rate will be positive.

The units associated with these rates are M/s or mol/(L · s), which allows the calculated rate to be independent of the scale of the reaction. Part of the unit must be the reactant or product on which it is based.

The relative rates for different participants in the reaction will be related by the coefficients in the balanced equation. To use a very simple example:

$$2\ H_2\ (g) + O_2\ (g) \rightarrow 2\ H_2O\ (l)$$

In this reaction, hydrogen gas will disappear at the same rate that water is formed, but oxygen will disappear at half the rate of hydrogen disappearance and water formation. This leads to a number of possible rates. To avoid this ambiguity, chemists define a *general rate* as:

$$general\ rate\ of\ reaction = \frac{change\ in\ concentration}{coefficient}$$

So, for the above reaction:

$$general\ rate\ of\ reaction = -\frac{1}{2} \times \frac{\Delta[H_2]}{\Delta t} = -\frac{\Delta[O_2]}{\Delta t} = \frac{1}{2} \times \frac{\Delta[H_2O]}{\Delta t}$$

When concentration is plotted versus time, we see average rates over a specific time period, but *instantaneous rates* can also be seen. An instantaneous rate is defined by the slope of a tangent to a curve that touches the curve at a specific time point. The instantaneous rate when time = 0, the very beginning of the reaction, is called the *initial rate*.

Workbook Example 13.1

PROBLEM:
In the following reaction:

$$4\ NH_3\ (g) + 3\ O_2\ (g) \rightarrow 2\ N_2\ (g) + 6\ H_2O\ (g)$$

the rate of formation of H_2O is 0.73 M·s^{-1} at 300 s. What is the rate of disappearance of NH_3?

SOLUTION:
With the rate of formation of water at , the stoichiometry of the balanced equation can be used to determine the rate of disappearance of NH_3.

$$\frac{0.73\ mol\ H_2O}{L\Delta s} \times \frac{4\ mol\ NH_3}{6\ mol\ H_2O} = \frac{0.49\ mol\ NH_3}{L\Delta s}$$

This rate is reported with a minus sign in front of it, –0.49 M·s^{-1}, because we are reporting the rate of *disappearance* of the reactant NH_3.

Workbook Problem 13.1

Given the following unbalanced equation:

$$HBr\ (g) \rightarrow H_2\ (g) + Br_2\ (g)$$

express the general rate of reaction with respect to all reactants and products. If the initial rate of disappearance of HBr is –3.26 × 10^{-3} M/s, what is the rate of appearance of bromine gas?

Strategy: The stoichiometry of the reaction allows relative reaction rates to be determined.

Step 1:   Balance the equation.

Step 2:   Write the reaction rates.

Step 3:   Determine the rate of formation of bromine from the rate of disappearance of HBr.

Section 13.2   Rate Laws and Reaction Order

The rate law for a reaction indicates its dependence on the concentration of each reactant. For a general reaction:

$$a\,\text{A} + b\,\text{B} \rightarrow \text{products,}$$
$$rate = k[A]^n[B]^m$$

In a **rate law**, k is a proportionality constant called the **rate constant**.

The exponents m and n must be experimentally determined because they are unrelated to the coefficients in the chemical reaction. However, these coefficients determine the **reaction order**, which is the sum of the exponents. So, a reaction in which $m + n = 2$ is said to be second order, a reaction in which $m + n = 3$ is said to be third order, and so on.

The exponents themselves are usually small positive integers, but in complex reactions they can be negative, zero, or even fractions.

- exponent = 1: rate depends linearly on the concentration of the reactant
- exponent = 0: rate is independent of the concentration of the reactant
- negative exponent: rate decreases as the concentration of the reactant increases

Workbook Problem 13.2

What are the units for zeroth-order, first-order, and second-order rate constants if the rate is expressed as M/s?

Strategy: Determine the units needed to cancel out to the rate of each reaction.

Step 1:   Zeroth-order reaction:

Step 2:   First-order reaction:

Step 3:   Second-order reaction:

Section 13.3   Method of Initial Rates: Experimental Determination of a Rate Law

Determining the reaction order experimentally is accomplished by measuring the initial rate of a reaction with differing initial concentrations. This is done carefully, so as to make it possible to see what effect changing each reaction has independently.

The following are the most common results:
- Doubling the concentration has no effect: the reaction is zero-order in that reactant.
- Doubling the concentration doubles the rate: the reaction is first order in that reactant.
- Doubling the concentration quadruples the rate: the reaction is second-order in that reactant.
- Doubling the concentration causes rate to increase eightfold: the reaction is third order in that reactant.

Initial rates are always used for these determinations because the reverse reaction will begin to occur, even if only slightly, as the reaction continues.

Only reactants and catalysts appear in the rate law.

Once the rate law is determined, it can be used to solve for the rate constant. Rate constants are temperature-dependent and are a characteristic of a reaction. The units of k, depend on the order of the reaction.

Workbook Example 13.2

PROBLEM:
For the following theoretical reaction:

$$2\,A + 3\,B \rightarrow A_2B_3$$

the following measurements were made:

Initial [A] (M)	Initial [B] (M)	Rate of formation of A_2B_3
0.1	0.1	$1\ M \cdot s^{-1}$
0.2	0.1	$2\ M \cdot s^{-1}$
0.1	0.2	$1\ M \cdot s^{-1}$

Determine the rate law, reaction order, and rate constant.

SOLUTION:
What this shows is that doubling the concentration of A doubles the rate, while doubling the concentration of B does not affect the rate.

From this, the rate law can be determined as

$$rate = k[A]^1 [B]^0 = k[A]$$

This is a first-order reaction.

To solve for k, the rate expression must be rearranged:

$$k = \frac{rate}{[A]} = \frac{1\,M \cdot s^{-1}}{0.1\,M} = 10\,s^{-1}$$

The unit of s^{-1} is characteristic of a first-order reaction.

Workbook Example 13.3

PROBLEM:
The following reaction

$$2\ NO\ (g) + 2\ H_2\ (g) \rightarrow N_2\ (g) + 2\ H_2O\ (g)$$

is first order in H_2 and second order in NO. Determine the rate law and the reaction order. How does the reaction rate change if the concentration of NO is doubled and the concentration of H_2 is held constant? How does the reaction rate change if the concentration of H_2 is cut in half and the concentration of NO is held constant?

SOLUTION:
Based on the information given, the rate law is

$$rate = k[NO]^2 [H_2]$$

The reaction is third order.

If the concentration of NO is doubled while the concentration of H_2 is held constant, the rate will quadruple. If the concentration of H_2 is halved, the rate will also be halved.

Workbook Problem 13.3

The following data were collected for this reaction:

$$2 NO_2 (g) + F_2 (g) \rightarrow 2NO_2F (g)$$

Exp	$[NO_2]$ (M)	$[F_2]$ (M)	Rate (M·s^{-1})
1	0.10	0.10	0.026
2	0.20	0.10	0.051
3	0.20	0.20	0.103

Determine the rate law from this data. What is the order of the reaction with respect to each reactant? What is the overall order of the reaction? Calculate the value of k.

Strategy: The rate law for the reaction is

$$rate = k[NO_2]^m [F_2]^n$$

To find m and n, compare the change in concentration with the change in rate. Verify your answer by using the data from any of the three experiments.

Step 1:   Determine the order of the reaction with respect to NO_2.

Step 2:   Determine the order of the reaction with respect to F_2.

Step 3:   Write the equation for the rate law.

Step 4:   Calculate the value of k.

Section 13.4   Integrated Rate Law: Zeroth-Order Reactions

A zeroth-order reaction is independent of the concentration of the reactants. Zeroth-order reactions are relatively uncommon and have the equation:

$$rate = -\frac{\Delta[A]}{\Delta t} = k[A]^0 = k$$

The Integrated rate law is

$$[A] = -kt + [A]_0$$

A plot of [A] versus time is a straight line for a zeroth-order reaction where the slope $= -k$, and the y-intercept $= [A]_0$.

Workbook Problem 13.4

A general zeroth-order reaction gave the following rate data:

Exp	[A] (M)	Rate (M·s^{-1})
1	0.10	0.026
2	0.20	0.026
3	0.40	0.026

Determine the rate constant and the amount of reactant remaining after 20 seconds with an initial concentration of 1.0 M.

Strategy: Substitute into the rate law for zeroth-order reactions, solve for [A] after 20 seconds.

Step 1: Solve for k.

Step 2: Substitute into the integrated rate law and solve for [A] after 20 seconds.

Section 13.5   Integrated Rate Law: First-Order Reactions

Using calculus, it is possible to convert a simple first-order rate law:

$$rate = -\frac{\Delta[A]}{\Delta t} = k[A]$$

to an integrated form:

$$\ln\frac{[A]_t}{[A]_0} = -kt$$

This is enormously valuable, because this concentration–time format allows us to make determinations about reaction progress.

Graphically, a plot of ln [A] versus time gives a straight line if the reaction is first order in A. The slope of the line will be equal to –k. This is indicative of a first-order reaction: a concentration vs. time plot will be curved, but a plot of ln A vs. time will be straight.

Workbook Example 13.4

PROBLEM:
When sucrose reacts with water, glucose is formed according to the reaction

$$C_{12}H_{22}O_{11} + H_2O \rightarrow 2\ C_6H_{12}O_6$$

This reaction follows first-order kinetics with respect to the sucrose. Calculate the value of k if it takes 3.00 hr for the concentration of sucrose to decrease from 0.0500 M to 0.0442 M. Determine the amount of time required for the reaction to go to 95% completion.

SOLUTION:
To calculate the value of k, we simply substitute the data given into the first-order integrated rate equation.

$$\ln\frac{0.0442}{0.0500} = -k(3.00\,h)$$
$$k = 4.11\times10^{-2}\ h^{-1}$$

To determine the amount of time needed for the reaction to go to 95% completion determine the concentration of sucrose at that point. The amount of sucrose remaining at 95% completion is 5% of the initial concentration:

$$0.05 \times 0.0500\ M = 0.0025\ M$$

This can be substituted into the integrated rate law with the rate constant previously calculated:

$$\ln\frac{0.0025}{0.0500} = -\left(4.11\times10^{-2}\ h^{-1}\right)t$$
$$t = 72.9\ h$$

Workbook Problem 13.5

In a first-order reaction, the concentration of the reactant goes from 0.100 M to 0.030 M over the course of 73 min. What is k, and how long will it take for the reaction to go to 80% completion?

Strategy: Substitute into the integrated rate law for a first-order reaction, then use k to solve for the time to 80% completion.

Step 1:   Solve for k.

Step 2:   Determine the concentration of the reactant at 80% completion.

Step 3:   Solve for *t* at 80% completion.

The half-life of a reaction is the time it takes for the reactant concentration to drop to one-half its initial value. From the integrated rate law:

$$\ln \frac{1}{2} = -k\, t_{1/2}$$
$$-0.693 = -k\, t_{1/2}$$
$$t_{1/2} = \frac{0.693}{k}$$

For a first-order reaction, half-life is a constant.

Workbook Example 13.5

PROBLEM:
If the reaction of sucrose and water has a rate constant of -4.11×10^{-2}, what is the half-life for the reaction?

SOLUTION:
Substitute the rate constant into the equation for the half-life of a first-order equation.

$$t_{1/2} = \frac{0.693}{4.11 \times 10^{-2}\, h^{-1}} = 16.9\, h$$

Workbook Problem 13.6

What is the half-life for the reaction described in workbook problem 13.5?

Strategy: Substitute into the half-life equation.

Section 13.6   Integrated Rate Law: Second-Order Reactions

Second-order reactions are dependent either on the concentration of a single reactant raised to the second power, $[A]^2$, or on two first-order reactants, $[A][B]$.

The rate law for a reaction of the simpler type is

$$rate = -\frac{\Delta[A]}{\Delta t} = k[A]^2$$

The integrated rate law for a second-order reaction is

$$\frac{1}{[A]_t} = kt + \frac{1}{[A]_0}$$

The half-life equation (taking the integrated rate law and set $[A]_t = 1/2 \; [A]_0$) is

$$t_{1/2} = \frac{1}{k[A]_0}$$

The half-life for a second-order reaction depends on both the rate constant and the initial concentration: as a result, each half-life for a second-order reaction is twice as long as the previous half-life.

A second-order reaction can be identified by the fact that a plot of $\frac{1}{[A]_t}$ versus time is a straight line; the slope is equal to the rate constant (k), and the y-intercept is equal to $\frac{1}{[A]_0}$.

Workbook Example 13.6

PROLBEM:
The reaction 2 NOBr (g) → 2 NO (g) + Br$_2$ (g) is a second-order reaction with respect to NOBr. The rate constant for this reaction is $k = 0.810 \; M^{-1} \cdot s^{-1}$ when the reaction is carried out at a temperature of 10 °C. If the initial concentration of NOBr = 8.3×10^{-3} M, how much NOBr will be left after a reaction time of 50 min? What is the half-life of this reaction?

SOLUTION:
We can solve for the amount of NOBr after 50 min by substituting into the integrated rate law for a second-order reaction.

$$\frac{1}{[NOBr]} = \left(0.81 \, M^{-1}s^{-1}\right)\left(50 \, min \times \frac{60 \, s}{1 \, min}\right) + \frac{1}{8.3 \times 10^{-3} \, M} = 2550 \, M^{-1}$$

$$[NOBr] = 3.92 \times 10^{-4} \, M$$

To determine the half-life, again substitute into the equation for the half-life of a second-order reaction:

$$t_{1/2} = \frac{1}{\left(0.810 \, M^{-1}s^{-1}\right)\left(8.3 \times 10^{-3}\right)} = 149 \, s$$

Workbook Problem 13.7

The following data were collected for the second-order reaction $NO_2 (g) \rightarrow NO (g) + O (g)$

Time (s)	0	100	200	300	400
$[NO_2]$	0.1000	0.0797	0.0662	0.0567	0.0495

Determine the rate constant, the time required for the reaction to reach 50% completion, and the time required to reach 95% completion.

Strategy: Use the integrated second-order rate law and experimental data to determine the rate constant, then solve for the first half-life (50% completion) and then for 95% completion using the integrated second-order rate law.

Step 1: Solve for k

Step 2: Use the equation for half-life to determine the time required for 50% completion.

Step 3: Determine the concentration of NO_2 at 95% completion (5% of NO_2 remaining).

Step 4: Substitute the concentrations into the integrated rate law, and solve for t.

Section 13.7   Reaction Rates and Temperature: The Arrhenius Equation

Reaction rates tend to double when reaction temperature is increased by 10 °C. This is consistent with the **collision theory** model of reactions. Increasing the kinetic energy of a reaction system by raising the temperature increases the energy with which the molecules collide, increasing the chances of a successful reaction.

Molecules must be correctly oriented at collision in order to produce a reaction, and an intermediate must form that is an aggregate of all the reactants. This aggregate is known as the **transition state**, and the energy required to form it is called the **activation energy**.

Evidence for the concept of an activation energy barrier comes from kinetic–molecular theory: only a tiny proportion of gas-phase collisions result in reactions. The proportion of the collisions that do result in reactions is described by the formula

$$f = e^{-\frac{E_a}{RT}}$$

As the temperature increases in a reaction, more molecules are available to collide at the energies necessary to generate a reaction. In fact, the distribution of these molecules according to both the kinetic–molecular theory and the dependence of reaction rate on temperature are exponential.

Also involved is the orientation of the reactants when they collide: the proportion of molecules colliding with the proper orientation is the **steric factor (p)**.

These factors are all brought together in the Arrhenius equation:

$$k = A e^{-\frac{E_a}{RT}}$$

The parameter A is called the **frequency factor,** or the pre-exponential factor. From this equation, the rate constant decreases as the activation energy (E_a) increases and increases as T increases.

Section 13.8   Using the Arrhenius Equation

If the rate constant is known at two different temperatures, the natural log of the equation can be taken:

$$\ln k = \ln A e^{-\frac{E_a}{RT}}$$

A plot of $\ln k$ versus $1/T$ gives a straight line, with a slope equal to $-E_a/R$. This equation can also be rearranged in order to calculate the activation energy:

$$\ln \frac{k_2}{k_1} = -\frac{E_a}{R}\left(\frac{1}{T_2} - \frac{1}{T_1}\right)$$

Workbook Example 13.7

PROBLEM:
The activation energy for the reaction $ClO_2F\ (g) \rightarrow ClOF\ (g) + O\ (g)$ is 186 kJ/mol. If the value of k is $6.76 \times 10^{-4}\ s^{-1}$ at 322 °C, what is the value of k at 150 °C?

SOLUTION:
Convert all temperatures to kelvins, and substitute into the Arrhenius equation:

$$\ln \frac{k_2}{6.76 \times 10^{-4}\ s^{-1}} = -\frac{1.86 \times 10^5\ J/mol}{8.314\ J/K \cdot mol}\left(\frac{1}{423\ K} - \frac{1}{595\ K}\right)$$

$$\ln \frac{k_2}{6.76 \times 10^{-4}\ s^{-1}} = -15.3$$

Taking the antilog of both sides gives us

$$e^{\ln\frac{k_2}{6.76\times10^{-4}\ s^{-1}}} = e^{-15.3}$$

$$\ln\frac{k_2}{6.67\times10^{-4}\ s^{-1}} = 2.27\times10^{-7}$$

$$k_2 = 1.51\times10^{-10}\ s^{-1}$$

Workbook Problem 13.8

If a reaction has a rate constant of $0.0123\ s^{-1}$ at 700 K and $0.00342\ s^{-1}$ at 250 K, what is the activation energy of the reaction, and what will the rate constant be at 500 K?

Strategy: Solve for E_a using the two-point Arrhenius equation, then use the same equation to solve for the rate constant at 500 K.

Step 1:   Solve for E_a using the two-point Arrhenius equation.

Step 2:   Solve for k at 500 K.

Section 13.9   Reaction Mechanisms

The mechanism is the sequence of molecular events that leads from reactants to products. Each step is known as an **elementary reaction** or an **elementary step**. Knowing reaction mechanisms makes it possible to increase the efficiency of industrial reactions.

Elementary reactions provide information about individual molecular events; the overall reaction provides stoichiometry, but no mechanism information.

A species that is formed then consumed in the course of reaction is known as a **reaction intermediate**. These do not appear in the overall equation.

Elementary reactions are classified based on their molecularity: the number of molecules or atoms on the reactant side. A unimolecular reaction involves only one reactant molecule, a bimolecular reaction involves energetic collisions between two reactant molecules, and a termolecular reaction (rare) involves collisions of three molecules or atoms.

The elementary steps must sum to give the overall equation.

Section 13.10   Rate Laws for Elementary Reactions

Although overall rate laws must be determined experimentally, the rate law for an elementary reaction will contain the concentration of each reactant raised to an exponent equal to its coefficient in the chemical equation.

For a unimolecular reaction, the rate will always be first order in the concentration of reactant, because the single molecule is being converted to products.

For a bimolecular reaction, the rate depends on the frequency of collisions between the reactant molecules, and thus on the concentration of those molecules.

This applies only to elementary reactions, not overall reactions.

Workbook Example 13.8

PROBLEM:
Write the rate law for the following elementary reaction:

$$2 \, H_2 \, (g) + O_2 \, (g) \rightarrow 2 \, H_2O$$

SOLUTION:
The rate law for an elementary reaction is equal to the concentration of the reactants raised to their stoichiometric coefficients:

$$rate = k\left[H_2\right]^2\left[O_2\right]$$

Section 13.11   Rate Laws for Overall Reactions

The experimentally observed rate law for an overall reaction depends on the reaction mechanism.

When the overall reaction occurs in two or more steps, the slowest step will determine the rate of the overall reaction. This is known as the **rate-determining step**.

The rate-determining step will determine the rate law for the overall reaction and will match the predicted rate law for the elementary reaction.

When a rate law for an overall reaction is proposed, the elementary reactions must sum to give the overall reaction, and the mechanism must be consistent with the observed rate law for the overall reaction.

To establish a reaction mechanism, the overall rate law must first be experimentally determined, then a series of elementary steps devised. The rate law can then be predicted based on the rate-determining step. If the observed and predicted rate laws agree, the proposed method is a possible one.

Workbook Example 13.9

PROBLEM:
Given the following reaction mechanism:

$$A_2 \rightarrow 2\,A$$
$$2\,A + 3\,H_2O \rightarrow 2\,AH_3 + 3/2\,O_2$$
$$2\,AH_3 + 4\,O_2 \rightarrow 2\,HAO_3 + 2\,H_2O$$

1. Determine the overall reaction.
2. Identify the reaction intermediates, and calculate the molecularity of each step.
3. Identify the rate law if the second step is the rate-determining step.

SOLUTION:
To determine the overall reaction, sum up the elementary reactions and cancel out the species that occur on both sides:

$$A_2 + 2\,A + 3\,H_2O + 2\,AH_3 + 4\,O_2 \rightarrow 2\,A + 2\,AH_3 + 3/2\,O_2 + 2\,HAO_3 + 2\,H_2O$$

Overall reaction: $A_2 + H_2O + 5/2\,O_2 \rightarrow 2\,HAO_3$

The reaction intermediates are the species that appear in the reaction mechanism but not in the overall reaction: A, AH_3.

The molecularity of the first step = 1 (unimolecular).
The molecularity of the second step = 5.
The molecularity of the third step = 6.
The rate law is

$$rate = k[A]^2[H_2O]^3$$

Workbook Problem 13.9

What rate law and overall reaction would be expected from the following mechanism:

$$2\,A \rightarrow A_2 \qquad \text{rate-determining step}$$
$$A_2 + B \rightarrow A_2B \qquad \text{fast step}$$

Strategy: Write the expected rate law based on the rate-determining step, and sum the reactions for the overall reaction.

Step 1:   Write the expected rate law.

Step 2:   Sum for the overall reaction.

Section 13.12   Catalysis

A **catalyst** is a substance that increases the rate of a reaction without being consumed in the reaction. Catalysts are very important in the chemical industry, where they increase reaction efficiencies, and in living organisms where enzymes (biological catalysts) facilitate biologically essential reactions.

Catalysts make a lower-energy mechanism available for reaction, either by increasing the frequency factor (for instance, by ensuring that reactants are brought together in the proper configuration to react) or, more commonly, by reducing the activation energy.

Section 13.13   Homogeneous and Heterogeneous Catalysts

Homogeneous catalysts exist in the same phase as the reactants—an aqueous enzyme, for example.

Heterogeneous catalysts are in a different phase than the reactants: a metal mesh that provides a reaction surface, for example. The mechanism for heterogeneous catalysis is not well understood, but seems to involve three steps:

- **Adsorption** of the reactants to the reaction surface
- Facilitation of the reaction
- **Desorption** of products from the reaction surface

Heterogeneous catalysts are the most important for industrial processes, in part because they are easily separated from the reactants and products. The catalytic converters in the exhaust systems of newer automobiles use heterogeneous catalysts to convert pollutants to water, carbon dioxide, nitrogen, and oxygen.

Putting It Together

The reaction mechanism for the reaction of triphenylphosphine, $P(C_6H_5)_3$, with $Ni(CO)_4$ is

$$Ni(CO)_4 \rightarrow Ni(CO)_3 + CO$$
$$Ni(CO)_3 + P(C_6H_5)_3 \rightarrow Ni(CO)_3[P(C_6H_5)_3]$$

a. What is the molecularity of each step?
b. Doubling the concentration of $Ni(CO)_4$ doubles the rate. However, doubling the concentration of $P(C_6H_5)_3$ does not affect the rate. What is the rate law for the reaction? Which step is the slow step in the reaction mechanism?
c. The rate constant for this reaction is $k = 9.3 \times 10^{-3}$ s^{-1} at 20 °C. If the initial concentration of $Ni(CO)_4$ is 0.30 M, how long will it take the reaction to be 60% complete?
d. How many grams of $Ni(CO)_3[P(C_6H_5)_3]$ will be formed if you start with 0.237 g $Ni(CO)_4$ and allow the reaction to proceed for two minutes?

Self–Test

This section is intended to test your knowledge of the material covered in this chapter. Think through these problems, and make certain you understand what they are asking. Make sure your answers make sense. Successful completion of these problems indicates that you have mastered the material in this chapter. You will receive the greatest benefit from this section if you use it as a mock exam, as this will allow you to determine which topics you need to study in more detail.

True–False
1. The rate of a reaction can change as the reaction proceeds.

2. The exponents in rate laws cannot be experimentally determined.

3. The order of a reaction is determined by adding up the coefficients in the balanced equation.

4. The rate of a reaction is not the same as the rate constant unless the reaction is of the zeroth order.

5. The rate of a first-order reaction is dependent on the concentration of a single reactant raised to the first power, $[A]^1$.

6. The half-life of a reaction is half of the original amount of a reactant.

7. The half-life of a second-order reaction is dependent on initial concentration.

8. Zeroth-order reactions are fairly common.

9. Reactions can occur in more than one step, and all the steps occur at the same rate.

10. High temperatures increase the likelihood that reactants will collide with sufficient energy to react.

11. Activation energy can be negative for exothermic reactions.

12. A catalyst is unchanged at the end of a reaction.

13. Biological reactions rarely involve catalysts.

14. Catalysts can be in a different phase than the rest of the reactant mixture.

Multiple Choice
15. The overall reaction order for the rate law, rate = $k[A]^2[B]$ is
    a. two   b. three   c. five   d. six

16. A reaction has the rate law, rate = $k[A]^2$. If the initial rate for this reaction is 0.15 mol/L·sec, and the concentration of A is doubled, the new initial rate for this reaction will be
    a. 0.45 mol/L·sec   b. 1.20 mol/L·sec
    c. 0.60 mol/L·sec   d. 1.35 mol/L·sec

17. If a plot of ln [A] versus time gives a straight line for a particular reaction, that reaction is
    a. first order   b. ½ order   c. second order   d. zeroth order
18. The reaction order for the elementary reaction step, A + B → products, is
    a. second order   b. third order
    c. unable to be determined from the   d. first order
        balanced equation.

19. For every collision that occurs in a reaction
    a. products are formed
    b. the particles have enough kinetic energy to overcome the potential energy barrier
    c. a small fraction of molecules will overcome the potential energy barrier to reaction
    d. none of the above

20. The exponents in a rate law indicate the
    a. dependence of the rate on time
    b. sum of the coefficients in a balanced equation
    c. time required for a certain amount of reactant to disappear
    d. order of the reaction

21. The half-life of a first-order reaction with $k = 3.23 \times 10^{-3}$ s^{-1} is
    a. unable to be determined
    b. 1.62×10^{-3} s
    c. 3.10×10^2
    d. 2.15×10^2 s

22. The half-life of a second-order reaction depends on
    a. the initial concentration
    b. only the rate constant
    c. the rate constant and the initial concentration
    d. none of the above

23. Reaction rates tend to double when the temperature is
    a. doubled
    b. increased by 10 °C
    c. increased by 100 K
    d. increased by 100 °C

24. An activated complex
    a. has more energy than either the reactants or products
    b. is the configuration of the atoms at the maximum of the potential energy barrier
    c. is a weak linking of all the atoms involved in reaction
    d. all of the above

Matching

25. Activation energy

26. Bimolecular reaction

27. Catalyst

28. Kinetics

29. Decay constant

30. Enzyme

31. Half-life

a. biological catalyst

b. study of rates and mechanisms of chemical reactions

c. having a different phase than the reaction

d. involving two molecules

e. not dependent on the concentration of reactants

f. first-order rate constant for radiodecay

g. slowest elementary reaction in an overall reaction mechanism

32. Heterogeneous catalyst

  h. factor related to orientation of molecules in collisions and their ability to react

33. Homogeneous catalyst

  i. occurring when the concentration of products is zero

34. Initial rate

  j. involving three particles

35. Molecularity

  k. the potential energy barrier to reaction

36. Rate-determining step

  l. reaction intermediate with the maximum potential energy

37. Steric factor

  m. time required for the concentration of a reactant to drop to one-half its initial value

38. Termolecular reaction

  n. species that lowers the activation energy without being consumed in the reaction

39. Transition state

  o. in the same phase as the reaction

40. Zeroth-order reaction

  p. the number of particles, whether atoms or molecules, involved in a reaction

Fill-in-the-Blank

41. The _____ _____ is defined as either the _____ in concentration of a reactant, or the _____ in the concentration of a _____.

42. Rate laws depend on the concentration(s) of reactants and/or a proportionality constant known as the _____ _____, abbreviated _____.

43. In a first-order reaction, doubling the concentration of a reactant multiplies the rate by a factor of ___. In a second-order reaction, doubling the concentration of a reactant multiplies the rate by a factor of _____. In a third-order reaction, doubling the concentration of a reactant multiplies the rate by a factor of _____. In a zeroth-order reaction, doubling the concentration of a reactant multiplies the rate by a factor of _____.

44. In experimental determinations of rate laws, initial rates of reaction are used because when the products reach a certain concentration, it is likely that the _____ _____ will begin to occur.

45. If a plot of log [A] versus time gives a straight line, the reaction is _____ with respect to A.

46. Reaction mechanisms can be made up of individual _____ _____, each of which will have its own rate. The overall rate of the reaction will be determined by the _____ _____ _____, which is the _____ step.

47. The Arrhenius equation can be used to calculate the _____ _____ of a reaction, the energy required to form a _____ _____ and from there to form products.

48. A catalyst can either increase the efficiency of collisions or make a _____ _____ reaction mechanism available.

49. When the temperature increases by 10 °C, reaction rates tend to _____.

50. Heterogeneous catalysts _____ reactants, facilitate reactions, then _____ products.

Problems

51. The following data for the reaction A + B → C was collected. Calculate an estimate of the initial rate and the rate between 30 and 75 sec.

[C] (mol/L)	Time (s)
0	0
0.005	15
0.010	30
0.015	45
0.017	60
0.019	75
0.020	90
0.021	105
0.022	120

52. The rate of disappearance of PH_3 for the reaction $4\ PH_3(g) \rightarrow P_4(g) + 6\ H_2\ (g)$ is 1.67×10^{-2} mol/L·sec. What is the rate of appearance of H_2?

53. The following data were collected for the reaction $2\ NO\ (g) + Cl_2\ (g) \rightarrow 2\ NOCl\ (g)$

[NO] (mol/L)	[Cl$_2$] (mol/L)	Initial Rate mol/(L·sec)
0.025	0.025	15.9
0.050	0.025	63.6
0.025	0.050	31.8

Determine the rate law from this data. What is the order of the reaction with respect to each reactant? What is the overall order of the reaction? Provide the rate law, and solve for k.

54. The following data were collected for the reaction $3 A + 2 B \rightarrow 2 D$

A] (mol/L)	[B] (mol/L)	Initial Rate mol/(L·sec)
0.150	0.150	6.70×10^{-2}
0.300	0.150	0.134
0.150	0.300	6.70×10^{-2}

Determine the rate law from this data. What is the order of the reaction with respect to each reactant? What is the overall order of the reaction? What is the rate law? Calculate the value of k. If the initial concentration of A = 0.542 M and the initial concentration of B = 0.830 M, what would the initial rate be?

55. The rate law for the decomposition of O_3 to O_2 is second order in ozone. The rate constant for this reaction equals 1.40×10^{-2} L/(mol·sec). If the initial concentration of ozone is 2.37 M, how much O_3 will be present after 12 hr?

56. The following data were collected for the reaction $2 HI\ (g) \rightarrow H_2\ (g) + I_2\ (g)$

[HI] (mol/L)	Time (min)
2.50	0
1.45	3
1.02	6
0.788	9
0.641	12
0.541	15
0.468	18
0.412	21
0.368	24
0.332	27
0.303	30

Determine the order of this reaction. Calculate the value of k and the half-life from the initial reaction.

57. The proposed reaction mechanism for a reaction is

$$NO_2\ (g) + Cl_2\ (g) \rightarrow NO_2Cl\ (g) + Cl\ (g)$$
$$Cl\ (g) + NO_2\ (g) \rightarrow NO_2Cl\ (g)$$

What is the overall reaction? What is the molecularity of each step? What intermediates are there? If the first step is the slow step in the mechanism, write the expected rate law. What will the order of the reaction be?

58. If the rate constant for a reaction is $k = 3.79 \times 10^{-5}$ L/mol·sec at 108 °C and the activation energy for the reaction is 11.8 kJ/mol, what is the value of A?

59. The activation energy is $E_a = 98.62$ kJ/mol for the reaction $2\ NOCl \rightarrow 2\ NO + Cl_2$
    If $k = 7.2 \times 10^{-5}$ s^{-1} at 67 °C, at what temperature will $k = 3.2 \times 10^{-3}$ s^{-1}?

Challenge Problem
60. For the reaction $C_2H_5I + OH^- \rightarrow C_2H_5OH + I^-$, $E_a = 79.7$ kJ/mol and A = 1.09×10^9 M^{-1}s^{-1} at 45°C. If the concentration of C_2H_5I is doubled while [OH$^-$] remains constant, the rate doubles. If the [OH$^-$] is doubled while [C_2H_5I] remains constant, the rate doubles. If a 250 mL solution of 0.475 g KOH in ethanol is mixed with a 250 mL solution of 1.378 g C_2H_5I in ethanol, what is the initial rate at 35 °C?

CHAPTER FOURTEEN

Chemical Equilibrium

Learning Objectives

As a result of reading and studying this chapter, you should be able to

Section 14.1   The Equilibrium State
1.   Describe the characteristics of a reaction in chemical equilibrium.

Section 14.2   The Equilibrium Constant K_c
2.   Write an equilibrium constant expression K_c for gas- and solution-phase reactions.
3.   Determine a new equilibrium constant expression when reactions are manipulated or added.
4.   Calculate the value of the equilibrium constant K_c given concentrations of reactant and products.
5.   Identify a reaction in equilibrium from a molecular representation and evaluate the equilibrium constant.

Section 14.3   The Equilibrium Constant K_P
6.   Write the equilibrium expression K_P in terms of partial pressures of reactants and products. Relate the equilibrium constants K_P and K_c.

Section 14.4   Heterogeneous Equilibria
7.   Write equilibrium constant expressions for reactions involving heterogeneous equilibria.

Section 14.5   Using the Equilibrium Constant
8.   Determine the extent of a reaction given the equilibrium constant.
9.   Predict the direction a reaction will shift to reach equilibrium given initial conditions.
10.   Calculate concentrations or partial pressures or products and reactants in equilibrium.
11.   Calculate concentrations of reactants and products in equilibrium when initial concentrations are unequal.

Sections 14.6 – 14.9   Factors that Alter the Composition of an Equilibrium Mixture: Le Châtelier's Principle (Concentration, Pressure, Volume, Temperature)
12.   Predict the direction a reaction at equilibrium will shift as a result of changes in concentration.
13.   Predict the direction a reaction at equilibrium will shift as a result of changes in volume or pressure.
14.   Predict the direction a reaction at equilibrium will shift as a result of changes in temperature.

Chapter Summary

This chapter introduces the concept of chemical equilibrium, the state reached when the concentration of reactants and products remains constant over time despite both forward and backward reactions continuing to occur. Equilibrium constants can be related to either concentrations or partial pressures, and equations calculated in different ways can be related to one another. These constants can be used to determine the final concentrations of reactants and products in an equilibrium mixture. When equilibria are stressed through changing conditions, the equilibrium will shift to minimize the stress. Changes in temperature, pressure/volume, and concentrations all shift the equilibrium mixture; adding a catalyst will speed the achievement of equilibrium, but does not change the equilibrium concentrations. Reaction conditions must be chosen carefully with the use of a catalyst. Kinetics and equilibrium are linked, and their constants can be compared.

The Chapter in Detail

Section 14.1   The Equilibrium State

Many reactions do not go to completion, but reach a **chemical equilibrium** at which the concentrations of reactants and products no longer change.

To avoid confusion, species on the left of an equation are always called *reactants* and those on the right are always called *products*, even though the direction of the reaction may be unclear. To indicate that a reaction can proceed in either direction, a double arrow is used.

As a reaction proceeds, there comes a point when the reverse reaction (right to left) proceeds at the same rate as the forward reaction. This occurs in all reactions, though some proceed so far toward completion that they are called "irreversible": the equilibrium mixture will contain almost all products and almost no reactants.

When equilibrium is reached, the reactions, forward and backward, continue to occur, but have the same rates: there is no net conversion of reactants to products.

Section 14.2   The Equilibrium Constant K_c

When reactions are permitted to go to equilibrium, the concentrations of reactants and products are related by the equilibrium constant, K_c.

For the generic reaction $a\,A + b\,B \rightleftharpoons c\,C + d\,D$, the equilibrium constant is

$$K_c = \frac{[C]^c [D]^d}{[A]^a [B]^b}$$

The equilibrium constant is temperature-dependent and unitless: all of the concentrations are considered to be divided by their standard state concentrations (1 M), which eliminates the units from the concentration values.

For the reverse reaction, the equilibrium constant will be the inverse of the constant for the forward reaction. This makes it important that the form of the equation be given.

Workbook Example 14.1

PROBLEM:

For the reaction CO (g) + 2 H$_2$ (g) $\rightleftharpoons$ CH$_3$OH (g), the equilibrium concentrations are [CO] = 2.38 M; [H$_2$] = 0.260 M; [CH$_3$OH] = 3.15 M at a temperature of 110 °C. Calculate the equilibrium constant for this reaction.

SOLUTION:
Write an equilibrium equation for the balanced chemical reaction:

$$K_c = \frac{[CH_3OH]}{[CO][H_2]^2}$$

Substitute the equilibrium concentrations and solve for K_c:

$$K_c = \frac{(3.15)}{(2.38)(0.260)^2} = 19.6$$

Workbook Problem 14.1

A weak acid dissociates according to the following equation:

$$HA\ (aq) \rightleftharpoons H^+ (aq) + A^- (aq)$$

What is the equilibrium constant at 20 °C if a 0.250 M solution is 1.73% ionized?

Strategy: Write the equilibrium equation for K_c, calculate concentrations of all species, and solve.

Step 1:   Write the equation for the equilibrium constant.

Step 2:   Determine the concentration of each species.

Step 3:   Solve for K_c.

Section 14.3   The Equilibrium Constant K_P

For gas-phase reactions, partial pressures can be used instead of molar concentrations. The ideal gas law can be rearranged to show the relationship between concentration (n/V) and pressure:

$$P_A = [A]RT$$

Then these terms are substituted into the equation for an equilibrium constant, the relationship between K_c and K_p can be shown to be

$$K_P = K_c (RT)^{\Delta n}$$

There Δn is the sum of the coefficients of the gaseous products minus the sum of the coefficients of the gaseous reactants.

Workbook Example 14.2

PROBLEM:
Calculate K_P for the reaction in the previous example.

SOLUTION:
The coefficients of the balanced equation can be used to calculate Δn. The sum of the coefficients on the products side is 1, while the sum of the coefficients on the reactant side is 3.
Δn = products – reactants = –2.

Substitute Δn and K_c into the equation, relating K_P to K_c:

$$K_P = (19.6)\left[\left(0.0821\frac{L \cdot atm}{K \cdot mol}\right)(383\,K)\right]^{-2} = 1.98 \times 10^{-2}$$

Workbook Problem 14.2

Determine the value of K_c and K_p for the reaction

$$PCl_5\,(g) \rightleftharpoons PCl_3\,(g) + Cl_2\,(g)$$

At equilibrium at 250 K, the concentrations measured for each reactant are $[PCl_5]_e$ = 2.35 M and $[PCl_3]_e$ = $[Cl]_e$ = 1.56 M.

Strategy: Write the equilibrium equation for K_c.

Step 1:   Solve for K_c using the molar concentrations given.

Step 2:   Determine Δn.

Step 3:   Solve for K_p, using the equation that expresses the relationship between K_c and K_p.

Section 14.4   Heterogeneous Equilibria

In **homogeneous equilibria**, all the reactants and products are in the same phase; in **heterogeneous equilibria**, reactants and products are in a mixture of phases.

Because the concentrations of solids and liquids are constant, they are not included in the equilibrium expression.

Workbook Example 14.3

PROBLEM:

Write the equilibrium expression for the reaction $H_2CO_3\ (s) \rightleftharpoons CO_2\ (g) + H_2O\ (l)$.

SOLUTION: This reaction generates only one product that is not a pure solid or liquid. As a result, the equilibrium constant will be

$$K_c = [CO_2]$$

Because this is a gas, K_p should be used:

$$K_P = K_c RT$$

Section 14.5   Using the Equilibrium Constant

Because of the form in which the equilibrium constant is written, its value provides information about the way the reaction proceeds at a specific temperature:

- If K_c is large, over 10^3, the reaction proceeds almost to completion. Products predominate over reactants.
- If K_c is small, under 10^{-3}, the reaction proceeds very little. Reactants will predominate over products in an equilibrium mixture.
- If the value of K_c is intermediate, there will be appreciable quantities of both reactants and products in an equilibrium mixture.

The equilibrium constant can also be used to predict the direction of reaction given the concentrations of reactants and products.

The reaction quotient, Q_c, is calculated in exactly the same way as the equilibrium constant expression, but uses concentrations that may or may not be equilibrium values. Comparing Q and K makes it possible to predict which way the reaction will go:

- $Q_c < K_c$; reaction goes from left to right, more products will be formed.
- $Q_c > K_c$; reaction goes from right to left, products will be converted to reactants.
- $Q_c = K_c$; reaction is at equilibrium, there will be no net change.

If all but one of the equilibrium concentrations is known, the last can be calculated algebraically using the equilibrium constant.

It is also possible to calculate equilibrium concentrations from initial concentrations using the equilibrium coefficient and the balanced chemical equation. To do this, the following steps are required:

- Write and balance the chemical equation.
- Make a table and list the initial concentrations of all reactants and products and the change expected. For the change, define the concentration of one of the species as x and use the balanced equation to define the changes in the other species in terms of x.
- Substitute the algebraic statements of final concentration into the equilibrium equation and solve for x. If this requires solving a quadratic equation, choose the answer that makes chemical sense (there are no negative concentrations).

Workbook Example 14.4

PROBLEM:
The following reaction has a K_c value of 4.18×10^{-9} at 425 °C:

$$2\ HBr\ (g) \rightleftharpoons H_2\ (g)\ +\ Br_2\ (g)$$

What is the position of the equilibrium? If the concentrations of all species present are [HBr] = 0.35 M; $[H_2]$ = $[Br_2]$ = 3.2×10^{-3} M, is the reaction mixture at equilibrium? If not, in which direction will the reaction proceed?

SOLUTION:
We can determine the position of the equilibrium by looking at the value of K_c. Since $K_c < 10^{-3}$, we know that reactants predominate over products and that the position of the equilibrium is to the left. This is reflected in the concentrations provided, but to determine if the reaction mixture is at equilibrium, we need to calculate Q_c and compare that value to K_c.

$$Q_c = \frac{[H_2][Br_2]}{[HBr]^2} = \frac{\left(3.2\times10^{-3}\,M\right)\left(3.2\times10^{-3}\,M\right)}{\left(0.35\,M\right)^2} = 8.4\times10^{-5}$$

Since $Q_c > K_c$, the reaction is not at equilibrium and will proceed from right to left.

Workbook Example 14.5

PROBLEM:

The reaction, $PCl_5 (g) \rightleftharpoons PCl_3 (g) + Cl_2 (g)$, has $K_c = 85.0$ at a temperature of 760 °C. Calculate the equilibrium concentrations of PCl_5, PCl_3, and Cl_2 if the initial concentration of PCl_5 is 3.50 M. Assume a volume of 1.00 L.

SOLUTION:

To solve this problem, follow the procedure outlined above:

The balanced equation is given.

The initial concentration of PCl_5 is 3.50 M. Let x be the concentration of PCl_5 that reacts on going to the equilibrium state. Since the mole to mole ratios are 1:1:1, if x amount of PCl_5 reacts, then x mol/L of PCl_3 and Cl_2 will be present at equilibrium. This can be summarized in the following table:

	$PCl_5 (g)$	$\rightleftharpoons$	$PCl_3 (g)$	+	$Cl_2 (g)$
Initial concentration (M)	3.5		0		0
Change (M)	$-x$		$+x$		$+x$
Equilibrium concentration (M)	$3.5 - x$		x		x

The equilibrium expression for the reaction is

$$K_{eq} = \frac{[PCl_3][Cl_2]}{[PCl_5]}$$

The equilibrium concentrations from the above table can be substituted into this expression, and the equation rearranged to facilitate the use of the quadratic equation.

$$85.0 = \frac{x^2}{3.5 - x}$$

$$x^2 + 85x - 298 = 0$$

Use the quadratic equation to solve for x:

$$x = \frac{-85 \pm \sqrt{(85)^2 - 4(1)(-298)}}{2}$$

$$x = 3.37, -88.3$$

A negative number does not make physical sense: the answer is 3.37.

Calculate the equilibrium concentrations:

$$[PCl_5] = 3.5 - 3.37 = 0.13 \text{ M}$$
$$[PCl_3] = [Cl_2] = 3.37 \text{ M}$$

Workbook Problem 14.3

The value of K_c for the reaction

$$C\ (s)\ +\ H_2O\ (g)\ \rightleftharpoons\ CO\ (g)\ +\ H_2\ (g)$$

is 3.0×10^{-2}. Determine the equilibrium concentration if the initial concentration of water is 11.75 M.

Strategy: Follow the previously outlined steps in your textbook:

Step 1:   Write the balanced equation for the reaction.

Step 2:   Make a table listing the initial concentration, the change in concentration, and the equilibrium concentration. (Let x = the amount of substance that reacts.)

Initial concentration (M)	
Change (M)	
Equilibrium concentration (M)	

Step 3:   From the balanced equation, write the equilibrium equation. Substitute the algebraic expressions for the equilibrium concentrations into the equilibrium equation, and solve for x. Use the quadratic equation if necessary.

Section 14.6   Factors That Alter the Composition of an Equilibrium Mixture: Le Châtelier's Principle

If a reaction does not proceed at room temperature and atmospheric pressure, the reaction conditions can be altered to force the reaction to proceed. There are four factors that alter the composition of an equilibrium mixture:

- Concentration of reactants or products
- Pressure and volume
- Temperature
- Catalysis

The fourth, addition of a catalyst, only affects the rate at which equilibrium is reached, not the composition of the equilibrium mixture.

If a stress—a change in concentration, pressure, volume, and/or temperature—is applied to a reaction mixture at equilibrium, the equilibrium will shift to minimize the stress. This is known as **Le Châtelier's principle**.

Section 14.7   Altering an Equilibrium Mixture: Changes in Concentration

When equilibrium is disturbed by the addition or removal of a reactant or product, a reaction will occur to *remove* an added substance and *replenish* a removed substance in order to restore equilibrium:

- Adding a reactant will drive the reaction toward products (to the right); adding a product will drive the reaction toward reactants (to the left).
- Removing a reactant will drive the reaction toward reactants (to the left); removing a product will drive the reaction toward products (to the right).

Mathematically, this works by changing Q_c: when a reactant is added or a product removed, Q_c becomes smaller: the reaction must shift right toward products, increasing the numerator again. When a reactant is removed, or a product added, Q_c becomes larger. To return to equilibrium, the reaction must shift left: products will be converted to reactants.

Workbook Example 14.6

PROBLEM:
Predict the results of adding $CaCO_3$, adding CO_2, and removing CO_2 to the following reaction:

$$CaCO_3 (s) \rightleftharpoons CaO (s) + CO_2 (g)$$

SOLUTION:

Adding $CaCO_3$ will not affect the equilibrium: solids do not appear in either K or Q.

Adding CO_2 will push the reaction to shift left toward reactants, thus reducing the amount of CO_2.

Removing CO_2 will push the reaction to shift right toward products, thereby increasing the amount of CO_2.

Workbook Problem 14.4

The following reaction is at equilibrium:

$$2 \, BrNO (g) \rightleftharpoons 2 \, NO (g) + Br_2 (g)$$

What would be the effect on the equilibrium if (1) BrNO was added, (2) BrNO was removed, (3) NO was added, and (4) Br_2 was removed.

Strategy: Use Le Châtelier's principle.

Step 1:   Determine the direction of reaction needed to minimize each stress.

Section 14.8   Altering an Equilibrium Mixture: Changes in Pressure and Volume

When the moles of products and moles of reactants are different, changing the volume and/or pressure can shift the reaction direction. An increase in the pressure will shift the reaction to the side with fewer moles of gas. A decrease in the pressure will shift the reaction in the direction of more moles of gas.

Increases in pressure (due to decrease in volume) are akin to increases in concentration for gases, as n/V increases.

- If a reaction has the same number of moles of gaseous products and reactants, changing the pressure will not affect the reaction.
- In heterogeneous equilibria, changes in pressure do not affect solids or liquids, because their volume is essentially independent of pressure.

These effects only are seen when pressure is changed as a result of changing volume. If pressure is increased by adding a nonreacting (inert) gas to the mixture, the equilibrium will not change.

Workbook Problem 14.5

The following reaction is at equilibrium:

$$N_2O_4\ (g) \rightleftharpoons 2\ NO_2\ (g)$$

What would be the effect on the equilibrium if the pressure was increased? If the pressure was decreased? If 10 L of neon were added to the reaction mixture at constant volume?

Strategy: Use Le Châtelier's principle.

Step 1:   Determine the direction of reaction needed to minimize each stress.

Section 14.9   Altering an Equilibrium Mixture: Changes in Temperature

As long as the temperature remains constant, the value of the equilibrium constant remains the same. Changes in pressure of concentration only change Q.

When the temperature changes, the equilibrium constant itself is altered:

- An exothermic reaction is less favored as temperature increases: the equilibrium constant decreases with increasing temperature. The heat added by the reaction will *increase*, not decrease, the stress on the equilibrium. As the temperature decreases, the equilibrium constant decreases: the reaction is contributing heat, thereby reducing the stress on the reaction.
- An endothermic reaction is more favored as the temperature increases: the equilibrium constant increases with increasing temperature, because the reaction absorbs some of the temperature increase. As the temperature decreases, the equilibrium coefficient decreases: the heat absorbed by the reaction *increases*, not decreases, the stress on the equilibrium.

Workbook Example 14.7

PROBLEM:
How will the following changes alter the equilibrium for the reaction?

$$C(graphite) + CO_2(g) \rightleftharpoons 2CO(g) \qquad \Delta H^\circ = +172.5 \text{ kJ}$$

a. CO is removed from the system.
b. CO_2 is removed from the system.
c. The volume of the container is increased.
d. Graphite is added to the system.
e. The temperature is raised.

SOLUTION:
a. When CO is removed from the system, the equilibrium shifts right.
b. When CO_2 is removed from the system, the equilibrium shifts left.
c. Increasing the volume of the container decreases the pressure in the container. The equilibrium will shift to the right to minimize the change.
d. Adding a solid to the system does not affect the equilibrium since solids are not included in the equilibrium expression.
e. Raising the temperature causes the equilibrium to shift right, because the reaction absorbs temperature.

Section 14.10   The Link between Chemical Equilibrium and Chemical Kinetics

Adding a catalyst lowers the activation energy for both the forward and reverse reactions. As a result, the reaction reaches equilibrium faster, but the equilibrium constant remains the same, and the equilibrium concentrations will be unaffected.

Although a catalyst does not affect the equilibrium, it can affect the reaction conditions chosen: if good yields are obtained slowly at a certain temperature, finding a catalyst can make the reaction more practical.

Workbook Problem 14.6

Consider the following reaction, which occurs in a catalytic converter in the presence of a palladium catalyst:

$$2CO\ (g) + O_2\ (g) \xrightarrow{\ Pd\ } 2CO_2\ (g) \qquad\qquad \Delta H° = -566\ kJ$$

What would be the effect on the concentration of CO_2 if the pressure was increased? If the pressure was decreased? If the palladium catalyst was removed? If the temperature was lowered? If the concentration of O_2 was increased?

Strategy: Use Le Châtelier's principle

Step 1:   Determine the direction of reaction needed to minimize each stress and its effect on $[CO_2]$.

If the reaction $A\ +\ B\ \rightleftharpoons\ C\ +\ D$ is a simple reaction with an elementary forward and reverse reaction, it has two rate constants:

- Rate of forward reaction = $k_f[A][B]$
- Rate of reverse reaction = $k_r[C][D]$

At equilibrium, these rates are equal to one another, and they can be related to the equilibrium constant:

$$k_f\left[A\right]\left[B\right] = k_r\left[C\right]\left[D\right] \quad \text{or} \quad \frac{k_f}{k_r} = \frac{[C][D]}{[A][B]} = K_c$$

It is the relative values of k_f and k_r that determine the composition of the equilibrium mixture. If the rate constant for the forward reaction is much larger than that of the reverse reaction, K_c is very large, and the reaction goes to completion. Because the reverse reaction is too slow to be detected, this is an irreversible reaction.

If instead the values of k_f and k_r are close to one another, there will be an equilibrium in which all species—reactants and products—will be present.

The temperature-dependence of equilibrium constants is also explained by the relationship derived above:

$$\frac{k_f}{k_r} = K_c$$

Because the kinetic constants are temperature-dependent and increase by different amounts when the temperature increases (as described by the Arrhenius equation), equilibrium is also temperature-dependent.

Workbook Example 14.8

PROBLEM:
Calculate the equilibrium constant K_c for the reaction

$$H_2O\ (l) \rightleftharpoons H^+\ (aq) + OH^-\ (aq)$$

given that $k_f = 2.4 \times 10^{-5}\ s^{-1}$ and $k_r = 1.3 \times 10^{11}\ M^{-1} \cdot s^{-1}$ and that the equilibrium expression is

$$K_c = \left[H^+\right]\left[OH^-\right]$$

SOLUTION:
To calculate K_c, use the equation

$$K_c = \frac{k_f}{k_r} = \frac{2.4 \times 10^{-5}\ s^{-1}}{1.3 \times 10^{11}\ s^{-1}} = 1.8 \times 10^{-16}$$

Putting It All Together

Calculate the pressure of all species at equilibrium for the reaction

$$2\ NO\ (g) + Br_2\ (g) \rightleftharpoons 2\ NOBr\ (g)$$

given an initial pressure of NO = 78.4 mm Hg and the initial pressure of Br_2 = 51.3 mm Hg. The total pressure at equilibrium = 120.5 mm Hg. Determine the value of K_p.

Self–Test

This section is intended to test your knowledge of the material covered in this chapter. Think through these problems, and make certain you understand what they are asking. Make sure your answers make sense. Successful completion of these problems indicates that you have mastered the material in this chapter. You will receive the greatest benefit from this section if you use it as a mock exam, as this will allow you to determine which topics you need to study in more detail.

True–False
1. All chemical reactions are reversible to some extent.

2. The equilibrium constant can be determined by running the same reaction at different temperatures.

3. The equilibrium constant has units of M^{-1}.

4. Equilibrium constants can be related to concentrations or to partial pressures.

5. A large equilibrium constant indicates a reaction with more reactants than products at equilibrium.

6. Increasing the pressure on a gas-phase reaction will drive the reaction in the direction of more moles of gas.

7. Adding more of a reactant pushes the reaction to the right toward products.

8. Catalysts speed the rate at which equilibrium is achieved.

9. The relative values of k_f and k_r determine the composition of the equilibrium mixture.

10. The value of K makes it possible to predict the direction of a reaction.

Multiple Choice

11. A state of chemical equilibrium is reached when
    a. the rate of the forward reaction is the same as the rate of the reverse reaction
    b. $Q = K$
    c. the concentrations of the products and reactants have reached constant value
    d. all of the above

12. The equilibrium constant is calculated as
    a. concentration of products over concentration of reactants
    b. concentration of products raised to stoichiometric coefficients over concentration of reactants raised to stoichiometric coefficients
    c. concentration of reactants over concentration of products
    d. concentration of reactants raised to stoichiometric coefficients over concentration of products raised to stoichiometric coefficients

13. In heterogeneous equilibria, the following are included in the equilibrium constant:
    a. solids, liquids, solutions, and gases   b. liquids, solutions, and gases
    c. solutions and gases   d. only gases

14. An equilibrium constant in the range of 10^{-3} to 10^3 indicates that
    a. both products and reactants will be appreciably present at equilibrium
    b. only products will be appreciably present at equilibrium
    c. only reactants will be appreciably present at equilibrium
    d. no information about final concentrations can be gained from this

15. The following stresses will result in a change in equilibrium concentrations:
    a. change of temperature   b. change of concentrations
    c. change in volume and pressure   d. all of the above

16. Which of the following is an expected response to stress on a reaction?
    a. Lowering the temperature on an exothermic reaction increases the concentration of products.
    b. Lowering the temperature on an endothermic reaction increases the concentration of products.
    c. Raising the temperature on an exothermic reaction increases the concentration of products.
    d. Raising the temperature on an endothermic reaction has no effect on the concentration of products.

17. A catalyst
    a. increases the concentration of products
    b. increases the percent yield of a reaction
    c. makes a reaction go to completion
    d. increases the speed at which equilibrium is obtained

18. K_p
    a. is the same as K_c
    b. is the reciprocal of K_c
    c. is the equilibrium constant calculated using partial pressures
    d. includes the concentration of pure liquids in the equilibrium expression

19. The reaction quotient, Q_c, is
    a. calculated in the same manner as K_c
    b. less than K_c at equilibrium
    c. greater than K_c at equilibrium
    d. equal to K_p

20. When a reactant is removed from a system at equilibrium
    a. the reaction shifts right
    b. the reaction shifts left
    c. the reactants are converted to products
    d. the denominator in the K_c expression becomes larger

Matching

21. Chemical equilibrium phase

22. Equilibrium constant, K_c

23. Equilibrium constant, K_p

24. Equilibrium mixture

25. Homogeneous equilibria

26. Heterogeneous equilibria

27. Reaction quotient, Q_c

28. Reversible reaction

a. constant in the equilibrium equation for gas-reactions

b. equilibrium in which all reactants and products are in the same phase

c. calculated like K_{eq}, but with nonequilibrium concentrations.

d. constant in the equilibrium equation for solutions

e. if a stress is applied to a reaction, the reaction occurs in the direction that relieves the stress

f. any state in which change in forward and reverse directions is equal in rate, leading to no net change in concentration of reactants and products

g. a state in which the concentration of products and reactants in a reaction no longer shows any net change

h. equilibrium in which reactants and products are in different phases

29. Dynamic state

    i. a reaction that can go forward toward products or back to reactants, depending on reaction conditions

30. Le Châtelier's principle

    j. combination of reactants and products with characteristic proportions

Fill–in–the–Blank

31. Chemical equilibrium is a _____ state in which the _____ of reactants and products is _____.

32. The equilibrium equation for a _____ equilibrium does not include concentrations of pure _____ or _____.

33. When an equilibrium mixture is subjected to a stress, it responds to _____ that stress. Adding more reactant will shift a reaction toward _____, while adding more products will shift a reaction toward _____. Changing the _____ affects the mixture by changing the _____ constants. Adding a _____ does not alter the equilibrium mixture, simply _____ the rate at which it is achieved.

34. Comparing K and Q makes it possible to determine the _____ of a reaction. When $K = Q$, the reaction is at _____. When $K < Q$, the reaction will proceed from _____ to _____. When $K < Q$, the reaction will proceed from _____ to _____.

35. Increasing the pressure on a gas-phase reaction will shift the equilibrium away from small molecules toward larger molecules. This has the result of _____ the pressure.

Problems

36. Write equilibrium expressions for the following balanced equations.
    a. $O_3\ (g)\ +\ Cl\ (g) \rightleftharpoons O_2\ (g)\ +\ ClO\ (g)$
    b. $2\ NO\ (g)\ +\ Br_2\ (g) \rightleftharpoons 2\ NOBr\ (g)$
    c. $2\ Cu\ (s) + O_2\ (g) \rightleftharpoons 2\ CuO\ (s)$
    d. $2\ N_2O\ (g) \rightleftharpoons 2\ N_2\ (g)\ +\ O_2\ (g)$

37. The partial equilibrium pressures for N_2, O_2, and NO in the reaction

$$N_2\ (g)\ +\ O_2\ (g) \rightleftharpoons 2\ NO\ (g)$$

are $p(N_2) = 0.27$ atm, $p(O_2) = 0.187$ atm, and $p(NO) = 0.045$ atm. Calculate K_p for this reaction.

38. Calculate K_c for the reaction in Question 37.

39. From the value of K_c, determine whether mainly products or mainly reactants exist at equilibrium for the following balanced equations:

    a. $CH_3Cl\ (aq) + OH^-\ (aq) \rightleftharpoons CH_3OH\ (aq) + Cl^-\ (aq)$ $K_c = 1 \times 10^{16}$

    b. $2SO_2\ (g) + O_2\ (g) \rightleftharpoons 2\ SO_3\ (g)$ $K_p = 3.3$

    c. $2\ HBr\ (g) \rightleftharpoons H_2\ (g) + Br_2\ (g)$ $K = 2 \times 10^{-19}$

40. For the reaction

$$PCl_5\ (g) \rightleftharpoons PCl_3\ (g) + Cl_2\ (g)$$

    $K_c = 33.3$ at 760 °C.   Is this system at equilibrium if $[PCl_5] = 2.43 \times 10^{-3}$ M, $[PCl_3] = [Cl_2] = 0.830$ M? If not, determine the direction the reaction must go to reach equilibrium.

41. Nitrogen dioxide is produced from the reaction of dinitrogen oxide and oxygen according to the equation

$$2\ N_2O\ (g) + 3\ O_2\ (g) \rightleftharpoons 4\ NO_2\ (g)$$

    At 25 °C, 0.0342 moles of N_2O and 0.0415 moles of O_2 are placed in a 1.00 L container and allowed to react. The $[NO_2]$ at equilibrium is 0.0298 moles. What are the equilibrium concentrations of N_2O and O_2? What is the value of K_c?

42. The equilibrium constant for the reaction

$$SO_2\ (g) + NO_2\ (g) \rightleftharpoons NO\ (g) + SO_3\ (g)$$

    is $K_c = 85.0$ at 460 °C. Calculate the concentrations of all species at equilibrium when the initial reaction mixture contains 0.100 M SO_2 and 0.100 M NO_2.

43. Consider the reaction

$$2\ SO_2\ (g) + O_2\ (g) \rightleftharpoons 2\ SO_3\ (g) \qquad \Delta H° = +197\ Kj$$

    What will be the direction of the reaction when the following stress is applied?
    a. addition of SO_3 b. decrease in temperature
    c. increase in volume   d. addition of N_2 gas
    e. addition of SO_2 gas

44. For the reaction $A + B \rightleftharpoons 2\ C$, $k_f = 6.8 \times 10^{-5}\ s^{-1}$ and $k_r = 7.2 \times 10^{-7}\ s^{-1}$. Calculate K.

45. Calculate the equilibrium concentration for all species present in the reaction

$$HCONH_2\ (g) \rightleftharpoons NH_3\ (g) + CO\ (g)$$

    if the initial concentration of formamide ($HCONH_2$) is 0.125 M. $K_c = 4.84$ at 127 °C.

46. For the reaction

$$2 \, H_2S \, (g) \; + \; CH_4 \, (g) \; \rightleftharpoons \; 4 \, H_2 \, (g) \; + \; CS_2 \, (g)$$

$K_c = 5.27 \times 10^{-8}$ at 700 °C. What is the position of equilibrium for this reaction? Is the reaction mixture at equilibrium when the concentrations of the reactants and products are $[CH_4] = 1.75$ M, $[H_2S] = 2.00$ M, $[H_2] = 0.450$ M, and $[CS_2] = 0.085$ M? If not, in which direction will the reaction proceed?

Challenge Problem

47. The following reaction was carried out in a 4.00 L vessel:

$$NO_2 \, (g) \; + \; NO \, (g) \; \rightleftharpoons \; N_2O \, (g) \; + \; O_2 \, (g)$$

Determine the number of moles of reactants and products present at equilibrium if the initial number of moles are 0.100 mol NO_2, 0.050 mol NO, 0.0125 mol N_2O, and 0.00125 mol O_2. Use a value of $K_c = 0.914$.

CHAPTER FIFTEEN

Aqueous Equilibria: Acids and Bases

Learning Objectives

As a result of reading and studying this chapter, you should be able to

Section 15.1   **Acid–Base Concepts: The Brønsted-Lowry Theory**
    1.   Identify Arrhenius acids and bases and Brønsted–Lowry acids and bases and conjugate acid–base pairs.

Section 15.2   **Acid Strength and Base Strength**
    2.   Predict the direction of a reaction based upon relative strengths of the conjugate acid–base pairs.

Section 15.3   **Factors That Affect Acid Strength**
    3.   Predict the relative strengths of binary acids (HA) and oxoacids (H_nYO_m) based on their chemical structure.

Section 15.4   **Dissociation of Water**
    4.   Calculate the concentration of $[H_3O^+]$ or $[OH^-]$ using the value of K_w.

Sections 15.5 – 15.6   **The pH Scale and Measuring pH**
    5.   Calculate the pH of a solution given $[H_3O^+]$ or $[OH^-]$.
    6.   Calculate $[H_3O^+]$ or $[OH^-]$ given the pH of a solution.

Section 15.7   **The pH in Solutions of Strong Acids and Strong Bases**
    7.   Calculate the pH of a strong acid or base solution.

Sections 15.8 – 15.10   **Equilibria in Solutions of Weak Acids**
    8.   Calculate the value of K_a given the initial concentration of a weak acid and its equilibrium pH.
    9.   Given the K_a value of a weak acid and its initial concentration, calculate the pH, the percent dissociation, and the concentration of all species present in solution.

Section 15.11   **Polyprotic Acids**
    10.   Calculate the pH and the concentration of all species present in a solution of a diprotic acid.
    11.   Visualize the species present when a diprotic acid dissociates.

Sections 15.12 – 15.13 **Equilibria in Solutions of Weak Bases and the Relation between K_a and K_b**
    12.   Calculate the pH and equilibrium concentrations in a solution of a weak base.
    13.   Relate K_a, K_b, pK_a, and pK_b for a conjugate acid–base pair.

Section 15.14   Acid–Base Properties of Salts
14.   Predict whether a salt is acidic, basic, or neutral and calculate the pH of the salt solution.

Section 15.15   Lewis Acids and Bases
15.   Identify the Lewis acid and Lewis base and use curved arrow notation to indicate donation of a lone pair of electrons in Lewis acid–base reactions.

Chapter Summary

The study of equilibrium carries over into the study of acids and bases. In this chapter, we discuss the three definitions of acids and bases and conjugate acids and base pairs. We also introduce reactions of strong and weak acids and bases and the concepts of pH and pK_a. We explain polyprotic acids and their reactions, as well as the identifications of acidic salts and basic salts and the pH of their solutions. Finally, we look at periodic variations in acid strength and how they can be used to predict the strength of various acids.

The Chapter in Detail

Section 15.1   Acid–Base Concepts: The Brønsted–Lowry Theory

The Arrhenius definition of acids and bases is the simplest, and we have used it in this book thus far. In the Arrhenius definition, acids are described as dissociating to form hydrogen ions and bases as dissociating to form hydroxide ions. This theory is limited to aqueous solutions, and it is unable to account for the basicity of substances—such as ammonia (NH_3)—that are bases without having any hydroxide ions to contribute.

The Brønsted–Lowry theory of acids and bases solves these problems by describing an acid as a *proton donor* and a base as a *proton acceptor*, with acid–base reactions being *proton transfers*.

Chemical species that differ only by a proton are called **conjugate acid–base pairs**.

HA indicates the acid and A⁻ the conjugate base. A conjugate base has one less proton than the acid.
BH⁺ is the conjugate acid of B; the conjugate acid has one more proton than the base.

Acid dissociation is when a proton is transferred to a solvent, for example:

$$HA + H_2O\ (l) \rightleftarrows A^-\ (aq) + H_3O^+\ (aq)$$

Base dissociation is when a proton is removed from a solvent, for example:

$$B + H_2O\ (l) \rightleftarrows BH^+\ (aq) + OH^-\ (aq)$$

All Brønsted–Lowry bases have one or more lone pairs of electrons. This makes it possible for them to pull protons off solvent molecules.

Workbook Example 15.1

PROBLEM:
Write the proton-transfer equilibria for the following acids or bases in aqueous solution, and identify the conjugate acid–base pairs in each one: CH_3COOH (acetic acid), PO_4^{3-} (phosphate ion – base)

SOLUTION:

$$CH_3COOH\ (l) + H_2O\ (l) \leftrightarrows CH_3COO^-\ (aq) + H_3O^+\ (aq)$$

Acid = CH_3COOH conjugate base = CH_3COO^-

$$PO_4^{3-}\ (aq) + H_2O\ (l) \leftrightarrows HPO_4^{2-}\ (aq) + OH^-\ (aq)$$

Base = PO_4^{3-} conjugate acid = HPO_4^{2-}

Workbook Problem 15.1

Write the proton-transfer reactions for nitrous acid (HNO_2) and basic sulfate (SO_4^{2-}). Identify the acid or base and its conjugate.

Strategy: Write the reaction with water, in which the acid protonates the water, and the base deprotonates the water.

Step 1:   Write the reactions in the forward direction.

Step 2:   Identify the conjugate acid and the conjugate base.

Section 15.2   Acid Strength and Base Strength

In an acid-dissociation reaction, the acid and water molecules compete for the loosely-attached hydrogen ion that defines the acid. If the water molecules are more basic than the acid ion, they will pull the hydrogen ion off the acid, and the acid will dissociate. If the acid ion (A^-) is more basic than the water molecules, the acid will not dissociate.

A strong acid dissociates completely in aqueous solution; therefore, it is also defined as a strong electrolyte. The remaining ion, A^-, is a weak base with only a negligible tendency to combine with a proton in aqueous solution. Strong acids include perchloric ($HClO_4$), hydrochloric (HCl), sulfuric (H_2SO_4), and nitric (HNO_3).

A weak acid only partially dissociates in aqueous solution and is therefore a weak electrolyte. The remaining ion, A^-, is a relatively strong conjugate base and will thus tend to pull protons back out of solution. Weak acids include acetic acid (CH_3COOH), nitrous acid (HNO_2), and hydrofluoric acid (HF).

Based on this information, we can conclude the following—a crucial point to remember:

The weaker the acid, the stronger its conjugate base.
The stronger an acid, the weaker its conjugate base.

Protons will always be transferred from a weaker base to a stronger base, so protons will transfer from a strong acid to the conjugate base of a weak acid.

A table of acids and their conjugate bases, arranged by strength, is on Page 609 of your textbook.

Workbook Example 15.2

PROBLEM:
Determine the direction of reactions involving hydrochloric acid, Cl⁻, acetic acid, and acetate ion.

SOLUTION:
Hydrochloric acid is a strong acid, making Cl⁻ a very weak base. Acetic acid is a weak acid, so acetate ions are a relatively strong conjugate base. As a result, hydrochloric acid will transfer protons to acetate ions. The reaction is

$$HCl + CH_3COO^- \ (aq) \leftrightarrows CH_3COOH \ (aq) + Cl^- \ (aq)$$

Workbook Problem 15.2

Write the proton-transfer reaction between HCl and CO_3^{2-} and the proton-transfer reaction between CN^- and H_3PO_4.

Strategy: Determine which species is the stronger acid and which is the stronger base, then transfer protons accordingly.

Step 1: Write the reactions, transferring protons from the stronger acid to the stronger base.

Section 15.3 Factors that Affect Acid Strength

The extent of dissociation of an acid is often determined by the strength and polarity of the HA bond: the weaker or more polar the HA bond is, the stronger the acid.

For binary acids, as the size of A⁻ increases down the periodic table, the strength of the HA bond is weakened; therefore, acidity increases down the groups. Within a row, the acidity will increase with increasing polarity. As A⁻ becomes more electronegative, acidity increases.

Oxoacids have the general formula H_nYO_m, where Y is a nonmetallic atom and n and m are integers. Y is always bonded to one or more hydroxyl (OH) groups, and may be bound to one or more oxygen atoms as well. The dissociation of an oxoacid involves the breaking of an OH bond, so anything that weakens these bonds or increases their polarity will increase the strength of the acid.

As Y becomes more electronegative, the strength of oxoacids increases. As the number of oxygens changes in an oxoacid (sulfuric acid to sulfurous acid, for example), an increase in the oxidation number of Y will increase acid strength. In other words, the more oxygens the central atom is trying to hold on to, the stronger the acid will be. H_2SO_4 is stronger than H_2SO_3, HNO_3 is stronger than HNO_2.

Workbook Example 15.3

PROBLEM:
Order the following by increasing acid strength:

a) HF, NH_3, H_2O;
b) NH_3, AsH_3, PH_3;
c) HIO_3, HIO, HIO_2, HIO_4;
d) H_2SeO_4, H_2TeO_4, H_2SO_4

SOLUTION:

a) Across a period, an increase in electronegativity gives rise to an increase in acidity: $NH_3 < H_2O < HF$.

b) Down a group, an increase in size gives rise to an increase in acid strength: $NH_3 < PH_3 < AsH_3$.

c) For oxoacids, acid strength increases with an increasing number of oxygen atoms: $HIO < HIO_2 < HIO_3 < HIO_4$.

d) For oxoacids that contain the same number of OH groups and the same number of O atoms, acid strength increases with increasing electronegativity of Y: $H_2TeO_4 < H_2SeO_4 < H_2SO_4$.

Section 15.4   Dissociation of Water

When acids dissociate, the acid releases a hydrogen ion, H^+. This is too reactive to exist loose in solution, so it associates with a water molecule to form a hydronium ion, H_3O^+.

The simplest hydrate of the proton is $[H(H_2O)]^+$, but others are also formed with the general formula $[H(H_2O)_n]^+$.

For practical purposes, H^+ (*aq*) and H_3O^+ (*aq*) are used interchangeably to represent a proton hydrated by an unspecified number of water molecules.

Water can act both as an acid and as a base, and it will dissociate in a reaction in which water acts as both acid and base:

$$H_2O\ (l) + H_2O\ (l) \leftrightarrows H_3O^+\ (aq) + OH^-\ (aq)$$

Water has a characteristic dissociation constant, abbreviated K_W. Water is omitted from the equilibrium expression because it is a pure liquid.

$$K_w = \left[H_3O^+\right]\left[OH^-\right] = 1.0 \times 10^{-14}$$

This equilibrium lies very far to the left: in pure water, only about 2 out of 10^9 molecules are dissociated.

The forward and reverse reactions are rapid, and in pure water at 25 °C,

$$\left[H_3O^+\right] = \left[OH^-\right] = 1.0 \times 10^{-7}\ M$$

In aqueous solutions, the relative values of $[H_3O^+]$ and $[OH^-]$ determine whether the solution is acidic, basic, or neutral:

- $[H_3O^+] = [OH^-]$ neutral solution $[H_3O^+] = 1.0 \times 10^{-7}$ M.
- $[H_3O^+] > [OH^-]$ acid solution $[H_3O^+] > 1.0 \times 10^{-7}$ M.
- $[H_3O^+] < [OH^-]$ basic solution $[H_3O^+] < 1.0 \times 10^{-7}$ M

Given the concentration of one ion, it is possible to calculate the concentration of the other using the equilibrium constant:

$$\left[H_3O^+\right] = \frac{1 \times 10^{-14}}{\left[OH^-\right]} \qquad\qquad \left[OH^-\right] = \frac{1 \times 10^{-14}}{\left[H_3O^+\right]}$$

Workbook Example 15.4

PROBLEM:
Calculate the molarity of OH^- in a solution with an H_3O^+ concentration of 0.0042 M. Is this solution acidic, basic, or neutral? What is the $[H_3O^+]$ in a solution with $[OH^-]$ of 9.35×10^{-7} M? Will the solution be acidic, basic, or neutral?

SOLUTION:
If we know the concentration of either the $[H_3O^+]$ or the $[OH^-]$, we can calculate the other from the relationship $K_W = [H_3O^+][OH^-]$.

$$\left[OH^-\right] = \frac{1 \times 10^{-14}}{0.0042\ M} = 2.4 \times 10^{-12} \qquad [H_3O^+] > [OH^-],\ \text{so this solution is acidic.}$$

$$\left[H_3O^+\right] = \frac{1.0 \times 10^{-14}}{9.35 \times 10^{-7}\ M} = 1.07 \times 10^{-8} \qquad [H_3O^+] < [OH^-];\ \text{this solution is basic.}$$

Workbook Problem 15.3

Determine the $[OH^-]$ in a solution with $[H_3O^+] = 2.3 \times 10^{-5}$ M, and the $[H_3O^+]$ of a solution with $[OH^-] = 5.73 \times 10^{-9}$ M. Are these solutions acidic, basic, or neutral?

Strategy: Use the water dissociation constant to determine the concentrations. A basic solution will have $[OH^-] > [H_3O^+]$, and an acid solution will have $[H_3O^+] > [OH^-]$.

Step 1:   Use K_w to calculate the ion concentrations.

Section 15.5   The pH Scale

The pH scale is a logarithmic scale used to express the hydronium ion concentration in a solution:

$$pH = -\log\left[H_3O^+\right]$$

$$\left[H_3O^+\right] = antilog\left(-pH\right) = 10^{-pH}$$

Note that only the digits to the right of the decimal point are significant in a logarithm: the number to the left of the decimal point is an exact number related to the integral power of 10.

As $[H_3O^+]$ increases, the pH decreases:

- Acidic solutions: pH < 7.
- Basic solutions: pH > 7.
- Neutral solutions: pH = 7.

A less-often-used scale is pOH:

$$pOH = -\log\left[OH^-\right]$$

Just as $K_w = [H_3O^+][OH^-] = 1 \times 10^{-14}$, the sum of pH + pOH always equals 14.

Workbook Example 15.5

PROBLEM:
Calculate the $[H_3O^+]$ and $[OH^-]$ for orange juice, which has a pH of 3.87.

SOLUTION:

$$\left[H_3O^+\right] = antilog\left(-pH\right) = 10^{-3.87} = 1.35 \times 10^{-4}\ M$$

$$\left[OH^-\right] = \frac{1 \times 10^{-14}}{1.35 \times 10^{-4}} = 7.41 \times 10^{-11}\ M$$

Workbook Example 15.6

PROBLEM:

Calculate the pH of a solution with a $[H_3O^+]$ of 3.56×10^{-7} M. Is this solution acidic, basic, or neutral?

SOLUTION:

$$pH = -\log\left(3.56 \times 10^{-7}\, M\right) = 6.45$$

The pH is less than 7: this solution is slightly acidic.

Workbook Problem 14.4

Calculate the pH of a solution whose $[OH^-] = 2.35 \times 10^{-6}$ M. Is the solution acidic, basic, or neutral?

Strategy: Calculate the $[H_3O^+]$, then determine pH.

Step 1: Calculate the $[H_3O^+]$, using the expression for the dissociation of water.

Step 2: Calculate the pH.

Step 3: Determine if the solution is acidic, basic, or neutral.

Section 15.6   Measuring pH

To measure pH, there are a number of acid–base indicators: compounds that change color in a specific pH range. They are weak acids with different colors in their acid and conjugate base forms and change color over a range of 2 pH units.

Particularly useful is a mixture of indicators that change color over the entire range of pH: this is known as a *universal indicator*.

For greater precision than is possible with such indicators, pH meters are electronic instruments that measure the pH-dependent electrical potential of the test solution.

Section 15.7   The pH in Solutions of Strong Acids and Strong Bases

Three of the strong acids are monoprotic, meaning that they dissociate completely to form one proton. Because these dissociate completely, $[H_3O^+] = [A^-] = $ initial concentration of the acid, and $[HA] = 0$. This makes the pH of a strong acid solution very straightforward to calculate:

$$pH = -\log[acid]$$

The most common strong bases are alkali metal hydroxides with the formula MOH. These also dissociate completely to form metal ions and hydroxide ions. The pH can be calculated from the $[OH^-]$.

Alkaline earth hydroxides have the formula $M(OH)_2$. When dissolved, they dissociate completely to form one metal ion and two hydroxide ions, but they are not as soluble as the alkali metal hydroxides.

Alkaline earth oxides are very strong bases because the O^{2-} ion is an even stronger base than the hydroxide ion. An O^{2-} ion will immediately pull a proton off a water molecule, forming two hydroxide ions: one from the oxide ion, now bound to a hydrogen, and the other from the split water molecule that is now missing a hydrogen:

$$O^{2-} + H_2O \rightarrow 2\ OH^-$$

Workbook Example 15.6

PROBLEM:
Calculate the pH of a 2.37×10^{-2} M HNO_3 solution.

SOLUTION:
HNO_3 is a strong acid that completely dissociates in water. Therefore, $[H_3O^+] = $ initial concentration of the undissociated acid.

$$pH = -\log\left[H_3O^+\right] = -\log\left[2.37 \times 10^{-2}\right] = 1.63$$

Workbook Example 15.7

PROBLEM:
Calculate the pH of a 2.37×10^{-6} M $Mg(OH)_2$ solution.

SOLUTION:
$Mg(OH)_2$ is a strong base that completely dissociates in water to form a metal cation and two hydroxide anions. Therefore, $[OH^-] = 2 \times$ initial concentration of the undissociated base.

$$\left[OH^-\right] = 2\left(2.37 \times 10^{-6}\ M\right) = 4.74 \times 10^{-6}\ M$$

in dilute aqueous solution, $[OH^-][H_3O^+] = 1 \times 10^{-14}$

$$\left[H_3O^+\right] = \frac{1\times10^{-14}}{4.74\times10^{-6}} = 2.11\times10^{-9} \; M$$

$$pH = -\log\left[H_3O^+\right] = -\log\left(2.11\times10^{-9} \; M\right) = 8.68$$

Workbook Problem 15.5

Calculate the pH of a 7.2×10^{-3} M KOH solution.

Strategy: Determine the $[H^+]$ from the K_w expression and then calculate pH.

Step 1: Determine the concentration of $[OH^-]$.

Step 2: Calculate the $[H_3O^+]$.

Step 3: Calculate the pH.

Section 15.8   Equilibria in Solutions of Weak Acids

A weak acid only partially dissociates in solution, allowing for an equilibrium expression, abbreviated K_a:

$$HA \; (aq) + H_2O \; (l) \leftrightharpoons H_3O^+ \; (aq) + A^- \; (aq)$$

$$K_a = \frac{\left[H^+\right]\left[A^-\right]}{\left[HA\right]}$$

Because it is a pure liquid, water is omitted from the equilibrium expression.

K_a values are experimentally determined: the higher the value, the stronger the acid.

Also sometimes used are pK_a values. We calculate pK_a in the same manner as we do for pH:

$$pK_a = -\log K_a$$

The higher the K_a is, the lower the pK_a will be. It is fair to note that the pK_a scale is used very heavily in organic chemistry, so some of you will be seeing this again!

Workbook Example 15.8

PROBLEM:
The pH of a 0.400 M solution of nicotinic acid ($HC_6H_4NO_2$) is 2.63. Determine the value of K_a for this acid.

SOLUTION:
To determine K_a, first write the balanced equation for the dissociation equilibrium and the equilibrium equation that defines K_a.

$$HC_6H_4NO_2\ (aq) + H_2O\ (l) \rightleftharpoons H_3O^+(aq) + C_6H_4NO_2^-\ (aq)$$

$$K_a = \frac{\left[H_3O^+\right]\left[C_6H_4NO_2^-\right]}{\left[HC_6H_4NO_2\right]}$$

To determine K_a, we need to know the concentrations of the species in the equilibrium mixture. We can determine the concentration of H_3O^+ from the pH.

$$\left[H_3O^+\right] = antilog\left(-pH\right) = 10^{-2.63} = 2.34\times10^{-3}\ M$$

Dissociation of one $HC_6H_4NO_2$ molecule produces one H_3O^+ and $C_6H_4NO_2^-$ ion; the H_3O^+ and $C_6H_4NO_2^-$ concentrations are equal.

$$[H_3O^+] = [C_6H_4NO_2^-] = 2.34 \times 10^{-3}\ M$$

The [$HC_6H_4NO_2$] concentration at equilibrium is equal to the initial concentration minus the amount of $HC_6H_4NO_2$ that dissociates.

$$[HC_6H_4NO_2] = 0.400 - (2.34 \times 10^{-3}) = 0.398\ M$$

$$K_a = \frac{\left[H_3O^+\right]\left[C_6H_4NO_2^-\right]}{\left[HC_6H_4NO_2\right]} = \frac{\left(2.34\times10^{-3}\right)\left(2.34\times10^{-3}\right)}{0.398} = 1.38\times10^{-5}$$

Section 15.9   Calculating Equilibrium Concentrations in Solutions of Weak Acid

Both equilibrium concentrations and pH can be calculated using the K_a or pK_a of a weak acid.

Although there are two reactions occurring, the autoionization of water and the partial dissociation of the weak acid, the autoionization of water can be disregarded when $K_a \gg K_w$ and the [H_3O^+] is attributed only to the acid. When these values are close together, however, both reactions must be considered.

When equilibrium problems involving acids are worked, it is often possible to make the simplifying assumption that so little of the acid dissociates that the initial and final concentrations are the same.

For the equation:

$$HA + H_2O \rightleftharpoons A^- + H_3O^+$$

the equilibrium expression is

$$K_a = \frac{\left[H_3O^+\right]\left[A^-\right]}{\left[HWA\right]}$$

When solving such an equation, it is frequently the case that the amount of dissociation is so small that [HA] before and after dissociation is essentially the same, which greatly simplifies the solution of the equilibrium expression. Proceed with caution, however! The assumption needs to be checked every time a problem is worked, by comparing the value obtained for x with the concentration of the acid. If [HA] $- x \neq$ [HA], the problem will need to be solved using the quadratic equation without simplifying assumptions.

Workbook Example 15.9

PROBLEM:
Calculate the pH and the concentration of all species present (H_3O^+, $C_4H_7O_2^-$, $HC_4H_7O_2$, and OH^-) and the pH in 0.355 M butyric acid ($HC_4H_7O_2$), which has $K_a = 1.5 \times 10^{-5}$.

SOLUTION:
Because $K_a \gg K_w$, it is possible to disregard the contribution of autoionization of water. The dissociation reaction is

$$HC_4H_7O_2\ (aq) + H_2O\ (l) \rightleftharpoons H_3O^+\ (aq) + C_4H_7O_2^-\ (aq)$$

Principal reaction	$HC_4H_7O_2\ (aq)$	$\rightleftharpoons$	H_3O^+	$+$	$C_4H_7O_2^-$
Initial concentration (M)	0.355		0		0
Change (M)	$-x$		$+x$		$+x$
Equilibrium concentration (M)	$0.355 - x$		$+x$		$+x$

Substitute the equilibrium concentrations into the equilibrium expression:

$$K_a = 1.5 \times 10^{-5} = \frac{\left[H_3O^+\right]\left[C_4H_7O_2^-\right]}{\left[HC_4H_7O_2\right]} = \frac{(x)(x)}{0.355 - x}$$

Assume that x is negligible compared with the initial concentration of the acid, so $0.355 - x \approx 0.355$. Using this value in the denominator, solve for x.

$$x^2 = \left(1.5 \times 10^{-5}\right)(0.355)$$

$$x^2 = 5.33 \times 10^{-6}\ M$$

$$x = \sqrt{5.33 \times 10^{-6}\ M} = 2.31 \times 10^{-3}\ M$$

The equilibrium concentrations are

$$[HC_4H_7O_2] = 0.355 - 0.00231 = 0.353 \text{ M}$$
$$[H_3O^+] = [C_4H_7O_2^-] = 2.31 \times 10^{-3} \text{ M}$$

$[OH^-]$ is obtained from the dissociation of water.

$$\left[OH^-\right] = \frac{K_w}{\left[H_3O^+\right]} = \frac{1.0 \times 10^{-14}}{2.31 \times 10^{-3} \text{ M}} = 4.33 \times 10^{-12} \text{ M}$$

$$pH = -\log\left(2.31 \times 10^{-3} \text{ M}\right) = 2.64$$

Workbook Problem 15.6

Determine the pH of a 1.63 M lactic acid ($HC_3H_5O_3$) solution. ($K_a = 1.4 \times 10^{-4}$)

Strategy: Follow the steps outlined previously in your textbook.

Step 1:   Determine the species present initially.

Step 2:   Write the proton-transfer reaction.

Step 3:   Make a table showing the principal reaction and the initial and equilibrium concentrations.

Step 4:   Substitute the equilibrium concentrations into the equilibrium equation, and solve for x. (Since $K_a < 1.0 \times 10^{-3}$, assume that x is negligible and that $1.63 - x \cong 1.63$.)

Step 5:   Calculate the equilibrium concentrations.

Step 6:   Calculate the pH.

Section 15.10 Percent Dissociation in Solutions of Weak Acids

Another way of expressing acid strength is in terms of *percent dissociation*. This is calculated as

$$Percent\ dissociation = \frac{[HA]_{dissociated}}{[HA]_{initial}} \times 100\%$$

The value depends on the concentration and type of acid: in general, the percent dissociation increases with an increasing K_a, and increases in more dilute solutions.

Workbook Example 15.10

PROBLEM:
Calculate the percent dissociation of 0.100 and 1.00 M vitamin C solutions. The molecular formula for ascorbic acid is $C_6H_8O_6$; the $K_a = 8.0 \times 10^{-5}$.

SOLUTION:
We need to know what the concentration is of the conjugate base of ascorbic acid; namely, $C_6H_7O_6^-$. To do this, we need to write the acid dissociation constant expression and solve for $C_6H_7O_6^-$:

$$K_a = \frac{[H_3O^+][C_6H_7O_6^-]}{[C_6H_8O_6]} = 8.0 \times 10^{-5}$$

Working with the first solution ($[C_6H_8O_6] = 0.100$ M):

$$K_a = \frac{x^2}{0.100}$$

$$x^2 = (0.100)(8.0 \times 10^{-5}) = 8.0 \times 10^{-6}$$

$$x = 0.00283\,M$$

This tells us that the concentration of dissociated ascorbic acid is 0.00283 M. Substituting this into the percent dissociation equation:

$$percent\ dissociated = \frac{0.00283\,M}{0.100\,M} \times 100\% = 2.83\%$$

For the next solution, the concentration of ascorbic acid is 1.00 M:

$$K_a = \frac{x^2}{1.00}$$

$$x^2 = (1.00)(8.0 \times 10^{-5}) = 8.0 \times 10^{-5}$$

$$x = 0.00894\,M$$

Substituting this value into the percent dissociated equation gives

$$percent\ dissociated = \frac{0.00894\,M}{1.00\,M} \times 100\% = 0.89\%$$

What this tells us is this: the more dilute the solution, the more we'll see molecules dissociate.

Workbook Problem 15.7

Calculate the percent dissociation of a 0.79 M benzoic acid solution. Benzoic acid has the molecular formula C_6H_5COOH and has a K_a value of 6.5×10^{-5}.

Strategy: To determine the percent dissociation, we need to first determine the concentration of the dissociated HA. This requires a determination of the equilibrium concentration of H_3O^+ and $C_6H_5COO^-$.

Step 1:   Determine the species present initially, and write the proton-transfer reaction.

Step 2:   Make a table showing the principal reaction, initial concentration, change in concentration, and the equilibrium concentration.

Step 3:   Substitute the equilibrium concentrations into the equilibrium expression, and solve for x.

Step 4:   Calculate the concentrations of the major species present.

Step 5:   Calculate the percent dissociation.

Section 15.11   Polyprotic Acids

Some acids, such as sulfuric acid (strong) and phosphoric acid (weak), have more than one proton than can dissociate. As a result, they have a stepwise dissociation, and each step has its own characteristic K_a.

It is more difficult to pull a hydrogen off of an already negatively charged ion, so the dissociation constants decrease dramatically:

$$K_{a1} > K_{a2} > K_{a3}$$

A diprotic acid solution contains a mixture of acids: H_2A, HA^-, and H_2O. The diprotic species is by far the strongest, and as a rule the first dissociation step is the only one that needs to be considered, because the hydronium ions contributed by the second dissociation (or the third in the case of phosphoric acid) are too few to change the pH calculated from the first dissociation.

The exception to the above rule is sulfuric acid, the only strong polyprotic acid. Sulfuric acid dissociates completely at the first dissociation step and has a $K_{a2} = 1.2 \times 10^{-2}$, stronger than the K_a of many weak acids. In this case, it is necessary to calculate the $[H_3O^+]$ generated by both dissociations.

Workbook Example 15.11

PROBLEM:
Calculate the pH of a 0.145 M sulfuric acid solution. K_{a1} = complete, $K_{a2} = 1.2 \times 10^{-2}$.

SOLUTION:
Because K_{a1} is complete, it means that $[H_3O^+]$ is equal to the concentration of $[HSO_4^-]$, which is 0.145 M.

To determine the concentration of H_3O^+ for the second dissociation:

$$K_{a2} = \frac{\left[H_3O^+\right]\left[HSO_4^-\right]}{\left[SO_4^{2-}\right]} = \frac{x^2}{0.145 - x} = 1.2 \times 10^{-2}$$

With a K_{a2} of this magnitude, it is unlikely that the value of x will be negligible. We need to solve for x using the quadratic formula:

$$x^2 + 0.012x = 0.00174 = 0$$

$$x = \frac{-0.012 \pm \sqrt{(0.012)^2 - 4(1)(-0.001740)}}{2(1)}$$

$$x = 0.036\,M$$

The total concentration of H_3O^+ is: 0.145 M + 0.036 M = 0.181 M.

$$pH = -\log(0.181M) = 0.74$$

Workbook Problem 15.8

Write the stepwise dissociation for arsenic acid (H_3AsO_4) and determine the pH of a 1.69×10^{-3} M solution. ($K_{a1} = 5.62 \times 10^{-3}$; $K_{a2} = 1.70 \times 10^{-7}$; $K_{a3} = 3.95 \times 10^{-12}$)

Step 1:   Write the stepwise dissociation for arsenic acid.

Step 2: Determine the principal reaction.

Step 3: Make a table showing the principal reaction, initial concentration, change in concentration, and equilibrium concentration of the reactant and products.

Step 4: Substitute the equilibrium concentrations into the equilibrium expression.

Step 5: Calculate the pH of the solution.

Section 15.12   Equilibria in Solutions of Weak Bases

Weak bases will accept a proton from water, leaving OH⁻ ions and the protonated conjugate acid of the base:

$$B \ (aq) + H_2O \ (l) \leftrightharpoons BH^+ \ (aq) + OH^- \ (aq)$$

The position of the equilibrium is characterized by the base-dissociation constant, K_b.

$$K_b = \frac{\left[OH^- \right]\left[BH^+ \right]}{[B]}$$

Many weak bases are amines, ammonia derivatives in which one of the hydrogen atoms is replaced by another chemical group: the basicity of these compounds is due to the lone pair of electrons on the nitrogen atom, which can be used to bind a proton.

Problems involving weak bases are solved just like those involving weak acids. The only difference is that [OH⁻] is being solved for in the equilibrium equation. To obtain the [H⁺] and the pH of the solution requires using the autoionization constant of water.

Workbook Example 15.12

PROBLEM:
Calculate the pH of a 0.10 M solution of methylamine, CH_3NH_2; $K_b = 3.7 \times 10^{-4}$.

SOLUTION:

$$K_b = 3.7 \times 10^{-4} = \frac{\left[CH_3NH_3^+\right]\left[OH^-\right]}{\left[CH_3NH_2\right]} = \frac{x^2}{0.10 - x}$$

K_b is within 3% of the concentration of the base; use the quadratic equation:

$$x^2 + 3.7 \times 10^{-4} x - 3.7 \times 10^{-5} = 0$$

$$x = \frac{-3.7 \times 10^{-4} \pm \sqrt{\left(3.7 \times 10^{-4}\right)^2 - 4(1)\left(-3.7 \times 10^{-5}\right)}}{2(1)} = 0.0059\,M$$

To solve for pH, calculate the concentration of H^+ using the autoionization constant of water:

$$\left[H^+\right] = \frac{1 \times 10^{-14}}{\left[OH^-\right]} = \frac{1 \times 10^{-14}}{0.0059\,M} = 1.69 \times 10^{-12}\,M$$

$$pH = -\log\left(1.69 \times 10^{-12}\right) = 11.8$$

Workbook Problem 15.9

Determine the pH of a 0.975 M trimethylamine, $(CH_3)_3N$, solution. $K_b = 6.5 \times 10^{-5}$.

Strategy: Solve using a table of equilibrium values.

Step 1:   Write the principal reaction.

Step 2:   Construct a table with concentrations of the reactant and products.

Step 3:   Substitute the equilibrium concentrations into the equilibrium expression, and solve for x.

Step 4:   Determine the equilibrium concentrations of the species present.

Step 5:   Calculate the pH of the solution.

Section 15.13   Relation between K_a and K_b

For a conjugate acid–base pair, K_a and K_b can be related through the ionization constant of water:

$$K_a \times K_b = K_w$$

As the strength of an acid increases, the strength of its conjugate base decreases; likewise, as the strength of a base increases, the strength of its conjugate acid decreases. This makes sense, especially since these values are always proportional to K_w.

Workbook Example 15.13

PROBLEM:
Calculate K_a for methylamine, where $K_b = 1.4 \times 10^{-4}$.

SOLUTION:

$$K_a = \frac{K_w}{K_b} = \frac{1 \times 10^{-14}}{1.4 \times 10^{-4}} = 7.14 \times 10^{-11}$$

Section 15.14   Acid–Base Properties of Salts

The pH of a salt solution is determined by the acid–base properties of the cations and anions that make up that solution. The salt(s) that are formed in an acid–base reaction will contain the conjugate acid and the conjugate base of the reactants. The stronger of the two will predominate:

- Strong acid + Strong base → Neutral solution
- Strong acid + Weak base → Acidic solution
- Weak acid + Strong base → Basic solution

Let's look at these one by one.

1) <u>Strong acid + strong base</u>: When a strong acid and a strong base react, they form a neutral salt; for example, the reaction of sodium hydroxide and hydrochloric acid forms sodium chloride. These are all strong electrolytes, meaning that they have no tendency to associate with anything else. The Cl^- ions in solution are not looking to pull hydrogens off water—they do not even combine with their Na^+ counterparts.

So, what would form neutral cations and anions? Neutral cations would form from Group 1A and 2A metals (except for beryllium), and NO_3^-, Cl^-, ClO_4^-, Br^-, and I^- would form neutral anions.

2) <u>Strong acid + weak base</u>: When a weak base reacts with a strong acid, the anion is neutral, as above, but the cation will be the conjugate acid of a weak base. This weak acid can react with water to form H_3O^+ ions, making the solution slightly acidic.

Small, highly charged metal cations are also acid in solution: the oxygens in the water molecules bind to them so tightly that the hydrogens can dissociate easily.

3) <u>Weak acid + strong base</u>: When a strong base reacts with a weak acid, the opposite occurs: the salts will have a neutral cation that does not wish to associate with anything and a weak base—the conjugate base of the weak acid.

When a weak base and a weak acid react, the pH of the solution will tend toward the stronger of the two compounds:

- K_a (for the cation) $> K_b$ (for the anion) → acidic solution
- K_a (for the cation) $< K_b$ (for the anion) → basic solution
- K_a (for the cation) $\approx K_b$ (for the anion) → neutral solution

---

Workbook Example 15.14

PROBLEM:
Determine whether aqueous solutions of the following salts are acidic, neutral, or basic. Write the hydrolysis reaction for those solutions that are acidic or basic. NH_4ClO_4, $RbCl$, $NaCH_3CO_2$, NH_4CO_3

SOLUTION:
To determine the acidity of an aqueous solution of a salt, we must determine if the salt is derived from a 1) strong acid/strong base reaction, 2) weak acid/strong base reaction, 3) strong acid/weak base reaction, or 4) weak acid/weak base reaction.

NH_4ClO_4: derived from the weak base NH_3 (aq) and the strong acid $HClO_4$ (aq). Acidic (NH_4^+ is the weak conjugate acid of NH_3).

$$NH_4^+ (aq) + H_2O \leftrightarrows NH_3 (aq) + H_3O^+ (aq)$$

$RbCl$: derived from the strong base $RbOH$ (aq) and the strong acid HCl (aq). Neutral.

$NaCH_3CO_2$: derived from the strong base $NaOH$ and the weak acid CH_3COOH. Basic.
($CH_3CO_2^-$ is the weak conjugate base of CH_3COOH.)

$$CH_3CO_2^- (aq) + H_2O \leftrightarrows CH_3COOH (aq) + OH(aq)$$

NH_4CO_3: derived from the weak base NH_3 and the weak acid HCO_3^- (aq). To determine the acidity of this solution we must calculate the K_a of the cation and the K_b of the anion of the salt and compare the two.

$$NH_4^+ \ K_a = \frac{1\times10^{-14}}{K_b \ NH_3} = \frac{1\times10^{-14}}{1.8\times10^{-5}} = 5.6\times10^{-10}$$

$$CO_3^{2-} \ K_b = \frac{1\times10^{-14}}{K_a \ HCO_3^-} = \frac{1\times10^{-14}}{5.6\times10^{-11}} = 1.8\times10^{-4}$$

Since the $K_b > K_a$, this solution has to be basic. The hydrolysis reaction is

$$CO_3^{2-} \ (aq) + H_2O \ (l) \rightleftharpoons HCO_3^- \ (aq) + OH^- \ (aq)$$

Workbook Example 15.15

PROBLEM:
Determine the pH of a 0.150 M solution of NH_4Cl.

SOLUTION:
To determine the acidity of an aqueous solution of a salt, the K_a for the acidic portion must be used.

From the last example problem, the K_a value of the ammonium cation is 5.6×10^{-10}.

$$5.6\times10^{-10} = \frac{[NH_3]\left[H_3O^+\right]}{\left[NH_4^+\right]} = \frac{x^2}{0.150}$$

$$x = 9.17\times10^{-6}$$

$$pH = -\log\left(9.16\times10^{-6}\right) = 5.04$$

Workbook Problem 15.10

Calculate the pH of a 0.137 M KOCl solution. The K_a for HOCl is 3.5×10^{-8}.

Strategy: Determine if KOCl produces an acidic, basic, or neutral aqueous solution, write the hydrolysis reaction for this salt, and determine the pH.

Step 1: Write the hydrolysis reaction for this salt.

Step 2: Write an equilibrium expression for this reaction.

Step 3: Construct a table.

Step 4: Substitute the equilibrium concentrations into the equilibrium expression.

Step 5: Calculate the pH of the solution.

Section 15.15   Lewis Acids and Bases

At this point in the story of acids and bases, we have been introduced to the Arrhenius theory of acids and bases, and we have worked extensively with the Brønsted–Lowry theory of acids and base. Another theory of acids and bases was developed by G.N. Lewis, of the Lewis dot structure fame. Lewis defined a base as an electron-pair *donor* and an acid as an electron-pair *acceptor*.

For bases, the Lewis and Brønsted–Lowry definitions agree, but for acids, the Lewis definition accommodates several situations that are not addressed by the Brønsted–Lowry definition, including cations and neutral molecules with vacant valence orbitals. The Lewis definition explains the acidity of Al^{3+} and CO_2, for example.

This theory is very useful when dealing with organic compounds. If you continue on to organic chemistry (which I hope you will!), you will be working with this in a lot more detail.

Putting It Together

Sulfanilic acid, a compound that is used in making dyes, is produced from the reaction of aniline and sulfuric acid in aqueous solution.

$$C_6H_5NH_2 \ (aq) + H_2SO_4 \ (aq) \rightarrow C_6H_4NH_2SO_3H \ (aq)$$

If you prepare 125 g of sulfanilic acid with an experimental yield of 90%, how much aniline will you use? ($d_{aniline} = 1.02$ g/mL). The K_a of sulfanilic acid is 5.9×10^{-4}. What is the pH of an aqueous solution prepared by dissolving 5.73 g of $NaC_6H_4NH_2SO_3H$ in 250 mL of water?

Self–Test

This section is intended to test your knowledge of the material covered in this chapter. Think through these problems, and make certain you understand what they are asking. Make sure your answers make sense. Successful completion of these problems indicates that you have mastered the material in this chapter. You will receive the greatest benefit from this section if you use it as a mock exam, as this will allow you to determine which topics you need to study in more detail.

True–False

1. The Brønsted–Lowry definition states that acids are electron donors and bases are electron acceptors.

2. A strong acid is more corrosive than a weak acid.

3. H^+ is unstable in solution and attaches to a water molecule to form a hydronium (H_3O^+) ion.

4. A solution with a pH of 7.9 is slightly basic.

5. The strength of an acid and the strength of its conjugate base are directly related: the stronger the acid, the stronger its conjugate base.

6. Dilute solutions of weak acids will have a higher percent dissociation.

7. For most polyprotic acids, two or more dissociations must be considered.

8. Salts can be acidic, basic, or neutral.

9. K_a can be used to calculate the K_b of a weak base.

10. $Mg(OH)_2$ is a weak base.

Multiple Choice

11. In the Arrhenius, Brønsted–Lowry, and Lewis definitions of an acid, acids are
    a. hydrogen ion generators, proton acceptors, and electron acceptors
    b. hydroxide acceptors, proton donors, and electron acceptors
    c. hydrogen ion generators, proton donors, and electron acceptors
    d. hydroxide acceptors, proton acceptors, and electron donors

12. The pH of a salt solution
    a. = 7 for all salts
    b. > 7 for salts derived from a strong acid and a strong base
    c. = 7 for salts derived from a weak acid and a weak base
    d. < 7 for salts derived from a strong acid and a weak base

13. A weak acid or base
    a. dissociates incompletely in solution
    b. has a pH close to 7
    c. dissociates fully in solution
    d. is always less corrosive than a strong acid or base

14. The concentrations of H_3O^+ and OH^- in pure water at 25 °C are
    a. 1×10^{-7} M, 1×10^{-14} M   b. 1×10^{-7} M, 1×10^{-7} M
    c. 1×10^{-14} M, 1×10^{-7} M   d. 1×10^{-14} M, 1×10^{-7} M

15. Solutions with $[H_3O^+] = 1\times10^{-5}$ M have a pH of
    a. 1×10^{-5} b. −5   c. 5   d. 9

16. The strength of a binary acid is affected by
    a. polarity of the HA bond
    b. the size of A–
    c. the strength of the HA bond
    d. all of the above

17. Acid–base indicators
    a. are strong acids
    b. are substances that change color in a specific pH range
    c. have the same colors in their acid (HIn) and conjugate base (In⁻) forms
    d. can be used to determine the exact pH of a solution

18. Which of the following is the strongest acid?
    a. $HClO_4$ b. $HClO_3$ c. $HClO_2$ d. $HClO$

Matching

19. Acid–base indicator

20. Acid dissociation constant

21. Arrhenius acid

22. Arrhenius base

23. Base dissociation constant

24. Brønsted–Lowry acid

25. Brønsted–Lowry base

26. Conjugate acid–base pair

27. Hydronium ion

28. Ion product for water

29. Lewis acid

30. Lewis base

31. Percent dissociation

32. pH

33. Polyprotic acid

34. Strong acid

35. Weak acid

a. K_a

b. H^+ donor

c. chemical species that differ only by one proton

d. molecule that dissociates in water to form OH^- ions

e. electron acceptor

f. a molecule with different colored forms at different pH values

g. K_b

h. dissociated/undissociated ×100

i. $-\log [H_3O^+]$

j. molecule that dissociates in water to release H^+ ions

k. H^+ acceptor

l. acid that is capable of more than one dissociation

m. ion in which H^+ is bonded to the oxygen atom of a solvent water molecule

n. electron donor

o. given by $K_w = \left[H_3O^+\right]\left[OH^-\right]$

p. acid that dissociates incompletely

q. acid that dissociates completely

Fill-in-the-Blank

36. Only a small fraction of _____ acid molecules transfer a _____ to water.

37. In acidic solutions, the concentration of _____ ions is greater than the

    concentration of _____ ions. In _____ solutions, these concentrations

    are _____ .

38. In general, the percent dissociation of an acid _____ with increasing K_a, and

    _____ with increasing concentration.

39. Polyprotic acids have a _____ dissociation in which each step has a much

    _____ K_a.

40. Amines are weak bases. They have _____ that can be donated to a bond

    with _____ .

Problems

41. What is the $[OH^-]$ for a solution of $Ca(OH)_2$ whose pH = 11.78?

42. Calculate the pH of a solution prepared by dissolving 1.83 g of $Ba(OH)_2$ in 150 mL of water.

43. The percent dissociation for a 7.35×10^{-3} M weak acid solution is 0.51%. Calculate the pH of this solution and determine the K_a for the acid.

44. Determine the K_a of histidine, a weak organic acid, if the pH of a 1.30×10^{-3} M solution is 6.03.

45. What is the pH of a 1.24 M solution of pyridine, a weak base whose formula is C_6H_5N? The base dissociation constant is 1.8×10^{-9}.

46. Calculate the concentration of all species present and the pH of a 0.125 M oxalic acid $(H_2C_2O_4)$ solution. The acid dissociation constants are $K_{a1} = 5.9 \times 10^{-2}$, $K_{a2} = 6.4 \times 10^{-5}$.

47. Calculate the pH of a 0.350 M solution of benzylamine. $(K_b = 2.14 \times 10^{-5})$

48. Determine a) K_b for the conjugate base of ascorbic acid $(K_a = 8.0 \times 10^{-5})$ and b) the K_a for the conjugate acid of hydrazine $(K_b = 8.9 \times 10^{-7})$.

49. 1.75 g of KCN is dissolved in 150 mL of water. What is the concentration of HCN (aq) at equilibrium? What is the pH of the solution? The acid dissociation constant is $K_a = 4.9 \times 10^{-10}$.

50. Determine which acid is stronger, and explain why.
    a. HBr (aq) or HI (aq) b. H_3PO_4 or H_3PO_3

Challenge Problem

51. A 25.0% by mass solution of H_3PO_4 has a density of 1.1667 g/mL. Calculate the pH and concentrations of all phosphate-containing species present.

CHAPTER SIXTEEN

Applications of Aqueous Equilibria

Learning Objectives

As a result of reading and studying this chapter, you should be able to

Section 16.1   Neutralization Reactions

1.   Write a balanced equation for a neutralization reaction, calculate the equilibrium constant, and determine whether the pH after neutralization is greater than, equal to, or less than 7.

Section 16.2   The Common-Ion Effect

2.   Calculate the effect of a common ion on concentrations, pH, and percent dissociation in a solution of a weak acid.

3.   Visualize the common-ion effect at the molecular level and predict the effect on pH and percent dissociation of a weak acid.

Sections 16.3–16.4   Buffer Solutions and the Henderson–Hasselbalch Equation

4.   Calculate the pH of a buffer solution and the change in pH on addition of a strong acid or a strong base.

5.   Use the Henderson–Hasselbalch equation to calculate the pH of a buffer solution and to prepare a buffer solution that has a given pH.

Sections 16.5–16.9   pH Titration Curves: strong acid–strong base; weak acid–strong base; weak base–strong acid; diprotic acid–strong base

6.   Calculate the pH at various points in a strong acid–strong base titration.

7.   Calculate the pH at various points in a weak acid–strong base titration.

8.   Calculate the pH at various points in a weak base–strong acid titration.

9.   Calculate the pH at various points in a diprotic acid–strong base titration.

10.   Visualize the molecular species present during a titration and interpret titration curves.

Sections 16.10 – 16.11   Solubility Equilibria for Ionic Compounds

11.   Write the equilibrium-constant expression for dissolution of an ionic compound, and calculate the value of its K_{sp}.

12.   Calculate ion concentrations and the solubility of an ionic compound from its K_{sp}.

Section 16.12   Factors That Affect Solubility

13.   Describe how the presence of molecular and ionic species affects the solubility of an ionic compound.

14.   Calculate solubility in a solution that contains a common ion.

15.   Use the formation constant, K_f, to calculate ion concentrations in a solution that contains a complex ion.

16.   Use the formation constant, K_f, to calculate the solubility of an ionic compound when the cation forms a complex ion.

Sections 16.13–16.15   Precipitation of Ionic Compounds and Separation of Ions in Qualitative Analysis

17.   Calculate the ion product, IP, for an ionic compound, and determine whether a precipitate will form when various solutions are mixed.

18.   Design a scheme for separating ions in a mixture based on selective precipitation.

Chapter Summary

Concepts of equilibrium can be applied to neutralization reactions between strong and weak acids and bases. The titration curves of each of these reactions have distinctive features that make it possible to determine the relative strength of the reactants. The common-ion effect also affects equilibrium mixtures. It is this corollary of Le Châtelier's principle that explains the functioning of buffers: solutions of weak acids and their conjugate bases that resist changes in pH. The pH of a buffer system will be close to the pK_a of the weak acid, and it can be calculated using the Henderson–Hasselbalch equation. The solubility product of an ionic compound, K_{sp}, is the equilibrium constant for its dissolution. This equilibrium constant can be converted to molar solubility and can be altered by changing the conditions surrounding the compound. Using knowledge of these constants and how they can be changed enables us to separate ions in solution through selective precipitation.

The Chapter in Detail

Section 16.1   Neutralization Reactions

When a strong acid and a strong base react, the net ionic equation for the reaction is

$$H_3O^+ \, (aq) + OH^- \, (aq) \rightarrow 2 \, H_2O \, (l)$$

When equal numbers of moles of acid and base are mixed together, $[H_3O^+]$ and $[OH^-] = 1 \times 10^{-7}$ M, the solution is neutral, and the reaction proceeds to completion. The equilibrium constant is the reciprocal of the ion–product constant for water. The reaction leaves behind a neutral salt, and the pH is 7.

When a weak acid and a strong base react, the net ionic reaction involves proton transfer from HA to the strong base:

$$HA \, (aq) + OH^- \, (aq) \rightarrow H_2O + A^- \, (aq)$$

The equilibrium constant for this reaction is obtained by combining the reactions for the dissociation of the weak acid and formation of water (reverse of the dissociation of water). As a result, the equilibrium constant for these reactions is

$$K_n = K_a \left(\frac{1}{K_w} \right)$$

The neutralization reaction will go to completion, but it will leave behind the conjugate base of the weak acid; therefore, the pH will be greater than 7.

When a strong acid and a weak base react, the net ionic equation involves proton transfer from the strong acid to the weak base:

$$H_3O^+ (aq) + B (aq) \rightarrow H_2O (l) + BH^+ (aq)$$

The equilibrium constant for this reaction is obtained by multiplying the equilibrium constant for the dissociation of the base with the formation of water (this is the reverse of the dissociation of water).

$$K_n = K_b \left(\frac{1}{K_w} \right)$$

The neutralization reaction will go to completion, but will leave behind the conjugate acid of the base. As a result, the pH will be less than 7.

When a weak acid and a weak base react, the reaction involves proton transfer from the weak acid to the weak base:

$$HA (aq) + B (aq) \leftrightarrows BH^+ (aq) + A^- (aq)$$

The equilibrium constant for the overall reaction will be the combination of the dissociation of the weak acid, the protonation of the weak base, and the formation of water:

$$K_n = \frac{K_a K_b}{K_w}$$

This type of reaction has far less tendency to proceed to completion than neutralizations involving strong acids or strong bases.

Workbook Example 16.1

PROBLEM:
Write the net ionic equation, and predict the pH for the following reactions:

$$HF (aq) + KOH \qquad\qquad HCl (aq) + CH_3NH_2 \text{ (a weak base)}$$

SOLUTION:

The first equation is the reaction of a weak acid with a strong base. The net ionic equation is

$$HF (aq) + OH^- (aq) \leftrightarrows H_2O (l) + F^- (aq)$$

The pH of this solution will be higher than 7 (basic), because a weak base is left in solution.

The second equation is the reaction of a strong acid with a weak base. The net ionic equation is

$$H_3O^+ (aq) + CH_3NH_2 \leftrightarrows H_2O (l) + CH_3NH_3^+ (aq)$$

The pH of this solution is expected to be less than 7 (acidic), since the conjugate acid of a weak base is a weak acid.

Workbook Problem 16.1

Write a balanced net ionic equation for the neutralization of nitrous acid by sodium hydroxide. Determine K_n and the position of equilibrium for this neutralization reaction. Predict the pH of the solution. The K_a for nitrous acid is 7.24×10^{-4}.

Strategy: Write the net reaction, determine K_n.

Step 1:   Write the net neutralization reaction by writing individual reactions.

Step 2:   Calculate K_n based on the equilibrium constants for the individual reactions in step 1.

Step 3:   Determine the position of equilibrium from the value of K_n.

Step 4:   Predict the p.

Section 16.2   The Common-Ion Effect

A solution of a weak acid and a salt of its conjugate base is an important mixture, because this type of mixture regulates the pH of biological systems.

The common-ion effect is a shift in equilibrium caused by adding more of an ion that is involved in an equilibrium. It is an example of Le Châtelier's principle: adding the salt of the conjugate base of an acid will push the equilibrium toward less dissociation of the acid.

To determine the properties of a solution prepared from a weak acid and a salt of its conjugate base, we need to know the properties of the various species. The principal reaction is the dissociation of the weak acid. The equilibrium calculations must be adjusted for the initial concentration of the conjugate base—this provides an initial concentration of A^-, and the dissociation of the weak acid will provide $[H_3O^+]$ and the change in $[A^-]$.

Workbook Example 16.2

PROBLEM:
Calculate the concentration of all species present, the pH, and the percent dissociation of nitrous acid in a solution that is 0.150 M HNO_2 and 0.075 M $NaNO_2$.

SOLUTION:
Because the salt is 100% dissociated, the species present initially are HNO_2, NO_2^-, Na^+, and H_2O. Na^+ is inert; HNO_2 is a weak acid ($K_a = 4.5 \times 10^{-4}$); NO_2^- is the conjugate base of a weak acid; and H_2O can be either an acid or a base. The principal reaction is transfer of a proton from HNO_2 to H_2O.

Principal reaction	$HNO_2\ (aq) + H_2O\ (l) \rightleftharpoons$	$H_3O^+\ (aq)$	$NO_2^-\ (aq)$
Initial concentration (M)	0.150	0	0.075
Change (M)	$-x$	$+x$	$+x$
Eq. concentration (M)	$0.150 - x$	$+ x$	$0.075 + x$

The common ion in this problem is NO_2^-. The equilibrium equation for the principal reaction is

$$K_a = 4.5 \times 10^{-4} = \frac{\left[H_3O^+\right]\left[NO_2^-\right]}{\left[HNO_2\right]} = \frac{(x)(0.075 + x)}{(0.150 - x)} = \frac{(x)(0.075)}{(0.150)}$$

Because K_a is so small, x is assumed to be negligible. The equilibrium is shifted to the left due to the common-ion effect.

$$x = \left[H_3O^+\right] = \frac{\left(4.5 \times 10^{-4}\right)(0.150)}{(0.075)} = 9.0 \times 10^{-4}$$

Note that the assumption concerning the size of x is justified.

$$pH = -\log\left(9.0 \times 10^{-4}\right) = 3.05$$

The percent dissociation of HNO_2 is

$$percent\ dissociation = \frac{\left[HNO_2\right]_{dissociated}}{\left[HNO_2\right]_{initial}} \times 100\% = \frac{9.0 \times 10^{-4}}{0.150} \times 100\% = 0.60\%$$

Workbook Problem 16.2

Calculate the pH and the percent dissociation of phosphorous acid in a solution that is 0.250 M H_3PO_3 and 0.175 M NaH_2PO_3. The K_a of phosphorous acid is $K_a = 1.0 \times 10^{-2}$.

Strategy: Write the reaction; determine the concentration of H_3O^+, $H_2PO_3^-$.

Step 1:   Write the overall reaction.

Step 2:   Make a chart of initial and equilibrium concentrations.

Step 3:   Solve for the equilibrium concentrations of H_3O^+, $H_2PO_2^-$.

Step 4:   Determine the pH of the solution.

Step 5:   Determine the percent dissociation of the acid.

Section 16.3   Buffer Solutions

Solutions like these, which contain a weak acid and its conjugate base or a weak base and its conjugate acid, are called **buffer solutions**. They are important to biological systems because they are able to resist changes in pH.

If a small amount of a base is added to one of these solutions, it will be neutralized by the weak acid. If a small amount of acid is added, it will be neutralized by the conjugate base.

For a buffer prepared from a weak acid, HA, and its conjugate base, B:

$$\left[H_3O^+\right] = K_a \frac{[HA]}{[B]}$$

Because the dissociations involved in these reactions are so small, these calculations can be done using initial concentrations.

Buffer capacity is a measure of how much acid or base a solution can absorb before the pH begins to be seriously affected. The take-home message of buffer capacity is this: the higher the concentration of the acid and base, the greater the buffer capacity. For the same concentration, a larger volume of solution will also have a greater buffer capacity.

Workbook Example 16.3

PROBLEM:
Which of these solutions will have the greatest buffer capacity? 1 M acetic acid, 0.1 M sodium acetate; 0.1M acetic acid, 1 M sodium acetate; 1 M acetic acid, 1 M sodium acetate; or 0.1 M acetic acid and 0.1 M sodium acetate?

SOLUTION:
The higher concentrations will have the greatest buffer capacities, so the solution that is 1 M in both species will have the highest buffer capacity.

Workbook Problem 16.3

Calculate the pH of a 1.0 L buffer solution containing 0.35 M HCOOH and 0.25 M HCOONa. Determine the pH of this solution after the addition of 0.10 mol HCl (assume no volume change). K_a for formic acid equals 1.8×10^{-4}

Strategy: The pH of the initial solution is determined from K_a and the equilibrium expression. The pH of the solution after addition of HCl is determined knowing the changes in concentration that occur after the addition of $[H_3O^+]$.

Step 1: Determine the principal reaction and equilibrium concentrations.

Step 2: Using the equilibrium equation, solve for $[H_3O^+]$.

Step 3: Solve for the pH.

Step 4: Write the neutralization reaction that occurs upon addition of HCl and set up a table showing the number of moles present before and after the addition of HCl.

Step 5: Determine the concentrations of the buffer components after neutralization occurs.

Step 6: Substitute the concentrations of the buffer components into the equilibrium expression and calculate the pH.

Section 16.4   The Henderson–Hasselbalch Equation

The equation developed above can be rearranged in logarithmic form called the Henderson–Hasselbalch equation:

$$pH = pK_a + \log \frac{[base]}{[acid]}$$

The Henderson–Hasselbalch equation helps us determine the pH of a solution at certain moments during a titration:

- When the ratio of base to acid is 1, the pH will be equal to the pK_a.
- When the pH is not equal to the pK_a, this equation can be used to determine the percent dissociation of the acid:

$$\log \frac{[base]}{[acid]} = pH - pK_a$$

- If the pH of a solution is two units above the pK_a of the weak acid, the pH of the solution is equal to the $pK_a + 2$.

$$\log \frac{[base]}{[acid]} = 2$$

$$\frac{[base]}{[acid]} = 1 \times 10^2 = \frac{100}{1}$$

This indicates that in a sample of molecules, there will be 100 base molecules for every one acid molecule:

$$percent\ dissociation = \frac{100\ base\ molecules}{101\ total\ molecules} \times 100\% = 99\%$$

This equation is also extremely useful for preparing buffer solutions. A weak acid with a pK_a within one pH unit of the desired pH of the solution can be used to make a solution that will hold the desired pH by adjusting the amount of base.

The pH of the final solution is determined only by the mole ratio of base to acid, not by the volume, but the volume will also determine the buffer capacity.

Workbook Example 16.4

PROBLEM:
You are performing an experiment that requires your solution be buffered at a pH of 4.32. Benzoic acid has a K_a of 6.5×10^{-5}. What proportion of benzoic acid to sodium benzoate would give a pH of 4.32?

SOLUTION:
The pK_a of benzoic acid is $-\log (6.5 \times 10^{-5}) = 4.19$.

$$\log\frac{[base]}{[acid]} = pH - pK_a = 4.32 - 4.19 = 0.13$$

$$\frac{[base]}{[acid]} = 10^{0.13} = 1.35$$

A good buffer for this pH would contain 1.35 moles of sodium benzoate for every mole of benzoic acid.

Workbook Problem 16.4

Determine the [sodium acetate]/[acetic acid] ratio for a buffer system with a pH = 5.75. The K_a of acetic acid is 1.8×10^{-5}. How much sodium acetate would you need to add to 1 L of a 0.01 M solution of acetic acid to make a buffer at this pH?

Strategy: Use the Henderson–Hasselbalch equation to calculate the [base]/[acid] ratio. This information provides the mole ratio of acetate to acetic acid.

Step 1:   Calculate the pK_a of acetic acid.

Step 2:   Use the Henderson–Hasselbalch equation to calculate the [base]/[acid] ratio.

Step 3:   Determine the mass of sodium acetate needed.

Workbook Problem 16.5

The pK_a of asparagine is 8.8. At what pH is asparagine 25% dissociated?

Strategy: Using the Henderson–Hasselbalch equation, calculate the log of the [base]/[acid] ratio, then determine the pH at which this value is obtained.

Section 16.5   pH Titration Curves

In an acid–base titration, a solution of a known concentration of base (or acid) is slowly added from a buret to a solution containing an unknown concentration of acid (or base).

This method is used to determine the equivalence point—the point at which stoichiometrically equivalent quantities of acid and base have been mixed together.

Titrations can be graphically represented as titration curves: a plot of the pH of the solution versus the volume of added titrant. These are useful in determining the equivalence point, which allows the experimenter to choose a suitable acid–base indicator.

Section 16.6   Strong Acid–Strong Base Titrations

Before any base is added, the $[H_3O^+]$ = concentration of strong acid. As base is added incrementally, it will neutralize the acid, leading to a slow rise in pH:

$$mmol\ H_3O^+\ after\ neutralization = mmol\ H_3O^+ initial - mmol\ OH^-\ added$$

$$[H_2O]\ after\ neutralization = \frac{mmol\ H_2O^+\ after\ neutralization}{total\ volume\ of\ acid\ and\ base}$$

At the equivalence point, all of the initial acid is neutralized, and the solution contains a neutral salt. The pH will be 7.00.

After the equivalence point, the pH will increase dramatically, because there is an excess of OH^- present.

$$mmol\ of\ excess\ OH^- = mmol\ of\ OH^-\ added - mmol\ of\ acid\ initially\ present$$

$$\left[OH^-\right] after\ neutralization = \frac{mmol\ excess\ of\ OH^-}{total\ volume\ of\ acid\ and\ base}$$

Workbook Problem 16.6

What is the pH that results when 50 mL of 0.1 M NaOH are added to 15 mL of 1.0 M HCl?

Strategy:	Determine the number of moles of H_3O^+ and OH^- present in the mixture and determine the pH based on the excess.
Step 1:	Calculate the moles of OH^- and the moles of H_3O^+.
Step 2:	Determine the excess quantity and the resultant $[H_3O^+]$.
Step 3:	Determine the pH of the solution.

Section 16.7   Weak Acid–Strong Base Titrations

Before any base is added, the pH can be calculated as usual for a solution of weak acid.

As base is added, but before the equivalence point is reached, the amount of conjugate base in the solution is equal to the amount of strong base added: the reaction between the weak acid and the strong base will proceed to completion.

$$Amount\ of\ base\ present\left[A^-\right] = mL\ strong\ base\ added \times \left[strong\ base\right]$$

$$\left[A^-\right] = \frac{mmol\ of\ A^-}{total\ mL\ of\ acid\ and\ base}$$

$$Amount\ of\ acid, HA, after\ neutralization = mmol\ of\ HA\ initially - mmol\ of\ A^-$$

$$\left[HA\right] after\ neutralization = \frac{mmol\ HA\ after\ neutralization}{total\ volume\ of\ acid\ and\ base}$$

At the equivalence point, all the weak acid has been neutralized, leaving behind a basic salt solution: the conjugate base of the weak acid.

$$mmol\ A^-\ at\ equivalence\ point = initial\ mmol\ of\ weak\ acid$$

$$\left[A^-\right] = \frac{initial\ mmol\ of\ HA}{total\ mL\ of\ acid\ and\ base}$$

The pH at the equivalence point will be the pH of the basic salt solution, and it will be greater than 7. After the equivalence point, the pH is determined by the excess OH^- from the strong base.

$$mmol\ OH^- added = mL\ of\ strong\ base \times [strong\ base]$$

$$mmol\ A^- present = mmol\ A^- at\ the\ equivalence\ point$$

$$mmol\ OH^- present = mmol\ OH^- added - mmol\ A^- present$$

$$\left[H_3O^+\right] = \frac{K_w}{\left[OH^-\right]}$$

The pH titration curve for a strong base and a weak acid has a different appearance than that of a strong base and a strong acid. Because of the buffering action of the weak acid/conjugate base mixture that develops before the equivalence point, the pH initially rises quickly, then levels off. Where pH = pK_a, the curve is flattest. Near the equivalence point, the pH increase is smaller than in a strong acid/strong base reaction, and the pH at the equivalence point is greater than 7.

As the K_a of a weak acid increases, the equivalence point gets more difficult to detect: in other words, the weaker the acid, the flatter the curve, because the initial pH is so close to neutral.

Workbook Example 16.5

PROBLEM:
25.0 mL of 0.200 M HF is titrated with 0.100 M NaOH. How many mL of base are required to reach the equivalence point? Calculate the pH at each of the following points: a) after addition of 10.0 mL of base, b) halfway to the equivalence point, c) at the equivalence point, and d) after addition of 65.0 mL of base.

SOLUTION:
To determine the number of mL of base needed to reach the equivalence point, we need to first calculate the number of mmol of HF present.

$$mmol\ HF = 25.0\,mL \times \frac{0.200\,mmol\ HF}{1\,mL} = 5.00\,mmol$$

We need 5.0 mmol of NaOH to reach the equivalence point, which means we need 50 mL of 0.100 M NaOH.

To calculate the pH at the different points in the titration, we need to use the steps that are outlined above.

(i) After addition of 10.0 mL of NaOH:

$$mmol\ F^- = 10.0\,mL \times \frac{0.100\,mmol}{1\,mL} = 1.00\,mmol$$

$$\left[F^-\right] = \frac{1.00\,mmol}{35.0\,mL} = 2.86 \times 10^{-2}\,M$$

$$mmol\ HF = 5.00\,mmol - 1.00\,mmol = 4.00\,mmol$$

$$\left[HF\right] = \frac{4.00\,mmol}{35.0\,mL} = 0.114\,M$$

$$pH = pK_a + \log\frac{\left[F^-\right]}{\left[HF\right]}$$

The K_a for HF is 3.5×10^{-4}; the pK_a is 3.46. Plugging this information into the Henderson–Hasselbalch equation:

$$pH = pK_a + \log\frac{\left[F^-\right]}{\left[HF\right]}$$

$$pH = 3.46 + \log\frac{\left[2.82\times10^{-2}\,M\right]}{\left[0.114\,M\right]} = 2.85$$

(ii) Halfway to the equivalence point:

At this point, 25.0 mL of NaOH have been added to the solution (because 50 mL is required to reach the equivalence point). Following the same steps as before:

$$mmol\ F^- = 25.0\,mL\times\frac{0.100\,mmol}{1\,mL} = 2.50\,mmol$$

$$\left[F^-\right] = \frac{2.50\,mmol}{50.0\,mL} = 0.0500\,M$$

$$mmol\ HF = 5.00\,mmol - 2.50\,mmol = 2.50\,mmol$$

$$\left[HF\right] = \frac{2.50\,mmol}{50.0\,mL} = 0.0500\,M$$

$$pH = 3.46 + \log\frac{\left[0.0500\,M\right]}{\left[0.0500\,M\right]} = 3.46$$

(iii) At the equivalence point:

At this point, all the HF has been neutralized, and the pH is determined by $[F^-]$.

$$mmol\ F^- = 50.0\,mL\times\frac{0.100\,mmol}{1\,mL} = 5.00\,mmol$$

$$\left[F^-\right] = \frac{5.00\,mmol}{75.0\,mL} = 0.0667\,M$$

Remember that F^- is the anion of a weak acid and gives a basic solution. The principal reaction is

$$F^-\ (aq) + H_2O\ (aq) \rightleftharpoons HF\ (aq) + OH^-\ (aq)$$

We can calculate the K_b for this reaction from the K_a:

$$K_b = \frac{K_w}{K_a} = \frac{1.0\times10^{-14}}{3.5\times10^{-4}} = 2.9\times10^{-11}$$

We can also solve for the $[OH^-]$:

$$K_b = \frac{\left[HF\right]\left[OH^-\right]}{\left[F^-\right]} = \frac{x^2}{6.67\times10^{-2}}$$

$$x = \left[OH^-\right] = 1.4\times10^{-6}\,M$$

$$\left[H_3O^+\right] = \frac{1.0\times10^{-14}}{1.4\times10^{-6}} = 7.1\times10^{-9}$$

$$pH = -\log\left(7.1\times10^{-9}\right) = 8.15$$

(iv) After addition of 65.0 mL of NaOH:

$$mmol\ OH^-\ added = 65.0\,mL\times\frac{0.100\,mmol}{1\,mL} = 6.50\,mmol$$

We calculated in the last part that the amount of fluoride ion present was 5.00 mmol. Using this in the equation:

$$mmol\ OH^-\ present = 6.50\,mmol - 5.00\,mmol = 1.50\,mmol$$

$$\left[OH^-\right] = \frac{1.5\,mmol}{90.0\,mL} = 1.67\times10^{-2}\ M$$

$$\left[H_3O^+\right] = \frac{1.0\times10^{-14}}{1.67\times10^{-2}} = 6.0\times10^{-13}$$

$$pH = -\log\left(6.0\times10^{-13}\right) = 12.22$$

Workbook Problem 16.7

What quantity of 0.10 M NaOH is required to titrate 100 mL of 0.050 M phenol ($K_a = 1.3 \times 10^{-10}$) to the equivalence point? Calculate the pH at the equivalence point.

Strategy: Determine the volume of NaOH required to provide equimolar amounts of base and acid; determine the concentration of conjugate base and pH.

Step 1:   Calculate the moles of phenol present and thus the moles of OH⁻ needed.

Step 2:   Determine the volume of sodium hydroxide solution required.

Step 3:   Determine the total volume of the resultant solution.

Step 4:   Solve for the concentration of salt.

Step 5:　　　Set up an equilibrium table.

Step 6:　　　Solve for K_b and [OH].

Step 7:　　　Solve for [H$_3$O] and pH.

Section 16.8　　Weak Base–Strong Acid Titrations

Before any acid is added, the pH can be calculated as usual for a solution of weak base.

As acid is added, but before the equivalence point is reached, the amount of conjugate acid in the solution is equal to the amount of strong acid added: the reaction between the weak base and the strong acid will proceed to completion.

$$Amount\ of\ acid\ present \left[BH^+ \right] = mL\ strong\ acid\ added \times \left[strong\ acid \right]$$

$$\left[BH^+ \right] = \frac{mmol\ of\ BH^+}{total\ mL\ of\ acid\ and\ base}$$

$$Amount\ of\ weak\ base\ after\ neutralization = mmol\ of\ B\ initially - mmol\ of\ strong\ acid\ added$$

$$\left[B \right] after\ neutralization = \frac{mmol\ B\ after\ neutralization}{total\ volume\ of\ acid\ and\ base}$$

At the equivalence point, all the weak base has been neutralized, leaving behind an acidic salt solution: the conjugate acid of the weak base.

$$mmol\ BH^+ at\ equivalence\ point = initial\ mmol\ weak\ base$$

$$\left[BH^+ \right] = \frac{initial\ mmol\ of\ B}{total\ mL\ of\ acid\ and\ base}$$

The pH at the equivalence point will be the pH of the acidic salt solution, and this will be less than 7. After the equivalence point, the pH is determined by the excess H$_3$O$^+$ from the strong acid.

The pH titration curve for a weak base and a strong acid has a different appearance than that of a strong base and a strong acid. Because of the buffering action of the weak acid/conjugate base mixture that develops before the equivalence point, the pH initially drops more quickly, then levels off where the pH equals the pK_a; the curve is flattest in this region. Near the equivalence point, the pH increase is smaller than in a strong acid/strong base reaction, and the pH at the equivalence point is less than 7.

Workbook Problem 16.8

A 40.00 mL solution of 0.375 M trimethylamine, $(CH_3)_3N$, is titrated with 0.100 M HCl.
a) How many mL of acid are required to reach the equivalence point? b) Calculate the initial pH, and the pH after c) the addition of 10.00 mL of acid, d) halfway to the equivalence point, e) at the equivalence point, and f) after addition of 100.00 mL of acid. The K_b of trimethylamine is 6.5×10^{-5}.

Strategy:　　　Use the method outlined above to calculate the pH at various points in the titration.

a)　mL of acid required to reach the equivalence point:
Step 1:　　　Determine the number of mmoles of $(CH_3)_3N$ in the initial solution.

Step 2:　　　Determine the mmol of HCl needed to react with the $(CH_3)_3N$ and the volume of 0.100 M HCl needed.

b)　Initial pH of the $(CH_3)_3N$ solution:
Step 1:　　　Determine the initial pH of the solution.

c)　pH of the solution after addition of 10.00 mL of acid.
Step 1:　　　Calculate the concentration of $(CH_3)_3N$ and its conjugate acid after the addition of 10.00 mL of HCl.

Step 2:　　　Determine the K_a and pK_a of the conjugate acid of $(CH_3)_3N$.

Step 3:   Using the Henderson–Hasselbalch equation, calculate the pH of the solution.

d) pH halfway to the equivalence point:
Step 1:   Use the Henderson–Hasselbalch equation to calculate the pH. Halfway to the
                 equivalence point, [base] = [acid].

e) pH at the equivalence point:
Step 1:   Calculate the concentration of the conjugate acid, using the initial moles of base,
                 and the total volume of solution.

Step 2:   Determine the pH of the solution.

f) pH after the addition of 200.00 mL of 0.100 M HCl:
Step 1:   Determine the number of mmoles of excess HCl.

Step 2:   Determine the concentration of excess HCl, and calculate the pH.

Section 16.9   Polyprotic Acid–Strong Base Titrations

These calculations share the same features of other calculations except that there are two equivalence points. Amino acids have two dissociable protons, and so they will react with two molar amounts of strong base. These titrations are done with protonated amino acids, so they begin at a low pH with the form H_2A^+.

Before the addition of any base, the pH is calculated as an equilibrium problem. The principal reaction is the dissociation of H_2A^+:

$$H_2A^+ \ (aq) \leftrightarrows HA \ (aq) + H_3O^+ \ (aq)$$

Adding strong base before the first equivalence point generates an H_2A^+/HA buffer solution. The concentrations can be calculated as in any weak acid–strong base titration, using the Henderson–Hasselbalch equation to calculate pH. Halfway to the first equivalence point, pH = K_{a1}.

At the first equivalence point, all the H_2A^+ is converted to HA. At this point, the principal reaction will be proton transfer between HA molecules:

$$2\ HA\ (aq) \leftrightarrows H_2A^+\ (aq) + A^-\ (aq) \qquad\qquad K_a = \frac{K_{a2}}{K_{a1}}$$

The pH is the average of pK_{a1} and pK_{a2}.

$$pH = \frac{pK_{a1} + pK_{a2}}{2}$$

For an amino acid, this is also the *isoelectric point*—the neutral form [HA] is at a maximum and the $[H_2A^+]$ and $[A^-]$ are both equal and very small. In biochemistry, the isoelectric point is useful for separating amino acids and proteins.

When strong base is added between the first and second equivalence points, the final proton is removed:

$$HA\ (aq) \leftrightarrows H_3O^+\ (aq) + A^-\ (aq)$$

This generates an HA/A$^-$ buffer solution. The Henderson–Hasselbalch equation can be used to calculate the pH.

At the second equivalence point, all of the HA is converted to A$^-$. $[A^-]$ is equal to the initial concentration of amino acid - $[H_2A^+]$. This is now the solution of a basic salt:

$$A^-\ (aq) + H_2O\ (l) \leftrightarrows HA\ (aq) + OH^-\ (aq) \qquad\qquad K_b = \frac{K_w}{K_{a2}}$$

Any strong base that is added beyond the second equivalence point will determine the pH.

Workbook Problem 16.9

What would the pH be if 25 mg of NaOH was added to 100 mL of a 0.050 M solution of the amino acid proline? (pK_{a1} = 1.952, pK_{a2} = 10.64) What is the isoelectric point of proline?

Strategy: Determine how far this amount of sodium hydroxide takes the titration: use the Henderson–Hasselbalch equation to determine pH. Calculate the isoelectric point from pK_{a1} and pK_{a2}.

Step 1:   Determine how many moles of proline and how many moles of sodium hydroxide are present.

Step 2:   Use the Henderson–Hasselbalch equation to determine the pH.

Step 3:   Calculate the isoelectric point.

Section 16.10   Solubility Equilibria for Ionic Compounds

When a sparingly soluble ionic compound dissolves, an equilibrium is set up between the solid and the dissolved ions in solution. The equilibrium constant for solubility is called the **solubility product constant**, or K_{sp}.

For the generic reaction

$$\text{M}_a\text{X}_b\,(s) \rightleftharpoons a\,\text{M}^{b+}\,(aq) + b\,\text{X}^{a-}\,(aq) \qquad K_{sp} = \left[M^{b+}\right]^a \left[X^{a-}\right]^b$$

Section 16.11   Measuring K_{sp} and Calculating Solubility from K_{sp}

The solubility constant is measured experimentally: the concentrations of an equilibrium mixture of species are measured, either by allowing the solid to come to equilibrium or by generating a solution by adding together soluble components to create a precipitate. The solubility constant is temperature-dependent and can be used to calculate molar solubility.

Workbook Example 16.6

PROBLEM:
Calculate the molar solubility of AgCl if the K_{sp} for AgCl is 1.8×10^{-10}.

SOLUTION:
The solubility equilibrium for AgCl is

$$\text{AgCl} \rightleftharpoons \text{Ag}^+\,(aq) + \text{Cl}^-\,(aq)$$

$$K_{sp} = \left[Ag^+\right]\left[Cl^-\right]$$

$$1.8 \times 10^{-10} = x^2$$

$$x = 1.34 \times 10^{-5}\,M$$

The molar solubility of AgCl is 1.34×10^{-5} M.

Workbook Example 16.7

PROBLEM:
The molar solubility of copper (I) chloride is 1.095×10^{-3} M. Calculate K_{sp} for CuCl.

SOLUTION:
The solubility equilibrium for AgCl is

$$CuCl \rightleftharpoons Cu^+ (aq) + Cl^- (aq)$$
$$K_{sp} = \left[Cu^+\right]\left[Cl^-\right]$$
$$K_{sp} = \left(1.095 \times 10^{-3}\right)^2$$
$$K_{sp} = 1.20 \times 10^{-6}$$

Workbook Problem 16.10

Calculate the K_{sp} for a solution of $CdCl_2$ prepared in pure water, given that the $[Cd^{2+}] = 1.10 \times 10^{-5}$ M.

Strategy: Write the balanced chemical equation for the solubility of $CdCl_2$, and use the stoichiometry and information given to calculate K_{sp}.

Step 1:   Write the solubility equilibrium expression for $CdCl_2$.

Step 2:   Determine the concentration of Cd^{2+} and Cl^- and calculate the K_{sp}

Workbook Problem 16.11

The K_{sp} of $Cu(OH)_2$ is 2.18×10^{-20}. Calculate the molar solubility of $Cu(OH)_2$.

Strategy: Write the solubility equilibrium for $Cu(OH)_2$ and the equilibrium expression. Calculate the molar solubility of $Cu(OH)_2$.

Step 1:   Let x be the number of mol/L of $Cu(OH)_2$ that dissolves. The saturated solution then contains x mol/L of Cu^{2+} and $2x$ mol/L of OH^-. Solve for K_{sp}.

Section 16.12   Factors That Affect Solubility

The common-ion effect will work to decrease the solubility of a slightly soluble ionic compound: adding more of an ion in the compound will tend to cause the common ions that are in solution to immediately precipitate back out.

If the compound contains a basic anion, the solubility will increase as the pH decreases.

The formation of complex ions, in which a Lewis base forms a coordinate covalent bond with the metal cation, will push the equilibrium in the direction of increased solubility by removing dissolved metal cations by forming complex ions.

The formation of complex ions is a stepwise process, but the equilibrium constant for the ion formation incorporates all of the steps. A high K_f indicates the formation of a stable complex ion.

Amphoteric oxides are metal hydroxides that will dissolve in both strongly basic and strongly acidic solutions. This occurs in basic solutions when excess OH^- ions convert the hydroxide into a complex ion in basic solution. In acid solutions, the OH^- is neutralized, and the metal cation becomes soluble.

Only oxides that can be converted to more soluble complex ions exhibit this behavior.

Workbook Example 16.8

PROBLEM:
Determine the molar solubility of $Cu(C_2O_4)$ ($K_{sp} = 2.87 \times 10^{-8}$ at 25 °C) in a 0.25 M $CuCl_2$ solution.

SOLUTION:
The solubility equilibrium expression is

$$Cu(C_2O_4) \leftrightarrows Cu^{2+} (aq) + C_2O_4^{2-} (aq)$$

The equilibrium expression for this reaction is

$$K_{sp} = \left[Cu^{2+} \right]\left[C_2O_4^{2-} \right]$$

Let x be the number of mol/L of $Cu(C_2O_4)$ that dissolves. We can now construct a table showing the equilibrium concentrations:

Solubility equilibrium	$Cu(C_2O_4)_2$ (s)	$\leftrightarrows$	Cu^{2+} (aq) +	$C_2O_4^{2-}$ (aq)
Initial concentration (M)			0.25	0
Equilibrium concentration (M)			0.25 + x	+ x

Given the value of K_{sp}, we can assume that x is negligible; it means that the equilibrium concentration of Cu^{2+} is approximately 0.25 M. Substituting these values into the equilibrium expression gives

$$2.87 \times 10^{-8} = (0.25)(x)$$

$$x = 1.15 \times 10^{-7} \ M$$

Workbook Problem 16.12

The initial pH of a solution containing $Sn(OH)_2$ was 9.45 after addition of excess NH_3. Determine the molar solubility of $Sn(OH)_2$ in this solution. For $Sn(OH)_2$, $K_{sp} = 5.4 \times 10^{-27}$.

Step 1:   Write the solubility equation and solubility expression.

Step 2:   Using the pH, determine the $[OH^-]$ concentration.

Step 3:   Let x be the number of mol/L of $Sn(OH)_2$ that dissolves. Construct a table showing the equilibrium concentrations.

Step 4:   Calculate the solubility of $Sn(OH)_2$.

Workbook Example 16.9

PROBLEM:
Write a balanced net ionic equation for the dissolution reaction between AgBr and $Na_2S_2O_3$, and calculate the equilibrium constant given K_{sp} (AgBr) $= 5.4 \times 10^{-13}$ and K_f {$Ag(S_2O_3)_2^{3-}$} $= 4.7 \times 10^{13}$.

SOLUTION:
The net ionic equation for the dissolution of AgBr in $Na_2S_2O_3$ is obtained by combining the solubility expression for AgBr and the formation expression for $Ag(S_2O_3)_2^{3-}$.

$$AgBr\ (s) \leftrightharpoons Ag^+\ (aq) + Br^-\ (aq) \qquad\qquad K_{sp} = 5.4 \times 10^{-13}$$

$$Ag^+\ (aq) + 2\ S_2O_3^{2-} \leftrightharpoons Ag(S_2O_3)_2^{3-} \qquad\qquad K_f = 4.7 \times 10^{13}$$

Adding the two equations together and canceling out what is common on both the reactant and product sides gives

$AgBr\ (s) \leftrightharpoons \cancel{Ag^+}\ (aq) + Br^-\ (aq)$	$K_{sp} = 5.4 \times 10^{-13}$
$\cancel{Ag^+}\ (aq) + 2\ S_2O_3^{2-}\ (aq) \leftrightharpoons Ag(S_2O_3)_2^{3-}\ (aq)$	$K_f = 4.7 \times 10^{13}$
$AgBr\ (s) + 2\ S_2O_3^{2-}\ (aq) \leftrightharpoons Ag(S_2O_3)_2^{3-}\ (aq) + Br^-\ (aq)$	$K = K_{sp} \times K_f = 25.4$

Section 16.13   Precipitation of Ionic Compounds

Solubility guidelines begin to describe the behavior of ions in solution, but a far more accurate method is the use of the **ion product** (**IP**).

For the salt, M_aX_b, $IP = [M^{b+}]^a[X^{a-}]^b$.

The ion product is calculated using initial concentration, not equilibrium concentrations: this is a reaction quotient, and it will determine which direction the reaction will go using the following guidelines:

- If $IP > K_{sp}$, the solution is supersaturated, and precipitation occurs.
- If $IP = K_{sp}$, the solution is saturated and at equilibrium.
- If $IP < K_{sp}$, the solution is unsaturated and precipitation will not occur.

Workbook Problem 16.13

Will a precipitate form when 100 mL of 0.75 M $Zn(NO_3)_2$ is mixed with 250 mL of 1.50 M Na_2CO_3?

Strategy: Write a metathesis (exchange) reaction for the reaction between $Zn(NO_3)_2$ and Na_2CO_3, and apply the solubility rules in Chapter 4 to determine if a precipitate will form.

Step 1:   Write the solubility equation and IP expression for the precipitate.

Step 2:   Calculate the IP, and compare its value to the K_{sp}. Will a precipitate form?

Section 16.14   Separation of Ions by Selective Precipitation

It is possible to separate a mixture of ions in solution by adding a reagent that will precipitate some of the ions but not others.

Using an acid solution, it is possible to separate insoluble and soluble metal sulfides. The equilibrium constant for this reaction is K_{spa} – the solubility constant in acid solution.

For the general equation

$$MS\ (s) + 2\ H_3O^+\ (aq) \leftrightharpoons M^{2+}\ (aq) + H_2S\ (aq) + 2\ H_2O\ (l)$$

$$K_{spa} = \frac{\left[M^{2+}\right]\left[H_2S\right]}{\left[H_3O^+\right]^2}$$

Using a reaction quotient:

$$Q_c = \frac{\left[M^{2+}\right]\left[H_2S\right]}{\left[H_3O^+\right]}$$

It is possible to adjust $[H_3O^+]$ so that $Q_c > K_{spa}$ for the more insoluble metal sulfide. The more insoluble sulfide will precipitate out, and the more soluble sulfide will stay in solution.

Section 16.15   Qualitative Analysis

Qualitative analysis is a procedure for identifying the ions that are present in an unknown solution.

In the traditional scheme of analysis for metal cations, 20 cations are separated into five groups for selective precipitation.

- Aqueous HCl will remove Ag, Hg, and Pb as insoluble chlorides (some Pb will remain).
- Bubbling H_2S through the acid solution will precipitate out the remaining Pb, as well as Cu, Hg, Cd, Bi, and Sn.
- Addition of NH_3 neutralizes the solution, precipitating out the remaining sulfates: Mn, Fe, Co, Ni, and Zn. Insoluble hydroxides of Al and Cr also precipitate out.
- $(NH_4)_2CO_3$ is now added to precipitate out Ca and Ba.
- $(NH_4)_2HPO_4$ will precipitate out Mg.

All that can remain in solution is K and Na. These can be identified using flame testing: Na imparts a persistent yellow color to a flame, while K imparts a transient violet flame.

The ions in each group can be separated with further analysis. Although there are now more sophisticated techniques for the analysis of metal ions in solution, this method is still an excellent technique for learning laboratory skills and learning about acid–base, solubility, and complex-ion equilibria.

Putting It Together

Aluminum phosphate is formed from the reaction of aluminum chloride and phosphoric acid. a) Write a balanced chemical equation for this reaction. b) If you begin with 95 g of aluminum chloride and 1.50 L of 0.75 M phosphoric acid, how many grams of aluminum phosphate will be produced? c) If you place 25.0 g of aluminum phosphate in water to produce a solution with a volume of 1.00 L, what are the equilibrium concentrations of Al^{3+} and PO_4^{3-}? ($K_{sp} = 1.3 \times 10^{-20}$) d) How does the addition of HCl affect the solubility of aluminum phosphate? e) If you mix 0.75 L of 4.00×10^{-3} M $AlCl_3$ with 1.25 L of 0.700 M Na_3PO_4, will a precipitate of aluminum phosphate form? If so, how many grams of $AlPO_4$ will form?

Self–Test

This section is intended to test your knowledge of the material covered in this chapter. Think through these problems, and make certain you understand what they are asking. Make sure your answers make sense. Successful completion of these problems indicates that you have mastered the material in this chapter. You will receive the greatest benefit from this section if you use it as a mock exam, as this will allow you to determine which topics you need to study in more detail.

True–False

1. Mixing 100 mL of 0.1 M HCl with 100 mL of 0.1 M NaOH will yield a solution with pH 7.

2. Mixing 100 mL of 0.1 M acetic acid with 100 mL of 0.1 M NaOH will yield a solution with pH < 7.

3. Adding a common ion will reduce the dissociation of a weak acid.

4. When OH^- ions are added to a buffer system, the equilibrium shifts toward undissociated acid molecules.

5. When pH = pK_a, all of the weak acid in a solution is dissociated.

6. A strong acid–strong base titration has an equivalence point at pH 7.

7. A strong acid–weak base titration has an equivalence point above pH 7.

8. Molar solubility and K_{sp} must be experimentally determined separately.

9. Metal ions in solution can be separated based on reactivity.

10. Na and K ions in solution can be separated by adding HCl to solution.

Matching

11. Amphoteric	a.	metal cation bonded to one or more small molecules
12. Buffer capacity	b.	solution of weak acid and conjugate acid that resists changes in pH
13. Buffer solution	c.	point in a titration at which molar equivalents have been added to solution
14. Common-ion effect	d.	exhibiting both acidic and basic properties
15. Complex ion	e.	relationship between pH and pK_a of a weak acid, calculated using the amount of dissociation.
16. Equivalence point	f.	procedure for identifying ions in solution
17. Formation constant	g.	shift in equilibrium caused by adding an ion already involved in the equilibrium

18. Henderson–Hasselbalch equation   h. K_{eq} for complex ions

19. Ion product   i. The amount of acid or base a buffer can absorb without a significant change in pH

20. Qualitative analysis   j. Q for K_{sp}

Fill-in-the-Blank
21. The neutralization of a weak acid with a _____ base, will go to _____ because OH⁻ has a great affinity for _____.

22. Solutions which contain a _____ acid and _____ base resist changes in _____ through shifts in _____ . These solutions are called _____ .

23. Complex ions are formed when _____ cations form _____ covalent bonds with small _____ like ammonia or water.

Problems
24. Write balanced net ionic equations, and predict the pH for reactions between
    a. HNO_3 and NaOH
    b. HCl and NH_2NH_2
    c. CH_3COOH (acetic acid) and NaOH

25. Calculate the pH of a solution containing 0.10 mol of NH_3 and 0.25 mol of NH_4Cl in 1.0 L. For ammonia, $K_b = 1.8 \times 10^{-5}$.

26. Determine the pH and concentration of all species present in a buffer solution containing 0.45 mol HClO and 0.25 mol NaClO in 1.0 L of solution. Determine the change in pH upon addition of 0.10 mol NaOH. Determine the change in pH on addition of 0.10 mol HCl. K_a for HClO is 3.5×10^{-8}.

27. Determine the ratio of lactic acid to lactate ion required for preparing a buffer solution whose pH is 4.75. The K_a for lactic acid is 1.4×10^{-4}. If the buffer is to be 0.750 M overall in these compounds, how many moles of each will be added per liter of solution?

28. Determine the change in pH that will occur on addition of 0.300 mol HCl (assuming no volume change) to a 1 L buffer solution prepared from 0.50 M HCOOH and 0.25 M HCO_2Na. The pK_a of formic acid is 3.74.

29. Calculate the pH of 75.0 mL of 0.0400 M HCl after addition of the following volumes of 0.1000M NaOH: a) 0.0 mL, b) 10.0 mL, c) 30.0 mL, d) 50.0 mL.

30. Calculate the pH of 50.0 mL of 0.0500 M acetic acid ($K_a = 1.8 \times 10^{-5}$) after addition of the following volumes of 0.1000 M NaOH: a) 0.0 mL, b) 10.0 mL, c) 25.0 mL, d) 30.0 mL.

31. The molar solubility of $CaCO_3$ is 7.07×10^{-5} M. Calculate the value of K_{sp}.

32. Calculate the molar solubility for $Ca_3(PO_4)_2$, given that the $K_{sp} = 2.1 \times 10^{-33}$.

33. Calculate the solubility of a solution of $CaSO_4$ that contains 0.50 M Na_2SO_4. ($K_{sp} = 7.1 \times 10^{-5}$)

34. Which of the following compounds are more soluble in acidic solution than in pure water? Why?
    a. AgBr
    b. Na_2S
    c. LiCN
    d. CaF_2

35. What are the concentrations of Zn^{2+} and $Zn(NH_3)_4^{2+}$ in a solution prepared by adding 0.50 mol of $Zn(NO_3)_2$ to 1.0 L of 4.0 M NH_3? ($K_f = 7.8 \times 10^8$)

36. Will a precipitate form when 350 mL of 0.75 M $CaCl_2$ is mixed with 200 mL of 1.50 M Na_3PO_4? $K_{sp} = 2.1 \times 10^{-33}$ for calcium phosphate.

37. Is it possible to separate Cu^{2+} from Fe^{2+} by bubbling H_2S through a 0.20 M HCl solution that contains 0.007 M Cu^{2+} and 0.007 M Fe^{2+}? K_{spa} for $CuS = 6 \times 10^{-16}$, K_{spa} for $FeS = 6 \times 10^2$, and $[H_2S] = 0.10$ M.

Challenge Problem

38. CuS has $K_{spa} = 6.0 \times 10^{-16}$. Determine the solubility of CuS in a buffer prepared from 0.45 M formic acid and 0.25 M sodium formate. $K_a = 1.8 \times 10^{-4}$ for formic acid.

CHAPTER SEVENTEEN

Thermodynamics: Entropy, Free Energy, and Equilibrium

Learning Objectives

As a result of reading and studying this chapter, you should be able to

Section 17.1   Spontaneous Processes
1.   Define a spontaneous process and classify various physical processes and chemical reactions as spontaneous or nonspontaneous.

Sections 17.2–17.4   Entropy and Spontaneous Processes
2.   Predict the sign of ΔS for various physical processes and chemical reactions.
3.   Determine the number of arrangements in a system and use Boltzmann's equation to calculate the entropy.
4.   Predict which state of substance has higher entropy.
5.   Calculate ΔS for the expansion or compression of an ideal gas at constant pressure.

Section 17.5   Standard Molar Entropies and Standard Entropies of Reaction
6.   Predict the relative entropies of two substances.
7.   Use standard molar entropies ($\Delta S°$) to calculate the standard entropy of reaction ($\Delta S°_{rxn}$).

Section 17.6   Entropy and the Second Law of Thermodynamics
8.   Describe how the changes in entropy and enthalpy of a system affect spontaneity of a reaction.
9.   Calculate values of ΔS_{sys}, ΔS_{surr}, and ΔS_{total} for a reaction, and use these values to determine whether the reaction is spontaneous. Estimate the temperature at which a reaction changes between spontaneous and nonspontaneous.

Section 17.7   Free Energy and the Spontaneity of Chemical Reactions
10.   Describe how the sign and magnitude of ΔS, ΔH, and temperature affect the value of ΔG and the spontaneity of a reaction.
11.   Calculate values of ΔS, ΔH, and ΔG for a reaction, and use these values to determine whether the reaction is spontaneous. Estimate the temperature at which a reaction changes between spontaneous and nonspontaneous.

Sections 17.8–17.9   Standard Free-Energy Changes and Standard Free Energies of Formation
12.   Define the standard free-energy change ($\Delta G°$) for a reaction, and calculate $\Delta G°$ from $\Delta H°$ and $\Delta S°$.
13.   Use values of $\Delta G°_f$ to determine if a compound is thermodynamically stable.
14.   Calculate $\Delta G°$ for a reaction from values of $\Delta G°_f$, determine if the reaction is spontaneous under standard conditions, and describe how temperature affects spontaneity.

Sections 17.10–17.11   Free Energy, Nonstandard Conditions, and Chemical
                        Equilibrium

15.   Calculate ΔG for a reaction under nonstandard-state conditions.
16.   Use the relationship between $\Delta G°$ and the equilibrium constant for a
       reaction, and be able to calculate one from the other.

Chapter Summary

Spontaneous processes are those that proceed toward equilibrium without any outside influence. Spontaneity is related to entropy, the amount of molecular randomness. The standard molar entropy of substances can be used to calculate the standard entropy of reaction, and the second law of thermodynamics states that in any spontaneous process, the total entropy of the system and its surroundings increases. Free energy is related to entropy through the equation $\Delta G = \Delta H - T\Delta S$. It is a state function that determines the spontaneity of a reaction. The relative importance of the enthalpy (ΔH) and entropy (ΔS) terms is determined by the temperature of a reaction. Free energy of reaction can be calculated either using the equation above, or the equation for free energies of formation. The free energy of a reaction is also mathematically related to the equilibrium constant, K, through the equation $\Delta G° = -RT \ln K$.

Section 17.1   Spontaneous Processes

A **spontaneous process** is one that proceeds on its own without any external influence or input of energy.

Spontaneous reactions always move toward equilibrium. Because of this, their spontaneity is dependent on the same things that the equilibrium is dependent on: temperature, pressure, and the composition of the reaction mixture.

When $Q < K$, the reaction proceeds in the forward direction. When $Q > K$, the reaction proceeds in the reverse direction.

Spontaneity of a reaction is not an indication of the speed of the reaction. The speed of a reaction is a function of kinetics and the height of the **activation energy** barrier.

Workbook Example 17.1

PROBLEM:
Determine which of the following processes are spontaneous and which are nonspontaneous.
a.   The cooling of a cup of tea.
b.   The smell of baking cookies being detectable by sensors throughout a house.
c.   The decomposition of table salt into sodium metal and chlorine gas.
d.   The reaction of S and H_2 to produce SH_2 if the concentration of H_2 is 1 M and the
     concentration of SH_2 is 5 M; the equilibrium constant is 7.8×10^5.

SOLUTION:
Both a. and b. are spontaneous processes and c. is a nonspontaneous process. To determine if d is spontaneous, we need to determine Q for the reaction. If $Q < K$, then the reaction is spontaneous.

If $Q > K$, then the reaction is nonspontaneous in the forward direction, but will be spontaneous in the reverse direction. The equilibrium expression for this reaction is

$$S\ (s) + H_2\ (g) \rightleftharpoons SH_2\ (g)$$

$$Q = \frac{[SH_2]}{[H_2]} = \frac{5\,M}{1\,M} = 5$$

$Q < K$; the reaction is spontaneous in the forward direction.

Section 17.2 Enthalpy, Entropy, and Spontaneous Processes: A Brief Review

Most spontaneous processes release heat, but not all do. For example, ice melts spontaneously, even though this is an endothermic process.

Enthalpy alone cannot account for spontaneity in a chemical reaction. The second factor that is involved is **entropy**: the tendency for a system to move toward maximum randomness.

Entropy, abbreviated S, is a state function, meaning that it is independent of path. When a solid melts, a liquid vaporizes, a gas is heated, or a solute dissolves, the freedom of movement of the individual particles, as well as the number of possible arrangements of those particles both increase; when this occurs entropy will also increase.

Dissolution of ionic solids is a slightly more complex case, which is why not all ionic solids are soluble. When ions are released into solution, they are also **hydrated**—surrounded by an orderly collection of water molecules that neutralizes their charge. This reduces the entropy of the water molecules. As a result, most ionic solids with +1 and –1 charges, such as NaCl, are soluble, while those with higher charges, and subsequently with more water molecules required for hydration, are not.

An increase in entropy is favorable for a reaction to proceed. As with all state functions, the change is calculated as final minus initial, as shown here for entropy:

$$\Delta S = S_{final} - S_{initial}$$

When ΔS is positive, entropy is favorable because disorder has increased. When ΔS is negative, entropy is unfavorable, because order has increased.

Workbook Example 17.2

PROBLEM:
Predict the sign of ΔS for
a. $Br_2\ (l) \rightarrow Br_2\ (g)$
b. $H_2O\ (l) \rightarrow H_2O\ (s)$
c. $NaCl\ (aq) \rightarrow NaCl\ (s)$
d. $Zn\ (s) + 2\ HCl\ (aq) \rightarrow H_2\ (g) + ZnCl_2\ (aq)$

SOLUTION:
a. ΔS is positive; gas molecules are less orderly than liquid molecules.
b. ΔS is negative; liquid molecules decrease in randomness when they freeze into a solid.
c. ΔS is negative; aqueous ions lose their randomness when they condense into a crystal.
d. ΔS is positive; gas is generated, which greatly increases the disorder of the system.

Section 17.3   Entropy and Probability

A random state is more likely than an ordered one. Although we are discussing here the state of things at the molecular level, we can see this in everyday life as well. It would be a rare thing indeed, for instance, to open a clothes dryer and find that the clothes have spontaneously folded themselves. It is much more likely that they are found in one of the multitude of disordered states that are available to them.

Ludwig Boltzmann proposed that the entropy of a particular state is related to the number of ways that this state can be achieved, and it can be calculated by the following formula:

$$S = k \ln W$$

where $k = R/N_A = 1.38 \times 10^{-23}$ J/K, and $\ln W$ = the number of available states.

For a perfect crystal, entropy will be 0, because there is only one way to achieve that state, and $\ln 1 = 0$. Molecules with strong **dipole moments** are more likely to form more orderly crystals. As a result, their entropy will be lower.

Gas expands spontaneously because the state of greater volume is more probable: the larger the volume, the higher the number of disordered states available to the gas particles.

For an ideal gas, it is possible to derive the following corollaries to the Boltzmann equation:

$$\Delta S = nR \ln \frac{V_{final}}{V_{initial}}$$

$$\Delta S = nR \ln \frac{P_{initial}}{P_{final}}$$

From these equations, it is possible to see that increasing the volume or decreasing the pressure both increase the entropy of a gas.

Workbook Problem 17.1

What is the entropy change if the volume of 3 g of hydrogen gas increases from 5 L to 20 L at a constant temperature? (R = 8.314 J/K)

Strategy: Determine the number of moles of hydrogen; calculate ΔS.

Step 1:   Calculate the moles of hydrogen.

Step 2:   Calculate ΔS.

Section 17.4   Entropy and Temperature

As the temperature of a substance increases, the overall kinetic energy of the system also increases.

This affects entropy in several ways. When temperature increases, molecular motion increases, randomness increases, and individual molecular energies occur across a wider spectrum, introducing another area of disorder. All of these events increase the entropy of a system.

A plot of entropy versus temperature will show a steady increase in entropy according to temperature, with large jumps for phase changes.

The third law of thermodynamics states that the entropy of a perfectly ordered crystalline substance at 0 K is zero.

Section 17.5   Standard Molar Entropies and Standard Entropies of Reaction

Standard molar entropies, $S°$ are available for many materials. A standard molar entropy is the entropy of 1 mole of the pure substance at 1 atm pressure and a specified temperature, usually 25 °C. The units of these values are in J/(K•mol). For the same substance, the gas-phase form will have the highest entropy, the solid phase the lowest entropy.

Within the same phase, larger molecules—which have the ability to adopt different conformations and thus can have internal entropy—have higher entropy than smaller molecules.

Entropies of reaction can be calculated from the standard entropies of the reactants and products by subtracting the sum of the entropies of the reactants from the sum of the entropies of the products. Just as with enthalpies of reaction, the individual values must be multiplied by the stoichiometric coefficients.

Workbook Example 17.3

PROBLEM:
Calculate the $\Delta S°_{rxn}$ for

$$2\,NO\,(g) + O_2\,(g) \rightarrow 2\,NO_2\,(g)$$

SOLUTION:
Remember that the equation we need to use is

$$\Delta S^o_{rxn} = \Delta S^o_{products} - \Delta S^o_{reactants}$$

By looking up each of the compounds (keep in mind that states of matter do matter!) in Appendix B of your text, we see that we have the following entropy values for the reactants and products:

$$S° \,(NO) = 210.7 \text{ J/mol·K}$$
$$S° \,(O_2) = 205.0 \text{ J/mol·K}$$
$$S° \,(NO_2) = 240.0 \text{ J/mol·K}$$

Substituting these values into the equation above, we get

$$\Delta S^o_{rxn} = \left[2\left(240.0\,\tfrac{J}{K \cdot mol}\right)\right] - \left[\left(205.0\,\tfrac{J}{K \cdot mol}\right) + 2\left(210.7\,\tfrac{J}{K \cdot mol}\right)\right] = -146.4\,\tfrac{J}{K \cdot mol}$$

Workbook Problem 17.2

Calculate the $\Delta S°_{rxn}$ for the combustion of hydrogen to form water vapor.

$$S° \, (H_2) = 130.6 \text{ J/mol·K}$$
$$S° \, (O_2) = 205.0 \text{ J/mol·K}$$
$$S° \, (H_2O) = 188.7 \text{ J/mol·K}$$

Strategy: Write the formula for the reaction; calculate ΔS.

Step 1: Write the chemical equation.

Step 2: Calculate ΔS.

Section 17.6 Entropy and the Second Law of Thermodynamics

Free-energy changes determine the spontaneity of a reaction; we calculate free energy by using the following equation:

$$\Delta G = \Delta H - T\Delta S$$

The value of the free energy can be interpreted this way: when the free energy (ΔG) > than 0, the reaction is nonspontaneous; when $\Delta G = 0$, the reaction is at equilibrium; and when $\Delta G < 0$, the reaction will occur spontaneously.

Remember that the first law of thermodynamics states that in any process, whether spontaneous or nonspontaneous, the total energy of a system and its surroundings is constant. You may remember that the first law of thermodynamics is really just a restatement of the law of conservation of energy, which says nothing about the spontaneity of a chemical reaction.

The second law of thermodynamics states that in any spontaneous process, the total entropy of a system and its surroundings will always increase. Therefore, we can use the following equation spontaneity:

$$\Delta S_{total} = \Delta S_{system} + \Delta S_{surroundings}$$

The value of ΔS_{total} can be interpreted this way: when ΔS_{total} is positive, the reaction is spontaneous; when ΔS_{total} is equal to 0, the reaction is at equilibrium; and when ΔS_{total} is less than 0, the reaction is nonspontaneous. All reactions proceed spontaneously in the direction that increases the total entropy of the system plus its surroundings.

At constant pressure, $\Delta S_{surroundings} = \dfrac{-\Delta H_{rxn}}{T}$. This makes it possible to rewrite the above equation as:

$$\Delta S_{total} = \Delta S_{system} - \dfrac{\Delta H_{rxn}}{T}$$

Workbook Example 17.4

PROBLEM:
Determine whether the following reaction is spontaneous at 25 °C by determining ΔS_{total}:

$$H_2\,(g) + Cl_2\,(g) \rightarrow 2\,HCl\,(g)$$

SOLUTION:
We need to calculate $\Delta S°$, ΔH_{rxn}, and ΔS_{total}. Appendix B of the textbook provides the following entropy values:

$S°\,(H_2) = 130.6$ J/mol·K $S°\,(Cl_2) = 223.0$ J/mol·K

$S°\,(HCl) = 186.8$ J/mol·K $\Delta H_f°(HCl) = -92.3$ kJ/mol

$$\Delta S°_{rxn} = \left[2\left(186.8\,\tfrac{J}{K\cdot mol}\right)\right] - \left[\left(130.6\,\tfrac{J}{K\cdot mol}\right) + \left(223.0\,\tfrac{J}{K\cdot mol}\right)\right] = 20.0\,\tfrac{J}{K\cdot mol}$$

$$\Delta H°_{rxn} = \left[2\left(-92.3\,\tfrac{kJ}{mol}\right)\right] - \left[\left(0\,\tfrac{kJ}{mol}\right) + \left(0\,\tfrac{kJ}{mol}\right)\right] = -184.6\,\tfrac{kJ}{mol}$$

$$\Delta S_{total} = \Delta S°_{rxn} - \dfrac{\Delta H_{rxn}}{T} = 20.0\,\tfrac{J}{K\cdot mol} - \dfrac{-184,600\,\tfrac{J}{mol}}{298\,K} = 639\,\tfrac{J}{K\cdot mol}$$

The overall entropy change is positive, so the reaction is spontaneous.

Workbook Problem 17.3

Calculate ΔS_{total} for the following reaction at 298 K, and determine if the reaction is spontaneous.

$$AgNO_3\,(aq) + NaBr\,(aq) \rightarrow AgBr\,(s) + NaNO_3\,(aq)$$

The following values may be used in this problem:

$S°\,(Ag^+) = 72.7$ J/mol·K $S°\,(Br^-) = 82.4$ J/mol·K

$S°\,(AgBr) = 107$ J/mol·K $\Delta H°\,(Ag^+) = 105.6$ kJ/mol

$\Delta H°\,(Br^-) = -121.5$ kJ/mol $\Delta H°(AgBr) = -100.4$ kJ/mol

Strategy: Calculate the values of ΔS_{system} and ΔS_{surr}.

Step 1:   Write the net ionic equation for this reaction.

Step 2:   Calculate ΔS_{system}.

Step 3:   Calculate ΔH_{rxn}.

Step 4:   Calculate ΔS_{total}.

Section 17.7   Free Energy and the Spontaneity of Chemical Reactions

In the last section, we discussed the following equation in some detail:

$$\Delta G = \Delta H - T \Delta S$$

There is a way to see how this equation is related to spontaneity. In order to understand it, we have to go back to the ΔS term, remembering that

$$\Delta S_{total} = \Delta S_{system} + \Delta S_{surroundings}$$

Since we know that $\Delta S_{surroundings} = -\Delta H/T$, we can also write the equation as

$$\Delta S_{total} = \Delta S_{system} - \frac{\Delta H_{rxn}}{T}$$

We multiply both sides of the equation by $-T$ so that we get

$$-T \Delta S_{total} = -T \Delta S_{system} + \Delta H$$

Rearranging the right side of the equation, we get

$$-T \Delta S_{total} = \Delta H - T \Delta S_{system}$$

Remember that $\Delta G = \Delta H - T \Delta S$! Replacing the right side of the equation with ΔG, we get

$$-T \Delta S_{total} = \Delta G$$

Based on this last equation, we can now conclude that if ΔG is

- less than 0 ($\Delta G < 0$), the reaction will be spontaneous.
- greater than 0 ($\Delta G > 0$), the reaction will be nonspontaneous.
- lequal to 0 ($\Delta G = 0$), the reaction mixture is at equilibrium.

In any spontaneous process at constant temperature and pressure, the free energy of the system decreases. Temperature can determine the relative importance of the enthalpy and entropy changes.

Using the free-energy equation, it is possible to determine the temperature at which a reaction becomes spontaneous by finding the temperature at which the reaction is at equilibrium. Keeping in mind that a reaction is at equilibrium when ΔG is equal to 0, we can rearrange the equation to get

$$T = \frac{\Delta H^{\circ}}{\Delta S^{\circ}}$$

Workbook Example 17.5

PROBLEM:
Pure chromium is obtained by reducing Cr_2O_3 with aluminum.

$$Cr_2O_3 \ (s) + 2 \ Al \ (s) \rightarrow 2 \ Cr \ (s) + Al_2O_3 \ (s)$$

Determine ΔH° and ΔS° for the reaction. Is the reaction spontaneous at 350 °C?

SOLUTION:
Using Appendix B in your textbook to get the ΔS° and ΔH°_f values that we need, we can solve for ΔH°_{rxn} and ΔS°_{rxn}:

$$\Delta H^{\circ}_{rxn} = \Delta H^{\circ}_f \left(Al_2O_3 \right) - \Delta H^{\circ}_f \left(Cr_2O_3 \right)$$

$$\Delta H^{\circ}_{rxn} = \left[1 \, mol \left(-1676 \, kJ\!\!\Big/\!\!_{mol} \right) \right] - \left[1 \, mol \left(-1140 \, kJ\!\!\Big/\!\!_{mol} \right) \right]$$

$$\Delta H^{\circ}_{rxn} = -536 \, kJ$$

$$\Delta S^{\circ}_{rxn} = S^{\circ}_f \left(Al_2O_3 \right) - S^{\circ}_f \left(Cr_2O_3 \right)$$

$$\Delta S^{\circ}_{rxn} = \left[\left(2 \, mol \, Cr \times 23.8 \, J\!\!\Big/\!\!_{k \cdot mol} \right) + \left(1 \, mol \, Al_2O_3 \times 50.9 \, J\!\!\Big/\!\!_{K \cdot mol} \right) \right] - \left[\left(2 \, mol \, Al \times 28.3 \, J\!\!\Big/\!\!_{K \cdot mol} \right) \right.$$
$$\left. + \left(1 \, mol \, Cr_2O_3 \times \left(-1058 \, kJ\!\!\Big/\!\!_{K \cdot mol} \right) \right) \right]$$

$$\Delta S^{\circ}_{rxn} = 1099.9 \, J\!\!\Big/\!\!_{K}$$

To determine if the reaction is spontaneous, we need to calculate ΔG.

$$\Delta G^{\circ} = \Delta H^{\circ} - T\Delta S^{\circ}$$

$$\Delta G^{\circ} = -536 \, kJ - \left(623 \, K \right) \left(1.0999 \, kJ\!\!\Big/\!\!_{K} \right) = -1221 \, kJ$$

Since the value for ΔG° is negative, the reaction is spontaneous.

Workbook Problem 17.4

Given the following information, what is the normal boiling point of methanol?

$$CH_3OH \ (l) \rightarrow CH_3OH \ (g)$$

$S° \ (l) = 127$ J/mol·K $S° \ (g) = 238$ J/mol·K

$\Delta H_f° \ (l) = -238.7$ kJ/mol $\Delta H_f° \ (g) = -201.2$ kJ/mol

Strategy: Solve for ΔS and ΔH; solve for the temperature at which $\Delta G = 0$.

Step 1:   Calculate ΔS.

Step 2:   Calculate ΔH.

Step 3:   Calculate T.

Section 17.8   Standard Free-Energy Changes for Reactions

The **standard free-energy change**, $\Delta G°$, is the change in free energy that occurs when reactants in their standard states are converted to products in their standard states. It is an extensive property, and refers to the number of moles in the equation.

A standard state is the pure form of a solid, liquid, or gas, or solute at 1 M, 1 atm pressure, and 25 °C.

$\Delta G°$ can be calculated from the standard enthalpy change, $\Delta H°$, and the standard entropy change, $\Delta S°$, using the equation $\Delta G° = \Delta H° - T\Delta S°$.

Workbook Problem 17.5

Determine $\Delta G°$ for the reaction

$$NaCl \ (aq) + AgNO_3 \ (aq) \rightarrow NaNO_3 \ (aq) + AgCl \ (s)$$

given the following information:

	$\Delta H°$ (kJ/mol)	$S°$ (J/mol·K)
AgCl (s)	−127.1	96.2
AgNO$_3$ (aq)	−101.8	219.1
NaNO$_3$ (aq)	−447.5	205.4
NaCl (aq)	−407.3	115.5

Strategy: From the information given, you can determine $\Delta G°$ from the equation
$\Delta G° = \Delta H° - T\Delta S°$.

Step 1: Determine $\Delta H°_{rxn}$.

Step 2: Determine $\Delta S°_{rxn}$.

Step 3: Calculate $\Delta G°_{rxn}$.

Section 17.9   Standard Free Energies of Formation

The **standard free energy of formation,** $\Delta G°_f$, of a substance is the free-energy change of formation of 1 mole of the substance in its standard state from the most stable form of its constituent elements in their standard states. For an element in its most stable form at 25 °C, $\Delta G°_f = 0$.

If $\Delta G°_f$ is negative, the substance is stable and will not decompose back into its elements. If $\Delta G°_f$ is positive, the substance is unstable and prone to decomposition. This says nothing about the rate of decomposition, however.

We can use $\Delta G°_f$ to calculate the standard free-energy changes for reactions:

$$\Delta G° = \Delta G_f^o\left(products\right) - \Delta G_f^o\left(reactants\right)$$

Workbook Example 17.6

PROBLEM:
Calculate $\Delta G°_{rxn}$ for the following reaction:

$$C_2H_4\ (g) + Cl_2\ (g) \rightarrow C_2H_3Cl\ (g) + HCl\ (g)$$

given the following information:

	$\Delta G°_f$(kJ/mol)
$C_2H_4\ (g)$	68.1
$Cl_2\ (g)$	0
$HCl\ (g)$	−95.3
C_2H_3Cl	51.9

SOLUTION:

$$\Delta G_{rxn}^{o} = \left[51.9\,^{kJ}\!/\!_{mol} + \left(-95.3\,^{kJ}\!/\!_{mol}\right) \right] - \left[\left(68.1\,^{kJ}\!/\!_{mol}\right) + 0\,^{kJ}\!/\!_{mol} \right] = -111.5\,kJ$$

Workbook Problem 17.6

Determine ΔG^{o}_{rxn} from ΔG^{o}_{f} for the reaction below. Is the reaction spontaneous? If not, determine the temperature at which the reaction becomes spontaneous.

$$CH_4\,(g) + 2\,Cl_2\,(g) \rightleftharpoons CCl_4\,(l) + 2H_2\,(g)$$

Strategy: Determine ΔG^{o}_{rxn}, then calculate ΔH° and ΔS° to determine the temperature at which the reaction becomes spontaneous. Will the reaction be spontaneous at 400 °C? 45 °C? –20 °C?

	ΔG° (kJ/mol)	ΔH° (kJ/mol)	S° (J/K·mol)
$CH_4\,(g)$	–50.8	–74.8	186.2
$Cl_2\,(g)$	0	0	223.0
$CCl_4\,(l)$	–65.3	–135.4	216.4
$H_2\,(g)$	0	0	130.6

Step 1: Calculate ΔG°_{rxn}; determine if the reaction is spontaneous.

Step 2: Determine the temperature at which the reaction will be spontaneous.

Step 3: Determine the spontaneity at the temperatures listed.

Section 17.10   Free-Energy Changes for Reactions Under Nonstandard-State Conditions

When reactants and products are present at nonstandard-state pressures and concentrations we use the following equation:

$$\Delta G = \Delta G^{\circ} + RT \ln Q$$

where Q is the reaction quotient and takes the same form as the equilibrium constant expression.

Workbook Example 17.7

PROBLEM:
Calculate ΔG for the formation of acetylene (C_2H_2) if the pressure of hydrogen is 415 atm, the pressure of acetylene is 0.1 atm, and the temperature is 25 °C.

SOLUTION:
The formula for this reaction is

$$2\,C\,(s) + H_2\,(g) \rightarrow C_2H_4\,(g)$$

$$\Delta G° = 209.2\,{}^{kJ}\!/_{mol}$$

$$\Delta G = 209200\,{}^{J}\!/_{mol} + 8.314\,{}^{J}\!/_{K\cdot mol}\,(298\,K)\ln\frac{0.1}{415} = 188600\,{}^{J}\!/_{mol} = 189\,{}^{kJ}\!/_{mol}$$

Workbook Problem 17.7

Consider the following reaction:

$$2\,NO\,(g) + O_2\,(g) \rightleftharpoons 2NO_2\,(g)$$

Determine the spontaneity under standard-state conditions and again under the following conditions: $P_{NO} = 0.150\,atm$, $P_{O_2} = 0.250\,atm$, $P_{NO_2} = 0.001\,atm$. Under which conditions is the reaction more spontaneous?

Strategy: Determine $\Delta G°_{rxn}$ and determine which reaction is more spontaneous.

	$\Delta G°$ (kJ/mol)
NO (g)	86.6
O₂ (g)	0
NO₂ (g)	51.3

Step 1: Calculate $\Delta G°_{rxn}$, determine if the reaction is spontaneous.

Step 2: Calculate ΔG.

Section 17.11   Free Energy and Chemical Equilibrium

As a reaction proceeds toward equilibrium, the value of Q changes: when there are no products, $Q = 0$, and $\ln Q$ is infinite. As the reaction proceeds, the total free energy decreases. When the reaction is mostly products, $Q \gg$, $\ln Q$ is greater than 0, and the reaction will proceed spontaneously in the reverse direction.

Free energy can be plotted against reaction progress: there will be a minimum between pure reactants and pure products; this minimum is the equilibrium mixture.

At equilibrium, $\Delta G° = 0$, and $Q = K$, allowing the following relationship to be derived:

$$\Delta G° = -RT \ln K$$

Workbook Example 17.8

PROBLEM:
Calculate K for the following reaction at 25 °C:

$$C_2H_6\ (g) + H_2\ (g) \rightarrow 2\ CH_4\ (g)$$

with the following data:
$\Delta G°$ for $CH_4 = -50.8$ kJ/mol $\Delta G°$ for $C_2H_6 = -32.9$ kJ/mol

SOLUTION:
Calculate $\Delta G°_{rxn}$:

$$\Delta G°_{rxn} = 2\left(-50.8\ {}^{kJ}\!/\!_{mol}\right) - \left(-32.9\ {}^{kJ}\!/\!_{mol}\right) = -68.7\ {}^{kJ}\!/\!_{mol}$$

Calculate K:

$$\ln K = -\frac{\Delta G°}{RT} = -\frac{-68,700\ {}^{J}\!/\!_{mol}}{\left[\left(8.314\ {}^{J}\!/\!_{K \cdot mol}\right)(298\,K)\right]} = 27.7$$

$$e^{\ln K} = e^{27.7} = 1.07 \times 10^{12}$$

Workbook Problem 17.8

Consider the following reaction:

$$I_2\ (g) + Cl_2\ (g) \rightleftharpoons 2ICl\ (g)$$

The equilibrium constant, K_p, for this reaction is 81.9 at 25 °C. What is the standard free-energy change for this reaction?

Strategy: Use the equation $\Delta G° = -RT \ln K$

Step 1: Solve for $\Delta G°$.

Putting It Together

At 37 °C (normal body temperature), $\Delta G° = -14.0$ kJ for the following reaction:

$$Hb-O_2\ (aq) + CO\ (g) \rightleftharpoons Hb-CO\ (aq) + O_2\ (g)$$

When a person is exposed to carbon monoxide gas, it binds preferentially to hemoglobin (Hb) molecules in the blood; Hb normally operates as an oxygen carrier. The free-energy change above shows that binding to carbon monoxide is favored over binding to oxygen, making even a small quantity of carbon monoxide dangerous. If the concentration of carbon monoxide is 10% of the concentration of oxygen, what percentage of hemoglobin molecules are bound to carbon monoxide?

Self–Test

This section is intended to test your knowledge of the material covered in this chapter. Think through these problems, and make certain you understand what they are asking. Make sure your answers make sense. Successful completion of these problems indicates that you have mastered the material in this chapter. You will receive the greatest benefit from this section if you use it as a mock exam, as this will allow you to determine which topics you need to study in more detail.

True–False
1. A spontaneous process occurs without any external energy and occurs quickly.

2. A reaction that is spontaneous and endothermic will have a large positive ΔS.

3. An increase in temperature always leads to an increase in entropy.

4. Standard entropies of reaction cannot be calculated without knowing the enthalpy of reaction.

5. A spontaneous reaction always increases the total entropy of the system plus the surroundings.

6. If ΔH is negative and ΔS is also negative, the reaction will be spontaneous at low temperatures.

7. The temperature of a reaction system determines the relative importance of the enthalpy and entropy terms.

8. The standard enthalpy of formation of a pure element is always negative.

9. ΔH determines whether a reaction is spontaneous or nonspontaneous.

10. The standard molar entropy of a pure substance is always positive.

Matching

11. Entropy

a. the total energy of a system and its surroundings is constant

12. First law of thermodynamics

b. the study of energy changes in chemical reactions

13. Free energy

c. the amount of molecular randomness in a system

14. Second law of thermodynamics

d. a process that proceeds on its own without any external influence

15. Spontaneous process

e. in any spontaneous process, the total entropy of a system and its surroundings always increases

16. Standard state

f. the capacity of a system to continue a chemical reaction.

17. Thermodynamics

g. the entropy of a perfectly ordered crystalline substance at 0 K is zero

18. Third law of thermodynamics

h. most stable form at 25 °C and 1 atm, 1 M for solutions

Fill-in-the-Blank

19. The expansion of a gas is a(n) _____ process that

_____ the entropy of the gas.

20. When an ionic solid dissolves, the entropy of the solid _____, but the entropy of the solvent _____.

21. As the _____ of a system increases, molecular _____ and _____ increase.

22. Standard molar entropies show that _____ molecules have higher entropy than _____ molecules, _____ have higher entropy than liquids, and liquids have higher entropy than _____.

23. Reactions can be made spontaneous by the _____ term, the _____ term, or both. What matters is their relative contributions to _____ _____. The relative contributions of the two terms are determined by the _____.

24. If a substance has a(n) _____ free energy of formation, it is thermodynamically _____. However, it may decompose imperceptibly slowly.

25. A reaction with a positive free energy will be spontaneous in the _____ direction.

Problems

26. Which has the higher entropy?
    a. An orderly dishwasher or a toy box
    b. Sugar crystals in a sugar bowl or sweetened tea
    c. 32 g O_2 gas at STP or 1 mol of O_2 gas at 273 K in a volume of 35.5 L

27. Predict the sign of ΔS for the following processes or reactions:
    a. an increase in the volume of a gas at constant temperature
    b. formation of solid products from aqueous reactants
    c. $MgO\ (s) + 2\ NH_4Cl\ (s) \rightarrow 2\ NH_3\ (g) + MgCl_2\ (s) + H_2O\ (l)$
    d. $N_2\ (g) + H_2\ (g) \rightarrow N_2H_4\ (l)$
    e. separation of $^{235}_{92}U$ from a mixture of $^{235}_{92}U$ and $^{238}_{92}U$.

28. Calculate the standard entropy of reaction for
    a. $2\ NH_4NO_3\ (s) \rightarrow 2\ N_2\ (g) + O_2\ (g) + 4\ H_2O\ (g)$

    $S°\ NH_4NO_3\ (s)\ = 151.1\ J/(mol•K)$ $S°\ N_2\ (g) = 191.5\ J/(mol•K)$
    $S°\ O_2\ (g) = 205.0\ J/(mol•K)$ $S°\ H_2O\ (g) = 188.7\ J/(mol•K)$

    b. $2\ S\ (s) + 3\ O_2\ (g) \rightarrow 2\ SO_3\ (g)$

    $S°\ S\ (s) =\ 31.8\ J/(mol•K)$ $S°\ O_2\ (g) = 205.0\ J/(mol•K)$
    $S°\ SO_3\ (g) = 256.6\ J/(mol•K)$

29. For a given reaction at room temperature, if $\Delta S°_{total} = 1.957 \times 10^3\ J/K$ and $\Delta S°_{rxn} = 515.7\ J/K$.

    Calculate $\Delta H°_{rxn}$.

30. Calculate ΔS if 1 mol of hydrogen gas at STP expands to a volume of 45.3 L at 273 K and 1 atm of pressure.

31. Determine the normal boiling point of ethanol given the following information:
    $S°\ (l)\ \ = 161\ J/mol \cdot K$ $S°\ (g) = 282.6\ J/mol \cdot K$
    $\Delta H_f°\ (l)\ = -277.7\ kJ/mol$ $\Delta H_f°\ (g) = -235.1\ kJ/mol$

32. Calculate ΔS_{total}, ΔS_{sys}, and ΔS_{surr} for the following reaction at 25 °C:

$$H_2\ (g) + Cl_2\ (g) \rightarrow 2\ HCl\ (g)$$

    $S°\ H_2\ (g) = 130.6\ J/(mol•K)$ $\Delta H_f°\ H_2\ (g) = 0\ kJ/mol$
    $S°\ Cl_2\ (g) = 223.0\ J/(mol•K)$ $\Delta H_f°\ Cl_2\ (g) =\ 0\ kJ/mol$
    $S°HCl\ (g) = 186.8\ J/(mol•K)$ $\Delta H_f°\ HCl\ (g) = -92.3\ kJ/mol$

33. Calculate the standard free energy of reaction from the standard free energies of formation for the following reaction:

$$C_2H_4\ (g) + 3\ O_2\ (g) \rightarrow 2\ CO_2\ (g) + 2\ H_2O\ (l)$$

    $\Delta G_f°\ C_2H_4\ (g) = 68.1\ kJ/mol$ $\Delta G_f°\ O_2\ (g) = 0\ kJ/mol$
    $\Delta G_f°\ CO_2\ (g) = -394.4\ kJ/mol$ $\Delta G_f°\ H_2O\ (l) = -237.2\ kJ/mol$

34. Calculate the free-energy change for the following reaction if the partial pressures are 0.5 atm for CO, 15 atm for CO_2, and 0.1 atm for O_2.

$$2\ CO\ (g) + O_2\ (g) \rightleftharpoons 2\ CO_2(g)$$

$\Delta G_f^\circ\ CO\ (g) = -137.2\ \text{kJ/mol}$ $\Delta G_f^\circ\ O_2\ (g) = 0\ \text{kJ/mol}$

$\Delta G_f^\circ\ CO_2\ (g) = -394.4\ \text{kJ/mol}$

35. Determine the equilibrium constant for the following reaction at 37 °C.

$$H_2CO_3\ (aq) \rightleftharpoons CO_2\ (aq) + H_2O\ (l)$$

$\Delta G_f^\circ\ H_2CO_3\ (aq)) = -623\ \text{kJ/mol}$ $\Delta G_f^\circ\ CO_2\ (aq) = -386.0\ \text{kJ/mol}$

$\Delta G_f^\circ\ H_2O\ (l) = -237.2\ \text{kJ/mol}$

36. Alkali metals are refined at high temperatures. What is ΔG_{rxn} for this process? At what temperature does it become spontaneous? What is the equilibrium constant for the reaction at this temperature?

$$2\ NaCl\ (s) \rightleftharpoons 2\ Na\ (s) + Cl_2\ (g)$$

	ΔG° (kJ/mol)	ΔH° (kJ/mol)	S° (J/mol·K)
NaCl (s)	−384.2	−411.2	72.1
Cl_2 (g)	0	0	223.0
Na (s)	0	0	51.2

Challenge Problem

37. Determine the value of K at 25 °C for the reaction

$$N_2\ (g) + O_2\ (g) \rightleftharpoons 2\ NO\ (g)$$

	ΔG° (kJ/mol)	ΔH° (kJ/mol)	S° (J/mol·K)
N_2 (g)	0	0	191.5
O_2 (g)	0	0	205.0
NO (g)	86.6	90.2	210.7

CHAPTER EIGHTEEN

Electrochemistry

Learning Objectives

As a result of reading and studying this chapter, you should be able to

Section 18.1	**Balancing Redox Reactions by the Half-Reaction Method**
1.	Assign oxidation numbers and classify half-reactions as oxidation or reduction reactions.
2.	Use the half-reaction method to balance a redox reaction.
Sections 18.2–18.3	**Galvanic Cells**
3.	Draw and interpret the sketch of a galvanic cell, label the anode and cathode to indicate the direction of electron and ion flow, and write balanced equations for the electrode and overall cell reactions.
4.	Use shorthand notation to represent a galvanic cell.
Section 18.4	**Cell Potentials and Free-Energy Changes for Cell Reactions**
5.	Interconvert between cell potentials and standard free-energy changes for electrochemical reactions.
Sections 18.5–18.6	**Standard Reduction Potentials**
6.	Use tabulated values of standard reduction potentials for half-reactions (Appendix D) to order substances by increasing strength as an oxidizing or reducing agent.
7.	Use standard reduction potentials for half-reactions to calculate a standard cell potential and determine spontaneity.
Sections 18.7–18.8	**Cell Potentials under the Nonstandard Conditions: The Nernst Equation**
8.	Calculate the cell potential under nonstandard-state conditions using the Nernst equation.
9.	Use the cell potential to calculate solution concentration.
Section 18.9	**Standard Cell Potentials and Equilibrium Constants**
10.	Use standard reduction potentials to calculate the equilibrium constant.
Sections 18.10–18.11	**Batteries and Corrosion**
11.	Describe batteries and fuel cells, and calculate values of $E°$, $\Delta G°$, and K for the cell reaction.
12.	Describe how corrosion occurs and methods used to prevent it.
Sections 18.12–18.14	**Electrolysis**
13.	Draw and interpret the sketch of an electrolytic cell, label the anode and cathode, indicate the direction of electron and ion flow, and write balanced equations for the electrode and overall cell reactions.

14.   Predict the products formed and the electrode and overall cell reactions when an aqueous solution of an ionic compound is electrolyzed.

15.   Relate the current, time, and amount of product produced in an electrolytic cell.

Chapter Summary

Electrochemistry is the area of chemistry concerned with the interconversion of chemical and electrical energy. Galvanic cells convert chemical energy to electrical energy. The standard cell potential of these reactions can be calculated from tables of standard half-reactions, and potentials at nonstandard-state conditions can be calculated using the Nernst equation. These potentials can also be used to calculate the free-energy changes and equilibrium constants of the reactions. The most common applications of these reactions are in batteries and fuel cells, with the same processes causing the much-less-productive process of corrosion. When the process is reversed, electrical energy can be used to drive nonspontaneous chemical changes. This process, known as electrolysis, is used to produce sodium, chlorine, and sodium hydroxide; it is also used in electrorefining and electroplating. The products obtained during electrolysis depend on the overvoltage required to drive the processes, not only on the standard voltages of the reactions involved. Quantitative calculations involving electrolytic processes require that the amount of current, and thus the number of electrons that flow into the reaction, be known.

The Chapter in Detail

Section 18.1   Balancing Redox Reactions by the Half-Reaction Method

By the nature of redox reactions, there is a reduction going on at the same time as an oxidation. Separating the reaction into two half-reactions provides another way to balance these complex equations.

In this method, the two portions of the equation are separated and balanced alone, just as the reactions above were balanced. To balance for charge, electrons must be added in.

Because electrons do not appear in the final equation, the proportions of the two half-reactions relative to each other have to be adjusted so that when the equations are added back together, the electrons on the product side and the reactant side cancel out.

Workbook Example 18.1

PROBLEM:
Using the half-reaction method, balance the following reaction, which takes place in an acidic solution:

$$Mn^{2+}(aq) + NaBiO_3(s) \rightarrow Bi^{3+}(aq) + MnO_4^-\ (aq)$$

SOLUTION:
1. Write two unbalanced half-reactions:

Oxidation: $Mn^{2+}(aq) \rightarrow MnO_4^-\ (aq)$ (Mn goes from +2 to +7)

Reduction: $NaBiO_3(s) \rightarrow Bi^{3+}(aq)$ (Bi goes from +5 to +3)

---

2. Balance each half-reaction for atoms other than H and O.

$$Mn^{2+}(aq) \rightarrow MnO_4^- (aq) \qquad\qquad NaBiO_3(s) \rightarrow Bi^{3+} (aq) + Na^+ (aq)$$

3. Add H_2O to balance-in oxygen and H^+ to balance-in hydrogen.

$$4\ H_2O\ (l) + Mn^{2+}(aq) \rightarrow MnO_4^- (aq) + 8\ H^+ (aq)$$
$$6\ H^+ (aq) + NaBiO_3(s) \rightarrow Bi^{3+} (aq) + Na^+ (aq) + 3\ H_2O\ (l)$$

4. Balance each reaction for charge.

$$4\ H_2O\ (l) + Mn^{2+}(aq) \rightarrow MnO_4^- (aq) + 8\ H^+ (aq) + 5e^-$$
$$2\ e^- + 6\ H^+ (aq) + NaBiO_3(s) \rightarrow Bi^{3+} (aq) + Na^+ (aq) + 3\ H_2O\ (l)$$

5. Make the electron count the same in both reactions.

$$(4\ H_2O\ (l) + Mn^{2+}(aq) \rightarrow MnO_4^- (aq) + 8\ H^+ (aq) + 5e^-) \times 2$$
$$(2\ e^- + 6\ H^+ (aq) + NaBiO_3(s) \rightarrow Bi^{3+} (aq) + Na^+ (aq) + 3\ H_2O\ (l)) \times 5$$
$$8\ H_2O\ (l) + 2\ Mn^{2+}(aq) \rightarrow 2\ MnO_4^- (aq) + 16\ H^+ (aq) + 10\ e^-$$
$$10\ e^- + 30\ H^+ (aq) + 5\ NaBiO_3(s) \rightarrow 5\ Bi^{3+} (aq) + 5\ Na^+ (aq) + 15\ H_2O\ (l)$$

6. Add the two half-reactions together, canceling anything that appears on both sides of the equation.

$$\cancel{8\ H_2O\ (l)} + 2\ Mn^{2+}(aq) + \cancel{10\ e^-} + 14\ \cancel{30}H^+(aq) + 5\ NaBiO_3(s) \rightarrow$$
$$2\ MnO_4^- (aq) + \cancel{16\ H^+} (aq) + \cancel{10e^-} + 5Bi^{3+} (aq) + 5\ Na^+ (aq) + 7\ \cancel{15}H_2O\ (l)$$

$$2\ Mn^{2+}(aq) + 14\ H^+ (aq) + 5\ NaBiO_3(s) \rightarrow$$
$$2\ MnO_4^- (aq) + 5\ Bi^{3+} (aq) + 5\ Na^+ (aq) + 7\ H_2O\ (l)$$

7. Check to make sure all atoms and charges are balanced.

Reactant side:	2Mn	14H	5Na	5Bi	15O	net charge: +18
Product Side:	2Mn	14H	5Na	5Bi	15O	net charge: +18

Workbook Problem 18.1

Using the half-reaction method, balance the following reaction, which takes place in a basic solution:
$$KMnO_4 + Na_2SO_3 \rightarrow MnO_2 + Na_2SO_4$$

Strategy: Follow the steps in the preceding worked example.

Step 1: Write two unbalanced half-reactions.

Step 2:   Balance each half-reaction for atoms other than H and O.

Step 3:   Add H_2O to balance-in oxygen and H^+ to balance-in hydrogen.

Step 4:   Balance each reaction for charge.

Step 5:   Make the electron count the same in both reactions.

Step 6:   Add the two half-reactions together, canceling anything that appears on both sides of the equation.

Step 7:   Make the solution basic by adding 1 OH^- to each side for every H^+.

Step 8:   Cancel water molecules that appear on both sides of the equation.

Step 9:   Check your answer to make sure both atoms and charges are balanced.

Section 18.2   Galvanic Cells

In galvanic cells, a spontaneous chemical reaction generates an electric current. In electrolytic cells, an electric current drives a nonspontaneous reaction.

In redox reactions, recall that oxidation is a loss of electrons (an increase in oxidation number), and reduction is a gain of electrons (a decrease in oxidation number).

To make it easier to see what is happening in these types of reactions, they are typically broken into half-reactions: one for the oxidation portion (with electrons as products) and another for the reduction portion (with electrons as reactants). The species that causes the oxidation and accepts the electrons is referred to as the *oxidizing agent*, while the species that gives up the electrons is referred to as the *reducing agent*.

These reactions can be carried out directly, as when a strip of zinc is placed in a beaker that contains copper ions; in this case, the copper is reduced by accepting electrons directly off the zinc metal, while the zinc is oxidized by donating electrons to the copper ions. Any enthalpy of reaction is lost into the solution.

Redox reactions can also be separated and run in a Daniell or other type of galvanic cell. In this situation, the oxidation and reduction portions of the reaction are run separately. The metals are immersed in a solution that contains a common anion and are connected by a wire through which electrons flow from the reducing agent to the oxidizing agent.

Specific names are given to the two metals, which are known as **electrodes**. The **anode** is the electrode at which oxidation takes place. It is defined as the negative (–) electrode and electrons are produced at this end. The **cathode** is the electrode at which reduction takes place. It is defined as the positive (+) electrode, and it consumes electrons.

The solutions are connected by a salt bridge, a U-shaped tube containing a gel suffused with an electrolyte that does not participate in the reaction. Neutrality of the solutions is maintained by the electrolyte in the salt bridge. Anions move toward the anode, and cations move toward the cathode.

The half-reaction at the anode and the half-reaction at the cathode must add together to give the overall reaction.

Workbook Example 18.2

PROBLEM:
Describe the construction of a galvanic cell based on the following reaction:

$$Ni^{2+} (aq) + Mg (s) \rightarrow Ni (s) + Mg^{2+} (aq)$$

SOLUTION:
Start by taking the overall cell reaction and breaking it into two half-reactions.

$$Ni^{2+} (aq) + 2\ e^{-} \rightarrow Ni (s)$$

$$Mg (s) \rightarrow Mg^{2+} (aq) + 2\ e^{-}$$

Looking at the two half-reactions, we find that the Ni^{2+} is being reduced and the Mg is being oxidized. Therefore, the anode compartment of our cell consists of a strip of magnesium metal immersed in a solution containing Mg^{2+} ions (such as magnesium nitrate). The cathode compartment consists of a strip of nickel immersed in a solution containing Ni^{2+} ions (such as nickel(II) nitrate). The two half-cells are connected to each other with a salt bridge and an external wire. Electrons flow through the wire from the magnesium anode to the nickel cathode. Anions move from the cathode compartment toward the anode, while cations migrate from the anode compartment toward the cathode.

Workbook Problem 18.2

Describe the construction of a galvanic cell based on the following reaction:

$$Pb^{2+} (aq) + Cd (s) \rightarrow Pb (s) + Cd^{2+} (aq)$$

Strategy: Break the overall reaction into half-reactions and design a galvanic cell, identifying the cathode and the anode.

Step 1:   Write the half-reactions, identifying the oxidizing agent and the reducing agent.

Step 2:   Identify solutions for each half-cell and a suitable inert electrolyte for the salt bridge.

Section 18.3   Shorthand Notation for Galvanic Cells

To make it easier to see what is going on in a galvanic cell, a shorthand notation is used. For the example above, the equation

$$Ni^{2+} (aq) + Mg (s) \rightarrow Ni (s) + Mg^{2+} (aq)$$

would be written

$$Mg \mid Mg^{2+}(aq) \parallel Ni^{2+} (aq) \mid Ni$$

In this shorthand, the single vertical line, $\mid$, represents a phase boundary—the solid electrode and the aqueous solution—while the double vertical line, $\parallel$, represents the salt bridge.

The anode half-cell (oxidation) is always written on the left, and the electrons flow to the right into the cathode half-cell. Reactants are written first, followed by products.

If a gas is involved (as when hydrogen is reduced or chlorine oxidized), an additional line will be present to identify the electrode:

$$Zn \mid Zn^{2+}(aq) \parallel H^{+} (aq) \mid H_2 (g) \mid Pt (s)$$

A very detailed version of the shorthand notation also includes ion concentrations and gas pressures.

Workbook Example 18.3

PROBLEM:
Give the shorthand notation for a galvanic cell that employs the overall reaction

$$Ni(NO_3)_2 \ (aq) + Cd \ (s) \rightarrow Ni \ (s) + Cd(NO_3)_2 \ (aq)$$

Give a brief description of the cell.

SOLUTION:
The two half-reactions for this overall reaction are

$$Ni^{2+} \ (aq) + 2 \ e^- \rightarrow Ni \ (s)$$

$$Cd \ (s) \rightarrow Cd^{2+} \ (aq) + 2 \ e^-$$

From these half-reactions, we know that nickel is being reduced and cadmium is being oxidized. Therefore, cadmium is the anode and nickel is the cathode. The cell notation is

$$Cd \ (s) \ | \ Cd^{2+} \ (aq) \ || \ Ni^{2+} \ (aq) \ | \ Ni \ (s)$$

This cell consists of a strip of cadmium metal as the anode dipping into an aqueous solution of $Cd(NO_3)_2$ and a strip of nickel metal as the cathode dipping into an aqueous solution of $Ni(NO_3)_2$. The two half-cells are connected by a salt bridge and a wire.

Workbook Problem 18.3

Given the shorthand notation below, determine the half-reactions occurring at the anode and cathode. Write the overall reaction for the cell and give a brief description of the cell's construction.

$$Mg \ (s) \ | \ Mg^{2+} \ (aq) \ | \ | \ Cr^{3+} \ (aq) \ | \ Cr \ (s)$$

Strategy: Based on the shorthand notation above, determine which species is the cathode and which is the anode, and write the half-reaction that occurs at each electrode.

Step 1: Write the half-reaction occurring at the anode. Remember, the reactant in each half-cell is written first, followed by the product.

Step 2: Write the half-reaction occurring at the cathode.

Step 3:   Combine the two half-reactions to obtain the overall cell reaction (recall that the equation must be balanced in electrons).

Step 4:   Describe the cell indicated in the shorthand notation.

Section 18.4   Cell Potentials and Free-Energy Changes for Cell Reactions

The chemical potential that pushes electrons away from the anode and toward the cathode is known as the **electromotive force**, the **cell potential** (E) or the **cell voltage**. The potential of a galvanic cell is defined as a positive quantity with units of volts (V).

The following relationships are found among the units of energy, electrical charge, and current:

$$1\,watt = 1\frac{Joule}{sec} = \frac{1\,Coulomb \times 1\,volt}{sec} = 1\,ampere \times 1\,volt$$

Watts are a measure of power, coulombs are a measure of charge, volts are a measure of potential, and amperes are a measure of current.

Cell potential is measured with a voltmeter: when the + and – terminals of the voltmeter are connected to cathode (+) and anode (–), a positive value will be obtained for the voltage. This makes it possible to use a voltmeter to determine which electrode is the cathode and which is the anode.

Cell potential and free-energy are both driving forces for chemical reactions. They are related by the following equation:

$$\Delta G = -nFE$$

where n = moles of electrons transferred in the reaction and F is the *Faraday constant*. The Faraday constant is the electrical charge per mole of electrons and is equal to 96,500 C/mol e⁻.

Note that ΔG and E have opposite signs: a spontaneous reaction is indicated by a negative free-energy change or a positive cell potential.

Standard cell potential, $E°$, is the cell potential when both reactants and products are in their standard states: solutes at 1 M concentration, gases at a partial pressure of 1 atm, solids and liquids in pure form, at 25 °C.

Workbook Example 18.4

PROBLEM:
If the standard cell potential for the following reaction is 2.37 V, what is the standard free-energy change? Write the shorthand notation for this reaction.

$$Al\ (s) + Fe^{3+}\ (aq) \rightarrow Al^{3+}\ (aq) + Fe\ (s)$$

SOLUTION:
Use the equation $\Delta G° = -nFE°$

F is 96,500 C/mol, and n can be inferred from the balanced equation. Three electrons are being transferred per mole of reactants, so $n = 3$.

$$\Delta G° = -3\left(96500\frac{C}{mol}\right)(2.37V)\left(1\frac{J}{C \cdot V}\right) = -686\,kJ\big/_{mol}$$

The shorthand notation for this reaction is $Al\ (s)|\ Al^{3+}\ (aq)\ \|\ Fe^{3+}\ (aq)\ |\ Fe\ (s)$

Workbook Problem 18.4

The reaction shown below in shorthand notation has a standard cell potential of 1.61 V. Write the equation for this reaction and calculate the standard free-energy change.

$$Mg\ (s)\,|\,Mg^{2+}\ (aq)\,|\,|\,Zn^{2+}\ (aq)\,|\,Zn\ (s)$$

Strategy: Based on the shorthand notation above, write the equation. Using the formula $\Delta G° = -nFE°$, calculate the standard free-energy change.

Step 1:   Write the balanced equation for the reaction.

Step 2:   Identify n and use $\Delta G° = -nFE°$ to calculate $\Delta G°$.

Section 18.5   Standard Reduction Potentials

The standard potential of a galvanic cell is the sum of the half-cell potentials:

$$E^o_{cell} = E^o_{oxid} + E^o_{red}$$

Only a potential *difference* can be measured, so to determine half-reaction potentials, a standard electrode must be used against which all other reactions can be measured.

The standard hydrogen electrode (S.H.E.) is assigned an arbitrary potential of exactly 0 V, and all other voltages are relative to this reaction:

$$2\,H^+\,(aq,\,1\,M) + 2\,e^- \rightarrow H_2\,(g,\,1\,atm) \qquad E° = 0.00\,V$$

The shorthand notation for S.H.E. is $H^+\,(1\,M)\,|\,H_2\,(1\,atm)\,|\,Pt\,(s)$, where the Pt electrode is in contact with both H_2 gas and $H^+\,(aq)$.

Standard potentials for half-cells can be determined by constructing a galvanic cell in which the half-cell of interest is paired up with the standard hydrogen electrode: standard oxidation potential is the half-cell potential for an oxidation half-reaction, and standard reduction potential is the half-cell potential for a reduction half-reaction.

As with free-energy changes, when the direction of a half-reaction is reversed, the sign of $E°$ must be reversed.

In a table of half-cell reaction, all the reactions are written as reductions. Oxidizing agents are on the reactant side, and reducing agents are on the product side. Reactions are listed in decreasing order of standard reduction potential, with the strongest oxidizing agents at the upper left and the strongest reducing agents at the lower right.

Section 18.6   Using Standard Reduction Potentials

Tables of standard reduction potentials summarize an enormous quantity of chemical information in a small space. They make it possible to arrange oxidizing or reducing agents in order of strength and predict the spontaneity of thousands of redox reactions.

To balance the equations in electrons, it may be necessary to place coefficients on the half-reactions. What this means is that it may be necessary to multiply half-reactions by some factor to ensure that electrons cancel. In this case, the values of $E°$ are *not* multiplied by the same factor: the voltage and free-energy depend on the number of electrons transferred, not on the stoichiometry of the atoms.

An oxidizing agent can oxidize any reducing agent that lies below it in the table. For the reaction to be spontaneous, $E°$ must be positive.

Workbook Example 18.5

PROBLEM:
Write the balanced net ionic equation, and calculate $E°$ for the following galvanic cell:

$$Al\,(s)\,|\,Al^{3+}\,(aq)\,||\,Cl_2\,(g)\,|\,Cl^-\,(aq)$$

SOLUTION:
Al (s) is the anode and therefore undergoes oxidation, while Cl_2 (s) is the cathode and undergoes reduction. The half-reactions and their cell potentials are

$$Al\,(s) \rightarrow Al^{3+}\,(aq) + 3\,e^- \qquad E° = +1.66\,V$$

$$Cl_2\,(g) + 2\,e^- \rightarrow Cl^-\,(aq) \qquad E° = +1.36\,V$$

(Notice that the sign for $E°$ for the Al/Al^{3+} half-reactions has been reversed.) To write the balanced net ionic equation, we need to make sure that the electrons cancel out on both sides. Therefore, we need to multiply the top reaction by 2 and the bottom reaction by 3.

$$2\ Al\ (s) \rightarrow 2\ Al^{3+}\ (aq) + 6\ e^- \qquad E° = +1.66\ V$$

$$3\ Cl_2\ (aq) + 6\ e^- \rightarrow 6\ Cl^-\ (aq) \qquad E° = +1.36\ V$$

Remember that $E°$ values are independent of the amount of reaction and are not multiplied by stoichiometric coefficients.

$$E^o_{cell} = E^o_{Al \rightarrow Al^{3+}} + E^o_{Cl_2 \rightarrow Cl^-} = +1.66V + 1.36V = 3.02V$$

Workbook Problem 18.5

Using the table of standard reduction potentials and without calculating the cell potential, determine if the following reactions are spontaneous.

$$Cu\ (s) + 2\ Ag^+\ (aq) \rightarrow 2\ Ag\ (aq) + Cu^{2+}\ (aq)$$

$$2\ MnO_4^-\ (aq) + 16\ H^+\ (aq) + 10\ F^-\ (aq) \rightarrow 2\ Mn^{2+}\ (aq) + 8\ H_2O\ (l) + 5\ F_2\ (g)$$

Strategy: Determine the position of the reducing agent relative to the position of the oxidizing agent, and based on those results, determine the spontaneity of the reactions.

Section 18.7   Cell Potentials Under the Nonstandard-State Conditions: The Nernst Equation

Like free-energy changes, cell potentials depend on temperature and on the composition of the reaction mixture. Remember from Chapter 17 that $\Delta G = \Delta G° + RT \ln Q$, and using the relationship from section 18.4 ($\Delta G = -nFE$), the following relationship can be derived:

$$-nFE = -nFE° + RT \ln Q$$

Dividing both sides by $-nF$ yields the Nernst equation:

$$E = E° - \frac{0.0592}{n} \log Q$$

This equation works at 25 °C, and the unit of E is in volts.

This equation makes it possible to calculate cell potentials under nonstandard-state conditions.

Workbook Example 18.6

PROBLEM:
Calculate E_{cell} for the following reaction:

$$2\ Fe^{3+}\ (aq) +\ Pb\ (s) \rightarrow 2\ Fe^{+}\ (aq) + Pb^{2+}\ (aq)$$

with the following concentrations: $[Pb^{2+}] = 0.15$ M, $[Fe^{3+}] = 0.50$ M, and $[Fe^{+}] = 0.10$ M.

SOLUTION:
The half-reactions for this equation are

$$Pb\ (s) \rightarrow Pb^{2+} + 2\ e^{-} \qquad\qquad E° = +0.13\ V$$

$$Fe^{3+}\ (aq) + 2\ e^{-} \rightarrow Fe^{+}\ (aq) \qquad\qquad E° = 0.77\ V$$

$$E^{o}_{cell} = 0.13V + 0.77V = 0.90V$$

The Nernst equation for this reaction is

$$E_{cell} = E^{o}_{cell} - \frac{0.0592}{2} \log \frac{\left[Pb^{2+}\right]\left[Fe^{+}\right]^{2}}{\left[Fe^{3+}\right]^{2}}$$

Substituting the information given and solving for E_{cell} gives:

$$E_{cell} = 0.90V - \frac{0.0592}{2} \log \frac{(0.15\,M)(0.10\,M)^{2}}{(0.50\,M)^{2}} = 0.97V$$

Workbook Problem 18.6

If $E_{cell} = 0.17$ V for the following reaction, what is the concentration of $[Sn^{4+}]$ ions? The concentrations are $[Cu^{2+}] = 0.63$ M, $[Sn^{2+}] = 0.72$ M.

$$Sn^{2+}\ (aq)\ + Cu^{2+}\ (aq) \rightarrow Sn^{4+}\ (aq) + Cu\ (s)$$

Strategy: Calculate the E^{o}_{cell} for the cell reaction, substitute the information given into the Nernst equation, and solve for $[Sn^{4+}]$.

Step 1: Calculate E^{o}_{cell}.

Step 2: Using E^{o}_{cell} and E, solve for $[Sn^{4+}]$.

Section 18.8 Electrochemical Determination of pH

Using the Nernst equation, it is possible to derive an equation to calculate the pH of a solution.

When the following reaction is considered with a hydrogen electrode in a solution of unknown pH

$$Pt\ (s)\ |\ H_2\ (1\ atm)\ |\ H^+\ (?\ M)\ |\ |\ reference\ cathode$$

the difference between E_{cell} and E_{ref} can be directly related to pH, by the following derived equation:

$$pH = \frac{E_{cell} - E_{ref}}{0.0592}$$

As a result of this equation, the pH of a solution can be measured by measuring E_{cell}. The awkwardness of the hydrogen electrode has resulted in its substitution with a glass electrode consisting of a silver wire coated with silver chloride immersed in a reference solution of dilute hydrochloric acid. The reference electrode is known as a *calomel* electrode: Hg_2Cl_2 in contact with liquid mercury and aqueous KCl.

Workbook Example 18.7

PROBLEM:
The following cell has a potential of 0.65 V. Calculate the pH of the solution in the anode compartment.

$$Pt\ (s)\ |\ H_2\ (g)\ (1\ atm)\ |\ H^+\ (pH = ?)\ |\ |\ Cl^-\ (aq)\ (1\ M)\ |\ Hg_2Cl_2\ (s)\ |\ Hg\ (l)$$

SOLUTION:
The cell reaction is

$$Hg_2Cl_2\ (s) + H_2\ (g) \rightarrow 2\ Hg\ (l) + 2\ Cl^-\ (aq) + 2\ H^+\ (aq)$$

$E°$ for the calomel electrode is 0.28 V:

$$E^o = E^o_{H_2 \rightarrow H^+} + E^o_{Hg_2Cl_2 \rightarrow Hg,Cl^-} = 0.00V + 0.28V = 0.28V$$

The pH can be calculated using the equation

$$pH = \frac{E_{cell} - E_{ref}}{0.0592}$$

$$pH = \frac{0.65V - 0.28V}{0.0592} = 6.25$$

Workbook Problem 18.7

If $E_{cell} = 0.62$ V for the following reaction, what is the pH in the anode compartment?

$$Cu^{2+}(aq) + H_2(g) \rightarrow 2\,H^+(aq) + Cu(s)$$

Strategy: Determine the half-reactions occurring and calculate E_{ref}. Substitute into the modified Nernst equation to determine pH of the solution ($Cu^{2+}(aq) + 2\,e^- \rightarrow Cu(s)\ E° = 0.34$).

Step 1:   Determine E_{ref}.

Step 2:   Calculate the pH of the anode solution.

Section 18.9   Standard Cell Potentials and Equilibrium Constants

The standard cell potential is related to the standard free-energy change for the reaction. These are described by the following reactions:

$$\Delta G° = -nFE°$$

and

$$\Delta G° = -RT \ln K$$

These equations can be set equal to each other and simplified:

$$E° = \frac{RT}{nF} \ln K = \frac{2.3030\,RT}{nF} \log K$$

$$E° = \frac{0.0592}{n} \log K$$

or

$$\log K = E° \frac{n}{0.0592}$$

This makes it possible to calculate equilibrium constants from cell potentials. This is convenient, as potentials are easily measured, but small concentrations are not as easily measured.

Compared with the equilibrium constants for acid–base reactions, those for redox reactions tend to be either very large or very small. In other words, redox reactions either go to completion or do not proceed at all. A positive value of $E°$ corresponds to $K > 1$, while a negative value of $E°$ corresponds to $K < 1$.

With the addition of this method, there are three different ways to determine the value of an equilibrium constant K.

- K from concentration data: $K = \dfrac{C^c\, D^d}{A^a\, B^b}$

- K from thermochemical data: $\ln K = -\dfrac{\Delta G°}{RT}$

- K from electrochemical data: $\ln K = \dfrac{nFE°}{RT}$

Workbook Example 18.8

PROBLEM:
Calculate the equilibrium constant for the following reaction at 25 °C.

$$2\ Fe^{3+}\ (aq) + 3\ Pb\ (s) \rightarrow 2\ Fe\ (s) + 3\ Pb^{2+}\ (aq)$$

SOLUTION:
The half-reactions for this reaction are

$$Fe^{3+}\ (aq) + 2\ e^- \rightarrow Fe\ (s) \qquad E° = 0.32\ V$$

$$Pb\ (s) \rightarrow Pb^{2+}\ (aq) + 2\ e^- \qquad E° = 0.13\ V$$

$$E^o_{cell} = 0.32 + 0.13 = +0.45V$$

The value of n for this reaction is 6. We can now solve for K.

$$\log K = \frac{nE°}{0.0592V} = \frac{(6)(0.45V)}{0.0592V} = 45.6$$

$$K = 10^{45.6} = 3.98 \times 10^{45}$$

Workbook Problem 18.8

Find K for the reaction:

$$H_2O_2\ (l)\ + 2\ H^+\ (aq) + 2\ Br^-\ (aq) \rightarrow 2\ H_2O\ (l) + Br_2\ (l)$$

Strategy: Determine the half-reactions occurring and calculate $E°$ from the standard potentials of the half-cell reactions. Calculate K from the equation $\log K = E° \dfrac{n}{0.0592}$.

Step 1:   Determine $E°$.

Step 2:   Calculate K from $E°$.

Section 18.10   Batteries

Galvanic cells can be linked in series, with the final voltage the sum of all the individual voltages. Battery features depend on the application, but in general, batteries should be rugged, compact, lightweight, and inexpensive and should provide stable power for a relatively long time.

Lead storage batteries have been generating power for automobiles for nearly a century. Six 2 V cells are connected to generate 12 V of power. Inside the battery, the anode is a series of grids packed with spongy lead, while the cathode is a series of grids packed with lead dioxide; both are dipped in a 38% solution of sulfuric acid.

Anode: $Pb\ (s) + HSO_4^-\ (aq) \rightarrow PbSO_4\ (s) + H^+\ (aq) + 2e^-$ $E° = 0.296$ V

Cathode: $PbO_2\ (s) + 3H^+\ (aq) + HSO_4^-\ (aq) + 2e^- \rightarrow PbSO_4\ (s) + 2H_2O\ (l)$ $E° = 1.628$ V

Overall: $Pb\ (s) + PbO_2\ (s) + 2H^+\ (aq) + 2HSO_4^-\ (aq) \rightarrow 2PbSO_4\ (s) + 2H_2O\ (l)\ E° = 1.924$ V

As the reaction proceeds, $PbSO_4$ adheres to the surface of the electrodes. By running power back through the battery, the $PbSO_4$ is driven off the surfaces. Eventually, this is no longer possible, and the battery must be recycled.

Dry-cell batteries are common household batteries. They originated with *Leclanché cells*, which were patented in 1866. Leclanché cells consist of a zinc metal can (the anode), an inert graphite rod surrounded by a paste of solid MnO_2 ,and carbon black (cathode), and a paste of NH_4Cl and $ZnCl_2$ in starch (electrolyte).

Leclanché cells have been displaced by alkaline dry cells, in which the acidic NH_4Cl solution (which corrodes the zinc) is replaced by $NaOH$ or KOH. This slows the corrosion, which increases battery life, power, and stability.

Anode: $Zn\ (s) + 2\ OH^-\ (aq) \rightarrow Zn(OH)_2\ (aq) + 2\ e^-$

Cathode: $2\ MnO_2\ (s) + H_2O\ (l) + 2\ e^- \rightarrow Mn_2O_3\ (s) + 2\ OH^-\ (aq)$

Nickel–cadmium, or "ni-cad," batteries are rechargeable: like lead storage batteries, the solid products of the reactions adhere to the surface of the electrodes. The anode is Cd, and the cathode is NiO(OH).

Anode: $Cd\ (s) + 2\ OH^-\ (aq) \rightarrow Cd(OH)_2\ (s) + 2\ e^-$

Cathode: $NiO(OH)\ (s) + H_2O\ (l) + e^- \rightarrow Ni(OH)_2\ (s) + OH^-\ (aq)$

Nickel–metal hydride batteries, or *NiMH*, batteries have replaced ni-cad batteries in many applications because cadmium is expensive and toxic—NiMH batteries are more environmentally friendly. They have the same voltage as ni-cad batteries, but have twice the energy density: half as much battery mass is required for the same voltage. The cathode is the same as in ni-cad batteries, but the anode is a special metal alloy that can absorb and release large amounts of H_2 at ordinary temperature:

Anode: $MH_{ab}\ (s) + OH^-\ (aq) \rightarrow M\ (s) + H_2O\ (l) + e^-$

Cathode: $NiO(OH)\ (s) + H_2O\ (l) + e^- \rightarrow Ni(OH)_2\ (s) + OH^-\ (aq)$

The overall reaction transfers hydrogen from the anode to the cathode.

NiMH batteries are used in many consumer products and also in hybrid automobiles, where they supply energy to the electric motor and are recharged when the brakes are applied, using some of the kinetic energy of the vehicle.

Lithium and *lithium-ion* batteries are lightweight, high-voltage, and rechargeable. They are used in cell phones, laptops, cameras, tools, and Tesla electric cars. With an atomic weight of 6.97 g/mol, less than 7 g of lithium can provide 1 mol of electrons. In addition, lithium has the highest standard oxidation potential of any metal.

The anode is lithium metal, the cathode is typically MnO_2, and the electrolyte is a lithium salt such as $LiClO_4$ in an organic solvent.

Anode: $x\ Li\ (s) \rightarrow x\ Li^+\ (soln) + x\ e^-$

Cathode: $MnO_2\ (s) + Li^+\ (soln) + x\ e^- \rightarrow Li_xMnO_2\ (s)$

Lithium ion batteries contain a lithiated graphite anode. Abbreviated Li_xC_6, this is graphite with lithium atoms inserted between the layers. The cathode is CoO_2, which is also able to incorporate lithium ions into its structure. The electrolyte is a solution of lithium salt in an organic solvent or a solid-state polymer that can transport lithium ions.

Anode: $Li_xC_6\ (s) \rightarrow x\ Li^+\ (soln) + 6\ C\ (s) + x\ e^-$

Cathode: $Li_{1-x}CoO_2\ (s) + x\ Li^+\ (soln) + x\ e^- \rightarrow LiCoO_2\ (s)$

There are safety concerns about lithium ion batteries that are being addressed by the development of alternative electrode materials.

Section 18.11   Corrosion

Corrosion is the oxidative deterioration of a metal. The conversion of iron to rust is a well-known example; it requires both oxygen and water and involves pitting of the metal surface and deposition of hydrated iron(III) oxide at a location separate from the pit.

A possible mechanism involves an electrochemical process in which iron is oxidized in one region of the surface while oxygen is reduced in another region. Step 1 is

Anode region: $Fe\ (s) \rightarrow Fe^{2+}\ (aq) + 2\ e^-$ $\qquad E° = 0.45\ V$

Cathode region: $O_2\ (g) + 4\ H^+\ (aq) + 4\ e^- \rightarrow 2\ H_2O\ (l)$ $\qquad E° = 1.23\ V$

Soluble Fe^{2+} ions migrate through the water droplets and encounter dissolved oxygen, leading to the further oxidation of Fe^{2+} to Fe^{3+}. Fe^{3+} then reacts with water in the final step, producing $Fe_2O_3 \cdot H_2O$, which is rust.

This mechanism also explains the increased corrosion of cars in the presence of road salt—salt in the water greatly increases its conductivity, allowing the reaction to proceed more quickly.

O_2 is able to oxidize all metals except a few, as can be seen by the fact that the O_2/H_2O half-reaction lies above the M^{n+}/M half-reaction for all but a very few metals, such as gold. In the case of many metals, including aluminum, magnesium, chromium, titanium, and zinc, the oxidation forms a hard coating that prevents further oxidation.

Corrosion of iron can be avoided by shielding it from oxygen and water. This can be done with paint, but is more effective with another metal. Coating steel with molten zinc is known as **galvanizing**, and it provides protection even if the zinc layer is scratched, because the zinc is preferentially oxidized and will reduce any oxidized iron.

Cathodic protection, a similar method, uses a *sacrificial anode*: a metal that is more easily oxidized is electrically connected to the steel or iron and corrodes instead of the iron. This method is used to protect ships and large buildings from corrosion.

Section 18.12   Electrolysis and Electrolytic Cells

In a galvanic cell, a spontaneous reaction generates a current. In an **electrolytic cell**, a current is used to drive a nonspontaneous reaction. The process of using an electric current to bring about chemical change is called **electrolysis**.

An electrolytic cell has two electrodes connected to a source of current and dipped into an electrolyte. The battery acts as an electron pump, removing electrons from the anode, moving electrons through the electrolyte, and pulling electrons into the cathode.

Electrolysis of molted NaCl: the negative electrode (cathode) attracts Na^+, which is reduced to Na (*l*). The positive electrode (anode) attracts Cl^-, which is oxidized to Cl_2 (*g*).

Anode:	$2\ Cl^-\ (l) \rightarrow Cl_2\ (g) + e^-$
Cathode:	$Na^+\ (l) + e^- \rightarrow Na\ (l)$

Electrolysis of aqueous NaCl: when an aqueous solution is used, the reactions may differ because water may be involved.

Cathode:	$Na^+\ (aq) + e^- \rightarrow Na\ (s)$	$E° = -2.71\ V$ *or*
	$2\ H_2O + 2\ e^- \rightarrow H_2\ (g) + 2\ OH^-\ (aq)$	$E° = -0.83\ V$
Anode:	$2\ Cl^- \rightarrow Cl_2\ (g) + e^-$	$E° = -1.36\ V$ *or*
	$2\ H_2O \rightarrow O_2\ (g) + 4\ H^+\ (aq) + 4\ e^-$	$E° = -1.23\ V$

At the cathode, water is split because that reaction is far less negative than the reduction of sodium, and hydrogen bubbles are seen. At the anode, however, chlorine gas is formed. This is due to a phenomenon known as *overvoltage*. A basic solution of NaOH is formed.

Overvoltage is the amount of voltage needed above the calculated standard reduction (or oxidation) potential for electrolysis to occur. It is needed when the rate of electron transfer is very slow. Only a small overvoltage is needed for the solution or deposition of metals—they are good conductors of electrons—but a large overvoltage is required for the formation of gases, especially O_2 or H_2. Current theories do not allow this to be predicted easily, so experimental evidence is needed when cell potentials are similar.

Electrolysis of water: if the electrolyte in solution is less easily oxidized and reduced than water, water will react at both electrodes:

Anode: $2 H_2O (l) \rightarrow O_2 (g) + 4 H^+ (aq) + 4 e^-$

Cathode: $4 H_2O (l) + 4 e^- \rightarrow 2 H_2 (g) + 4 OH^- (aq)$

Equal quantities of $H^+ (aq)$ and $OH^- (aq)$ are formed, so the solution will remain neutral.

Overall cell reaction: $2 H_2O (l) \rightarrow O_2 (g) + 2 H_2 (g)$

Workbook Example 18.9

PROBLEM:
Predict the half-cell reactions when a solution of aqueous KCl is electrolyzed. Write the overall cell reaction.

SOLUTION:
The possible half-reactions are

Cathode: $K^+ (aq) + e^- \rightarrow K (s)$ $E° = -2.93$ V or
$2 H_2O (l) + 2 e^- \rightarrow H_2 (g) + 2 OH^- (aq)$ $E° = -0.83$ V

Anode: $2 Cl^- (aq) \rightarrow Cl_2 (g) + 2 e^-$ $E° = -1.36$ V or
$2 H_2O (l) \rightarrow O_2 (g) + 4 H^+ (aq) + 4 e^-$ $E° = -1.23$ V

At the cathode, water will be reduced, generating hydrogen gas. Remember that at the anode, the overvoltage needed to generate chlorine gas is less than that needed to generate oxygen gas, so chlorine gas will be produced, despite its higher oxidation potential.

Overall reaction: $2 H_2O (l) + 2 Cl^- (aq) \rightarrow H_2 (g) + Cl_2 (g) + 2 OH^- (aq)$

Workbook Problem 18.9

Predict the half-cell and overall reaction for the electrolysis of an aqueous Li_2SO_4 solution.

Strategy: Determine the possible half-reactions and their standard voltages to predict the reactions at the cathode and anode, then determine the overall reaction.

Step 1: Determine the possible reactions at the cathode and their standard voltages.

Step 2:   Determine the possible reactions at the anode and their standard voltages.

Step 3:   Predict which reactions will occur and calculate the overall reaction.

Section 18.13   Commercial Applications of Electrolysis

Sodium is produced commercially in a *Downs cell* by electrolysis of a molten mixture of $NaCl$ and $CaCl_2$. The addition of the $CaCl_2$ lowers the melting point of the mixture by more than 200 °C. The liquid Na produced at the cathode is less dense than the mixture and can be drawn off from the top. The Downs cell process requires extremely high currents, so manufacture is often near hydroelectric power plants.

Chlorine and sodium hydroxide are produced by the electrolysis of aqueous sodium chloride in what is known as the *chlor–alkali industry*. The anode reaction, which produces Cl_2 gas, is separated from the cathode compartment by a membrane that permits the passage of sodium ions. When water is hydrolyzed at the cathode, a sodium hydroxide solution is produced.

Aluminum is produced in the *Hall–Héroult* process, which involves the electrolysis of a molten mixture of Al_2O_3 and cryolite (Na_3AlF_6) at 1000 °C in a cell with graphite electrodes. At the cathode, molten aluminum is produced and sinks to the bottom of the cell. At the anode, O_2 gas is formed. This reacts with the electrode to generate CO_2 gas, so frequent replacement of the anode is necessary. The reactions at the electrodes are not well understood, and a great deal of current is required for the process, making aluminum production the largest single consumer of electricity in the United States.

Electrorefining is the purification of a metal by means of electrolysis. For example, copper can be purified using an impure copper anode and a pure copper cathode. Copper ions move away from the impure electrode and deposit on the pure electrode, leaving other metals either dissolved in the electrode solution or on the bottom of the container as *anode mud*.

Electroplating is a similar process in which one metal is coated on the surface of another using electrolysis. The cathode is the carefully cleaned object to be plated, while the solution in the electrolytic cell contains ions of the metal to be deposited.

Section 18.14   Quantitative Aspects of Electrolysis

The amount of substance produced at an electrode by electrolysis depends on the quantity of electrons passed through the cell. This is directly related to the stoichiometry of the reaction: 1 mol of electrons are needed to reduce 1 mol of sodium ions to sodium metal.

Moles of electrons passed through a cell can be calculated from electric current and time by the following formulas:

$$Charge\ (C) = Current\ (A) \times Time\ (s)$$

$$Moles\ of\ e^- = charge\,(C) \times \frac{1\,mol\,e^-}{96500\,C}$$

Figure 18.20 (page 793 in your text) gives a sequence that can be used to calculate the amount of product produced by passing a current through an electrolytic cell for a fixed period of time. The figure is reproduced below.

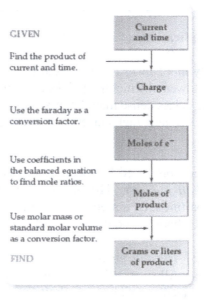

Workbook Example 18.10

PROBLEM:
How many grams of Cl_2 gas will be produced in the electrolysis of molten NaCl by a current of 5.75 A for 50.0 min? How many grams of sodium will be produced in the same time period?

SOLUTION:
(Remember that a coulomb is an A·s and that an ampere is C/s.)

$$2\ Cl^- \rightarrow Cl_2 + 2\ e^-$$

Moles of electrons = 2

$$\frac{5.75\,C}{s} \times 50.0\,min \times \frac{60\,s}{1\,min} \times \frac{1\,mol\,e^-}{96500\,C} \times \frac{1\,mol\,Cl_2}{2\,mol\,e^-} \times \frac{70.9\,g\,Cl_2}{1\,mole\,Cl_2} = 6.34\,g\,Cl_2$$

$$\frac{5.75\,C}{s} \times 50.0\,min \times \frac{60\,s}{1\,min} \times \frac{1\,mol\,e^-}{96500\,C} \times \frac{1\,mol\,Na}{1\,mol\,e^-} \times \frac{23.0\,g\,Na}{1\,mole\,Na} = 4.11\,g\,Na$$

Workbook Problem 18.10

The Dow process isolates Mg metal from seawater. The final step in this process involves the electrolysis of molten $MgCl_2$ to metal. How long would it take to produce 12 kg of Mg (s) at a current of 15 A? How many liters of Cl_2 gas would be produced (at STP) in the same amount of time?

Strategy: Write the electrolysis reaction and consider the conversion process to calculate time.

Step 1: Treat the electrons as reactants in the chemical equation, and solve for the time required to produce 12 kg of magnesium. Remember, the conversion factor of 96,500 C (or A·s) per mole of electrons.

Step 2: Use the information calculated above to determine the volume of Cl_2 gas.

Putting It Together

Determine $E°$ for the following reaction:

$$O_2 (g) + 4 H^+ (aq) + 4 Br^- (aq) \rightarrow 2 H_2O (l) + 2 Br_2 (l)$$

Is this reaction spontaneous? If not, why? A buffer containing 0.35 M sodium formate ($NaCHO_2$) and 0.20 M formic acid is added to adjust the pH of the reaction. Assume [Br^- = 1.0 M]. Determine E after the addition of the buffer. Now is the reaction spontaneous?

Self–Test

This section is intended to test your knowledge of the material covered in this chapter. Think through these problems, and make certain you understand what they are asking. Make sure your answers make sense. Successful completion of these problems indicates that you have mastered the material in this chapter. You will receive the greatest benefit from this section if you use it as a mock exam, as this will allow you to determine which topics you need to study in more detail.

True–False

1. In galvanic or voltaic cells, a spontaneous reaction occurs.

2. A salt bridge allows electrons to flow between the two halves of a galvanic cell.

3. If the cell potential of a reaction is positive, the free-energy change of the reaction is also positive.

4. When balancing electrochemical equations, $E°$ is never multiplied by the stoichiometric coefficients.

5. E for a reaction at nonstandard conditions can only be determined by experiment.

6. Cell potential and pH have a linear relationship.

7. The voltage of galvanic cells is added when they are connected in series to form a battery.

8. Direct methanol fuel cells have been proposed for use in small consumer electronics.

9. Corrosion of any metal can be prevented by contact with another metal above it in the list of standard reduction potentials.

10. The results of the electrolysis of an aqueous salt solution can be predicted using only standard half-cell voltages.

Multiple Choice

11. For the following galvanic cell, Ni (s) | Ni^{2+} (aq) || Br$^-$ (aq) | Br$_2$ (l) | Pt (s),
    a. the cathode is Ni (s)
    b. the electrons flow from the Ni (s) electrode to the Pt (s) electrode
    c. the Ni^{2+} ions flow to the anode
    d. electrons flow from the Pt (s) electrode to the Ni (s) electrode

12. The reaction carried out in a galvanic cell is
    a. spontaneous, and therefore has a positive $E°$ value
    b. nonspontaneous, and therefore has a positive $E°$ value
    c. spontaneous, and therefore has a negative $E°$ value
    d. nonspontaneous, and therefore has a negative $E°$ value

13. For the net reaction Cr (s) + Fe^{3+} (aq) → Cr^{3+} (aq) + Fe (s)
    a. Cr (s) is the reducing agent   b. Fe (s) is the anode
    c. Fe^{3+} (aq) is the reducing agent   d. Cr (s) undergoes reduction

14. Given the following reduction half-cells:
    $$PbO_2 (s) + 3\,H^+ (aq) + HSO_4^- (aq) + 2\,e^- \rightarrow PbSO_4 (s) + 2\,H_2O (l) \qquad E° = 1.628\ V$$
    $$Cr_2O_7^{2-} (aq) + 14\,H^+ (aq) + 6\,e^- \rightarrow 2\,Cr^{3+} (aq) + 7\,H_2O (l) \qquad E° = 1.33\ V$$
    $$SO_4^{2-} (aq) + 4\,H^+ (aq) + 2\,e^- \rightarrow H_2SO_3 (aq) + H_2O (l) \qquad E° = 0.17\ V$$
    $$2\,CO_2 (g) + 2\,H^+ (aq) + 2\,e^- \rightarrow H_2C_2O_4 (aq) \qquad E° = -0.49\ V$$

    a. the strongest oxidizing agent is PbSO$_4$ (s)
    b. PbSO$_4$ (s) will spontaneously react with CO$_2$ (g)
    c. the strongest oxidizing agent is PbO$_2$ (s)
    d. H$_2$C$_2$O$_4$ (aq) will not spontaneously react with PbO$_2$ (s)

15. In the electrolysis of molten BaI$_2$,
    a. the Ba^{2+} ions migrate toward the anode
    b. the I$^-$ ions migrate toward the anode
    c. water undergoes oxidation at the anode
    d. the system becomes increasingly basic

16. In the table of standard reduction potentials
    a. the strongest reducing agents are located in the bottom left of the table
    b. the strongest reducing agents are located in the top left of the table
    c. the strongest reducing agents are located in the lower right of the table
    d. the strongest oxidizing agents are located in the bottom left of the table

17. In an electrolytic cell
    a. the anode has a positive sign
    b. reduction occurs at the anode
    c. the anode has a negative sign
    d. oxidation occurs at the cathode

18. In an electrolytic cell
    a. anions migrate toward the cathode
    b. cations migrate toward the anode
    c. ions migrate through a salt bridge
    d. a current forces nonspontaneous reactions to occur

19. Alkaline dry cells
    a. contain an electrolyte of a moist paste of NaOH and $ZnCl_2$
    b. have a cathode in which steel is in contact with HgO in an alkaline medium
    c. contain an electrolyte of a moist paste of NH_4Cl and $ZnCl_2$
    d. are rechargeable nickel–cadmium batteries

20. The amount of substance produced at an electrode by electrolysis depends
    a. on the quantity of reactant present
    b. on the quantity of charge passed through the cell
    c. on the spontaneity of the reaction
    d. on the size of the electrolytic cell

Fill-in-the-Blank
21. Car batteries are _____ batteries, while common household batteries are

    _____.

22. The oxidative deterioration of a metal is called _____. In the case of steel,

    this can be prevented by coating with zinc, also called _____ for small objects,

    or with the use of another metal as a(n) _____ for buildings or ships.

23. The refining of _____ and _____ metals require very large quantities of

    _____. As a result, these processes are generally located near _____.

24. Standard half-cell potentials are written as _____, and are listed in

    _____ tendency to occur as written, and _____ tendency to occur in reverse.

    The "zero" on the scale is the _____.

25. The _____ of a reaction can be calculated from the standard

    _____, but these tend to be either very _____ or very _____,

    indicating that these reactions either go to _____, or not at all.

Matching

26. Anode

27. Cathode

28. Corrosion

29. Electrolysis

30. Electroplating

31. Electrorefining

32. Galvanizing

33. Overvoltage

34. Salt bridge

a. the process of using an electric current to bring about a chemical change

b. using current to deposit a thin layer of one metal on another

c. electrode at which reduction occurs

d. covering one metal with another to prevent corrosion

e. the voltage above the standard potential needed to speed a reaction.

f. electrode at which oxidation occurs

g. mechanism for allowing ion flow in a galvanic cell.

h. oxidative deterioration of a metal

i. using current to purify impure metals

Problems

35. Describe the galvanic cell that uses the following reaction:

$$Ni^{2+} (aq) + Mg (s) \rightarrow Ni (s) + Mg^{2+} (aq)$$

Write the half-reactions, identify the anode and cathode, and provide the shorthand notation.

36. Determine standard reaction potentials for the following unbalanced reactions and indicate whether they are spontaneous in the forward or reverse direction:
    a. $2 Al (s) + 3 Pb^{2+} (aq) \rightarrow 2 Al^{3+} (aq) + 3 Pb (s)$
    b. $Br_2 (l) + 2 Cl^- (aq) \rightarrow 2 Br^- (aq) + Cl_2 (g)$
    c. $Ni (s) + Pb^{2+} (aq) \rightarrow Ni^{2+} (aq) + Pb (s)$
    d. $Fe (s) + HNO_3 \rightarrow Fe^{3+} (aq) + NO_3^- (aq) + H_2 (g)$

37. Calculate E for the following cell, $Mg (s) | Mg^{2+} (aq) \| Fe^{3+} (aq) | Fe (s)$, under the following conditions:
    a. standard conditions   b. $[Mg^{2+}] = 0.1$ M, $[Fe^{3+}] = 2.0$ M
    c. $[Mg^{2+}] = 2.0$ M, $[Fe^{3+}] = 0.1$ M

38. Given the following reaction:

$$Cu^{2+} (aq) + 2 Ag (s) + 2 Br^- (aq) \rightarrow Cu (s) + 2 AgBr (s) \qquad E° = 0.27 \text{ V}$$

    Calculate $\Delta G°$ and K.

39. $E°$ for a galvanic cell in which Zn^{2+} is reduced to $Zn (s)$ is +1.61 V. What is the potential at the anode? What metal is oxidized at the anode? Write the half-reaction that occurs at the anode.

40. If the following cell has a potential of 0.14 V, what is the pH in the anode compartment?

$$\text{Pt } (s)| \text{ H}_2 \text{ (1 atm)} | \text{H}^+ (aq)\| \text{Ni}^{2+} \text{ (1 M)}| \text{Ni } (s)$$

41. Which of these metals, if coated onto iron, would prevent corrosion? Explain your answer.
    a. Sn   b. Mn   c. Cu

42. Calculate the K_{sp} for AgCl (s) using the table of standard reduction potentials.

43. Write the reactions that occur at the anode and cathode in the electrolysis of molten NaBr. How much current would be required to produce sodium metal at the rate of 3 g/hour?

44. If 3 A of current are applied to a copper object immersed in a solution of AgNO$_3$ and copper, how much time will be required to deposit 15 g of silver?

45. An acidic solution of NaMnO$_4$ can be used to oxidize zinc metal. If 3.27 g of zinc are to be oxidized in 250 mL of solution, what is the minimum number of grams of NaMnO$_4$ that need to be dissolved?

46. How many liters of O$_2$ are produced at 356 mm Hg when 2.97×10^3 C are passed through water at 25 °C?

Challenge Problem

47. The K_{sp} for lead(II) sulfate is 1.8×10^{-8}. The standard reduction potential for Pb^{2+}/Pb is $E° = -0.126$ V. Use this information to determine $E°$ for the reaction

$$\text{PbSO}_4 + 2 \text{ e}^- \rightarrow \text{ Pb } (s) + \text{SO}_4^{2-} (aq)$$

CHAPTER NINETEEN

Nuclear Chemistry

Learning Objectives

As a result of reading and studying this chapter, you should be able to

Sections 19.1–19.2		**Nuclear Reactions and Radioactivity**
	1.	State the different kinds of radioactive decay and the results of each.
	2.	Write balanced nuclear equations.
Section 19.3		**Nuclear Stability**
	3.	Predict the type of radioactive decay for a given isotope.
	4.	Write a balanced nuclear reaction for a radioactive decay series.
Section 19.4		**Radioactive Decay Rates**
	5.	Relate the half-life and decay constants.
	6.	Calculate the amount of radioactive isotope remaining after a given amount of time.
	7.	Relate decay rates to the decay constant, the half-life, or the amount remaining.
Section 19.5		**Energy Changes in Nuclear Reactions**
	8.	Calculate the mass defect and binding energy of a nucleus.
	9.	Calculate the mass and energy change in a nuclear reaction.
	10.	Calculate the mass and energy change of a chemical reaction.
Section 19.6		**Nuclear Fission and Fusion**
	11.	Identify key aspects of fission and fusion reactions and their role in nuclear power and weapons.
	12.	Calculate the energy change associated with a fission or fusion reaction.
Section 19.7		**Nuclear Transmutation**
	13.	Write balanced equations for nuclear transmutation reactions.
Section 19.8		**Detecting and Measuring Radioactivity**
	14.	Calculate the radiation intensity in a radioactive sample and convert between units of radiation intensity.
Section 19.9		**Applications of Nuclear Chemistry**
	15.	Calculate the age of an object using radioisotopic dating.

Chapter Summary

Nuclear chemistry is the study of the properties and reactions of atomic nuclei. The stability of an element is related to its proton/neutron ratio, and the binding energy of the nucleus can be calculated from this ratio. Some elements can undergo fission, either spontaneously or by being induced, as in nuclear reactors. Other elements can fuse together under extreme conditions to form larger nuclei. All of the transuranium elements and many other isotopes have been made by nuclear transmutation, a process by which the nuclei of one element are bombarded with particles to increase their size. Radiation varies in its ability to damage tissues; a number of different units are used for radiation exposures depending on the reason for the measurement. There are numerous practical uses of nuclear chemistry, from radiocarbon dating of ancient objects to therapeutic and imaging uses in medicine.

The Chapter in Detail

Section 19.1   Nuclear Reactions and Their Characteristics

Back in chapter 2, we discussed that elements in the periodic table are defined by the number of protons their nuclei contain. This is also known as their **atomic number**, Z. The elements are arranged in order of atomic number, which resolved a few discrepancies that occurred when they were arranged by mass. For example, tellurium ($Z = 52$, mass $= 127.6$) and iodine ($Z = 53$, mass $= 126.9$). Iodine behaves very much like a halogen, but if the periodic table is arranged by mass, it will not fall in the halogens column. The discovery of atomic numbers set this straight.

In a neutral (uncharged) atom, the number of protons will be the same as the number of electrons.

Most atomic nuclei also contain neutrons. The **mass number** of an atom, A, is found by adding the atomic number to the number of neutrons, N. The mass of electrons is so small compared to these other particles that they can be ignored.

$$A = Z + N$$

Elements are defined by their protons: any variation in the number of protons and you will have a different element. But the number of neutrons in the nucleus can change without changing the identity of the element because it has very little effect on the chemical properties of an atom. Atoms with different amounts of neutrons are called **isotopes**.

Isotope symbols are written using the following conventions, where X indicates the elemental symbol:

$$_Z^A X$$

From this, it is possible to calculate the number of neutrons ($A - Z$). It is not difficult to remember which of these numbers is which: the smaller will always be the atomic number, the larger will always be the mass number. Sometimes the atomic number is omitted, because it can be taken from the periodic table, and does not vary.

Workbook Example 19.1

PROBLEM:
Determine the number of protons, neutrons, and electrons in the following isotopes:

$$^{71}_{31}Ga \qquad ^{191}_{77}Ir \qquad ^{88}_{39}Y$$

SOLUTION:
The subscript is the atomic number (Z), and the superscript is the mass number (A).
Z = the number of protons = the number of electrons, and $A - Z$ = the number of neutrons.

$^{71}_{31}Ga$: $\quad Z = 31$; number of protons = number of electrons = 31

$\qquad\qquad A - Z = 71 - 31 = 40$; number of neutrons = 40

$^{191}_{77}Ir$: $\quad Z = 77$; number of protons = number of electrons = 77

$\qquad\qquad A - Z = 191 - 77 = 114$; number of neutrons = 114

$^{88}_{39}Y$: $\quad Z = 39$; number of protons = number of electrons = 39

$\qquad\qquad A - Z = 88 - 39 = 49$; number of neutrons = 49

Workbook Problem 19.1

Oxygen is the most abundant element by mass in the biosphere. All classes of biomolecules, proteins, sugars, fats, and nucleic acids contain oxygen, and oxygen makes up almost 90% of the mass of the world's oceans. Oxygen is made up of three stable isotopes, with mass numbers 16, 17, and 18. Write the standard symbols for these isotopes.

Strategy: Use the periodic table to determine the atomic number and the number of neutrons in each isotope. Give the standard symbol for each.

Workbook Problem 19.2

The metalloid X has been used as a cosmetic since at least 3000 B.C. When bound to sulfur atoms in a 2:3 ratio, it is known as *kohl*. An atom of the metal contains 51 protons, and its most stable isotope contains 70 neutrons. Identify the metal, and write the symbol in standard format.

Strategy: Determine the atomic number from the information given.

Section 19.2   Radioactivity

Experiments since 1896 have shown that many nuclei are **radioactive**; they undergo a spontaneous decay and release some sort of radiation in order to become stable. In 1897, Ernest

Rutherford (of the gold foil experiment fame) discovered three forms of the radiation that are released by spontaneous decay: alpha particles, beta particles, and gamma rays. Since then, two more types of radiation have been discovered: positron emission and electron capture.

In **alpha (α) radiation**, a particle is released that is positively charged and has the same mass-to-charge ratio as a helium nucleus. If you have a radioactive element that decays and gives off an alpha particle, the resulting new nucleus has an atomic number that's two protons less than what you started with, and the atomic mass is four units less than what you started with. We can write a reaction that describes what happened; for example, let's take a look at the alpha decay of francium-221:

$$^{221}_{87}Fr \rightarrow {}^{4}_{2}He + {}^{217}_{85}At$$

This type of reaction is called a **nuclear reaction**. In this type of reaction, we need to make sure that the atomic masses and atomic numbers on each side are equal.

One final word before we move on to the next particle. There are two ways to represent an alpha particle in a nuclear equation: we can write it as a helium nucleus that was spun off the equation, or we could write it with the alpha particle instead of He: $^{4}_{2}\alpha$. Either way is correct.

Workbook Example 19.2

PROBLEM:
Smoke detectors used in homes and apartments contain americium-241, which undergoes alpha decay. When alpha particles collide with air molecules, charged particles are produced that generate an electrical current. If smoke particles enter the detector, they interfere with the formation of charged particles in the air, and the electrical current is interrupted. This causes the alarm to sound and warns the occupants of a potential fire. Complete the following nuclear equation for the decay of americium-241:

$$^{241}_{95}Am \rightarrow ? + {}^{4}_{2}\alpha$$

SOLUTION:
We need to figure out the missing atomic number and atomic mass. The atomic number tells us the number of protons, which, in turn, tells us the identity of the resulting nucleus.

Looking at the nuclear equation, we have an atomic mass of 241 on the left side and a mass of 4 on the right-hand side.   Subtracting 4 from 241 gives us a mass of 237; this is the mass of the new nucleus.

Doing the same thing with atomic numbers, we have 95 on the left side and 2 on the right side. Subtracting 2 from 95 gives us 93; the new nucleus has an atomic number of 93. Looking up 93 on the periodic table tells us the new nucleus (nuclide) is neptunium, $^{237}_{93}Np$.

Inserting the nuclide into the equation gives us:

$$^{241}_{95}Am \rightarrow {}^{237}_{93}Np + {}^{4}_{2}\alpha$$

Workbook Problem 19.3

Write a balanced nuclear equation for the alpha decay of polonium-214.

Strategy: Use the periodic table to write the nuclide symbol of polonium-214 and the alpha particle. Using this information, write an incomplete nuclear equation and solve for the missing particle.

Step 1:   Write the nuclide symbols for polonium-214 and the alpha particle.

Step 2:   Write an incomplete nuclear equation representing the alpha decay of polonium-214.

Step 3:   Determine the missing mass number.

Step 4:   Determine the missing atomic number and determine the symbol of the new nucleus.

Step 5:   Write the complete balanced nuclear equation.

Another radioactive particle is the **beta (β) particle,** which is an electron that is being given off by the nucleus. The nuclide symbol is $_{-1}^{0}e$ or $_{-1}^{0}\beta$. An example nuclear reaction of this is the beta decay of cobalt-60:

$$_{27}^{60}Co \rightarrow {}_{28}^{60}Ni + {}_{-1}^{0}\beta$$

Gamma (γ) radiation has no mass; it is just high-energy electromagnetic radiation. Most of the time, gamma radiation accompanies alpha and beta emission in order to release energy, but it is often not shown when we write nuclear reactions.

A different type of radioactive decay is **positron emission.** This occurs when a proton in the nucleus changes into a neutron plus an ejected positron—the complete opposite of an electron, with no mass but a positive charge. We write this as $_{+1}^{0}e$ or $_{+1}^{0}\beta$. For example, the positron emission of aluminum-24 is written:

$$_{13}^{24}Al \rightarrow {}_{12}^{24}Mg + {}_{+1}^{0}\beta$$

Finally, another type of radioactive emission involves **electron capture**, a process in which the nucleus of an atom captures one of the surrounding electrons, by which it converts a proton into a neutron. When this happens, the mass stays the same, but the atomic number goes down by one unit. An example of this is the electron capture by polonium-204:

$$^{204}_{84}Po \rightarrow ^{\ 0}_{-1}e + ^{204}_{85}At$$

Workbook Problem 19.4

Write the balanced nuclear equation for the bombardment of nickel-58 by a proton, which produces a radioactive isotope and an alpha particle.

Strategy: Use the periodic table to write the nuclide symbols given in the problem. Using this information, write an incomplete nuclear equation and solve for the missing particle.

Step 1:   Write the incomplete nuclear equation.

Step 2:   Determine the missing atomic mass.

Step 3:   Determine the missing atomic number and the identity of the new nucleus.

Step 4:   Write the complete balanced nuclear equation.

Section 19.3   Nuclear Stability

What makes some nuclei unstable? In order to answer the question, we need to look at the neutron-to-proton ratio in the nucleus and the forces holding the nucleus together.

When the 3600 known isotopes of all the elements are plotted on a graph with the number of neutrons on the *y*-axis and the number of protons on the *x*-axis, we get something called the *band of nuclear stability*. Figure 19.2 (on page 813 of your textbook) shows this band of nuclear stability; anything outside of this band is considered to be unstable and will decay spontaneously.

A few generalizations can be made about this band of nuclear stability:

- Every element in the periodic table has at least one radioactive isotope.
- Hydrogen is the only element whose most abundant isotope (hydrogen-1) contains more protons (1) than neutrons (0).

- The ratio of neutrons to protons gradually increases, giving a curved appearance to the band of stability.
- All isotopes heavier than bismuth-209 are radioactive, even though they may decay slowly and be stable enough to occur naturally.

It turns out that that many radioactive nuclei cannot reach a stable nucleus with just one decay reaction; a series of nuclear decays, called **a radioactive decay chain**, will occur until an unstable nucleus becomes stable,

Section 19.4   Radioactive Decay Rates

The decay of radioactive nuclei is a first-order process in which the number of radioactive nuclei in a sample can be used in lieu of the concentration of a reactant. The rate constant, k, is here called the **decay constant**.

The integrated rate law for radioactive decay is

$$\ln \frac{N_t}{N_0} = -kt$$

where N_t is the number of radioactive nuclei remaining at a time, t, and N_0 is the original number of radioactive nuclei at time t=0.

The half-life and decay constant are related by the following equations:

$$t_{1/2} = \frac{\ln 2}{k} \qquad\qquad k = \frac{\ln 2}{t_{1/2}}$$

Workbook Example 19.3

PROBLEM:
If the decay constant of technetium-93 is 0.254 h^{-1}, what is its half-life?

SOLUTION:
Substitute the decay constant into the above equation:

$$t_{1/2} = \frac{\ln 2}{0.254\,h^{-1}} = 2.73\,h$$

Workbook Example 19.4

PROBLEM:
If the half-life of technetium-96 is 4.3 days, what is its decay constant?

SOLUTION:
Substitute the half-life into the above equation:

$$k = \frac{\ln 2}{4.3\,days} = 0.161\,days^{-1}$$

Workbook Problem 19.5

The half-life for the decay of C-14 is 5730 years. What is the decay constant? Since this is a first-order reaction, how old is a piece of wood in which 23% of the C-14 has decayed?

Strategy: Substitute into the equation relating half-life to the decay constant; use the integrated first-order rate law.

Step 1: Solve for k.

Step 2: Substitute into the integrated rate law and solve for t after 23% decay.

Section 19.5   Energy Changes in Nuclear Reactions

The ratio of neutrons to protons required for stability varies with the mass of the element. Lighter elements require a 1:1 ratio of neutrons to protons, but more neutrons are needed for heavier elements.

The more massive an atom, the more neutrons are needed to minimize the repulsions of the protons: neutrons serve as nuclear "glue."

The energy released when a nucleus is created cannot be measured, because the activation energy required to force the particles together is so high that temperatures rivaling those in the interior of the Sun are needed.

Although the energy changes cannot be measured, they can be calculated using Einstein's famous equation:

$$\Delta E = \Delta m c^2$$

Using the known masses of neutrons and protons, it is possible to calculate the **mass defect** of a nucleus. The mass of a nucleus is always less than the mass of the individual particles. This lost mass is a direct measure of the **binding energy** holding the nucleus together. The larger the binding energy, the more stable the nucleus.

The energy changes involved in nuclear reactions are millions of times greater than those involved in typical processes.

To make it easier to compare the stability of different nuclei, energies are usually expressed per **nucleon** in mega-electron volts, MeV. To covert mega-electron volts to joules, the conversion factor is 1 MeV = 1.6×10^{-13} J.

If binding energy is plotted against atomic number, a stability peak is found at Fe-56. Nuclear stability is increased when lighter nuclei fuse together or heavier nuclei break apart.

The result of these calculations is the discovery that mass and energy are not independently conserved, but the combination of the two is conserved. For non-nuclear reactions, the effect is too small to be measured and can thus be disregarded.

Workbook Example 19.5

PROBLEM:
Determine the mass defect and binding energy for chlorine-35 in both J/mol and MeV/nucleon. The mass of ^{35}Cl is 34.968 85 amu.

SOLUTION:
First, calculate the total mass of the nucleons and the mass defect:

Mass of 17 protons =	(17)(1.007 28 amu)	= 17.123 76 amu
Mass of 18 neutrons =	(18)(1.008 66 amu)	= 18.155 88 amu
Mass of 17 protons and 18 neutrons		= 35.279 64 amu

Mass of chlorine-35 atom	= 34.968 85 amu
Mass of electrons = (17) (5.486 × 10^{-4})	= 0.009 33 amu
Mass of chlorine-35 nucleus	= 34.959 52 amu

Mass defect: 35.279 64 amu – 34.959 52 amu= 0.320 12 amu

Convert this mass to grams:

$$0.32012\,amu \times \frac{1.6605 \times 10^{-24}\,g}{1\,amu} = 5.3156 \times 10^{-25}\,g$$

Use the Einstein equation to convert the mass defect to binding energy:

$$\Delta E = 5.3156 \times 10^{-25}\,g \times \frac{1\,kg}{1000\,g} \times \left(\frac{3.00 \times 10^8\,m}{1\,s}\right)^2 = 4.7840 \times 10^{-11}\,J$$

Convert to J/mol:

$$\frac{4.7840 \times 10^{-11}\,J}{1\,nucleus} \times \frac{6.02 \times 10^{23}\,nuclei}{1\,mol} = 2.881 \times 10^{13}\,J\big/mol$$

Convert to MeV/nucleon:

$$\frac{4.7840 \times 10^{-11}\,J}{1\,nucleus} \times \frac{1\,MeV}{1.60 \times 10^{-13}\,J} \times \frac{1\,nucleus}{35\,nucleons} = 8.54\,MeV\big/nucleon$$

Workbook Example 19.6

PROBLEM:
Calculate the mass change in g/mol when N atoms combine to form N_2 molecules.

$$2\,N \rightarrow N_2 \qquad \Delta E = -945.4 \text{ kJ/mol}$$

SOLUTION:
With a known energy change, rearrange the Einstein equation to solve for Δm:

$$\Delta m = \frac{\Delta E}{c^2}$$

This is an amount far too small to be measured.

Workbook Problem 19.6

Calculate the binding energy of a Xe-130 nucleus in Mev/nucleon. The mass of a ^{130}Xe nucleus is 129.903 amu.

Strategy: Calculate the mass defect and convert to energy using the Einstein equation.

Step 1: Determine the number of nucleons and their mass, subtract the mass of the electrons from the atomic mass, and calculate the mass defect in amu.

Step 2: Convert the mass defect to grams.

Step 3: Determine ΔE and convert to MeV/nucleon.

Section 19.6 Nuclear Fission and Fusion

Because the maximum stability is found in the middle of the periodic table, there is stability to be gained by the **fusion** of light nuclei and the **fission** of heavy nuclei.

Nuclear Fission
Certain large nuclei shatter into smaller pieces of roughly equal size when bombarded with neutrons. The fission of a given nucleus does not occur in the same way each time: more than 800 products have been identified from the fission of U-235, for example.

When neutrons are released as part of a nuclear fission in addition to those used to begin the reaction, a **chain reaction** can occur in which the reaction continues even after the external supply of neutrons is cut off.

In a small sample, many of the neutrons will escape without initiating further fission, but if there is a sufficient amount of reaction—known as a **critical mass**—it will become a self-perpetuating chain reaction.

Workbook Example 19.7

PROBLEM:
How much energy is released in both J and kJ/mol for the neutron-induced fission of U-235 (atomic mass = 235.043 92 amu) to form Te-140 (atomic mass = 139.938 85 amu) and Kr-93 (atomic mass = 92.931 27 amu)? The mass of a neutron is 1.008 66 amu.

SOLUTION:
To calculate the energy difference, the mass difference must be calculated and converted into energy.

Write and balance the equation: the overall masses need to balance!

$$^{235}_{92}U + {}^{1}_{0}n \rightarrow {}^{140}_{52}Te + {}^{93}_{36}Kr + 2\,{}^{1}_{0}n$$

mass of reactants = 235.04392 amu + 1.00866 amu = 236.05258 amu
mass of products = 139.93885 amu + 92.93127 amu + 2 (1.00866 amu) = 234.88744 amu

Calculate the mass difference:

$$reactants - products = 1.16514\, amu$$

Convert mass to kg:

$$1.16514\, amu \times \frac{1.6605 \times 10^{-24}\, g}{1\, amu} \times \frac{1\, kg}{1000\, g} = 1.93471 \times 10^{-27}\, kg$$

Use the Einstein equation to convert the mass difference to energy:

$$\Delta E = \left(1.93471 \times 10^{-27}\, kg\right)\left(3.00 \times 10^8\, \frac{m}{s}\right)^2 = 1.74124 \times 10^{-10}\, J$$

Convert to J/mol:

$$\frac{1.74124 \times 10^{-10}\, J}{1\, nucleus} \times \frac{6.022 \times 10^{23}\, nuclei}{1\, mol} = 1.049 \times 10^{14}\, J/_{mol}$$

Nuclear Reactors

Controlled fission reactions can be used to generate electricity. Uranium fuel is placed in a pressurized containment vessel surrounded by water and *control rods* made of boron and cadmium are added. The water slows the escape of neutrons, and the control rods absorb neutrons to regulate the speed of the fission reaction: they are raised to speed the reaction, allowing more neutrons into the fuel, and lowered to slow the reaction. The energy generated is used to heat circulating coolant, which produces steam that drives a turbine to produce electricity.

Naturally occurring uranium is a mixture of 99.3% non-fissionable ^{238}U and 0.07% fissionable ^{235}U. Reactor fuel is compressed UO_2 pellets that have been enriched to 3% ^{235}U and encased in zirconium.

The amount and concentration of nuclear fuel in a reactor is not enough to sustain a nuclear explosion: the worst-case scenario is when uncontrolled fission melts the reactor and the containment vessel, releasing radioactivity into the environment. Newer reactors have *passive safety* designs that automatically slow runaway reactors.

Thirty-one countries now use nuclear energy to supply some of their electricity needs. The primary issue holding back further use of nuclear energy is safe disposal of the wastes: ^{90}Sr takes 600 years to decay to safe levels, and ^{239}Pu takes 20,000 years to decay to safe levels.

Nuclear Fusion

Very light nuclei also release enormous amounts of energy when they undergo *fusion*: it is hydrogen fusion reactions that power the Sun and other stars. Hydrogen is cheap and plentiful, and fusion products are nonradioactive and nonpolluting, but with a temperature of approximately 40,000,000 K required to initiate the process, the technical barriers to making this a viable source of power are formidable, to say the least.

Workbook Problem 19.7

In Oklo, Gabon, a natural nuclear reactor operated in uranium ore for hundreds of thousands of years. Part of the evidence for this is the presence of ^{136}Xe and ^{99}Ru in the surrounding rocks. Write and balance an equation for the fission of ^{235}U resulting in these isotopes, and calculate the energy generated by this reaction in J/mol. The mass of ^{235}U is 235.043 92 amu, the mass of ^{136}Xe is 135.9072195 amu, the mass of ^{99}Ru is 98.905 939 amu. One neutron is needed to induce the fission, and one neutron is produced by the fission.

Strategy:	Balance the equation, calculate the mass defect, and convert to energy using the Einstein equation.
Step 1:	Write and balance the equation.
Step 2:	Calculate the mass defect in amu and convert to kilograms.
Step 3:	Determine ΔE and convert to J/mol.

Workbook Problem 19.8

Hydrogen bombs utilize the reaction between deuterium (^{2}H – 2.014 10 amu) and tritium (^{3}H – 3.016 05 amu) that produces ^{4}He (4.002 60 amu) and a neutron (1.008 66 amu). Calculate the energy generated by the reaction per mol of ^{4}He produced.

Strategy: Calculate the mass defect and convert to energy using the Einstein equation.

Step 1:   Determine the number of nucleons and their mass, and calculate the mass defect in amu.

Step 2:   Convert the mass defect to grams.

Step 3:   Determine ΔE and convert to J/mol.

Section 19.7   Nuclear Transmutation

Of the 3600 known isotopes, only 300 occur naturally. The remainder have been made by **nuclear transmutation**, the change of one element into another. This is often accomplished by bombardment of an atom with a high-energy particle such as a proton, a neutron, or an α-particle. In the collision, an unstable nucleus is momentarily generated, and a new element can be produced.

All of the transuranium elements—those with atomic numbers above 92—have been produced using transmutation.

The reactions and subsequent decays can be quite complex: because the elements can be produced in a variety of ways, they can also decay in a variety of ways.

Workbook Example 19.8

PROBLEM:
Ernest Rutherford was the first to accomplish nuclear transmutation, but the husband and wife team of Frederic and Irene Joliot-Curie were not far behind; they bombarded Al-27 with α-particles to produce P-30. Write and balance an equation for this reaction. What particle is also produced in this reaction?

SOLUTION:
According to the periodic table, Al has an atomic number of 13, and phosphorus an atomic number of 15. An α-particle is a helium nucleus: it has a mass of 4, and an atomic number of 2.

$$^{27}_{13}Al + ^{4}_{2}\alpha \rightarrow ^{30}_{15}P + ?$$

The number of protons adds up, but the number of neutrons does not: a neutron must also be produced by the reaction:

$$^{27}_{13}Al + ^{4}_{2}\alpha \rightarrow ^{30}_{15}P + ^{1}_{0}n$$

Workbook Problem 19.9

If U-238 is bombarded with C-12, the result is a new element and six neutrons. Write and balance the equation, and determine the identity of the element produced using the periodic table.

Strategy: Determine the overall mass, subtract out the neutrons, and determine the identity and mass of the element produced.

Section 19.8   Detecting and Measuring Radioactivity

Radiation is invisible, but it can be measured and detected by instruments, and its effect on biological systems can be readily observed.

Radiation intensity is measured in a number of different ways.

Decay events are measured as the *becquerel* (1 Bq = 1 disintegration/s) and the *curie* (1 Ci = 3.7×10^{10} disintegration/s). The curie is related to the decay rate of 1 g of radium.

Energy absorbed per kilogram of tissue is measured as the *gray* (1 Gy = 1 J/kg tissue) and the *rad*, which stands for *radiation absorbed dose*. A rad is one-hundredth of a Gy (1 rad = 0.01 Gy).

Tissue damage units take into account the type of radiation. The *sievert* (Sv) accounts for this. One Gy of α particles causes twenty times the damage of 1 Gy of γ rays, but 1 Sv of α particles causes the same damage as 1 Sv of γ rays. The *rem* (which stands for *roentgen equivalent for man*) is equivalent to one-hundredth of a sievert (1 rem = 0.01 Sv).

The seriousness of radiation exposures varies with the type of radiation, the energy of the radiation, the length of exposure, and whether the source is external or internal.

Coming from an external source, X rays and γ rays are the most harmful to the human body because they penetrate clothing and skin. They move at the speed of light and have 1000 times the penetrating power of α particles. Several inches of lead are required to stop them.

Coming from an internal source, α particles and β particles are more dangerous, because all their energy is absorbed by surrounding tissues. Alpha particles move at one-tenth the speed of light and can be stopped by the top layer of skin or a few pieces of paper. β particles are lighter and faster—they move at nine-tenths the speed of light and require a block of wood or heavy protective clothing to stop them from penetrating the body.

The biological effects of radiation exposure can be dire, but the usual exposure of an average person is only about 120 mrem annually. Transient health effects begin to be seen between 25 and 100 rem, where white cell counts begin to be temporarily decreased. The lethal dose of radiation is 600 rem—5000 times higher than most people's annual exposure.

About 70% of yearly exposures are natural, from rocks and cosmic rays, with the remaining 30% coming from medical procedures such as X rays.

Section 19.9   Applications of Nuclear Chemistry

Dating with Radioisotopes
Radiocarbon dating depends on the slow production of C-14 in the upper atmosphere by neutron bombardment of nitrogen. From this reaction, $^{14}CO_2$ diffuses into the lower atmosphere, where it is taken up by plants and animals. While the organisms are living, the ratio of ^{14}C to ^{12}C is constant, and is the same as that in the atmosphere. When an organism dies, the ^{14}C decays and is not replenished. By measuring the $^{14}C/^{12}C$ ratio, the approximate age of biological remains can be determined using the half-life of ^{14}C, which is 5715 years.

Variations on this technique use the fact that uranium-238 decays to lead-206 with a half-life of 4.47×10^9 years, making the dating of uranium-containing rocks possible. Potassium-40 decays with a half-life of 1.25×10^9 years to yield argon-40. The age of a rock can be estimated by crushing it and comparing the amount of ^{40}Ar and the amount of ^{40}K.

Workbook Example 19.9

PROBLEM:
A skull hoped to be ancient was found to have a ^{14}C decay rate of 4.7 disintegrations/min. What age is implied by this if living organisms have a decay rate of 15.3 disintegrations/min, and the half-life of ^{14}C is 5715 years?

SOLUTION:
Radioactive decay is a first-order process, so the time since the beginning of the reaction can be determined by calculating using the integrated rate law:

$$\ln \frac{N_t}{N_0} = -0.693 \left(\frac{t}{t_{1/2}} \right)$$

Substituting:

$$\ln \frac{4.7}{15.3} = -0.693 \left(\frac{t}{5715 \ years} \right)$$

$$\frac{\left(\ln \frac{4.7}{15.3} \right)(5715 \ years)}{-0.693} = 9734 \ years$$

The skull is approximately 10,000 years old.

Medical Uses of Radioactivity

In vivo procedures are those that take place inside the body to assess the function of a particular organ or body system. A radiopharmaceutical is administered, and its path in the body analyzed.

Therapeutic procedures are those in which radiation is used to kill diseased tissue. External radiation can be directed at tumors, often using γ rays from a Co-60 source. Although the radiation is directed as carefully as possible, patients typically develop some symptoms of radiation sickness. Internal radiation treatment is much more selective and can target tissues very specifically, as when I-131, which is selectively taken up by the thyroid, is used in the treatment of thyroid disease.

Imaging procedures provide diagnostic information by analyzing the distribution of radioisotopes introduced into the body. Depending on the use, a diseased area might concentrate the isotope, showing up as a "hot" spot, or fail to take up the isotope, thus showing up as a "cold" spot.

Putting It Together

47.0 % of the chlorine atoms in a 78.3 mg sample of sodium perchlorate are ^{36}Cl, which is a radioactive isotope of chlorine. The half-life of ^{36}Cl is 3.0×10^5 yr. How many disintegrations/second are produced by this sample?

Self–Test

This section is intended to test your knowledge of the material covered in this chapter. Think through these problems, and make certain you understand what they are asking. Make sure your answers make sense. Successful completion of these problems indicates that you have mastered the material in this chapter. You will receive the greatest benefit from this section if you use it as a mock exam, as this will allow you to determine which topics you need to study in more detail

True–False
1. Radioactive decay is a second-order process.

2. Lighter elements require more neutrons for stability, but heavier elements have a 1:1 ratio of protons and neutrons.

3. When nucleons come together to form a nucleus, a small amount of mass is gained.

4. When elements undergo neutron-induced fission, they tend to break into pieces of roughly equivalent size.

5. The control rods in a reactor absorb excess neutrons to slow the reaction.

6. Nuclear fusion has no toxic by-products, but is unlikely to be a practical source of power in the near future.

7. Given their positions in the periodic table, a nuclear transmutation reaction is more likely to turn gold into lead than lead into gold.

8. Alpha particles are most damaging from an internal source.

9. The curie measures only disintegrations per second, and it says nothing about tissue damage.

10. Radioisotopes can be used for disease treatment, imaging, and in vivo studies.

Multiple Choice

11. All of the following are nuclear processes *except*
    a. alpha emission   b. electron capture
    c. beta modification   d. transmutation

12. Nuclei are most stable when they
    a. have a 1:1 ratio of neutrons to protons, if heavy
    b. have fewer neutrons than protons
    c. are ionized
    d. are ^{56}Fe

13. The mass defect in nuclei is the result of
    a. inaccurate masses of subatomic particles
    b. the conversion of mass to energy to hold the nucleus together
    c. the law of conservation of mass
    d. nuclear transmutation

14. The process of nuclear fission
    a. can self-perpetuate as a chain reaction   b. generates very little energy
    c. generates only ^{4}He as a product   d. b and c

15. In a reactor, nuclear fission is used to generate power by
    a. generating electrons.
    b. converting water to steam, which turns a turbine.
    c. driving a piston with repeated small explosions.
    d. none of the above.

16. The control rods in a reactor
    a. give off electrons   b. generate neutrons
    c. absorb neutrons   d. emit alpha particles

17. Nuclear transmutation is a process that can be used to
    a. generate new elements and isotopes   b. generate electricity
    c. diagnose disease   d. date ancient objects

18. Which of the following provides information about tissue damage?
    a. becquerel   b. rem
    c. sievert   d. b and c

19. Which of the following correctly ranks these types of emissions by speed, fastest to slowest?
    a. alpha particle, gamma ray, beta particle   b. beta particle, alpha particle, X ray
    c. gamma ray, beta particle, alpha particle   d. X-ray, alpha particle, gamma ray

20. Radiocarbon dating can only be used to determine the ages of
    a. meteorites   b. formerly living objects
    c. uranium-containing rocks   d. all of the above

Fill–in–the–Blank

21. Neutrons appear to function as a nuclear _____ that overcomes the _____ to _____ repulsions that might otherwise cause a(n) _____ to fly apart.

22. The energy to hold the nucleus together cannot be directly measured, but can be calculated using the _____ _____ and the equation _____.

23. Binding energy per nucleon reaches a(n) _____ in the middle of the periodic table. Therefore, small nuclei can increase their stability through _____, and large nuclei can increase their stability through _____ or radioactive _____.

24. In nuclear _____, an atom is bombarded with any of a number of high-energy_____. In the collision, an unstable _____ is momentarily formed, and a new _____ is produced.

25. The effects of radiation on a biological system vary according to a large number of variables. The most damage comes from large, slow particles such as _____ and _____ particles as a(n) _____ source. Also damaging are the fastest moving radiation, _____ rays and _____ rays, which do the most damage as _____ sources. Large, slow particles are _____ by skin. The fast-moving rays require several inches of_____ to stop them.

26. Radiopharmaceuticals are used in a number of ways. They can be used in _____ procedures in which they are administered and monitored, in therapeutic procedures in which they are used to target _____ tissue (_____ radiation therapy is always more specific and targeted than _____ radiation therapy), and in _____, in which the isotope's location or absence is used to assist in diagnosis.

Matching

27. Binding energy

    a. a reaction in which the nucleus of an element splits into two nearly equal pieces

28. Chain reaction

    b. the loss in mass when nucleons come together to form a nucleus

29. Critical mass

    c. the joining of nuclei in a reaction

30. Fission

    d. the amount of material necessary for a nuclear reaction to be self-sustaining

31. Fusion

    e. the energy that holds nucleons together

32. Mass defect   f.   change of one element into another

33. Nuclear transmutation   g.   a self-sustaining reaction whose product initiates further reaction

Problems

34. Calculate the binding energy of europium-130 (mass = 129.963 57 amu), europium-153 (mass = 152.921 23 amu), and europium-167 (mass = 166.953 21amu) in MeV/nucleon. From these data, which is the most stable isotope? Which is the least stable?

35. Calculate the change in mass for the following reaction in g/mol:
$$2\,O \rightarrow O_2 \qquad \Delta E = -498.4 \text{ kJ/mol}$$

36. How much energy in J/mol is released in the following fission reaction?
$$^{235}U + {}^{1}n \rightarrow {}^{92}Kr + {}^{141}Ba + 3\,{}^{1}n$$
The mass of ^{235}U is 235.0439 amu, the mass of ^{92}Kr is 91.9262 amu, the mass of ^{141}Ba is 140.9144 amu, and the mass of a neutron is 1.008 66 amu.

38. How much energy is released in the following fusion reaction?
$$^{3}H + {}^{3}H \rightarrow {}^{4}He + 2\,{}^{1}n$$
The mass of tritium is 3.016 049 amu, the mass of ^{4}He is 4.002 60 amu, and the mass of a neutron is 1.008 66 amu.

39. Write chemical reactions for the following nuclear transmutations:
    a. Be-9 is bombarded with hydrogen to form Li-6 and an alpha particle.
    b. Pu-239 reacts with an alpha particle to produce an unknown element and a neutron.
    c. Pb-208 is collided with another element to form Hs-265 and a neutron.

40. The half-life for zirconium-96 is 6.3×10^{26} s. What is the decay constant for this radioactive element?

41. What is the age of a bone fragment that shows an average of 3.5 disintegrations per minute per gram of carbon? The carbon in living organisms undergoes an average of 15.3 disintegrations per minute per gram, and the half-life of C-14 is 5715 y.

42. The decay of U-235 to Pb-207 takes place with a half-life of 704 million years. If an asteroid that originally contained no lead is found to contain 56 g of Pb-207 and 27 g of U-235, how old is it?

Challenge Problem

43. In an experimental form of cancer treatment known as boron neutron capture therapy (BNCT), boron-10 is injected into a cancer patient, where it binds preferentially to tumor cells. A beam of neutrons is then directed at the tumor, and the boron-10 disintegrates into lithium. Only one type of radiation is produced by the reaction, and only one neutron is absorbed in the reaction. Write and balance the equation for this reaction, and explain the rationale behind this therapeutic method.

CHAPTER TWENTY

Transition Elements and Coordination Chemistry

Learning Objectives

As a result of reading and studying this chapter, you should be able to

Sections 20.1–20.2   **Electron Configurations and Properties of Transition Elements**

1. Write electron configurations for transition metals and predict the number of unpaired electrons.
2. Describe the trends in physical and chemical properties of transition metal elements across a period. Explain the reason for the trends.

Section 20.3   **Oxidation States of Transition Elements**

3. Relate the oxidizing and reducing strength of a transition metal ion to its oxidation state and location in the periodic table.

Section 20.4   **Chemistry of Selected Transition Elements**

4. Predict the products and write balanced equations for the common laboratory reactions of first-series transition metals, their ions, and their compounds.

Sections 20.5–20.6   **Coordination Compounds and Ligands**

5. Identify Lewis acids, Lewis bases, ligands, and donor atoms in a coordination complex.
6. Determine the formula of a coordination complex, identify the oxidation state and coordination number of the metal atom, and draw the structure of the complex.
7. Classify a ligand as mono-, bi-, tri-, tetra-, or hexadentate, based on its chemical structure.

Section 20.7   **Naming Coordination Compounds**

8. Determine the systematic name of a coordination complex from its formula or structure.
9. Determine the formula of a coordination complex from its systematic name or structure.

Sections 20.8–20.9   **Isomers and Enantiomers**

10. Classify isomers according to the scheme given in Figure 20.16.
11. Draw diastereomers for octahedral and square planar complexes.
12. Classify a coordination complex as chiral or achiral and draw enantiomers for chiral compounds.

Section 20.10   **Color of Transition Metal Complexes**

13. Relate the color of a metal complex, the wavelength of light it absorbs, and the energy difference between its ground and excited state.

Section 20.11   **Bonding in Complexes: Valence Bond Theory**

14.   Describe the bonding in a metal complex in terms of valence bond theory, including orbital diagrams, the hybrid orbitals used by the metal atom, and the number of unpaired electrons.

Section 20.12   **Crystal Field Theory**

15.   Draw crystal field energy-level diagrams for octahedral, tetrahedral, and square complexes. Classify ligands as strong or weak-field and predict (or account for) color and magnetic properties of complexes in terms of crystal field theory.

Chapter Summary

This chapter examines the properties and chemical behavior of transition-metal compounds, with particular emphasis on coordination compounds. Beginning with a review of the electronic configurations of the transition elements and how these configurations affect the properties and oxidation states of the ions formed, the chemistry of chromium, iron, and copper are examined in detail. Coordination chemistry is then explored, including structural properties, isomerization, color, and magnetism of the compounds. Two different theories, valence bond theory and crystal field theory that account for these properties are then discussed.

The Chapter in Detail

Section 20.1   Electron Configurations

Note: For convenience in this chapter, the orbitals will be listed by principal quantum number, because the $n + 1$ s electrons are lost first when transition metals ionize.

d orbitals begin at atomic number 21, scandium, and continue through to zinc. The filling proceeds according to Hund's rule, with the $n + 1$ s shells filling first, followed by one electron in each of the d orbitals, thereafter doubling up the electrons.

There are two exceptions to this pattern: Cr and Cu are both able to gain stability by varying from this rule. Cr places one electron in each of the orbitals for a completely half-filled configuration: $4s^1 3d^5$. Copper pulls one electron from the $4s$ orbital to gain a fully filled d orbital and leaves a half-filled s orbital: $4s^1 3d^{10}$.

Exceptions from the expected orbital-filling pattern always result in completely half-filled subshells or completely filled shells. The nd and $n + 1s$ orbitals have very similar energies, allowing electrons to move between them if it is energetically favorable to do so.

Two half-filled subshells minimize electron–electron repulsions, as does a completely filled d subshell with a half-filled s subshell.

When electrons are lost, the effective nuclear charge increases. The d orbitals experience a steeper drop in energy as the effective nuclear charge increases, making them quickly lower in energy than the s orbital. As a result, the s orbital empties first.

The designation of transition metals as groups 1B–8B was developed because these groups have analogous electron structures to the elements in groups 1A – 8A. Copper, in group 1B has Cu^+ as its most common ion, while Zn, in group 2B, will most commonly form a Zn^{2+} ion, like the main-group metals in group 2A. However, these designations are not as straightforward as those of the main-group elements, and they should be used with care.

Workbook Example 20.1

PROBLEM:
Write the electron configuration of Cu^{3+}. Would you expect this to be a stable species?

SOLUTION:
Writing the electron configuration for neutral copper gives

$$[Ar]\ 3d^{10}\ 4s^1$$

Removing three electrons to make Cu^{3+} gives

$$[Ar]\ 3d^8$$

because the s orbital will empty first. This will not be a stable species, as it has a high effective nuclear charge and two half-filled d orbitals. The most stable ion for copper will be Cu^+, with electron configuration $[Ar]\ 3d^{10}$.

Workbook Problem 20.1

Write the electron configurations for the following species:

Ti   Mn in MnO_4^- Ni^{2+}

Strategy: Write the expected electron configuration for the neutral element, determine the charge, and remove electrons accordingly.

Step 1:   Write the electron configuration of the neutral atom.

Step 2:   Determine the oxidation state of the metal.

Step 3:   Remove the necessary electrons.

Section 20.2   Properties of Transition Elements

As metals, all these elements are malleable, ductile, lustrous, and good conductors of heat and electricity. The sharing of d, as well as s, electrons gives rise to stronger metallic bonding in these elements than is seen in the group 1A and 2A metals, which makes those metals harder and more dense, with higher melting and boiling points.

Melting points increase as the number of unpaired d electrons available for metallic bonding increases and then decrease as the d electrons pair up and become less available for bonding, meaning that the melting points reach a maximum in the middle of each series.

Atomic radii and densities demonstrate a similar pattern: as the unpaired d electrons increase to a half-filled orbital, the atomic radii decrease and densities increase. As the electrons begin to pair up, the electron repulsions increase and the effective nuclear charge decreases, so the atomic radii will increase and the density will decrease.

The effects on atomic radii are more subtle in the transition metals than they are in the main-group elements. The similar atomic radii among the transition metals account for their ability to blend with one another into an alloy.

Across the f-block lanthanide elements, the increase in effective nuclear charge is almost exactly balanced by the increase in size expected from adding an entire quantum shell, resulting in smaller-than-expected atomic radii. Because of this **lanthanide contraction**, the elements in the third transition series have almost the same radii as the elements in the second transition series. This means that the third series, which includes Os, the most dense metal, are unusually dense.

The ionization energies of transition metals increase from left to right across a series, due to an increase in the effective nuclear charge and the corresponding decrease in atomic radius. The $E°$ for the **oxidation** potential of the first transition series is positive for every metal except Cu, meaning that these metals are oxidized more readily than H_2 gas is oxidized to H^+.

These metals can be oxidized by simple acids that lack an oxidizing anion, but the oxidation of Cu (s) requires a stronger oxidizing agent such as HNO_3.

The general trend for $E°$ is the same as that for ionization energies. The easier a metal is to oxidize, the easier it is to ionize.

Section 20.3   Oxidation States of Transition Elements

Transition metals exhibit a variety of oxidation states, which can be lower than their group number. All of the first series except scandium will form +2 cations, corresponding to the emptying of the $4s$ shell.

The first series can also lose $3d$ electrons to form ions with greater charges. The increased energy required to remove additional electrons is balanced by the larger $\Delta G°$ of hydration of the more highly charged cations. The most highly charged species are found in combination with the most electronegative elements, typically O.

For the group 3B–7B metals, the group number is the highest possible oxidation state, corresponding to the loss of all valence s and d electrons. This is not true for the 8B elements, however, as the loss of all the valence electrons is energetically prohibitive.

Transition metal ions in a high oxidation state are good oxidizing agents, while early transition metal ions in a low oxidation state are good reducing agents, but divalent ions of the later metals are poor reducing agents because of the larger effective nuclear charge.

Section 20.4   Chemistry of Selected Transition Metals

Chromium

Chromium is obtained from chromite ore ($FeO·Cr_2O_3$ or $FeO·Cr_2O_4$). If it is reduced with carbon, ferrochrome, an alloy used to make stainless steel, is produced. To obtain pure chromium, Cr_2O_3 is reduced with aluminum. Solid chromium is used to electroplate metal objects with a protective coating: it takes a high sheen and is corrosion-resistant because it develops a hard invisible coating of Cr_2O_3.

In aqueous solution, chromium will exist in +2, +3, and +6 oxidation states, with +3 the most stable.

When chromium metal reacts with acid in the absence of air, +2 chromium ions form a bright blue complex with six water molecules. When atmospheric O_2 is present, these ions are quickly oxidized to the more stable +3 ion. Although the $Cr(H_2O)_6^{3+}$ complex is violet, Cr^{3+} solutions are typically green because other anions replace some of the water molecules.

In basic solution, chromium(III) precipitates as chromium(III) hydroxide, a pale green solid that will dissolve in both acid and base. Chromium(II) hydroxide is a typical hydroxide that dissolves in acid but not in base, and chromium IV forms chromic acid, $CrO_2(OH)_2$ (also written H_2CrO_4), which is a strong acid. The higher the oxidation state of chromium, the more polar the OH bond and the greater the acidity.

In the +6 oxidation state, the two most important species are chromate, CrO_4^{2-}, in basic solution and dichromate, $Cr_2O_7^{2-}$, in acidic solution. Dichromate is a powerful oxidizing agent used in analytical chemistry.

Iron

Iron is the fourth-most abundant element in the Earth's crust and is immensely important in human civilization and in living systems. Alloys of iron with V, Cr, and Mn make steels that are less readily corroded and harder than iron alone.

The most important iron ores are hematite, Fe_2O_3, and magnetite, Fe_3O_4, which are reduced with coke in a blast furnace to give pure Fe.

The most important oxidation states of iron are +2 and +3.

Reaction with an acid that lacks an oxidizing anion (HCl, for example) in the absence of air will yield the iron (II) ion $Fe(H_2O)_6^{2+}$. The oxidation ceases at this point, because the standard potential for $Fe^{2+} \rightarrow Fe^{3+}$ is positive. In air, the iron ions slowly oxidize to Fe^{3+}.

Reaction with an acid that has an oxidizing anion directly yields Fe^{3+} ions.

The reaction of Fe^{3+} (aq) with a base yields $Fe(OH)_3$, which is exceptionally insoluble and forms when the pH rises above 2. It is the red-brown rust stain found in old iron sinks and bathtubs.

Copper

Copper is relatively rare, but can be found in the elemental state. Its most important ores are sulfides, including chalcopyrite, $CuFeS_2$. Sulfides are concentrated, separated from the iron, and converted to copper(I) sulfide, which can then be oxidized to form copper and sulfur dioxide. The resultant metal is 99% pure, but can be further purified by electrolysis.

Copper has high conductivity and negative oxidation potential, making it valuable for electrical wiring and corrosion-resistant water pipes. It also can be used to form alloys such as brass, Cu and Zn, and bronze, Cu and Sn.

Prolonged environmental exposure will eventually lead to oxidation: moist air and CO_2 lead to the development of $Cu_2(OH)_2CO_3$. Subsequent reaction with the highly dilute sulfuric acid in acid rain generates $Cu_2(OH)_2SO_4$, the green patina that develops on bronze.

Copper has two primary oxidation states: +1 (cuprous) and +2 (cupric). However, the $E°$ for $Cu^+ \rightarrow Cu^{2+}$ is less negative than that for the reaction of $Cu \rightarrow Cu^+$. As a result, anything strong enough to oxidize copper to Cu(I) ions is also strong enough to oxidize Cu(I) to Cu(II).

Cu^+ (aq) will disproportionate in solution:

$$2\,Cu^+\,(aq) \rightarrow Cu\,(s)\,+\,Cu^{2+}\,(aq) \qquad E° = +0.37\text{ V}, K = 1.8 \times 10^6$$

As a result of this large equilibrium constant, Cu^+ is not an important species in aqueous solution, although it does exist in solid compounds. When Cl^- is present, the disproportionation is reversed, and the precipitation of CuCl shifts the reaction to the left.

The more common Cu^{2+} is found in the blue aqueous $Cu(H_2O)_6^{2+}$. If aqueous ammonia is added to this ion, it will generate a blue precipitate of copper(II) hydroxide. If more ammonia is added, it will redissolve and form the dark blue complex ion $Cu(NH_3)_4^{2+}$.

Section 20.5   Coordination Compounds

A **coordination compound** is a compound in which a central metal ion is attached to a group of surrounding molecules or ions by coordinate covalent bonds. The molecules or ions surrounding the central metal ion are called **ligands**, while the atoms that are directly attached to the metal ion are the **ligand donor atoms**. Formation of these compounds is a Lewis acid–base interaction: the electron donors (Lewis bases) are the ligands, which always have lone pairs, and the electron acceptor (acting as a Lewis acid) is the central metal ion.

Some coordination compounds are salts containing a complex cation or anion along with enough ions of opposite charge to give a compound that is electrically neutral overall. The complex ion is enclosed in brackets when it is written to show that the other ions are simply balancing charges, while the complex ion is a discrete structural unit. The term **metal complex** refers to both neutral coordination compounds and the salts of ionic coordination compounds.

The number of ligand donor atoms surrounding the metal ion is the **coordination number**. The most common are 4 and 6, but others are also well known. The coordination number depends on the size, charge, and electron configuration of the central atom, as well as the size and shape of the ligands.

The characteristic shape of a metal complex depends on the metal ion's coordination number. A coordination number of 2 indicates a linear molecule, while a coordination number of 4 can be either tetrahedral or square planar. A coordination number of 6 will be octahedral.

The charge on a metal complex is equal to the oxidation state of the metal plus the sum of the charges on the ligands. If two are known, the third can be found.

Workbook Example 20.2

PROBLEM:
Silver forms a neutral complex with two ammonia molecules and a nitrate ion. What is the oxidation state of the metal, and what is the formula of the complex?

SOLUTION:
The charge on the complex is the sum of all the charges. The ammonia molecules are neutral, and nitrate has a charge of −1. For the complex to be neutral, the oxidation state of silver must be +1.

The formula of the complex is $[Ag(NH_3)_2]NO_3$.

Workbook Problem 20.2

Determine the oxidation state and coordination number of the metal in the following complexes:

$AgCl_2^-$ $[MnCl_4]^{2-}$ $[Fe(CN)_6]^{4-}$ $[Co(NH_3)_4Br_2]Br$

Strategy: Because we are given the charge and ligands, it is possible to determine the oxidation state of the metals. The coordination number is the number of ligand attachments.

Step 1:   Determine the oxidation state of the metals.

Step 2:   Determine the coordination numbers.

Section 20.6   Ligands

Ligands can be classified as **monodentate** or **polydentate** depending on the number of electron pairs that bond to the metal atom. Water and ammonia are monodentate ligands, binding through the O and N atoms. Glycinate ($NH_2CH_2COO^-$) is a bidentate ligand that binds through both the N and O.

Chelating agents are polydentate ligands. $EDTA^{4-}$ (Ethylenediamenetetraacetate) is a **hexadentate** ligand that will bind to metals. When the chelating agent is bound to a metal, it forms a **chelate ring**. A complex that contains one or more such rings is called a metal **chelate**.

$EDTA^{4-}$ forms particularly stable complexes, and can be used to hold metal ions in solution. It can be used in treating lead poisoning, preventing the precipitation of Pb in tissues by forming a chelate that is excreted by the kidneys. $EDTA^{4-}$ is often added to foods such as commercial salad dressing to remove metals that could catalyze the oxidation of oils and cause the dressing to become rancid.

Naturally occurring chelators are essential components of many biomolecules. For example, the heme group in hemoglobin is a planar, tetradentate ligand that binds an iron(II) ion. This complex reversibly binds an O_2 molecule to carry oxygen through the blood.

Section 20.7   Naming Coordination Compounds

If the compound is a salt, name the cation first and then the anion.

In naming a complex ion or a neutral complex, name the ligands first, in alphabetical order (ignoring prefixes), and then the metal. Anionic ligands end in –o. The complex name is one word. For more than one ligand of a particular type, use Greek prefixes (*di-, tri-, tetra,* etc.) to indicate the number of ligands.

If the name of a ligand itself contains a Greek prefix, put the ligand name in parentheses and use an alternate prefix (*bis-, tris-, tetrakis-*).

Use a Roman numeral in parentheses immediately following the name of the metal to indicate the oxidation state of the metal.

In naming the metal, use the ending –*ate* if the metal is in an anionic complex.

Anionic Ligand	**Ligand Name**	**Neutral Ligand**	**Ligand Name**
Bromide, Br^-	Bromo	Ammonia, NH_3	Ammine
Carbonate, CO_3^{2-}	Carbonato	Water, H_2O	Aqua
Chloride, Cl^-	Chloro	Carbon monoxide, CO	Carbonyl
Cyanide, CN^-	Cyano	Ethylenediamine, en	Ethylenediamine
Fluoride, F^-	Fluoro		
Glycinate, gly^-	Glycinato		
Hydroxide, OH^-	Hydroxo		
Oxalate, $C_2O_4^{2-}$	Oxalato		
Thiocyanate, SCN^-	Thiocyanate (S ligand)		
	Isothiocyanate (N ligand)		

Workbook Example 20.3

PROBLEM:
Name the following compounds:

$[Cr(H_2O)_5Cl](NO_3)_2$　　　　$Na_2[OsCl_4NO_3]^{3-}$　　　　$[Co(gly)NO_3]Cl$

SOLUTION:
Using the rules for naming coordination complexes:

$[Cr(H_2O)_5Cl](NO_3)_2$　　The ligands are *aqua* and *chloro*. With 5 H_2O, 1 Cl^-, and 2 charge-balancing NO_3^- ions, the charge on chromium is +3. This is pentaaquachlorochromium(III) nitrate.

$Na_2[OsCl_4NO_3]^{3-}$　　The ligands in this compound are *chloro* and *nitrato*. With 2 Na^+, 4 Cl^-, and 1 NO_3^-, and an overall charge of −3, the charge on osmium must be +6. Since the complex is an anion, the ending on osmium is changed to *ate*. This is sodium tetrachloronitratoosmate(VI).

$[Co(gly)NO_3]Cl$　　The ligands in this compound are *glycinate* and *nitrato*. With gly^-, NO_3^- and a balancing Cl^-, the charge on cobalt is +3. The compound is glycinonitratocobalt(III) chloride.

Workbook Example 20.4

PROBLEM:
Write formulas for the following compounds: hexaquanickel(II) chloride, ammonium diaquatribromochlorovanadate(III), sodium hexanitratochromium(III).

SOLUTION:
Determine the formula for the compounds:

The aqua ligand is H_2O. The formula is $[Ni(H_2O)_6]Cl$
The ammonium cation precedes the formula. $NH_3[V(H_2O)_2Br_3Cl]$
The nitrato cation is NO_3^-. $Na_3[Cr(NO_3)_6]$

Section 20.8   Isomers

In coordination chemistry, it is possible for compounds to have the same formula but a different arrangement of constituent atoms. This leads to different physical and chemical properties.

Constitutional isomers are isomers that have different connections among their constituent atoms.

Linkage isomers arise when a ligand can bond to a metal through either of two different donor atoms. For example, NO_2^- can use N or O as a donor atom.

Ionization isomers differ in the anion that is bonded to the metal ion. For instance, $[AgCl_2]NO_3$ has different properties than $[AgClNO_3]Cl$, but the same type of atoms.

Stereoisomers have the same connections among atoms but have a different arrangement of the atoms in space.

Diastereoisomers, or *geometric isomers*, have different relative orientations of their metal–ligand bonds. In a *cis* isomer, identical ligands occupy adjacent corners of the square in a square planar complex. In a *trans* isomer, identical ligands are across from one another in a square planar complex.

These isomers have different properties due to their different geometries, which affect dipoles and polarities in the isomer.

Square planar complexes of the type MA_2B_2 and MA_2BC (M = metal ion; A, B, C = ligands) can exist as *cis–trans* isomers. Tetrahedral complexes cannot have isomers, because all four corners are adjacent.

Octahedral complexes of the type MA_4B_2 or MA_4BC can also exist as diastereoisomers: two B ligands can be either on adjacent or opposite corners of the octahedron.

Workbook Example 20.5

PROBLEM:
Of the following compounds, which have geometric isomers? (Assume square planar or octahedral geometry.)

$Co(H_2O)_2NO_3Cl$ $[Fe(CN)_6]^{4-}$ $Ni(CO)_2Cl_2$

SOLUTION:
Determine the basic structure of the compounds and whether isomers are possible.

$Co(H_2O)_2NO_3Cl$ has the basic structure MA_2BC—it will have isomers
$[Fe(CN)_6]^{4-}$ has six identical ligands—it will not have isomers
$Ni(CO)_2Cl_2$ has the basic structure MA_2B_2—it will have isomers

Section 20.9   Enantiomers and Molecular Handedness

Enantiomers are molecules or ions that are nonidentical mirror images of one another; they are said to have different handedness, or *chirality*. They are also called *optical isomers* because of their opposite effects on polarized light.

Chiral objects do not have a symmetry plane; **achiral** objects do have a symmetry plane.

Certain molecules and ions are chiral. If in order of size, the ligands spiral to the right (clockwise), the enantiomer is referred to as "right-handed." If the ligands spiral to the left (counterclockwise) the enantiomer is referred to as "left-handed."

Enantiomers have identical properties except for their reactions with other chiral substances and their effect on plane-polarized light: a solution of one enantiomer will rotate light to either the right or left, and the other enantiomer will rotate the light the same amount in the other direction.

A **racemic mixture** is a 50:50 mixture of the two enantiomers. This will have no net effect on polarized light, because the two rotations cancel each other out.

Section 20.10   Color of Transition Metal Complexes

The color of transition metal complexes depends on the identity of the metal and the ligands. The colors will vary for the same metal with different ligands; they will also vary for different metals with the same ligand.

Metal complexes absorb light by undergoing an electronic transition from the lowest energy state (E_1) to a higher energy state (E_2). The wavelength absorbed depends on the energy difference between the two energy levels:

$$\Delta E = E_2 - E_1$$
$$\Delta E = h\nu = \frac{hc}{\lambda}$$

where h is Planck's constant, ν is the frequency, and λ is the wavelength.

Absorbance is the measure of the amount of light absorbed by a substance. The **absorption spectrum** is a plot of absorbance versus wavelength. The observed color is complementary to the color absorbed; for instance, if a complex strongly absorbs green light, it will appear orange.

Workbook Example 20.6

PROBLEM:
If a compound strongly absorbs at a frequency of 1.23×10^{15} s^{-1}, what is the difference between the ground state and the excited state of this compound in kJ/mol?

SOLUTION:
Use the formula $\Delta E = h\nu$ to determine the energy of the absorbed photons, then convert to kJ/mol.

$$\Delta E = h\nu = \left(6.626 \times 10^{-34}\ J \cdot s\right)\left(1.23 \times 10^{15}\ s^{-1}\right) = 8.15 \times 10^{-19}\ J\big/photon$$

$$\frac{8.15 \times 10^{-19}\ J}{1\ photon} \times \frac{6.02 \times 10^{23}\ photons}{1\ mole} = 491\ kJ\big/mole$$

Workbook Problem 20.3

If a coordination compound strongly absorbs light with a wavelength of 550 nm, what is the difference between the ground and excited states of the complex in kJ/mol of complex?

Strategy: Using the formula $\Delta E = \dfrac{hc}{\lambda}$, determine the energy of the absorbed photons, then convert to kJ/mol.

Step 1:   Determine the energy of the photons absorbed.

Step 2:   Convert to kJ/mol.

Section 20.11   Bonding in Complexes: Valence Bond Theory

In **valence bond theory**, bonding results when a filled ligand orbital containing a pair of electrons overlaps a vacant hybrid orbital on the metal ion to produce a coordinate covalent bond.

The geometry of these complexes is related to the hybrid orbitals used. Linear complexes have sp orbitals, tetrahedral complexes use sp^3 orbitals. Octahedral complexes have six equivalent hybrid orbitals that can be formed in either from d^2sp^3 or sp^3d^2. All are 90° from one another. Square planar complexes also use d orbitals in dsp^2 hybrids.

Transition metal complexes can be **paramagnetic**, containing unpaired electrons and attracted by magnetic fields, or **diamagnetic**, containing only paired electrons and weakly repelled by magnetic fields. The number of unpaired electrons can be measured by the force exerted on the complex by a magnetic field.

Metals can form complexes described as **high-spin complexes**, in which the d electrons are arranged to maximize unpaired electrons and minimize doubly occupied orbitals, or **low-spin complexes**, in which d electrons are arranged to maximize filled orbitals and minimize unpaired electrons.

Workbook Example 20.7

PROBLEM:
$[Co(NH_3)_6]^{3+}$ is a low-spin complex. Draw an electron diagram to demonstrate this.

SOLUTION:
Determine which hybrid orbitals are being used to minimize spin.

The free Co^{3+} ion has the valence $[Ar]3d^6$. This is an octahedral complex, so will use either d^2sp^3 or sp^3d^2 hybrid orbitals.

Co^{3+} :

$$\underset{3d}{\uparrow \ \uparrow \ \uparrow \ \uparrow \ \uparrow} \qquad \underset{4s}{\_\_} \quad \underset{4p}{\_\_ \ \_\_ \ \_\_} \qquad\qquad d^6$$

$[Co(NH_3)_6]^{3+}$:

$$\underset{3d}{\uparrow \ \uparrow \ \uparrow\downarrow \ \uparrow\downarrow \ \uparrow\downarrow} \quad \underset{4s}{\uparrow\downarrow} \quad \underset{4p}{\uparrow\downarrow \ \uparrow\downarrow \ \uparrow\downarrow} \qquad\qquad d^2sp^3$$

or

$[Co(NH_3)_6]^{3+}$:

$$\underset{3d}{\uparrow \ \uparrow \ \uparrow \ \uparrow \ \_\_} \quad \underset{4s}{\uparrow\downarrow} \quad \underset{4p}{\uparrow\downarrow \ \uparrow\downarrow \ \uparrow\downarrow} \quad \underset{4d}{\uparrow\downarrow \ \uparrow\downarrow \ \_\_ \ \_\_ \ \_\_} \qquad sp^3d^2$$

The d^2sp^3 configuration has fewer unpaired electrons and is also lower energy overall. Both support the choice of this configuration.

Workbook Problem 20.4

$[Co(Cl)_6]^{3-}$ is a high-spin complex. Draw an electron diagram to demonstrate this.

Strategy: Draw the orbital diagram and determine how to maximize unpaired electrons.

Step 1: Determine the electron configuration of the metal alone.

Step 2: Identify the possible hybridization schemes.

Step 3: Draw electron diagrams with both hybridizations.

Step 4:   Choose the hybridization that maximizes electron spin.

Section 20.12   Crystal Field Theory

Crystal field theory views the bonding in complexes as arising from electrostatic interactions and considers the effect of the ligand charges on the energies of the metal ion *d* orbitals. It is able to account for the colors and magnetic properties of transition metal complexes, as well as the difference between the high- and low-spin complexes.

In crystal field theory, there are no covalent bonds, no sharing of electrons, and no hybrid orbitals. Interactions with neutral ligands are ion–dipole.

In octahedral complexes, metal is positively charged and the ligands are negatively charged. The anions adopt the configuration that allows them to get as far away from each other as possible. The metal–ligand attraction is stronger than the ligand–ligand repulsions, so the complex is stable.

Ligand charges repel negatively charged *d* electrons, causing orbital energies to be higher in the complex: the lobes of the orbitals that are closest to the ligands are higher in energy than those that fall between ligands. This causes **crystal field splitting**, represented by the Greek letter Δ.

Crystal field splitting energy, Δ, corresponds to wavelengths in the visible region of the spectrum, providing an explanation for the colors of the complexes. The value of Δ can be calculated from the wavelength of the absorbed light.

The effect that ligands have on Δ allows ligands to be ranked in terms of the energy changes they cause. This is known as the **spectrochemical series**. **Weak-field ligands** produce relatively small values of Δ and low-spin complexes, while **strong-field ligands** produce high values of Δ and high-spin complexes.

If the value of Δ is greater than the value of *P*, the energy required to put two electrons in one orbital, it is more likely to be with a strong-field ligand and the electrons will pair up in the lower-energy orbital. If the value of Δ is less than that of *P*, the electrons will spread out through the available orbitals.

This is only an issue for complexes in which there are four to seven *d* electrons, known as d^4-d^7 complexes. For others, only one ground-state configuration is possible, and the compounds are colorless.

In tetrahedral complexes, the energy splitting for the *d* orbitals is different. In tetrahedral complexes, none of the ligands points directly at the *d* orbitals. As a result, *P* is almost always higher than Δ, and so tetrahedral complexes tend to be high-spin.

In square planar complexes, the energy is similar to that of octahedral complexes, with the *z*-axis ligands missing. This leads to a large energy gap between the *d* orbitals that are pointing toward ligands and those that are not, favoring low-spin complexes with the higher-energy orbitals vacant.

Putting It Together

The amount of iron in ore can be determined by dissolving a sample in a nonoxidizing acid, reducing all of the Fe^{3+} to Fe^{2+} and titrating with potassium dichromate. The reaction is the oxidation of the Fe^{2+} to Fe^{3+} and the reduction of the dichromate ion to Cr^{2+} in an acidic environment. Determine the mass percent of iron in a 1.513 g sample of ore if 33.28 mL of 0.090 M potassium dichromate is needed to reach the end point in a titration.

Self–Test

This section is intended to test your knowledge of the material covered in this chapter. Think through these problems, and make certain you understand what they are asking. Make sure your answers make sense. Successful completion of these problems indicates that you have mastered the material in this chapter. You will receive the greatest benefit from this section if you use it as a mock exam, as this will allow you to determine which topics you need to study in more detail.

True–False
1. Electron configurations are dependent on both orbital energies and electron–electron repulsions.

2. All of the transition elements are metals, and all are solids.

3. Densities of transition elements have a minimum density in the center of the row.

4. Transition metals are rarely found in more than one oxidation state.

5. Complexes with a coordinate number of 4 are either tetrahedral or square planar.

6. Ligands connect to metal atoms by one or more ligand donor atoms.

7. The oxidation state of the transition metal is not given as part of the name of a coordination compound.

8. Optical activity in an enantiomer is the result of chirality.

9. Complexes containing the SCN (thiocyanate) ion can have linkage isomers.

10. Crystal field theory assumes that all interactions between metal and ligand are covalent.

Multiple Choice
11. In a coordinate covalent bond,
    a. no electrons are shared
    b. a lone pair of electrons becomes a bonding pair
    c. metals bind to other metals
    d. ligands ionize

12 The multiple oxidation states of transition elements originate in
    a. the varying sizes of the elements
    b. increasing metallic character with ionization
    c. the closeness in energy of the *s* orbitals and the *n*–1 *d* orbitals
    d. transition elements do not have multiple oxidation states

13. All of the following are properties of transition metals:
    a. malleable, ductile, good conductors of heat and electricity
    b. malleable, ductile, poor conductors of heat and electricity
    c. brittle, ductile, good conductors of heat and electricity
    d. malleable, dull, good conductors of heat and electricity

14. From top to bottom in the transition metals, the following property varies almost not at all:
    a. ionization energy   b. electron configuration
    c. acidity   d. atomic size

15. Chromium is used to electroplate other metals. It is useful for this because
    a. it does not oxidize
    b. it is dull and grey, discouraging thieves
    c. It forms a hard, invisible layer of oxidation that protects against further corrosion
    d. none of the above

16. Ligands in coordination compounds are always
    a. anions   b. cations
    c. Lewis bases   d. Lewis acids

17. Numerous spatial arrangements are possible for coordination compounds. The shape of these compounds is determined by
    a. the hybridization of the d orbitals   b. the coordination number
    c. the polarity of the ligand   d. the group number of the metal

18. Which of the following will have geometric isomers?
    a. MA_2B_4 b. MA_6
    c. MA_4B_2 d. a and c

19. The observed color of transition metal complexes is
    a. the maximum absorbed wavelength
    b. complementary to the absorbed wavelength
    c. the absorbed wavelength rotated 90°
    d. none of the above

20. In contrast to valence bond theory, crystal field theory is able to account for
    a. geometry of complexes   b. color and spin
    c. paramagnetism   d. all of the above

Fill-in-the-Blank

21. Transition elements exhibit a variety of _____ states. Ions in which this is

    high are good _____ agents, while ions in which this is low are good

    _____ agents.

22. _____ compounds are those in which a central metal atom is   connected to

    surrounding _____ by _____ _____ bonds.

23. Many compounds exist as _____ , which have the same formula, but different

    arrangements of the atoms. These can be _____ in which the connectivity is different, or _____ in which the connections are the same, but have a different spatial arrangement.

24. The _____ of the light absorbed by a transition metal complex can be used to calculate the _____ required for its most common transition and can also be used to predict its observed _____.

25. _____ _____ _____ describes metal complexes in terms of overlapping _____ orbitals. _____ _____ _____ describes metal complexes as purely _____ in nature.

Matching

26. Absorption spectrum

27. Achiral

28. Bidentate ligand

29. Chelate

30. Chiral

31. *Cis* isomer

32. Coordination number

33. Crystal field splitting

34. Enantiomer

35. High-spin complex

36. Ligand

37. Ligand donor atom

38. Low-spin complex

a. a ligand that binds using two donor atoms

b. having handedness

c. the number of ligand donor atoms surrounding a metal ion

d. isomer in which identical ligands or groups are opposite one another

e. complex in which there is a maximum number of unpaired electrons

f. a cyclic complex formed by a metal and a polydentate ligand

g. isomer in which identical ligands or groups are adjacent to one another

h. a plot of the amount of light absorbed versus wavelength

i. atom attached directly to the metal

j. optically inactive mixture of enantiomers in a 50:50 ratio

k. the energy difference between two sets of *d* orbitals

l. ligand that has strong crystal field splitting

m. lacking handedness

39. Racemic mixture

    n. stereoisomers that are mirror images of one another

40. Strong-field ligand

    o. ligand that has weak crystal field splitting

41. *Trans* isomer

    p. complex in which there are a maximum number of paired electrons

42. Weak-field ligand

    q. molecule or ion that bonds to a central atom in a complex

Problems

43. Write the valence electron configurations for the following atoms and ions:
Mn^{5+}, Mo, Cr^{3+}, Fe^{2+}.

44. Identify the following neutral elements from the following electron configurations:
$[Ar]\ 4s^2 3d^5$, $[Kr]\ 5s^2 4d^7$, $[Xe]6s^2 5d^1$.

45. Arrange the following from the strongest oxidizing agent to the strongest reducing agent:
          Co   Cu   Mn   Zn

46. Arrange the following from the strongest oxidizing agent to the strongest reducing agent:
          Fe_2O_3 CoO MnO_4^- CrO_3^-

47. Identify the coordination number of each of the following compounds:
       $[HgCl_3]^-$ $[Mo(CN)_8]^{4-}$ $Fe(CO)_5$ $[CuCl_2]^-$

48. For the following compounds identify the ligands and their donor atoms. Determine the coordination number and the oxidation state of the metal and the charge on any complex ion. Determine if the compound can display geometric, linkage, or optical isomerism.
a. $[Co(H_2O)_4(NH_3)_2]Cl_3$ b. $[Co(NH_3)_4(H_2O)Br]NO_3$ c. $[Cr(NH_3)_4(SCN)_2]$
d. $(NH_4)_3[Fe(ox)_3]$ e. $[Co(en)_3]I_3$

49. Name the compounds in question 48.

50. Write formulas for the following compounds:
a. tetraamminediaquachromium(III) chloride
b. ethylenediaminedithiocyanatocopper(II)
c. pentacarbonylchloromanganate(I)
d. potassium tetracyanochloroferrate(II) dehydrate
e. tetraaquacopper(II) sulfate

51. Use valence bond theory to explain the bonding in $[Zn(NH_3)_4]^{2+}$ (tetrahedral complex) and $[Ni(CN)_4]^{2-}$ (square planar). What hybrid orbitals are used by the metal? Are the complexes low-spin or high-spin complexes?

52. Why is the crystal field splitting in a tetrahedral complex approximately half of the crystal field splitting in an octahedral complex?

Challenge Problem

53. Three different isomers with the formula $CrCl_3 \cdot 6\,H_2O$ exist. Two are green in color; one is violet. These isomers are labeled *A, B,* and *C*. When a 0.500 g sample of isomer *A* reacts with a dehydrating agent, the resulting mass is 0.432 g. When 50 mL of a 0.125 M solution of isomer *A* is titrated with excess $AgNO_3$, 0.269 g of AgCl are produced.

When a 0.500 g sample of isomer *B* reacts with a dehydrating agent, the resulting mass is 0.466 g. Reaction of a 50 mL solution of 0.125 M isomer *B* with excess $AgNO_3$ produced 0.539 g of AgCl. When a 0.500 g sample of isomer *C* reacts with a dehydrating agent, the resulting mass is 0.500 g. When 50 mL of a 0.125 M solution of isomer *C* is titrated with $AgNO_3$, 0.809 g of AgCl are produced. Determine the formula of each isomer.

CHAPTER TWENTY-ONE

Metals and Solid-State Materials

Learning Objectives

As a result of reading and studying this chapter, you should be able to

Section 21.1   **Sources of the Metallic Elements**

1. Relate the formulas of metal-containing minerals to the location of the metals in the periodic table.

Section 21.2   **Metallurgy**

2. Predict the relative magnitude of standard reduction potentials from an element's position in the periodic table and determine the best method for extracting a metal from an ore.
3. Describe the concentration, reduction, and refining aspects of metallurgical processes, and write balanced equations for the reactions that occur when a metal is produced from its ores.

Section 21.3   **Iron and Steel**

4. Describe and balance reactions that occur in a blast furnace and the basic oxygen process in the production of steel.
5. Apply principles of thermodynamics to various reactions in the production of iron.

Section 21.4   **Bonding in Metals**

6. Describe the electron-sea model of metals, and explain how it accounts for their properties.
7. Describe the molecular orbital theory for metals (band theory), draw molecular orbital energy-level diagrams for metals, and use band theory to account for their properties.
8. Describe the advantages of each of the two bonding theories in predicting properties of metals.

Sections 21.5–21.6   **Semiconductors and Applications**

9. Describe the bonding in semiconductors and account for their properties in terms of band theory.
10. Classify doped semiconductors as *n*-type or *p*-type.
11. Describe some semiconductor applications.
12. Predict relative band-gap energies based on periodic trends in atomic size and electronegativity.
13. Relate the wavelength of light emitted or absorbed by a semiconductor to the band gap in its band theory energy-level diagram.

Section 21.7   **Superconductors**

14. Write and balance a reaction for synthesizing a superconductor by hydrolysis of metal ethoxides.

15. Describe the appearance of a plot of electrical resistance versus temperature for a superconductor.
16. Identify coordination numbers of bonds in superconductors.

Sections 21.8 – 21.9   Ceramics and Composites
17. Describe the different types of bonding in ceramics and account for their properties in terms of the bonding model.
18. Describe the synthesis of ceramics, and write balanced equations for the reactions that occur in the sol-gel synthesis method.
19. Classify composites as ceramic–ceramic, ceramic–metal, or ceramic–polymer, and account for some of their properties.

Chapter Summary

In this chapter, we'll look at both metals and solid-state materials. Most metals are found on Earth as minerals: the science and technology of extracting them from their natural sources, known as *ores*, is called *metallurgy*. There are two models for describing the bonding and properties of metals, the electron-sea model and band theory, which you'll learn about in this chapter. Band theory also accounts for the electrical conductivity of metals, as well as the behavior of semiconductors, both with and without doping materials. We also introduce superconductors, ceramics, and composite materials.

The Chapter in Detail

Section 21.1   Sources of the Metallic Elements

The most common minerals in the Earth's crust are silicates and aluminosilicates, but because they are difficult to concentrate and reduce, they are not important sources of metals for industry. Oxides and sulfides such as hematite (Fe_2O_3), rutile (TiO_2), and cinnabar (HgS) are much more valuable. Substances such as these, from which metals can be produced economically and profitably, are called **ores**.

The chemical composition of the most common ores correlates with their location in the periodic table: early transition metals on the left side of the *d* block generally occur as oxides (group 3B as phosphates), while the more electronegative metals on the right side tend to occur as sulfides with a more covalent character.

Only Au and the platinum-group metals (Ru, Os, Rh, Ir, Pd, and Pt) are unreactive enough to occur as free metals.

Because *s*-block oxides are very basic and thus unable to exist around CO_2 and SiO_2, *s*-block metals occur as carbonates and silicates. Na and K exist as chlorides.

Section 21.2   Metallurgy

Ores are complex mixtures of metal-containing minerals and economically worthless material called **gangue** that consists of sand, clay, and other impurities. **Metallurgy** is the science of extracting metals from their ores and making alloys with them.

The process of extracting metal from ores is a three-step process: (1) concentration of the ore and chemical treatment prior to reduction if needed, (2) reduction of the mineral to the free metal, and (3) refining or purification of the metal.

Concentration and chemical treatment of ores

Various separation methods are used to extract minerals from the gangue that exploit the different properties of each. For example, density differences are used to separate gold from the less-dense silt it is found in. Magnetic properties can be used to separate magnetite iron ore from the nonmagnetic gangue.

Metal sulfides ores are concentrated by **flotation**, a process in which powdered ore is mixed with water, oil, and detergent. Ionic silicates sink into the water, while the less-polar sulfides float in the oil and detergent at the top of the tank.

Chemical separation methods include the *Bayer process*, in which amphoteric aluminum oxides are dissolved out of impurities using hot NaOH, and **roasting**, in which sulfide minerals are converted to oxides by heating the sulfides in air. In modern mineral roasting facilities, the SO_2 produced in the process is converted to sulfuric acid.

Reduction

Concentrated ore is reduced to the free metal either by chemical reduction or by electrolysis. The method used depends on the activity of the metal.

The most active metals, Au, and Pt, are found in nature in uncombined form.

Cu, Ag, and Hg are more active and are typically found in sulfide ores that are easily reduced by roasting: the sulfur is oxidized to SO_2, and the metal is reduced.

More active metals, such as Cr, Zn, and W, can be reduced by reacting their oxides with C, H, or an even more active metal such as Na or Mg. Carbon is the cheapest choice, but is unsuitable for metals that form stable carbides, such as tungsten, which is reduced using hydrogen.

The most active metals must be electrolytically reduced. Li, Na, and Mg are obtained by electrolysis of molten chlorides.

Refining

Purification methods for reduced ores include distillation, chemical purification, and electrorefining.

Zinc can be refined by distillation. Nickel is purified by the **Mond process**, a chemical method in which CO is passed over impure nickel to form $Ni(CO)_4$, which is then decomposed at higher temperatures on pellets of pure nickel.

Zirconium is similarly purified by conversion into ZrI_4, which is decomposed on a tungsten or zirconium filament at high temperature.

Copper is purified by electrorefining, an electrolytic process in which Cu is oxidized to Cu^{2+} at an impure Cu anode, and Cu^{2+} from aqueous $CuSO_4$ is reduced to Cu at a pure Cu cathode.

Workbook Problem 21.1

Copper occurs in a mineral called covellite that has the formula CuS. Write balanced equations for the roasting of covellite, its reaction with sulfuric acid, and its electrolytic purification.

Strategy: Write and balance equations for purification of the metal from the mineral ore.

Step 1:   Write the equation for roasting of covellite.

Step 2:   Write the equation for the reaction of the resultant oxide with sulfuric acid.

Step 3:   Write the equation for the electrolytic purification of the solution.

Section 21.3   Iron and Steel

The metallurgy of iron is of special importance, because iron is the major constituent of steel, the most widely used metal.

Iron is produced by carbon monoxide reduction of Fe ore in a blast furnace. The overall reaction is
$$Fe_2O_3 \, (s) + 3 \, CO \, (g) \rightarrow 2 \, Fe(l) + 3 \, CO_2 \, (g)$$

Limestone is also added to the mix: in the high temperatures of the furnace, it decomposes to CaO, which reacts with SiO_2 and other acidic oxides. These reactions produce **slag,** which consists mainly of calcium silicate and floats on top of the molten iron.

The impure iron obtained is called *cast iron* or *pig iron*, which is brittle and contains about 4% elemental carbon and smaller amounts of other impurities formed in the reducing atmosphere of the furnace.

To further purify the cast iron and convert it to steel, the **basic oxygen process** is used. The molten iron is exposed to a jet of pure oxygen in a furnace lined with basic oxides. The acidic oxides that form react with the basic oxides to form a slag that can be poured off. This produces steel with about 1% carbon and very small amounts of P and S.

The composition of liquid steel is monitored by chemical analysis, and the amounts of oxygen and impure steel are continuously varied to achieve the desired concentrations of impurities.

Hardness, strength, and malleability of steel depend on chemical composition, rate of cooling, and subsequent heat treatment. For example, stainless steel is a corrosion-resistant alloy that contains up to 30% chromium as well as smaller amounts of nickel.

Section 21.4   Bonding in Metals

Metals have some universal characteristics—they are malleable, ductile, lustrous, and good conductors of heat and electricity.

Two theoretical models are used to explain these properties: the *electron-sea model* and the *molecular orbital theory*.

Electron Sea Model of Metals

Metals do not have enough valence electrons to bond to adjacent molecules; their crystal structures contain delocalized electrons that belong to the structure as a whole. This acts as an electrostatic glue that holds the metal cations together.

This is a simple qualitative explanation for the properties of metals: electrons are mobile and therefore free to move from a negative electrode to a positive electrode when subjected to an electrical potential. Mobile electrons are also able to carry kinetic energy (heat) from one part of the crystal to another. Metals are malleable and ductile because no local bonds are broken when the solid is deformed. The more valence electrons there are, the harder the metal.

Molecular Orbital Theory for Metals

The number of molecular orbitals (MOs) formed is the same as the number of atomic orbitals combined. As the number of atoms increases, the difference in energy between successive MOs decreases, allowing the orbitals to merge into an almost continuous band of energy levels.

For this reason, molecular orbital theory in metals is referred to as **band theory**. The bottom half of the band consists of filled bonding MOs, while the top half consists of empty antibonding MOs.

In the absence of a current, electrons are equally likely to be traveling in one direction as another. Therefore, energy levels in a band occur in **degenerate** pairs; one set of energy levels applies to electrons moving one way and the other applies to electrons moving in the opposite direction within the metal. If an electrical potential is applied, there is a reason for electrons to move into the higher energy antibonding orbitals.

If there are no vacant orbitals in the band, there are no orbitals available for electrons to pass through, making the material an electrical insulator.

Main-group metals have *s* and *p* subshells sufficiently close in energy that the *s* and *p* bands overlap, resulting in a partially filled composite band and electrical conductivity.

Transition metals have overlapping *d* and *s* bands, which allows six MOs per metal atom, three bonding and three antibonding. As a result, the metals with six valence electrons (bonding orbitals completely filled) are the hardest and have the highest melting points.

> **Workbook Problem 21.2**
>
> Rank Mo, Cd, and Y in order of their expected melting points based on band theory.
>
> ***Strategy:***   Determine the number of bonding and antibonding electrons; rank in order by maximum bonding electrons and minimum antibonding electrons.
>
> *Step 1:*   Determine the numbers of electrons of each type.
>
> *Step 2:*   Rank in order of expected bonding strength.

Section 21.5   Semiconductors

A semiconductor has electrical conductivity intermediate between that of a metal and that of an insulator. In metals, there is no energy gap between the highest occupied orbitals and the lowest unoccupied orbital: electrons can move freely between these energy levels. In insulators, there is a large energy gap between the highest occupied orbitals and the lowest unoccupied orbital: electrons cannot move between these energy levels.

The bonding MOs are also referred to as the **valence band**. The higher-energy antibonding MOs are the **conduction band**. The difference in energy between the two is the **band gap**. Metals have no band gap, insulators have an insurmountable band gap, and semiconductors fall between the two.

In a semiconductor, a few electrons will have the energy to jump across the band gap. Conductivity increases with increasing temperature, as the number of high-energy electrons increases. This is the opposite of metals: as their temperature increases, conductivity drops, because the lattice of bonds is disrupted. This provides a convenient way to distinguish between the two types of material.

Doping is a process in which the conductivity of a semiconductor is increased by adding small amounts of impurities. Doping can increase the conductivity of a semiconductor by a factor of about 10^7.

If an impurity is introduced that has more electrons than the semiconductor, the extra electrons will reside in the conduction band. Because *negative* electrons are being added, these are called *n*-type semiconductors.

If an impurity is introduced that has fewer electrons than the semiconductor, there will be vacancies in the valence band that allow for increased conductivity. Because these can be thought of as adding *positive* holes in the valence band, these are *p*-type semiconductors.

Workbook Problem 21.3

Identify the following as *n*- or *p*-type semiconductors:
   a. Silicon doped with gallium
   b. Germanium doped with antimony
   c. Silicon doped with arsenic
   d. Germanium doped with phosphorus

Strategy: Determine the number of valence electrons in the semiconductor and compare to the number of valence electrons in the dopant. Because all of the elements are main group, their group numbers can be used.

Step 1:   Determine the numbers of valence electrons in each element.

Step 2:   Determine whether the dopant has more or fewer valence electrons than the semiconductor.

Step 3:   Identify the doped semiconductors as *n*- or *p*-type.

Section 21.6   Semiconductor Applications

Doped semiconductors are essential components in modern solid-state electronic devices.

Diodes

Diodes convert alternating current to direct current by permitting current to flow in one direction only. They consist of a *p*-type semiconductor in contact with an *n*-type semiconductor to give what is known as a *p–n* junction.

If the *n*-type side is connected to the negative terminal of a battery, current will flow through the junction: the electrons can find their way into the conductance band and across the gap into the valence band. If the *p*-type side is connected to the negative terminal of a battery, the positive holes move toward the battery, and the electrons in the conductance band across the gap are repelled. In this case, no current will flow.

Light-Emitting Diodes (LEDs)

If there is a difference in energy levels between the conduction band and the valence band, electrons will be able to fall from one to the other, releasing energy as light. The energy of the light emitted is roughly equivalent to the size of the band gap ($E_g = hv = hc/\lambda$).

LEDs are made of three to five semiconductors. These are 1:1 compounds of group 3A and 5A elements. They have the same basic structure as pure Si or Ge, because they have an average of four valence electrons.

Because these mixtures form a continuous series of solid solutions, the size of the band gap, and thus the color of the LED, can be "tuned" by varying the composition of the mixture. The red light so commonly seen in LEDs is the result of a solid solution of $GaP_{0.4}As_{0.6}$ with a band gap of 181 kJ/mol.

LEDs are smaller, brighter, longer lived, more energy efficient, and faster than incandescent bulbs.

Diode Lasers
The word *laser* is an acronym for *l*ight *a*mplification by *s*timulated *e*mission of *r*adiation. Diode lasers produce light in the same way as LEDs, but the light is more intense, highly directional, and all of the same frequency and phase.

The essential features of a diode laser are a very high forward bias and a laser cavity that allows emitted light to bounce back and forth, stimulating a cascade of electrons and holes that amplifies the amount of light produced.

Diode lasers are used in laser pointers, barcode readers, CD players, and fiber optic data systems.

Photovoltaic (Solar) Cells
Photovoltaic cells work in the opposite direction of an LED. Whereas an LED converts the energy difference of the band gap to light, a photovoltaic cell converts light to electricity.

When light shines on the *p–n* junctions, the photons from the light excite electrons from the valence band of the *p*-type semiconductor into the conduction band of the *n*-type semiconductor.

If the photovoltaic cell is part of an electrical circuit, light can flow in, charging the system, and the resultant energy can be used to power a device.

Current cells are only 20% efficient, so research is focusing on increasing efficiency and reducing cost.

Transistors
Transistors consist of *n–p–n* or *p–n–p* junctions that control or amplify electrical signals in modern integrated circuits. Vast numbers of these devices can be packed into tiny spaces, increasing the speed and decreasing the size of modern electronics.

Workbook Problem 21.4

If an LED has a band gap of 178 kJ/mol, what is the wavelength of the resulting light?

Strategy: Convert the band-gap energy to photon energy, and then to wavelength.

Step 1:   Determine the energy of an individual photon.

Step 2:   Convert the photon energy to wavelength.

Section 21.7   Superconductors

A **superconductor** is a material that loses all electrical resistance below a characteristic temperature, called the **superconducting transition temperature** (T_c). Below that temperature, the material becomes a perfect conductor: once an electric current is started, it will flow indefinitely without loss of energy.

Superconductors were discovered in 1911, but only superconductors with relatively low temperatures were known. In 1986, $Ba_xLa_{2-x}CuO_4$ was demonstrated to have a T_c of 35 K, and soon after that, other copper-containing oxides were found to have even higher superconducting transition temperatures. The record as of this writing is 138 K.

This was unexpected, because most metal oxides, nonmetallic inorganic solids called *ceramics,* are insulators. In this field, experimental data exceeds the ability of scientists to explain it: there is no generally accepted theory of superconductivity in ceramic superconductors.

If a superconductor is cooled to below T_c and a magnet is lowered toward it, the magnet and superconductor repel each other, causing the magnet to levitate. Magnets induce a supercurrent in the superconductor that generated a magnetic field opposite to that of the magnet. This is called the *Meissner effect*, and it is already being used to operate a high-speed train in Shanghai, China.

Superconducting magnets are also used in MRI instruments and particle accelerators, but these uses currently require cooling with liquid helium, which is expensive and requires cryogenic equipment.

Once the T_c of a material goes above 77 K, it can be cooled using liquid nitrogen, an abundant refrigerant that is cheaper than milk.

Currently available high-temperature superconductors are brittle powders with high melting points, not easily made into wires and coils needed for electrical equipment. However, superconducting thin films are being used as microwave filters, and superconducting wires as long as 1 km are now available.

High-temperature superconductors are also being developed based on fullerene (C_{60}) molecules. K_3C_{60} is a metallic conductor at room temperature and a superconductor at 18 K.

Fluorine-doped lanthanum oxide iron arsenide also shows promise, superconducting at 26 K, with substitution of the lanthanum by other elements, raising the temperature to 77 K.

Section 21.8   Ceramics

Ceramics are inorganic, nonmetallic, nonmolecular solids, including both crystalline and amorphous materials, such as glasses. Traditional silicate ceramics are made by heating aluminosilicate clays to high temperatures, while **advanced ceramics**—materials that have high-tech engineering, electronic, and biomedical applications—include oxide ceramics such as alumina (Al_2O_3) and nonoxide ceramics such as silicon carbide and silicon nitride.

Oxide ceramics are named by adding an –*a* suffix in place of the –*um* of the metal.

Compared to metals, ceramics have higher melting points; are stiffer, harder, and more resistant to wear and corrosion; maintain their strength at high temperatures; and are less dense than steel. This makes them attractive lightweight, high-temperature materials for replacing metal components in aircraft, space vehicles, and automobiles.

Nonoxide ceramics are covalent network solids with highly directional covalent bonds. This prevents planes of atoms from sliding over one another when the solid is subjected to the stress of a load or an impact. Because the solid cannot change shape, as a metal would, the bonds instead give way. This brittleness is the primary drawback of these materials. The bonding is largely ionic in oxide ceramics, but the material is still brittle.

Ceramic processing, the series of steps that leads from raw material to the finished ceramic object, determines the strength and the resistance to fracture of the product.

Sintering is a process done below the melting point of a material, in which the particles of the powder are welded together without completely melting. During the process, crystal grains grow larger and the density of the material increases as the void spaces between particles disappear. High-purity, fine powders that are tightly compacted prior to sintering are needed to prevent impurities and leftover voids that lead to cracking and material failure.

In the **sol–gel method** for preparing high-purity fine powders, a metal oxide powder is synthesized from a metal alkoxide (a compound derived from a metal and an alcohol). This forms a colloidal dispersion called a *sol*, consisting of extremely fine particles that can then be dehydrated and linked with oxygen bridges. As this cross-linking occurs, the sol becomes a more rigid material called a *gel*.

Oxide ceramics have many uses. Alumina is used in spark plugs, dental crowns, and the heads of artificial hips, as well as being the substrate for circuit boards. Silica is used to make the heat-resistant tiles that protect the space shuttle on re-entry into the atmosphere. Military armor is now also ceramic rather than steel, reducing weight and allowing the armor to shatter and deflect bullets.

Section 21.9   Composites

To counteract the brittleness of ceramic, a ceramic powder can be mixed, prior to sintering, with a second ceramic material to combine the properties of both components. This second compound is often in the form of *whiskers* (tiny fiber-shaped particles) and fibers.

Fibers and whiskers increase the strength and fracture-resistance of composite materials, because most of the chemical bonds are aligned along the fiber axis, giving the fibers great strength. The fibers can deflect cracks, preventing them from moving cleanly in one direction or bridging cracks to hold them together.

Composites can also be made of different materials. For example, **ceramic–metal composites,** or **cermets**, and **ceramic–polymer composites**. These materials have a high strength-to-weight ratio, making them ideal for aerospace and other applications.

Ceramic fibers used in composites are usually made by high-temperature methods. For example, carbon fiber is made by heating polyacrylonitrile to 1400–2500 °C, converting it to graphite. Silicon carbide chains are made using a similar method.

Putting It Together

The presence of manganese in steel can be determined by the following titration procedure:

1. A steel sample is dissolved in an acidic solution that oxidizes the manganese to the permanganate ion.
2. The permanganate ion is reduced to Mn^{2+} by reaction with an excess of iron(II) sulfate.
3. The unreacted Fe^{2+} is oxidized to Fe^{3+} by reaction with potassium dichromate.

Determine the percent by mass of manganese in a 0.450 g sample of steel if 22.4 mL of 0.0100 M potassium dichromate is needed to react with the Fe^{2+} remaining after 50.0 mL of 0.0800 M iron (II) sulfate reacts with the steel sample.

Self–Test

This section is intended to test your knowledge of the material covered in this chapter. Think through these problems, and make certain you understand what they are asking. Make sure your answers make sense. Successful completion of these problems indicates that you have mastered the material in this chapter. You will receive the greatest benefit from this section if you use it as a mock exam, as this will allow you to determine which topics you need to study in more detail.

True–False
1. Most silicate minerals are valuable sources of metals.

2. Less electronegative metals form ionic oxides, while more electronegative metals form sulfides with more covalent character.

3. Roasting is used to convert sulfides to more easily reduced oxides.

4. Iron is produced by reduction with NO_2 in a blast furnace.

5. The electron-sea model of metal bonding accounts for the superconductivity of some materials.

6. Valence band theory accounts for the different conductivities of metals and semiconductors.

7. Doping of semiconductors increases their conductivity.

8. Heating a semiconductor reduces its conductivity.

9. Diodes can convert alternating current to direct current.

10. Ceramics are ideal for aerospace applications because they are strong, malleable, and light.

Multiple Choice

11. A commercially viable, naturally occurring source of refinable metal is referred to as a(n)
    a. mineral   b. gangue   c. ore   d. flotation

12. Refining metals requires the following general steps in this order:
    a. concentration, oxidation, purification
    b. reduction, concentration and purification
    c. purification, oxidation, and concentration
    d. concentration, reduction, purification

13. Which of the following metals is found free in nature?
    a. aluminum   b. gold   c. copper   d. tungsten

14. Carbon is an inexpensive reducing agent, but cannot be used with metals that form
    a. carbides   b. alloys   c. ceramics   d. superconductors

15. Models of metallic bonding must account for metals'
    a. malleability and ductility   b. dull appearance
    c. electrical conductivity   d. a and c

16. Semiconducting metals conduct electricity, but not efficiently. This is as a result of
    a. impurities   b. a large band gap
    c. paramagnetism   d. no available p orbitals

17. Doping silicon with which of the following would produce a p-type semiconductor?
    a. boron   b. phosphorus   c. nitrogen   d. arsenic

18. Light-emitting diodes
    a. can be tuned to different wavelengths   b. are energy efficient
    c. a and b   d. none of the above

19. Superconductors
    a. are strongly magnetic   b. operate best at high temperatures
    c. have no electrical resistance below T_c d. are easily made into wires

20. Most ceramics are
    a. malleable   b. insulators   c. shiny   d. paramagnetic

Fill-in-the-Blank

21. The electron-sea model of metal bonding accounts for the _____ and

_____ of metals. These two abilities of metals to be shaped come from the

fact that bonds and electrons are not _____. As a result, metals are not

_____: they will deform rather than shatter.

22. In the absence of an electrical potential, there is no net electric _____ in a metal. If a potential is applied, electrons are able to access the _____ band, the _____ molecular orbitals that are _____ in the absence of potential.

23. Superconductors are fascinating, poorly understood materials with some interesting properties. _____ the T_c, the material loses all electrical _____.

24. The challenge for researchers is to find materials that can be cooled using liquid _____, a common and inexpensive coolant that requires no specialized equipment.

25. Advanced ceramics have properties that make them superior to _____ for many applications. They are lighter due to their lower _____, electrical _____, and resistant to _____. Their durability is limited in one respect: they are not _____, and will shatter under extreme stress.

26. Composite materials can consist of mixtures of any of the following: _____, _____, carbon or boron _____, and _____.

Matching

27. Alloy

28. Band gap

29. Band theory

30. Ceramic

31. Conduction band

32. Doping

33. Electron-sea model

34. Flotation

35. Gangue

a. molecular orbital theory for metals

b. mineral deposit from which a metal can be economically extracted

c. antibonding molecular orbitals in a semiconductor

d. material that has conductivity intermediate between metals and insulators

e. economically worthless material that an ore is found in.

f. crystalline inorganic constituent of rocks

g. energy difference between bonding MOs in the valence band and the nonbonding MOs in the conductance band

h. heating a mineral in air

i. process in which the particles of a powder are welded together without completely melting

36. Metallurgy

    j. addition of a small amount of an impurity to increase the conductivity of a semiconductor

37. Mineral

    k. science and technology of extracting metals from ore

38. Ore

    l. an inorganic, nonmetallic, nonmolecular solid

39. *p*-type semiconductor

    m. solid solution of two or more metals

40. Semiconductor

    n. method of preparing ceramics involving synthesis of metal oxide powder from metal alkoxide

41. Roasting

    o. by-product of iron production consisting mainly of calcium silicate

42. Sintering

    p. model that visualizes metals as cations surrounded by delocalized electrons

43. Slag

    q. temperature below which all electrical resistance is lost

44. Sol–gel method

    r. semiconductor doped with a material that has fewer electrons than necessary for bonding

45. Superconducting transition temperature

    s. bonding molecular orbitals in a semiconductor

46. Superconductor

    t. process that exploits differences in the ability of water and oil to wet the surfaces of mineral and gangue

47. Valence band

    u. material that loses all electrical resistance below a certain temperature

Problems

48. The manganese ore rhodochrosite has the formula $MnCO_3$. It is refined by heating it in the presence of oxygen to produce manganese(IV) oxide and drive off carbon dioxide, then reacting it with aluminum metal to reduce it to elemental manganese. Write and balance the equations for these reactions. If all reactions occur with 100% efficiency, how many kg of elemental manganese can be obtained from 1500 kg of pure rhodochrosite? How many kg of aluminum metal will be needed for this reaction?

49. Cobalt, chromium, and copper are all period 4 transition metals. Using molecular orbital theory, rank them in order of melting point.

50. Silicon and germanium are both semiconductors. Phosphorus, arsenic, and boron are all doping agents. Determine which type of semiconductor is formed by all of the combinations.

Challenge Problem

51. A concentration technique used to extract gold from low-grade ores is *cyanidation*. This technique involves reaction of the cyanide ion with the ore to produce $Au(CN)_2^-$ ions, which are then reduced with zinc to produce pure gold. Is it possible to use cyanidation to remove silver from the ore argentite, Ag_2S? $K_f [(Ag(CN)_2^-)] = 1 \times 10^{21}$. $K_{sp} (Ag_2S) = 6 \times 10^{-51}$. Could cyanidation be used to remove silver from horn silver, AgCl? $K_{sp} (AgCl) = 1.8 \times 10^{-10}$.

CHAPTER TWENTY-TWO

The Main-Group Elements

Learning Objectives

As a result of reading and studying this chapter, you should be able to

Section 22.1

A Review of General Properties and Periodic Trends

1. Use the periodic table to predict the properties of the main-group elements and their compounds.

Section 22.2

Distinctive Properties of the Second-Row Elements

2. Account for differences in the structure and properties of second-row elements.

Section 22.3

Group 1A: Hydrogen

3. Describe the preparation of elemental hydrogen.
4. Write and balance a chemical equation for the reaction of ionic hydrides with water.
5. Classify binary hydrides and describe their bonding behavior and properties.

Sections 22.4–22.5

Groups 1A and 2A: Alkali and Alkaline-Earth Metals

6. Use periodic trends to predict properties of Group 1A and 2A elements.
7. Write and balance reactions of alkali and alkaline-earth metals with halogens, oxygen, and water.
8. Write electrode reactions and perform calculations in the electrolysis of molten salts to produce alkali and alkaline-earth metals.

Section 22.6

Group 3A: Elements

9. Explain why the properties of boron differ from other 3A elements.
10. Describe the structure and bonding in diborane.

Sections 22.7–22.8

Group 4A: Carbon and Silicon

11. Draw electron-dot structures and predict hybrid orbitals and geometry for molecules and ions in main-group compounds containing elements from Group 4A.
12. Describe the structure and properties of the allotropes of carbon.
13. Use the shorthand notation for a silicate anion to represent its structure and interpret the notation to find the formula and charge of a silicate anion.

Chapter Summary

The main-group elements demonstrate strong periodic trends in ionization energy, electronegativity, atomic radius, and metallic character. The second-row elements have some unique properties as a result of their small size and high electronegativities. The families will be examined as groups, with particular attention given to the unique properties of hydrogen, boron, carbon, silicon, nitrogen, phosphorus, and sulfur.

The Chapter in Detail

Section 22.1   A Review of General Properties and Periodic Trends

Overall, the main group elements are divided into metals on the left of the periodic table, nonmetals on the right of the periodic table, and semimetals with intermediate properties along a stair-step line dividing them.

From left to right across the periodic table, the effective nuclear charge (Z_{eff}) increases, because the additional electrons are being added into a shell that does not completely shield the additional nuclear charge.

As a result of the increased Z_{eff}, electrons are more strongly attracted to the nucleus. Ionization energy and electronegativity increase to the right, atomic radius and metallic character decrease.

From top to bottom down the periodic table, atomic radius increases as additional electron shells are occupied. Because the valence electrons are farther from the nucleus, ionization energy and electronegativity decrease and metallic character increases.

The more metallic an element, the more likely it is to form ionic compounds with nonmetals. Binary hydrides can be metallic, in which case they are ionic solids, or nonmetallic, in which case they are gases, liquids, or low-melting-point solids. Similar trends are seen with oxides: metallic oxides are high-melting-point solids, while nonmetal oxides are gases or volatile liquids at room temperature.

Workbook Problem 22.1

Predict which of these pairs of elements has more metallic character:

           C or Si   Be or Li   Se or Br   As or I

Strategy: Examine the position of the elements on the periodic table.

Step 1:   Predict the metallic character of the elements based on their position on the periodic table.

Section 22.2   Distinctive Properties of the Second-Row Elements

The properties of second-row elements differ markedly from those below them in the same periodic group, because they have especially small sizes and high electronegativities. This causes the elements in this row to be more nonmetallic: they form mainly covalent molecular compounds, with a maximum of four covalent bonds, because they do not have access to d orbitals.

Hydrogen bonding is limited to N, O, and F.

The small size of second-row elements allows for the formation of multiple bonds when $2p$ orbitals overlap to form π bonds. $3p$ orbitals are more diffuse, leading to longer bond distances and poor π overlap. For example, O_2 is double-bonded, while S_8 consists of single-bonded, crown-shaped rings.

Section 22.3   Group 1A: Hydrogen

Hydrogen was first isolated by Henry Cavendish, an English chemist who showed that acid acting on metals generated a low-density flammable gas.

The name *hydrogen*, which means "water former," was given by French chemist Antoine Lavoisier, who noted that it combines with oxygen to form water.

Hydrogen is a colorless, odorless, and tasteless gas that primarily exists as a nonpolar, diatomic molecule. Its lack of polarity and small size lead to very weak intermolecular forces, and low melting and boiling points, but it has the highest dissociation energy of any diatomic element.

Hydrogen is thought to be approximately 75% of the mass of the universe, but it is very rare in the Earth's atmosphere because the Earth's gravity is not strong enough to hold it. Combined with other elements, it is the ninth most abundant element in the Earth's crust and oceans.

Isotopes of Hydrogen
Protium (hydrogen-1) is the lightest and most common form of hydrogen. It comprises 99.985% of atoms in naturally occurring hydrogen.

Deuterium (hydrogen-2, or D), is also known as *heavy hydrogen*, and is present in small amounts (0.015% of atoms), while *tritium* (hydrogen-3,or T) is radioactive and is present only in trace amounts.

All three isotopes have the same electron configuration and thus the same chemical behavior. However, their very different masses cause quantitative differences in their properties, known as **isotope effects**. These are greater for hydrogen than for any other element because the mass differences between isotopes are so great.

For hydrogen isotopes, we can see a trend: the heavier the isotope, the higher the melting point and the boiling point. The same is true of the isotopes when bound into water. When bonded in water, the DO and TO bonds are stronger than HO bonds, which make it possible to separate the isotopes.

The effect of isotopic mass on reaction rates is called a *kinetic–isotope effect*. D_2O can be separated from H_2O because its stronger bonds separate more slowly. As water is electrolyzed, the remaining molecules are enriched in D_2O. Reducing water from 2400 L to 83 mL yields 99% pure D_2O. About 150 metric tons of D_2O are manufactured in the United States per year for use as a coolant and moderator in nuclear reactors.

Workbook Problem 22.2

Benzene has the formula C_6H_6. What is the molecular mass of C_6D_6? What is the percent increase in the molecular mass when benzene is made with deuterium instead of protium? Make the same calculations for H_2O and D_2O, then qualitatively describe the changes expected for the melting and boiling points of C_6D_6 compared to C_6H_6. Would more or less energy be released from burning C_6D_6 compared to C_6H_6?

Strategy: Calculate the molecular masses of the four molecules, calculate the percent differences, and describe the kinetic–isotopic effects.

Step 1: Calculate the molecular weights for the four molecules.

Step 2: Calculate the percent difference in the two pairs.

Step 3: Predict the expected properties.

Workbook Problem 22.3

What volume of D_2O can be purified from 25 L of water?

Strategy: Set up a ratio to calculate the volume; choose an appropriate unit for the answer.

Preparation and Uses of Hydrogen
Electrolysis of water produces hydrogen of 99.95% purity, but it is impractical for large-scale production due to the large amount of energy required: 286 kJ/mol of energy for 1 mol (2g) of hydrogen.

Small-scale hydrogen production can be accomplished in the lab by reacting dilute acid with an active metal such as zinc.

Large-scale industrial preparation methods use inexpensive reducing agents to extract the oxygen from steam. Steam–hydrocarbon reforming, the most important method for producing industrial hydrogen, is a three-step process.

In the first step, a mixture of steam and methane are reacted at high temperature in the presence of a nickel catalyst to produce *synthesis gas*, so called because it can also be used to synthesize liquid fuels:

$$H_2O\ (g)\ +\ CH_4\ (g)\ \rightarrow\ CO\ (g)\ +\ 3\ H_2\ (g)$$

In the second step, the CO from the synthesis gas is mixed with more steam and then passed over a metal oxide catalyst at 400 °C. This **water–gas shift reaction** removes the toxic carbon monoxide and produces more hydrogen gas:

$$CO\ (g)\ +\ H_2O\ (g)\ \rightarrow\ CO_2\ (g)\ +\ H_2\ (g)$$

In the third step, the gas mixture is passed through a basic aqueous solution, which dissolves the carbon dioxide into carbonate ion:

$$CO_2\ (g)\ +\ 2\ OH^-\ (aq)\ \rightarrow\ CO_3^{2-}\ (aq)\ +\ H_2O\ (l)$$

95% of H_2 is produced and consumed in the same place. The largest single consumer is the Haber process for synthesizing ammonia:

$$N_2\ (g)\ +\ 3\ H_2\ (g) \rightarrow 2\ NH_3\ (g)$$

Large quantities are also used in the synthesis of methanol:

$$CO\ (g)\ +\ H_2\ (g)\ \rightarrow\ CH_3OH\ (l)$$

This is an industrial solvent and is used in making formaldehyde, a precursor to plastics.

Workbook Problem 22.4

Write a balanced equation for the production of synthesis gas from ethane, C_2H_6. If each step in the process has a 60% yield, how many liters of hydrogen gas at 25 °C and 1 atm can be produced from 15 kg of ethane?

Strategy: Write and balance the equation; determine the quantities of reactants and products at each step. Use the ideal gas law to calculate the volume of hydrogen gas.

Step 1:   Write and balance the equation for the formation of synthesis gas.

Step 2:   Determine the moles of reactants entering the first step and the moles of products produced.

Step 3:   Multiply the moles of products by 60%, and calculate the products of the second step.

Step 4:   Multiply the moles of products by 60%, and calculate the products of the third step.

Step 5:   Multiply the moles of products by 60%, and determine the volume of the H_2 gas.

Reactivity of Hydrogen

With its one valence electron, hydrogen has properties similar to both the alkali metals and the halogens. Like the alkali metals, it can ionize to form H^+ ($E_i = 1312$ kJ/mol). Like the halogens, it can share its electron to form covalent compounds.

Complete ionization of hydrogen is only possible in the gas phase. In liquids or solids, a bare proton is too reactive to exist by itself, so it will attach to a molecule with a lone pair of electrons.

Also like the halogens, hydrogen will accept an electron ($E_{ea} = -73$ kJ/mol) from an active metal to make an ionic hydride.

$$\text{Gain electron} — \text{H}^- \qquad\qquad E_{ea} = -73 \text{ kJ/mol}$$

Because of the strength of the H–H bond, hydrogen is relatively unreactive. But in the presence of oxygen, an explosive reaction can occur that requires as little as 4% hydrogen in a mix of gases. This is the potentially dangerous and highly exothermic reaction:

$$2 \text{ H}_2 \text{ (g)} + \text{O}_2 \text{ (g)} \rightarrow 2 \text{ H}_2\text{O} \text{ (l)} \qquad\qquad \Delta H = -572 \text{ kJ}$$

Workbook Problem 22.5

What is the change in volume involved when 15 g of hydrogen and 100 g of oxygen react at STP? What volume of water is produced? What volume of gas is left over?

Strategy: Determine the initial volume, the moles of each gas, and the quantity of gas left over (assume that the volume of the water produced is negligible). How many kilojoules are released in this reaction?

Step 1: Determine the moles of each gas and the limiting reagent.

Step 2: Calculate the moles of excess reagent remaining after the reaction.

Step 3: Calculate the volume change of the reaction.

Step 4: Calculate the kilojoules released by the reaction.

Binary Hydrides

Binary hydrides contain hydrogen and just one other element and can be ionic, covalent, or metallic.

The alkali metals and Ca, Sr, and Ba form ionic hydrides. They can be formed by direct reaction at 400 °C. Binary hydrides are salt-like, with the alkali metal hydrides forming a face-centered cubic cell like that of sodium chloride.

The hydride ion is an electron donor—a Brønsted–Lowry base—and therefore a good reducing agent. When dissolved in water, the hydride ion reduces water to generate H_2 gas and OH^- ions.

Covalent hydrides are molecules in which hydrogen is covalently bonded to another nonmetal. These tend to be small molecules with relatively weak intermolecular forces, and so are gases or volatile liquids at room temperature. Many are quite familiar—H_2O, NH_3, and CH_4 are all covalent hydrides.

Metallic hydrides will also form with large metal atoms, especially the lanthanides, actinides, and certain *d*-block transition metals. When these large metal atoms are compressed together, the hydrogen atoms—it is not known whether they ionize—pack into the interstitial spaces. These compounds can be stoichiometric, for example, UH_3, or not, as in the case of $ZrH_{1.9}$.

The properties of these metallic hydrides depend on their composition, which is also dependent on the partial pressure of H_2 gas in the surroundings. This makes these compounds of potential interest as hydrogen-storage devices.

Workbook Problem 22.6

Write a balanced equation for the reaction of sodium hydride with water. How many liters of hydrogen gas at STP will be evolved if 3.65 g of sodium hydride is allowed to react with excess water?

Strategy: Write and balance the equation, determine the moles of sodium hydride, and use the balanced equation to calculate the moles of gas and the standard molar volume to determine liters of hydrogen gas.

Step 1:   Write and balance the equation.

Step 2:   Determine the moles of sodium hydride and the moles of hydrogen evolved.

Step 3:   Determine the volume of hydrogen evolved.

Section 22.4   Group 1A: Alkali Metals

Description: Alkali metals have the lowest ionization energies of any elements, because they lose an ns^1 electron to gain a noble gas configuration. As a result, they are among the most powerful reducing agents in the periodic table.

Alkali metals are *metallic*: they are shiny, malleable, and good conductors of electricity. In contrast to the more common metal elements, alkali metals are soft enough to cut with a knife, have low melting points and densities, and are so reactive that they are stored under oil. They are not found in nature as metals, but as salts.

Production: Lithium and sodium are produced by *electrolysis*: the chloride salt is melted and an electric current passed through it while it is held at high temperature.

Potassium, rubidium, and cesium are produced by reduction at high temperatures, using sodium to reduce potassium and calcium to reduce rubidium and cesium. Although this might seem to contradict the activity series, the high temperatures allow tiny amounts of the desired metals to be produced; removing them from the reaction mixture shifts the equilibrium toward production of the desired metal.

Reactions: Alkali metals react with halogens to form *halides*. The reactivity of the alkali metals increases as the ionization energy decreases: cesium is the most reactive, lithium the least reactive.

Reactions with oxygen vary with the size of the ion: lithium produces a simple oxide, Li_2O. Sodium produces peroxide, Na_2O_2, in which the anion is O_2^{2-}. Potassium produces a superoxide, KO_2, in which the anion is O_2^-. This last is particularly valuable in spacecraft for its ability to react with moisture and carbon dioxide to generate oxygen.

Alkali metals take their name from their reaction with water to yield hydrogen gas and an alkali metal hydroxide. These reactions are vigorous enough to split water molecules, and, as with the halide reactions above, they increase in vigor with decreasing ionization energy. The reaction of lithium and water generates bubbles; the reactions of other alkali metals result in flames as the heat generated burns the hydrogen gas that has been generated:

$$2 \text{ M } (s) + 2 \text{ H}_2\text{O } (l) \rightarrow 2 \text{ MOH } (aq) + \text{H}_2 (g) + \text{heat}$$

A similar reaction occurs when alkali metals are added to liquid ammonia. These reactions produce hydrogen gas and a dissolved metal amide:

$$2 \text{ M } (s) + 2 \text{ NH}_3 (l) \rightarrow 2 \text{ MNH}_2 (aq) + \text{H}_2 (g)$$

Liquid ammonia will dissolve alkali metals at temperatures below –33 °C to form metal ions and dissolved electrons. These solutions are very strong reducing agents.

Section 22.5   Group 2A: Alkaline-Earth Metals

Description: Alkaline-earth metals are able to lose two ns^2 electrons to gain a noble gas configuration, but they have slightly higher ionization energies than the alkali metals. However, they do follow the same reactivity trend, in which the larger members of the group, with lower ionization energies, are the most reactive.

These elements are harder than the alkali metals and have higher densities and melting points. Despite their lower reactivity, they are also found in nature only as salts.

Production: Pure alkaline-earth metals are produced by chemical or electrolytic reduction. Beryllium is produced by reduction of beryllium fluoride with magnesium, and magnesium is prepared by electrolysis of its melted chloride. Calcium, strontium, and barium are all made by high-temperature reduction with aluminum metal.

Reactions: Alkaline-earth metals react with halogens to form halides with the form MX_2 and with oxygen to form oxides with the expected form MO, though strontium and barium will also form peroxides. Beryllium and magnesium are relatively unreactive, but will burn with a bright flame. The other alkaline-earth metals are all reactive enough that they are best protected from air.

With the exception of beryllium, the alkaline-earth metals will react with water to form hydroxides, but with much less vigor than the alkali metals. Magnesium will react only at temperatures exceeding 100 °C, calcium and strontium will react slowly at room temperature, and barium, with its low ionization energy, reacts vigorously at room temperature.

Workbook Problem 22.7

What is the predicted outcome of the following reactions? Balance the reactions that occur.

$$Sr\ (s) + O_2\ (g) \qquad Ca\ (s) + H_2O\ (l) \qquad Be\ (s) + H_2O\ (l)$$

Strategy: Based on the reactivity of the metals and the expected reactions:

Step 1:   Determine what reaction, if any, will occur.

Step 2:   If reactions occur, write their balanced equations.

Section 22.6   Group 3A: Elements

The group 3A elements are boron, aluminum, gallium, indium, and thallium. These are the first of the p block elements; they have the valence electron configuration $ns^2\,np^1$ and an oxidation state of +3. As expected, metallic character increases down the group: boron (described below) is the only semimetal; the others behave as metals. Aluminum (also described below) is the most commercially important member of the group.

Gallium is in its liquid form from 29.2 °C to 2204 °C. Gallium arsenide is a semiconductor used in making diode lasers for laser printers, CD players, and fiber-optic devices.

Indium is also used in making semiconductor devices such as transistors and *thermistors*, electrical-resistance thermometers.

Thallium is extremely toxic and has no commercial uses.

Boron
Boron is relatively rare, but occurs in concentrated deposits of borate minerals.

Pure crystalline boron is obtained by the following reaction:

$$2 \ BBr_3 \ (g) \ + \ 3 \ H_2 \ (g) \ \rightarrow \ 2 \ B \ (s) \ + \ 6 \ HBr \ (g)$$

It is a strong, high-melting-point substance that is chemically inert at room temperature, making it a desirable component in high-strength composite materials.

Boron halides are highly reactive, volatile covalent compounds that consist of trigonal planar BX_3 molecules. They have a vacant $2p$ orbital that allows them to behave as Lewis acids: they will form adducts with Lewis bases by accepting a share of a nonbonded pair of electrons.

Boron will also react with metal fluorides to form the tetrahedral BF_4^- anion, which is a catalyst in industrially important organic reactions.

Boron hydrides (boranes) are volatile, molecular compounds with formulas B_nH_m, the simplest of which is diborane, B_2H_6. Diborane is of interest because of its unusual structure: rather than having a structure similar to ethane (C_2H_6), it has an *electron-deficient* structure in which two BH_2 groups are connected by bridging hydrogens; therefore there are only 12 valence electrons. The unusual B–H–B bonds are much longer than the terminal B–H bonds.

Aluminum
The most abundant metal in the Earth's crust, aluminum is very difficult to refine: until an economical method for its manufacture was developed in 1886, it was considered a precious metal.

Ruby and sapphire are impure forms of Al_2O_3, the red color of rubies coming from chromium impurities and the blue of sapphires from tin impurities.

Aluminum metal is purified from bauxite, $Al_2O_3 \cdot x \ H_2O$. The Bayer process, which takes advantage of the fact that this is an amphoteric oxide, removes the Al_2O_3 and the Haber process then electrolyzes the aluminum out of a mixture of Al_2O_3 and cryolite.

Aluminum is a reducing agent that loses three electrons to form Al^{3+} ions. It reacts vigorously with halogens to form halides with formula AlX_3. AlF_3 and $AlCl_3$ have extended crystal structures, but $AlBr_3$ and AlI_3 both form dimers with bridging halogens that are similar to diborane.

When exposed to air, aluminum oxidizes immediately to form a thin, hard oxide coating that protects the underlying metal from oxygen or water. However, aluminum will still react with acid or base to form hydrogen gas and either Al^{3+} ions in acid solution, or aluminate ions ($Al(OH)_4^-$ in basic solutions.

Section 22.7   Group 4A: Carbon

Carbon, silicon, germanium, tin, and lead are important both in industry and in living organisms, and they exemplify the increase in metallic character down the periodic table.

Carbon is a nonmetal that is present in all plants and animals; it is an essential part of biological molecules.

Silicon is a semimetal that makes up numerous silicate minerals, is the second-most abundant element in the Earth's crust, and with germanium is used to make solid-state electronic devices. Germanium is a high-melting-point brittle semiconductor with the same structure as diamond and silicon.

Tin and lead are both soft, malleable, low-melting-point metals that have been known since ancient times. Tin has two allotropes: malleable, silvery metallic *white tin* and brittle, semiconducting *gray tin*. White tin is the most stable allotrope at room temperature, but when kept below 13 °C for long periods, it crumbles into gray tin, in what is known as "tin disease." Lead is found only in metallic form.

Group 4A has the valence electron configuration ns^2np^2, giving +4 as the most common oxidation state. The chlorides of the group 4A elements are all sp^3 covalently bonded volatile molecular liquids. An oxidation state of +2 is also possible for Sn and Pb in solution, which generally leads to ionic bonding. There are no +4 aqueous ions: these species instead exist as complex ions such as $Sn(OH)_6^{2-}$.

Carbon
Elemental carbon forms a number of allotropes:

Diamond has an sp^3-bonded covalent network in which each C atom forms a tetrahedral array of σ bonds. This forms the hardest known substance, as well as the element with the highest melting point: approximately 9000 K at a pressure of 6–10 million atm. The tight localization of the electrons makes diamond an electrical insulator.

Graphite has a two-dimensional, sheet-like structure in which sp^2 hybrid orbitals form trigonal planar σ bonds to three neighboring C atoms, while the remaining p orbital forms a delocalized π bond with its neighbors. These delocalized electrons make graphite useful as an electrode: its conductivity is 10^{20} times greater than that of diamond.

Gases can get in between the sheets of graphite, which are held together only by dispersion forces. The sheets can slide over one another, making graphite useful as a lubricant.

Graphene is a two-dimensional array of hexagonally arranged carbon atoms just one atom thick. It is very strong and flexible and is an excellent conductor, making it a possible material for electronics and composite materials.

Fullerene is a nearly spherical allotrope of C_{60} that can be prepared by vaporizing graphite in a helium atmosphere. As a molecular substance, it is soluble in nonpolar solvents, unlike graphite and diamond.

Derivatives of fullerene include compounds in which other atoms are attached to the C_{60} cage and compounds in which metal atoms are trapped in the C_{60} cage.

Similar compounds include the egg-shaped C_{70} molecule and *carbon nanotubes*. Carbon nanotubes can conduct electricity 10 times better than copper and will probably replace silicon in electronic devices at some point.

Carbon forms more than 40 million compounds, most of which are organic. Some of the inorganic compounds of carbon are as follows:

Oxides of carbon include carbon monoxide and carbon dioxide:

Carbon monoxide is a colorless, odorless, toxic gas that forms when carbon or hydrocarbons are burned in limited oxygen. The toxicity of CO arises from the fact that it binds 200 times more tightly to hemoglobin than oxygen. Thus, even low levels reduce the blood's ability to deliver oxygen to the tissues.

Carbon dioxide is a colorless, odorless, nonpoisonous gas. It is produced when fuels burn in an excess of O_2, is an end product of metabolism, is a by-product of the yeast-catalyzed fermentation of sugar in the manufacture of alcoholic beverages, and is produced when metal carbonates are heated or reacted with acids.

Carbon dioxide has numerous uses, including providing the "fizz" in beverages and fighting fires. CO_2 is denser than air, so it settles over small fires like a blanket, thereby smothering them. Solid CO_2 is known as "dry ice." It sublimes at –78 °, and is used primarily as a refrigerant.

Supercritical carbon dioxide, abbreviated scCO₂, is a form of carbon dioxide that exists above 31 °C and 73 atmospheres. It is neither true liquid nor true gas, and it is nontoxic, nonflammable, nonpolar, and easily recovered, making it useful as a replacement for organic solvents. It is already used to decaffeinate tea and coffee and in "green" dry cleaning, with other uses being investigated such as coating drug particles and manufacturing computer chips.

Carbonates also come in two forms: carbonates, which contain the CO_3^{2-} ion, and hydrogen carbonates, or bicarbonates, which contain the HCO_3^- ion.

Carbonates include Na_2CO_3, also known as soda ash. It is used in glassmaking, while $Na_2CO_3 \cdot$ 10 H_2O, or washing soda, is used in laundering textiles: the carbonate ions precipitate cations from hard water, as well as producing OH^- ions that remove grease.

Sodium hydrogen carbonate (sodium bicarbonate), $NaHCO_3$, is also called baking soda. It reacts with acidic substances in foods to produce bubbles of CO_2 that cause dough to rise.

Hydrogen cyanide, HCN, is a highly toxic volatile substance that is weakly acidic in solution, dissociating to form CN^- ions. These are known as *pseudohalides* because they behave like Cl^-, precipitating out Ag ions. In forming complex ions, cyanide behaves as a Lewis base, bonding through the lone pair of electrons on carbon.

The toxicity of HCN and cyanides is due to the strong and irreversible binding of CN^- to the iron (III) atom contained in cytochrome oxidase, an important enzyme in cellular metabolism. With this enzyme unable to function, cellular energy production ceases and death occurs rapidly.

CN^- ions are able to react with ores containing Au and Ag. When the cyanide salts are reduced with zinc, the precious metal can be recovered.

Carbides are binary compounds of carbon in which the carbon atom has a negative oxidation state. They can be ionic carbides of active metals (CaC_2), interstitial carbides of transition metals (Fe_3C), or covalent network carbides (SiC).

Workbook Problem 22.8

Determine the oxidation state of carbon in the following compounds:

| CO | CO_2 | C_2O_3 | CO_3^{2-} | HCN | CaC_2 |

Strategy: Use the rules for assigning oxidation numbers to determine the oxidation state of carbon.

Step 1:   Determine the oxidation state of carbon.

Section 22.8   Group 4A: Silicon

Silicon is a hard, gray, semiconducting solid that melts at 1414 °C. It has a diamond-like structure, but does not have a graphite-like allotrope because the silicon atoms are too large for effective overlap of π orbitals.

In nature, silicon is generally found in silica and silicate minerals. To generate elemental silicon, silica sand is reduced by carbon to form silicon and carbon monoxide. To make the ultrapure silicon needed for electronics, silicon is converted to $SiCl_4$, purified by fractional distillation, and reduced with hydrogen gas to form pure silicon and HCl. This can then be further purified by zone refining, in which a rod of silicon is slowly melted from top to bottom. Impurities concentrate in the molten zone and are dragged to the bottom of the rod.

Silicates are ionic compounds that contain silicon oxoanions and metal cations. The basic building block is the SiO_4 tetrahedron. Alone, it is the orthosilicate ion SiO_4^{4-}, but more commonly, there is an oxygen shared between units, leading to a large number of possible mineral structures, including cyclic anions, chains, layers, and extended three-dimensional structures. This bonding flexibility means that silicates include such varied mineral types as fibrous asbestos, sheets of mica, and emeralds.

Aluminosilicates occur when partial substitution of the Si^{4+} ion with Al^{3+} occurs. Among these, *feldspar*s are the most common of all minerals. In *zeolites*, the tetrahedra are joined together in an open structure with a three-dimensional network of cavities that can only be entered by small molecules, making them useful as molecular sieves to separate small molecules from larger ones and as catalysts in industrial processes, including the manufacture of gasoline.

Workbook Problem 22.9

The formula unit of garnet has three $SiO_4{}^{4-}$ units balanced by Ca^{2+} and Al^{3+} cations. How is the formula unit of garnet written?

Strategy: From the information provided, determine the overall silicate structure, remembering that oxygens are shared in the tetrahedral units. Balance the charge on the silicate using the ions in a correct ratio.

Step 1:   Determine the silicate structure.

Step 2:   Determine how many of each cation is needed to balance the charge on the silicate unit.

Step 3:   Write the formula unit.

Section 22.9   Group 5A: Nitrogen

Nitrogen, phosphorus, arsenic, antimony, and bismuth exhibit the expected trends down the periodic group—increasing atomic size, decreasing ionization energy and electronegativity, and acid–base properties of oxides. N and P oxides are acidic, As and Sb oxides are amphoteric, and bismuth oxides are basic.

The valence configuration of the group is ns^2np^3, which allows for a variety of oxidation states. Nitrogen and phosphorus will exhibit all oxidation states between –3 and +5, while As and Sb will exhibit both +3 and +5 oxidation states, but the +5 oxidation state is less stable as atomic size increases.

Sb^{3+} and Bi^{3+} are found in salts, but there are no simple cations of N or P.

As, Sb, and Bi are found in sulfide ores and are used in making various metal alloys. Arsenic is used in pesticides and semiconductors, and bismuth is present in some pharmaceuticals, such as Pepto-Bismol.

Nitrogen

Nitrogen is a colorless, odorless, tasteless gas that makes up 78% of Earth's atmosphere. It can be separated from liquid air by fractional distillation.

Nitrogen gas can be used as a protective inert atmosphere in manufacturing processes, while liquid nitrogen is commonly used as a refrigerant.

A large amount of energy is required to break the N≡N bond, so reactions involving N_2 typically have a high activation energy and/or an unfavorable equilibrium constant. As temperatures increase, equilibrium shifts to the right. The many compounds formed by nitrogen are conveniently classified by oxidation state.

Ammonia (NH_3) is the starting material for industrial synthesis of other nitrogen compounds. It is synthesized by the Haber process, which requires high temperatures, high pressures, and catalaysis.

Colorless, strong-smelling, and gaseous at room temperature, NH_3 molecules have a lone pair of electrons, making them polar, trigonal pyramidal, water soluble, and easily condensed, because they are able to hydrogen bond. This ability also makes NH_3 an excellent solvent for ionic compounds. It is a Brønsted–Lowry base and reacts with acids to yield ammonium salts.

Hydrazine ($H_2N–NH_2$) is an ammonia derivative in which one H is replaced by an *amino* (NH_2) group. It is prepared by reacting ammonia with OCl^-.

A poisonous, colorless liquid that smells like ammonia, hydrazine is explosive in the presence of oxidizing agents, including air, and is used as a rocket fuel: when it reacts with N_2O_4 (*l*), the two liquids generate large volumes of nitrogen gas and water vapor. It is safely handled in aqueous solution, where it is a weak base and versatile reducing agent.

Oxides of nitrogen: although many exist, we describe only three here:

- N_2O, or nitrous oxide, is a colorless, sweet-smelling gas used as a propellant. It is also used as a dental anesthetic, because it is mildly intoxicating in small doses, leading to its common name of "laughing gas." It is produced by gently heating molten ammonium nitrate.
- NO, or nitric oxide, is a colorless gas that can be produced in small quantities by reacting copper metal with dilute nitric acid. It is important in biological processes, where it transmits nerve impulses, kills harmful bacteria, and dilates blood vessels to increase blood flow.
- NO_2, or nitrogen dioxide, is a toxic, reddish-brown gas that can be produced by reacting copper with concentrated nitric acid. It is paramagnetic and tends to dimerize, forming an N–N bond between two units.

HNO_2, or nitrous acid, is produced when NO_2 reacts with water. This is a disproportionation reaction in which nitrogen goes from +4 in NO_2 to +3 in nitrous acid. It also goes to +5 in HNO_3, nitric acid, at which point it tends to again disproportionate into nitric oxide and nitric acid.

HNO_3, or nitric acid, is a strong acid primarily used to manufacture ammonium nitrate for fertilizers, as well as explosives, plastics, and dyes. It is produced by the multistep Ostwald process, in which ammonia is oxidized to nitric oxide, nitric oxide is oxidized to nitrogen dioxide, and nitrogen dioxide is disproportionated in water. Further removal of water generates concentrated nitric acid, which is 15 M and 68.5% HNO_3 by mass. It often has a slight yellow color due to the presence of small amounts of NO_2 produced by decomposition.

Nitric acid is a stronger oxidizing agent than H^+ alone and will oxidize relatively inactive metals, but the product depends upon the nature of the reducing agent and the reaction conditions.

Aqua regia is a 3:1 volume ratio of concentrated HCl and concentrated HNO_3. It is an even stronger oxidizing agent and will oxidize even inactive metals, such as gold, which do not react with either component separately.

Workbook Problem 22.10

Upon heating, ammonium nitrate decomposes to produce nitrous oxide (N_2O) and water. What volume of nitrous oxide, collected over water, at a total pressure of 705 mmHg and 22°C, can be produced from 5.3 g ammonium nitrate? (The vapor pressure of water at 22°C is 24 mmHg.)

Strategy: After writing the balanced chemical reaction, determine the volume of N_2O from the pressure of N_2O using the ideal gas law. You must use Dalton's law to determine the pressure of N_2O.

Step 1: Write the balanced chemical equation.

Step 2: Calculate the pressure of N_2O, using the total pressure and the vapor pressure of water.

Step 3: Calculate the number of moles of N_2O produced from 3.5 g NH_4NO_3.

Step 4: Calculate the volume of N_2O using the ideal gas law.

Section 22.10 Group 5A: Phosphorus

Phosphorus is found in phosphate rock as $Ca_3(PO_4)_2$ and in fluorapatite as $Ca_5(PO_4)_3F$. Apatites are minerals with the general formula 3 $Ca_3(PO_4)_2 \cdot CaX_2$ ($X^- = F^-$ or OH^-).

Phosphorus is also important in living systems, and it is the sixth most abundant element in the human body. Tooth enamel is almost pure hydroxyapatite, $Ca_5(PO_4)_3OH$, and bones contain hydroxyapatite and collagen. Phosphate is also an essential part of the backbone of nucleic acids and the phospholipids that are the major components of cell membranes.

Industrial production of phosphorus involves heating phosphate rock, coke, and silica sand at 1500 °C. The gaseous P_4 tetrahedra produced are condensed by passing the mixture through water.

Two main allotropes of phosphorus exist:

- White phosphorus is toxic, waxy, white solid with discrete P_4 tetrahedra. It is nonpolar, has a low melting point, and is highly reactive due to an unusual bonding structure that requires "bent" overlap of p orbitals. If exposed to air, white phosphorus will burst into flame.
- Red phosphorus can be produced by heating white phosphorus without air. It is a polymeric form, which is nontoxic, less soluble, and less reactive than white phosphorus.

Phosphorus will form compounds with all oxidation states from –3 to +5, but +3 and +5 are most common. It is less electronegative than nitrogen, so is more likely to be found in a positive oxidation state.

Phosphine (PH_3) is a colorless, extremely poisonous gas. It is neutral in solution and is easily oxidized, burning in air to form phosphoric acid, H_3PO_4.

Halides will form when phosphorus reacts with any halogen. When halogen is limited, the halides have the formula PX_3. When halogen is in excess, PX_5 is formed. These compounds are gases, volatile liquids, or low-melting-point solids.

Oxides of phosphorus follow a similar pattern. When oxygen is limited, P_4O_6 forms, but when oxygen is in excess, P_4O_{10} is formed. Both are acidic oxides that react with water to form acids. P_4O_{10} has such a high affinity for water that it is used as a drying agent for gases and organic solvents.

Phosphorous acid, H_3PO_3, is a weak diprotic acid in which only two of the three hydrogens are attached to oxygen.

Phosphoric acid, H_3PO_4, is a low-melting, colorless crystalline solid. In the laboratory, it is a syrupy, aqueous solution that is 82% H_3PO_4 by mass. For use as a food additive, it is manufactured by burning pure liquid phosphorus in the presence of steam. For fertilizer, an impure version is made by treating phosphorus rock ($Ca_3(PO_4)_2$) with sulfuric acid. When this is reacted again with phosphorus rock, the result is called triple superphosphate.

Section 22.11   Group 6A: Oxygen

Oxygen, sulfur, selenium, tellurium, and polonium exhibit the expected trends down the periodic group. O and S are typical nonmetals. Se and Te are largely nonmetallic and are both semiconductors. Polonium is a radioactive metal found in trace amounts in uranium ores.

The valence configuration of the group is $ns^2 np^4$, which makes the –2 oxidation state the most common. As metallic character increases, this oxidation state is increasingly unstable, and S, Se, and Te are commonly found in positive oxidation states, especially +4 and +6.

Commercial uses of Se, Te, and Po are limited. Se is used in making red glass and in photocopiers, Te is used in alloys to improve machinability, and Po is a heat source for space equipment and a source of alpha particles.

Elemental Oxygen
Oxygen is the most abundant element on the surface of our planet and is crucial to human life. Approximately 25% of all biological molecules contain oxygen.

Small amounts of O_2 can be prepared in the laboratory by the thermal decomposition of an oxoacid salt, like potassium chlorate, $KClO_3$

$$2 \, KClO_3 \, (s) \rightarrow 2 \, KCl \, (s) + 3 \, O_2 \, (g)$$

Oxygen can also be prepared on an industrial scale by undergoing the fractional distillation of liquefied air. More than 66% of the oxygen produced industrially is used in making steel. Oxygen is also used in sewage treatment and in paper bleaching.

Oxygen Compounds
Oxygen can obtain an octet configuration by reacting with active metals from Group 1A or 2A, such as lithium:

$$4 \, Li \, (s) + O_2 \, (g) \rightarrow 2 \, Li_2O \, (s)$$

Oxygen can also form covalent oxides with nonmetals like hydrogen, carbon, sulfur, and phosphorus:

$$C \, (s) + O_2 \rightarrow CO_2 \, (g)$$

Acid–Base Properties of Oxides
Binary compounds with oxygen (with a –2 oxidation state) are called **oxides** and can be acidic, basic, or amphoteric.

Section 22.12   Group 6A: Sulfur

Sulfur accounts for 0.016% of the Earth's crust by mass, has numerous allotropes, and is found in large underground deposits and as a component of numerous minerals. In biology, it is a component of proteins.

Rhombic sulfur is the most stable form. It is a yellow crystalline solid that contains crown-shaped S_8 rings. Above 95 °C, monoclinic sulfur is more stable. In this allotrope, the S_8 rings have a different packing arrangement than the crystals of rhombic sulfur.

Molten sulfur is a fluid, straw-colored liquid at just above its melting point, but between 160 °C and 195 °C, the S_8 rings open and form long chains that can contain more than 200,000 S atoms. In this state, the sulfur becomes viscous and dark reddish-brown. Above 195 °C the chains break, and viscosity declines. If molten sulfur is dropped in water, it freezes in a disordered state known as plastic sulfur.

Sulfur forms numerous compounds:

- Hydrogen sulfide (H_2S) is a colorless gas with a strong odor of rotten eggs. It is very toxic and dulls the sense of smell, making exposure all the more dangerous. It can be produced in the lab by treating iron(II) sulfide with dilute sulfuric acid or by hydrolysis of thioacetamide for use in qualitative analysis. It is a weak diprotic acid and a mild reducing agent.
- Sulfur dioxide and sulfur trioxide (SO_2 and SO_3) are the most important oxides of sulfur. Sulfur dioxide is produced by burning sulfur and is a colorless, toxic gas with a pungent odor. It is toxic to microorganisms, so is used to sterilize wine and dried fruit. In the atmosphere, sulfur dioxide is oxidized to sulfur trioxide, which dissolves in rainwater to

form sulfuric acid. Both species are acidic, but little sulfurous acid (H_2SO_3) is found in solution: SO_2 will dissolve without reacting.

- Sulfuric acid is the world's most important industrial chemical. It is manufactured by the **contact process**: S is burned to SO_2, SO_2 is oxidized to SO_3 with a catalyst, and the SO_3 is dissolved in sulfuric acid, rather than water, to make more sulfuric acid because SO_3 dissolves slowly in water. Concentrated sulfuric acid is 18 M and 98% H_2SO_4 by mass.

- Sulfuric acid is a strong acid at the first proton and dissociates at the second proton with a K_{a2} of 1.2×10^{-2}. It forms two series of salts: hydrogen sulfates, with a hydrogen as one of the cations ($NaHSO_4$), and sulfates (Na_2SO_4). In dilute solutions at room temperature, sulfuric acid behaves like HCl, oxidizing metals above H in the activity series. Hot, concentrated H_2SO_4 is a stronger oxidant than sulfuric acid and will come apart to re-form SO_2. Sulfuric acid is used to manufacture fertilizers and in many other industrial processes.

Section 22.13   Group 7A: Halogens

Description: The halogens are diatomic nonmetals, characterized by their gaining electrons in chemical reactions. They have large negative electron affinities and large positive ionization energies. Halogens are too reactive to occur free in nature: hence the name, which comes from the Greek roots *hals* (salt) and *gennan* (to make).

Production: Halogens are produced by oxidation of their anions. Fluorine and chlorine are produced by electrolysis, while bromine and iodine are prepared by oxidation of their anions by chlorine.

Reactions: Halogens are among the most reactive elements in the periodic table, with fluorine being the most reactive: it will form compounds with every other element in the periodic table except He, Ne, and Ar. Halogens become less reactive as they get larger, for exactly the same reason that metals gain reactivity as they get larger: outer electron shells are more weakly held, so elements that are gaining rather than losing electrons will become less reactive as their size increases.

Metal halides can be formed from every metal, predictably with main-group metals and less-predictably with transition metals, some of which will form more than one halide.

Hydrogen and halogens react to form hydrogen halides: fluorine reacts explosively and spontaneously, chlorine reacts in the presence of UV light or a spark, and bromine and iodine react very slowly. Hydrogen halides will dissolve in water to form **acids**. Hydrofluoric acid, though it is the only weak acid among the hydrogen halides, is also one of the few substances that can be used to etch glass.

Halogens have a valence configuration of ns^2np^5. They gain an electron to ionize and share one electron in molecular compounds in which they have an oxidation state of -1.

Numerous positive oxidation states are also available for the larger halogens—Cl, Br, and I—allowing them to share valence electrons with oxygen. The oxidation states available are $+1$, $+3$, $+5$, or $+7$, and the general formula for an oxoacid is HXO_n, where the value of n depends on the oxidation state of the halogen.

The higher the oxidation state of the halogen, the stronger the acid. Acidic protons are bonded to the oxygen (despite how the formula is typically written), and these are all strong oxidizing agents.

Hypohalous acids are formed when halogens dissolve in water, disproportionating to form an aqueous −1 ion and a +1 oxidation state in HOX:

$$X_2 \ (g, \ l, \text{ or } s) + H_2O \ (l) \rightarrow HOX \ (aq) + H^+ \ (aq) + X^- \ (aq)$$

Equilibrium lies to the left in this reaction, but shifts toward products in basic solution, generating OX^- ions. The most important salt of these is $NaOCl$, sodium hypochlorite, which is a strong oxidizing agent sold in 5% solution as chlorine bleach.

Further disproportionation of OCl^- to ClO_3^- is slow, but can be produced by the reaction of Cl_2 with hot NaOH. Chlorate salts are used as weed killers and oxidizing agents: $KClO_3$ is used in matches, fireworks, and explosives, and reacts with organic matter.

Sodium perchlorate can be converted to perchloric acid by reaction with HCl: anhydrous perchloric acid is a colorless, shock-sensitive liquid that decomposes explosively on heating. It is a powerful and dangerous oxidizing agent that is powerful enough to oxidize silver and gold. Even its salts are powerful oxidants: ammonium perchlorate is used as a rocket fuel. Perchlorate ions are toxic, because they prevent uptake of iodide by the thyroid, but the source of environmental exposures is unclear.

Iodine forms multiple perhalic acids as a result of its large size:

- Paraperiodic acid, H_5IO_6, is a weak, polyprotic acid obtained from evaporating HIO_4 solutions.
- Metaperiodic acid, HIO_4, is a strong monoprotic acid produced when H_5IO_6 loses water.

Section 22.14   Group 8A: Noble Gases

Description: Noble gases are unreactive gases. They have complete outer valence shells and complete energy levels. They become more reactive as they become larger, because their outer electrons are less tightly held than those closer to the nucleus. Even the larger noble gases will react only with fluorine, the most electronegative of all the elements.

Putting It Together

When sulfur dioxide reacts with aqueous hydrogen sulfide, elemental sulfur is produced. Write a balanced equation for the reaction. If a metric ton of coal containing 3.3% sulfur by mass is burned, how much hydrogen sulfide is needed to remove the sulfur dioxide? How many kilograms of elemental sulfur are produced? What volume of sulfur dioxide gas is produced at 150 °C and 740 mm Hg?

Self–Test

This section is intended to test your knowledge of the material covered in this chapter. Think through these problems, and make certain you understand what they are asking. Make sure your answers make sense. Successful completion of these problems indicates that you have mastered the material in this chapter. You will receive the greatest benefit from this section if you use it as a mock exam, as this will allow you to determine which topics you need to study in more detail.

True–False

1. The periodic table is divided into metals on the left, nonmetals on the right, and semimetals in the center.

2. Metallic character decreases from top to bottom in a column.

3. Second-row elements have unique bonding properties because of their small sizes and high electronegativities.

4. It is common to find double (π) bonds in third- and fourth-row elements.

5. Boranes can be connected with bridging hydrogen atoms.

6. Carbon allotropes include nonconducting covalent solids and sheets with delocalized electrons that will conduct electricity.

7. Silicates are major components of the Earth's crust and often consist of lengthy chains of double-bonded SiO_4^{4-} tetrahedra.

8. Nitrogen is obtained in pure form by fractional distillation of liquefied air.

9. White phosphorus is made up of tetrahedra of P atoms with strained, unstable bonds.

10. When it is liquefied, sulfur decreases steadily in viscosity to its boiling point.

Multiple Choice

11. Which of the following are characteristic of metals?
    a. brittle
    b. low ionization energy
    c. form covalent hydrides
    d. readily become anions

12. Because of their small size, second-row elements are able to form a maximum of _____ bonds.
    a. 2
    b. 3
    c. 4
    d. 5

13. Boron halides are able to act as Lewis acids because
    a. the empty *p* orbital can accept a lone pair from another atom
    b. they readily give up hydrogen ions in solution
    c. they are so reactive that they split water molecules
    d. all of the above

14. Nitrogen can exist in which of the following oxidation states?
    a. −5, −4, −3, 0, +1,+2
    b. −3, −2, −1, 0
    c. +1, +2, +3, +4, +5
    d. −3, −2, −1, 0, +1, +2, +3, +4, +5

15. Phosphorus is found in
    a. bones and teeth
    b. cell membranes
    c. DNA and RNA
    d. all of the above

16. 6A elements have the electron configuration
    a. ns^2np^2
    b. ns^2np^3
    c. ns^2np^4
    d. ns^2np^5

17. The halogen that is capable of the most flexibility of bonding is
    a. F   b. Cl   c. Br   d. I

18. The most electronegative elements in the periodic table are the
    a. halogens   b. noble gases   c. metals   d. boranes

19. The only elements that will form triple bonds are
    a. N and P   b. C and N   c. Br and I   d. O and N

20. Which of the following correctly identifies the highest abundances
    a. crust: silicon, atmosphere: oxygen, ocean: phosphorus
    b. crust: silicon, atmosphere: nitrogen, ocean: oxygen
    c. crust: phosphorus, atmosphere: oxygen, ocean: nitrogen
    d. crust: calcium, atmosphere: oxygen, ocean: hydrogen

Fill-in-the-Blank

21. From left to right across the periodic table, effective nuclear charge_____,

    ionization energy _____, atomic radius _____, and metallic

    character _____.

22. Carbon has a number of allotropes, ranging from the tetrahedral covalent solid

    _____, to the sheets of _____, to the remarkable spheres of

    _____.

23. When phosphorus is manufactured, an unstable allotrope known as

    _____ is produced. It is highly reactive because of the

    _____ pi bonds in the P_4 tetrahedra. If it is heated in the absence of

    air, it is converted to the more stable, less toxic, polymeric form known as

    _____.

24. Silicates are typically found as chains of rings with a basic unit of _____.

25. The properties of sulfur are dramatically temperature dependent. At room temperature, sulfur

    is a _____. When heated, it melts into a _____, then becomes

    more viscous as the _____ break and form _____. When these begin

    to break down, the _____ of the liquid begins to decrease with temperature.

Matching

26. Aluminosilicate   a. commercial process for making nitric acid from ammonia

27. Borane   b. purification technique in which a heater melts a narrow zone at the top of a rod of a material, then sweeps slowly down the rod, carrying impurities

28. Carbide

   c. silicate mineral in which partial substitution of Al^{3+} for Si^{4+} has occurred

29. Contact process

   d. any compound of boron and hydrogen

30. Ostwald process

   e. an ionic compound containing silicon oxoanions along with other cations

31. Silicate

   f. commercial process for making sulfuric acid from sulfur

32. Zone refining

   g. binary compound in which carbon has a negative oxidation state

Problems

33. For the following pairs of elements, determine which has the more metallic character.
    a. Na or Rb   b. In or Ga   c. Cs or Sr

34. For the following pairs of elements, determine which has the more ionic hydride.
    a. K or Na   b. Sn or In   c. Sb or As

35. What are the general trends down a group in the periodic table for ionization energy, atomic radius, electronegativity, and basicity of oxides?

36. Describe the unique bonding in diborane.

37. Describe four of the allotropes of carbon in terms of their bonding, structure, and properties.

38. Give an example of a nitrogen-containing compound for each common oxidation state exhibited by nitrogen.

39. Give the name of the halogen oxoacid HXO_n (X = Cl, Br, or I and n = 1, 2, 3, and 4).

40. Give the chemical equation for the production of NO_2 from nitric acid and copper.

Challenge Problem

41. Calcium carbide reacts with water to form acetylene (C_2H_2). If 15.0 g of calcium carbide react with 7.45 g of water, how many grams of acetylene are produced? If the resulting acetylene is burned to form carbon dioxide and water, how much energy is produced? (Assume 100% efficiency for all reactions.)

    Possibly useful information:

    $$\Delta H_f\ C_2H_2\ (g) = 227.4\ \text{kJ/mol}$$
    $$\Delta H_f\ CO_2\ (g) = -393.5\ \text{kJ/mol}$$
    $$\Delta H_f\ H_2O\ (g) = -241.8\ \text{kJ/mol}$$

CHAPTER TWENTY-THREE

Organic and Biological Chemistry

Learning Objectives

As a result of reading and studying this chapter, you should be able to

Section 23.1   **Organic Molecules and Their Structures: Alkanes**
1.   Represent the chemical structure of alkanes as condensed structures or line drawings.
2.   Identify and draw isomers of alkanes

Section 23.2   **Families of Organic Compounds: Functional Groups**
3.   Identify and name functional groups in organic molecules.
4.   Represent molecules with functional groups using line drawings.

Section 23.3   **Naming Organic Compounds**
5.   Convert between the IUPAC name and condensed formula or line drawing of alkanes and unsaturated compounds.

Section 23.4   **Carbohydrates: A Biological Example of Isomers**
6.   Classify a monosaccharide as an aldose or a ketose and draw a typical open-chain structure.
7.   Classify the type of isomerism in monosaccharides.

Section 23.5   **Valence Bond Theory and Orbital Overlap Pictures**
8.   Draw orbital overlap pictures to describe bonding in organic compounds.
9.   Identify and draw cis–trans isomers.

Section 23.6   **Lipids: A Biological Example of Cis–Trans Isomerism**
10.   Recognize and draw the structure of triacylglycerols.
11.   Convert between chemical structure and name of lipids.
12.   Relate lipid structure to physical properties.

Section 23.7   **Formal Charge and Resonance in Organic Compounds**
13.   Use curved arrows to depict electron movement to draw resonance structures and specify formal charges.

Section 23.8   **Conjugated Systems**
14.   Draw orbital overlap pictures for conjugated systems and identify localized and delocalized electrons.
15.   Identify conjugated systems and the hybridization of atoms included in the system.

Section 23.9   **Proteins: A Biological Example of Conjugation**
16.   Identify amino acids, show how they join together to form peptides, and name small peptides.

Section 23.10   **Aromatic Compounds and Molecular Orbital Theory**
     17.   Predict if an organic compound is aromatic.

Section 23.11   **Nucleic Acids: A Biological Example of Aromaticity**
     18.   Identify the bases found in the nucleic acids, draw structures
           showing how different components of a nucleic acid are joined,
           and write the DNA sequence that is complementary to a
           specified strand.
     19.   Draw resonance structures to show that DNA bases are aromatic.

Chapter Summary

Organic chemistry is the study of carbon compounds. In this chapter, we introduce you to the vast array of compounds and reactions in organic chemistry. We will describe naming of compounds, varying structures with multiple bonds, cyclic structures and functional groups, and the most basic reactions of these molecules. We will then provide a basic introduction to biochemical energetics and the main classes of biomolecules. Proteins, carbohydrates, lipids, and nucleic acids are discussed in terms of their basic building blocks, functions, and properties.

The Chapter in Detail

Section 23.1   Organic Molecules and Their Structures: Alkanes

There are a multitude of single-bonded compounds that contain only carbon and hydrogen. They are known as **hydrocarbons** and belong to a family of single-bonded organic molecules called **alkanes**.

Each carbon can form four bonds, either to hydrogen or to another carbon. There are only three possible structures for alkanes with one to three carbons, but at four carbons, branched-chain hydrocarbons become possible. This greatly increases the number of possible compounds. At six carbons, for example, there is one straight-chain alkane with no branches and four branched-chain alkanes.

Compounds such as these, with the same formula but different arrangements of atoms are called isomers. Despite their identical numbers and types of atoms, their differing structures give them different properties, both physical and chemical.

To reduce the complexity of drawing these molecules, condensed structures are used in which most single bonds are omitted, with only vertical bonds to branches shown:

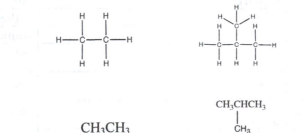

become

                                               CH_3CHCH_3
                                                     |
                 CH_3CH_3 CH_3

The condensed structure provides information about connectivity only. As molecules get larger, they are more and more able to adopt different conformations, but their connectivity remains the same. As a result, all of the following are different representations of the same molecule:

$$CH_3CH_3$$
$$|$$
$$CH_3CHCHCH_2CH_3$$

$$CH_3CHCHCH_2CH_3$$
$$/ |$$
$$CH_3CH_3$$

$$CH_3CH_2CHCHCH_3$$
$$/ \backslash$$
$$CH_3 \ CH_3$$

$$CH_3 \ CH_3$$
$$\backslash /$$
$$CH_3CH_2CHCHCH_3$$

$$CH_3$$
$$|$$
$$CH_3CH_2CHCHCH_3$$
$$|$$
$$CH_3$$

$$CH_3$$
$$\backslash$$
$$CH_3CHCHCH_2CH_3$$
$$|$$
$$CH_3$$

These are all 2,3-dimethylpentane.

Single bonds are free to rotate, so there are an infinite number of possible *conformations* for organic molecules. Most alkanes will have an extended zig-zag shape due to the tetrahedral arrangement of sp^3 orbitals.

Workbook Example 23.1

PROBLEM:
Draw the straight-chain alkane C_8H_{18} as a condensed structure.

SOLUTION:
Remembering that all carbons in an alkane must have four bonds, the structure is

$$CH_3CH_2CH_2CH_2CH_2CH_2CH_2CH_3$$

Section 23.2   Families of Organic Compounds: Functional Groups

Organic compounds can be classified into families according to structural features, and members of a given family also have similar reactivity. Alkanes, alkenes, alkynes, and the aromatic rings known as arenes contain only carbon–carbon bonds. Others contain bonds to nitrogen or oxygen: some contain bonds to halogens as well.

The following is a condensed version of the much more extensive chart on Page 985 of your book: for additional information, refer to that chart as well.

Family Name	Functional Group	Name Ending
Alkanes	Single bonds	-ane
Alkenes	One or more C=C bonds	-ene
Alkynes	One or more C≡C bonds	-yne
Arenes	C_6 ring, alternating C-C and C=C bonds	None
Alcohols	-C-OH	-ol
Ether	-C-O-C-	-ether
Amine	-C-N-	-amine
Aldehyde	O ‖ -C-C-H	-al
Ketone	O ‖ -C-C-C-	-one
Carboxylic acid	O ‖ -C-C-OH	-oic acid
Ester	O ‖ -C-C-O-C-	-oate
Amide	O ‖ -C-C-N-	-amide

Workbook Problem 23.1

Identify the functional groups in the following compounds:

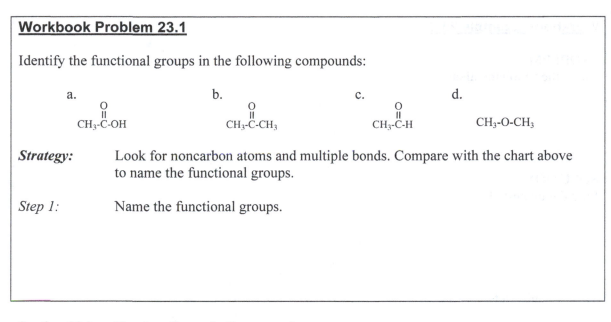

a.	b.	c.	d.
O	O	O	
‖	‖	‖	
CH_3-C-OH	CH_3-C-CH_3	CH_3-C-H	CH_3-O-CH_3

Strategy: Look for noncarbon atoms and multiple bonds. Compare with the chart above to name the functional groups.

Step 1: Name the functional groups.

Section 23.3   Naming Organic Compounds

Because of the large numbers and complexity of organic molecules, the International Union of Pure and Applied Chemistry (IUPAC) has come up with a standard method for naming organic molecules.

In this system, a molecule is named with a prefix that indicates its substituents, a parent name that indicates the number of carbons in the longest chain, and a suffix that indicates the chemical family of the molecule (these are the name endings in the chart above).

Straight-chain alkanes are names using Greek prefixes for those with chains of five carbons or more, thus, *pent*ane, *hex*ane, and so on. For historical reasons, chains with one to four carbons are named using unique prefixes:

    1 carbon = *meth-*   2 carbons = *eth-*   3 carbons = *prop-*   4 carbons = *but-*

Branched-chain alkanes are named using the following four steps:

- Name the parent chain. This requires finding the longest continuous carbon chain in the molecule, which may require taking a nonlinear path.
- Number the carbon atoms in the main chain, beginning at the end *nearest* the first branch point. The first branch point is at the lowest possible number.
- Identify and number the branching substituent. Sometimes, there will be more than one substituent on a carbon. In this case, assign the same number twice. Hydrocarbon substituents are called alkyl groups and are named using the same prefixes as the main chain, but with the suffix *–yl*. In other words, a four-carbon substituent is a *butyl* group.
- Write the name as a single word. Use hyphens to separate prefixes and commas to separate numbers if there are multiple substituents. If there are multiple identical substituents, name them as *di-* or *tri-*. *Trimethyl-*, for instance, indicates that there are three methyl group substituents. If there are multiple nonidentical substituents, name them in alphabetical order, ignoring any prefixes: *ethyl* comes before *dimethyl*.

Workbook Example 23.2

PROBLEM:
Name the following alkane:

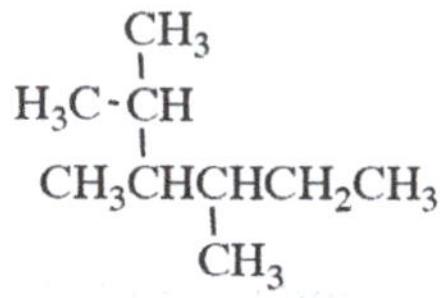

$$CH_3$$
$$H_3C\text{-}CH$$
$$CH_3CHCHCH_2CH_3$$
$$CH_3$$

SOLUTION:
Find the longest chain:

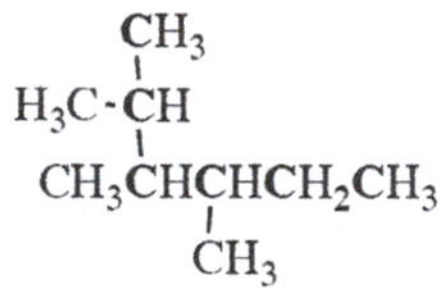

$$CH_3$$
$$H_3C\text{-}CH$$
$$CH_3CHCHCH_2CH_3$$
$$CH_3$$

This is a substituted *hexane*.

Number the chain so as to give the smallest numbers to the substituents:

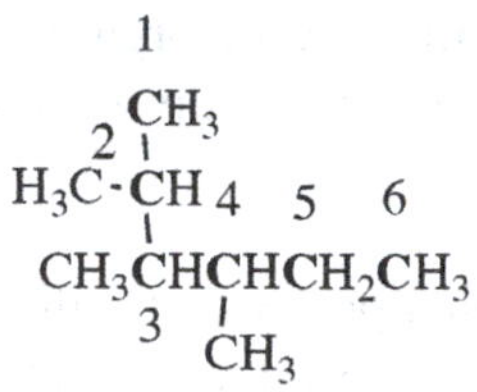

1
$$^2 CH_3$$
$$H_3C\text{-}CH\ 4\ \ 5\ \ 6$$
$$CH_3CHCHCH_2CH_3$$
$$3\ \ CH_3$$

There are *methyl* groups attached to carbons 2, 3, and 4. (Notice that if the chain had been numbered backward, these would be carbons 3, 4, and 5—always choose the lower numbers.)

The molecule's name is 2,3,4-trimethylhexane.

More About Alkyl Groups
Methyl and ethyl groups have only one possible attachment point, so where they attach is irrelevant. But with a three-carbon propyl group, the attachment can be at either end or in the middle, which gives an *isopropyl* group. Four-carbon butyl groups can be attached at either end or at carbon-2 to form *sec-butyl*. But with as few as four carbons, the substituent itself can be branched to form *isobutyl*, in which the substituent attaches at the end of the chain, or *tert-butyl*, in which the substituent attaches in the middle of the chain. The prefixes *sec-* and *tert-* indicate the number of carbons attached: *sec* indicates that there are two carbons attached to the branching carbon, and *tert-* indicates that there are three.

Workbook Problem 23.2

Name the following molecules:

a.
$$CH_3$$
$$CH_3CH_2CHCCH_3$$
$$CH_3\ CH_3$$

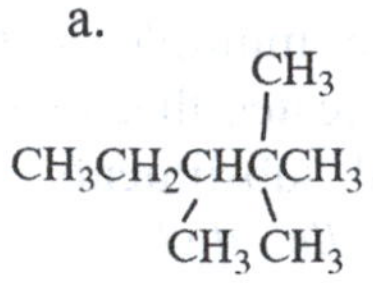

b.
$$CH_2CH_3$$
$$CH_3CH_2CH_2CCH_2CH_2CH_3$$
$$CH_2CH_2CH_3$$

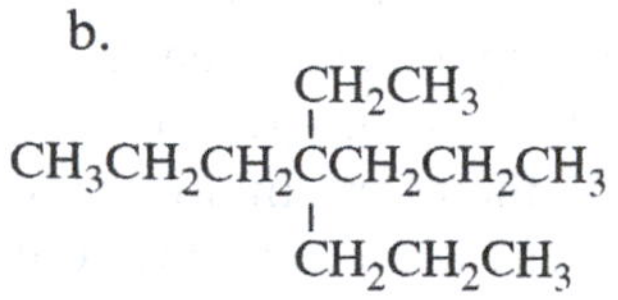

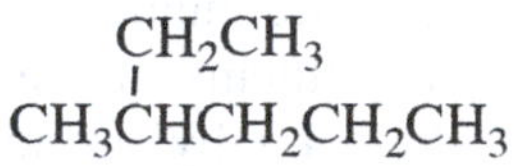

c.
$$CH_2CH_3$$
$$CH_3CHCH_2CH_2CH_3$$

Strategy: Use the naming strategy detailed above to identify the molecules.

Step 1:   Identify and name the parent chain.

Step 2:   Number the chain, beginning at the end closest to the first branch point.

Step 3:   Identify and number the substituents.

Step 4:   Write the name of the molecule as a single word.

Workbook Problem 23.3

Draw the following molecules:
- a. 4-isopropylheptane
- b. 2-methyl-3-ethylhexane
- c. 2,2-dimethylbutane
- d. 2,3-dimethyl-4-propyloctane

Strategy: Draw the carbon backbones, and add substituents in the proper places.

Step 1:   Draw the carbon backbones.

Step 2:   Add the substituents, subtracting hydrogens as necessary.

Unsaturated Organic Compounds: Alkenes and Alkynes

Alkenes are molecules that contain one or more double bonds, while alkynes contain one or more triple bonds. These are known as unsaturated hydrocarbons; saturated hydrocarbons are those that contain as many hydrogens bonded to the molecule as possible.

Alkenes are named by numbering the chain beginning closest to the double bond. If the double bond is in the center of the chain, numbering is begun closest to the substituent. The smallest alkenes are often named as –*ylene*; ethylene, for example, instead of ethene.

Isomers of alkenes can occur not only because of the position of the bond within the chain but also as a result of the position of substituents around the bond. Double bonds cannot rotate the way single bonds can because of their pi-bonds, which leads to the formation of **cis–trans** isomers. In *cis* isomers, identical substituents are on the same side of the double bond, while in *trans* isomers, identical substituents are on opposite sides of the double bond.

Workbook Example 23.3

PROBLEM:
Name the following compound:

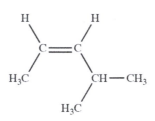

SOLUTION:
The longest chain has five carbons, and numbering begins at the left to minimize the number of the double bond. The substituents are both on the same side of the bond, and there is a methyl group on carbon 4. This is cis-4-methyl-2-pentene.

Alkynes are similar in many respects to alkenes and are named using the suffix –*yne*. The smallest alkyne, *ethyne*, is often called *acetylene*. The geometry of triple bonds is linear, so cis-trans isomerization does not occur.

The most common reactions of alkenes and alkynes are addition reactions in which a reagent adds to the multiple bond. Alkenes can be converted into alkanes by the addition of hydrogen, into dihalides by adding halogens, or into alcohols by adding water:

$$\text{Alkene} + H_2 \rightarrow \text{Alkane}$$

$$\text{Alkene} + X_2 \rightarrow \text{1,2-Dihaloalkane}$$

$$\text{Alkene} + H_2O \rightarrow \text{Alcohol}$$

Workbook Problem 23.4

Draw 4-ethyl-6-methyl-2-heptyne.

Strategy: Draw the carbon backbone, and add substituents in the proper places.

Step 1:   Draw the carbon backbone.

Step 2:   Add the substituents, subtracting hydrogens as necessary.

Workbook Problem 23.5

What are the reaction products from 3-hexene and

      a.   water in the presence of an acid
      b.   chlorine
      c.   hydrogen in the presence of a metal catalyst

Strategy: Draw the carbon backbone, and determine the possible products of the addition reactions.

Step 1:   Draw the carbon backbone.

Step 2:   Add the substituents.

Other Functional Groups

Alcohols have a hydroxyl (-OH) group in place of a hydrogen. Alcohols can form hydrogen bonds just like water. This makes small alcohols soluble in water and gives them higher boiling points than alkanes.

Alcohols are named by the point of attachment and are numbered beginning at the end of the chain closest to the alcohol group.

Methanol (CH_3OH) is also called wood alcohol. It is toxic, causing blindness in low doses and death in larger amounts. It is an important starting material for producing formaldehyde and acetic acid.

Ethanol is produced by fermentation of grains and sugars and is sometimes called grain alcohol. It is the alcohol present in alcoholic beverages. Enzymes in yeast break down sugars to make ethanol and CO_2.

Other alcohols include 2-propanol, also called isopropy alcohol or rubbing alcohol, which is a disinfectant; 1,2-ethanediol, also called ethylene glycol, used in automobile antifreeze; 1,2,3-propanetriol, also called glycerol, which is used as a moisturizing agent; and phenol, which is used in the manufacture of nylon, epoxy, and resins.

Ethers can be described as water derivatives in which both hydrogens have been replaced by organic substituents. They are relatively inert chemically, so they are often used as reaction solvents. Diethyl ether was used as an anesthetic agent for many years but has now been replaced by other less flammable alternatives.

Amines are organic derivatives of ammonia. They are names using the suffix *–amine* to describe the substituents on the nitrogen. Like ammonia, amines are weak bases that can use the lone pair of electrons on the nitrogen to accept a proton and form ammonium salts. These salts are more soluble than the neutral compounds.

Many drugs are amines: they become more soluble in body tissues if they are converted to their ammonium salts,

Carbon=oxygen double bonds are known generally as **carbonyl groups**. All biomolecules, most pharmaceuticals, and many synthetic polymers contain these groups. Carbonyl groups are polar, as a result of the electronegative oxygen atom, but can be even more polar if there is an additional substituent on the carbonyl carbon.

Aldehydes and ketones contain a carbonyl group at the end of a carbon chain for aldehydes and in the middle of a carbon chain for ketones. They are present in many biologically important compounds, as well as in small molecules such as formaldehyde, or, using the IUPAC naming system, methanal, which is used as a biological preservative and sterilizing agent, as well as a starting material for plastics manufacture. A common ketone is acetone, or propanone, a widely used organic solvent.

Carboxylic acids contain a $-\overset{\overset{\text{O}}{\|}}{\text{C}}-\text{OH}$ group and are found throughout the plant and animal kingdoms. Vinegar is primarily acetic or ethanoic acid, and long-chain carboxylic acids are constituents of all fats and oils. Systematic names replace the ending of the alkane parent chain with *–oic acid*.

These compounds are very weak acids, but will react with alcohols to produce an ester and water: the –OH of the acid and the H of the alcohol group combining to form water, leaving a bond through the oxygen of the alcohol.

Esters contain a $-\overset{\overset{\text{O}}{\|}}{\text{C}}-\text{O}-\text{C}$ functional group. Esters are common in pharmaceuticals, in polymers such as polyester, and in fragrance molecules.

The most common reaction of esters is conversion into carboxylic acids. They undergo a hydrolysis reaction in which a water molecule is added into the ester to form a carboxylic acid and water. Note: this is the reverse of the reaction described above for carboxylic acids. This reaction is catalyzed by both acid and base.

When this hydrolysis is base-catalyzed, it is also called *saponification*: soap is made by hydrolysis of long-chain esters in animal fat and is a mixture of the sodium salts of carboxylic acids.

Esters are named by identifying first the alcohol derivative and then the acid derivative, ending in *–ate*.

Amides contain a $-\overset{\displaystyle O}{\overset{\|}{C}}-N$ functional group. Amide bonds are the links that hold proteins together, as well as some polymers, particularly nylon. Amides are also found in some pharmaceuticals.

Amides are neutral because of the strong influence of the adjacent oxygen. This pulls the lone pair of electrons tightly to the nitrogen, preventing them from reacting.

Acid- or base-catalyzed hydrolysis of an amide generates a carboxylic acid and an amine. This is an important part of the digestion of protein.

Section 23.4   Carbohydrates: A Biological Example of Isomers

Carbohydrates occur in all living organisms. The name originated with carbon, with the empirical formula $C(H_2O)$, which led to the idea that it was a hydrate of carbon. Although that idea was discarded, the name persisted, and it is now used for the large class of hydroxyl-containing aldehydes and ketones that we commonly call *sugars*.

Monosaccharides, or *simple sugars,* are carbohydrates that cannot be broken into smaller molecules by acid hydrolysis. An *aldose* contains an aldehyde carbonyl group, while a *ketose* contains a ketone carbonyl group. The *–ose* suffix designates a sugar.

Monosaccharides are typically found in a ring form on which a hydroxyl group near one end of the chain combines with the carbonyl at or near the other end of the chain. When this happens, more than one form is possible depending on whether the following hydroxyl is above (β) or below (α) the ring. Crystalline forms of monosaccharides are generally in the α-form, but in solution, all three forms, α, β, and linear, are found.

Workbook Example 23.4

PROBLEM:
Classify the following monosaccharides:

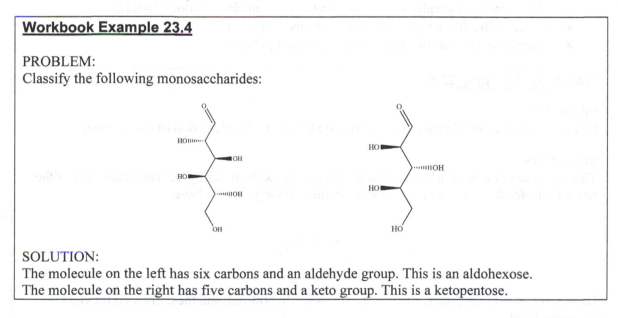

SOLUTION:
The molecule on the left has six carbons and an aldehyde group. This is an aldohexose.
The molecule on the right has five carbons and a keto group. This is a ketopentose.

Polysaccharides are formed when monosaccharides join together. Sucrose is a *disaccharide* formed of one glucose and one fructose molecule. Cellulose, the structural material in plants, is made up of thousands of β-glucose molecules joined together in an immensely long chain while starch is made up of thousands of α-glucose units. This small difference in configuration makes starches—found in rice, beans, and potatoes—digestible to us, and cellulose indigestible.

All living organisms do work, both physical and chemical. In animals, the energy for these processes comes from their food, which is broken down in interconnected reactions collectively called **metabolism**. Food molecules are oxidized to carbon dioxide, water, and energy. The reactions that break down larger molecules into smaller ones release energy and are known as **catabolism**. The reactions that build larger molecules from smaller ones absorb energy and are known as **anabolism**.

The four stages of catabolism:

- *Stage one* is digestion: large molecules are broken down into smaller molecules such as simple sugars, fatty acids, and amino acids.
- *Stage two* is the conversion of these small molecules into two-carbon $CH_3C=O$ acetyl groups that are connected to the carrier molecule *coenzyme A* through the carbonyl carbon. The resultant molecule is *acetyl coenzyme A* or *acetyl CoA*. This is an intermediate in the breakdown of all types of food molecules.
- *Stage three* is the oxidation of acetyl group in the *citric acid cycle* to yield carbon dioxide, water, and energy. This energy is used in stage four.
- *Stage four* is the *electron transport chain*. This takes the energy released by the citric acid cycle and uses it to make adenosine triphosphate (ATP), the "energy currency" of the cell.

Catabolic reactions make ATP, anabolic reactions "spend" it.

Section 23.5   Valence Bond Theory and Orbital Overlap Pictures

Valence bond theory uses an orbital overlap picture to show how electron pairs are shared in a chemical bond. In Chapter 8, you learned how to:

- Determine the hybridization of a central atom in an electron-dot structure.
- Describe the difference between a sigma and a pi bond.
- Sketch orbitals that overlap to form sigma and pi bonds.

Workbook Example 23.5

PROBLEM:
Draw the structure of ethanal (CH_3-CHO) and identify the hybridization of each carbon.

SOLUTION:
The -al ending of ethanal indicates that we are dealing with an aldehyde. The ethan- part of the name indicates that we have two carbons. Putting this together we have:

Based on this structure, the carbon on the left is sp^3-hybridized, and the carbon on the right is sp^2-hybridized.

Section 23.6   Lipids: A Biological Example of Cis–Trans Isomerism

A lipid is an organic molecule that dissolves in nonpolar solvent. The fact that lipids are defined by a physical property means that they exist in a variety of forms, but they all contain large hydrocarbon portions.

Animal fats and vegetable oils are the most abundant lipids. Fats and oils are both *triacylglycerols*, or *triglycerides*. These are esters of glycerol (1,2,3-propanetriol) with three long-chain carboxylic acids called *fatty acids*.

As shown below, fatty acids can be saturated, unsaturated, and *polyunsaturated:*

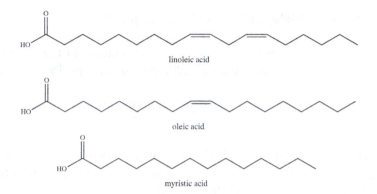

linoleic acid

oleic acid

myristic acid

It is the different properties of these fatty acids that lead to the different properties of fats and oils. Fats have primarily saturated fatty acids that pack together more efficiently and therefore have higher melting points. Oils contain primarily unsaturated fatty acids, but can be hydrogenated to yield saturated fats.

Section 23.7   Formal Charge and Resonance in Organic Compounds

Back in Chapter 7, we talked about formal charge and resonance. Let's revisit these now:

- **Formal charge** is the number of valence electrons in an isolated atom minus the number of valence electrons assigned to an atom in an electron-dot structure.

We calculate formal charge by using this formula:

$$\text{Formal charge} = \left(\begin{array}{c} \textit{Number of} \\ \textit{valence electrons} \\ \textit{in free atoms} \end{array} \right) = \frac{1}{2} \left(\begin{array}{c} \textit{number of} \\ \textit{bonding} \\ \textit{electrons} \end{array} \right) - \left(\begin{array}{c} \textit{number of} \\ \textit{nonbonding} \\ \textit{electrons} \end{array} \right)$$

- Different **resonance** forms of a substance differ only in the placement of bonding and nonbonding electrons. The connections between atoms and the relative positions of the atoms remain the same.

Section 23.8   Conjugated Systems

Conjugated systems occur when p orbitals are connected in compounds with alternating single and double bonds. When two pi-bonds are separated by a sigma-bond, we say that that system is conjugated.

In line drawings, electrons appear to be localized in bonds between two atoms, but in a conjugated system, the electrons are delocalized, or associated with more than one or two atoms in a molecule. Remember that molecular orbital theory describes delocalization very well.

In valence bond theory, the major principle is that atoms will adopt the hybridization that makes the molecule the most stable by lowering the overall energy. We follow two principles to make the most stable structure of a molecule:

- Conjugation is immensely stabilizing, so always look for conjugated systems first.
- If an atom is not conjugated, it will adopt a stable hybridization according to VSEPR theory.

Section 23.9   Proteins: A Biological Example of Conjugation

Proteins are major components of biological systems. Fifty percent of the dry weight of a human body is protein: our bodies contain more than 150,000 proteins that perform multiple functions.

Proteins can be structural, they can carry messages as hormones, and they can catalyze reactions as enzymes.

Proteins are polymers made up of individual units called *amino acids*. These small molecules contain an amide group, a central carbon, and a carboxylic acid group. They are linked together by amide bonds, which, in proteins, are referred to as *peptide bonds*.

Two amino acids linked together are known as a *dipeptide*; three as a *tripeptide*. Up to 100 amino acids linked together is known as a *peptide*, with the term *protein* reserved for chains that are even longer.

Twenty amino acids occur naturally, all of which have the basic structure below, with a variety of –R groups attached to the central carbon, known as the α-carbon. These are therefore also referred to as α-amino acids. Each is known by its name and a three-letter abbreviation (Serine = Ser, for example). There is a chart of all 20 on Page 1012 of the textbook.

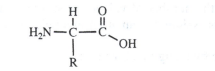

The –R group is also known as a *side chain*. Of the 20 amino acids, humans can synthesize only 11, with the remaining 9 *essential amino acids* required from dietary sources. Fifteen of the 20 have neutral side chains, two are acidic, and three are basic. Of the neutral amino acids, these can again be divided into polar and nonpolar or *hydrophobic* (nonpolar, literally, "water hating") and *hydrophilic* (polar, or "water-loving").

The number of possible combinations increases rapidly with the number of amino acids: with only three amino acids, six different tripeptides can be formed. All combinations will have an *N-terminal amino acid* with a free NH_2 group and a *C-terminal amino acid* with a free CO_2H group. By convention, sequences are written N-terminus on the left to C-terminus on the right.

Section 23.10   Aromatic Compounds and Molecular Orbital Theory

Aromatic molecules are given that name because the original members of this group were fragrant substances found in natural sources. However, it was soon realized that these behaved chemically differently than other organic molecules.

Aromatic now refers to molecules containing six-membered rings represented as having alternating single and double bonds. The reality of these molecules is that the carbon molecules are sp^2-hybridized, with a delocalized pi-bond that spreads over the entire molecule, generating a resonance structure.

Benzene (C_6H_6) is the most common aromatic molecule:

Substituted aromatic molecules are named using the suffix –*benzene*. Ethylbenzene, for example, has a single ethyl group. (It doesn't need a number, because all ring positions are equivalent.) If the benzene ring is itself a substituent of a larger molecule, it is given the name *phenyl*.

Disubstituted benzenes are named based on the relative positions of their substituents. *Ortho*-substituted benzenes have substituents directly next to one another on the ring, *meta*- substituted molecules have substituents separated by one carbon, and *para*-substituted benzenes have groups directly across the ring from one another.

Aromatic compounds typically undergo substitution reactions, because the benzene structure is so stable: one of the hydrogens on the ring is substituted by another group.

When benzene reacts with nitric acid in the presence of sulfuric acid, a *nitro* group is added to the ring. Nitrobenzene is a starting point for preparing dyes as well as explosives.

When halogens are added to a benzene ring in the presence of an iron-based catalyst (FeX_3, where X is the halogen being added), it yields a single-substituted benzene and HX.

Workbook Problem 23.6

Draw the following molecules:

        o-dichlorobenzene   *p*-methylnitrobenzene   *m*-ethylmethylbenzene

Strategy: Draw the benzene ring, and add substituents in the proper orientations.

Workbook Problem 23.7

Provide the possible products from the following reactions:

a.   *o*-dimethylbenzene with nitric acid in the presence of sulfuric acid (two possible products)
b.   *m*-bromochlorobenzene with chlorine in the presence of FeCl₃ (four possible products)

Strategy: Draw the reactant, and add the new substituent in all places that give a unique configuration.

Section 23.11   Nucleic Acids: A Biological Example of Aromaticity

Deoxyribonucleic acid (DNA) and ribonucleic acid (RNA) are the carriers of genetic information. Coded in an organism's nucleic acids is all the information needed to produce the many thousands of proteins required by that organism.

Nucleic acids are polymers built of nucleotide units. Nucleotides are built of nucleosides, which are amine bases linked to aldopentose sugars (ribose for RNA and deoxyribose for DNA) and phosphoric acid.

DNA and RNA share three amine bases:

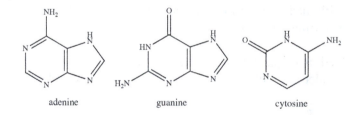

adenine   guanine   cytosine

Only DNA contains thymine:

thymine

Only RNA contains uracil:

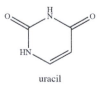

uracil

In both DNA and RNA, the amine base is bonded to the C1′ of the sugar, and the phosphoric acid is bonded to the C5′ carbon of the sugar. The "prime" on the C indicates a sugar carbon. Other numbered carbons are in the amine base. When the nucleotides are joined to make polymers, the phosphates join to the C3′ of the next nucleoside.

DNA sequences are listed beginning at the C5′ end, using the first letters of the nucleotides: TACG. DNA sequences from different tissues of the same species have the same proportions of each nucleotide, but sequences from different species may have very different proportions.

There are always equal amounts of A and T and G and C, regardless of species. In the Watson–Crick model of DNA structure, two complementary strands of DNA hydrogen-bond to each other, with A forming two hydrogen bonds with T, and G forming three hydrogen bonds with C. The two strands then coil around one another to form a double helix.

DNA is primarily found in the nuclei of cells coated with proteins and wound into *chromosomes*. Each chromosome contains several thousand **genes**, a segment of DNA coding for a specific protein. DNA is the storage mechanism for genetic information, while RNA is used to make proteins.

Replication is the process by which additional copies of DNA are made. The helix unwinds, and the bases are exposed. A new complementary strand is made for each half of the DNA, leading to two identical copies.

Transcription is the process by which the instructions in the DNA are copied for use. A single-stranded RNA copy of one of the two DNA strands is made, with uracil in the RNA taking the place of thymine. Once completed, the RNA separates from the template, and the DNA (with thymine replacing the uracil) helix reforms.

Translation is the conversion of the information contained in the nucleic acid template to protein. *Messenger RNA, or mRNA,* molecules travel to ribosomes where the RNA is read and the protein made. Each three-nucleotide sequence codes for a different amino acid. For instance, the sequence C-U-G codes for the amino acid leucine. A *transfer RNA, or tRNA,* has a complementary sequence and brings the next amino acid in to be joined to the chain. At the end of the mRNA there is a "stop" sequence that brings translation to a close.

Workbook Problem 23.8

Determine the mRNA sequence that would be produced from the following DNA sequence:

AGCTTGACACGTGTCACTAGA

How many amino acids would be coded for by this mRNA?

Strategy: Determine the complementary bases, and then determine the number of codons.

Putting It Together

Butyl butyrate is responsible for the distinctive scent of bananas. The percent composition of this ester is 67% carbon, 11% hydrogen, and 22% oxygen. Reaction with water yields butanol and an acid. The molar mass of the acid is 144 g/mol. What is the structure of butyl butyrate?

Self–Test

This section is intended to test your knowledge of the material covered in this chapter. Think through these problems, and make certain you understand what they are asking. Make sure your answers make sense. Successful completion of these problems indicates that you have mastered the material in this chapter. You will receive the greatest benefit from this section if you use it as a mock exam, as this will allow you to determine which topics you need to study in more detail.

True–False
1. Different isomers will have different physical and chemical properties.

2. A double or triple bond is not considered a functional group.

3. Whether an alkene is cis or trans has very little practical significance.

4. Alcohols are able to participate in hydrogen bonding.

5. Organic carboxylic acids are strong acids.

6. The links between amino acids are amide bonds, but when they are in peptides they are referred to as peptide bonds.

7. Lipids are defined by their physical properties, not by their structure.

8. Monosaccharides can be aldoses or ketoses.

9. All polysaccharides are readily digested to form glucose monomers.

10. Nucleic acids are the primary structural components of plant cell walls.

Multiple Choice
11. In naming alkanes, the parent chain is
    a. the longest straight chain
    b. the longest continuous chain
    c. numbered to give the maximum numbers for substituents
    d. b and c

12. When alkenes react with halides
    a. one halogen adds to the side of the double bond
    b. the double bond is completely broken, and two molecules are formed
    c. one halogen adds to each carbon of the double bond
    d. a strong acid is needed as a catalyst

13. A three-carbon alkane is called
    a. propane   b. triane   c. propene   d. butane

14. The smallest cycloalkane without stress in the bonds is
    a. propane   b. butane   c. pentane   d. hexane

15. Aromatic rings
    a. contain six-membered carbon rings   b. have delocalized *p* electrons
    c. are flat   d. all of the above

16. Alcohols and ethers can both be considered water derivatives, but they differ in that
    a. ethers have double bonds to oxygen
    b. alcohols have one hydrogen of the water substituted, ethers have both
    c. ethers contain nitrogens
    d. alcohols have double bonds to oxygen

17. Which of the following are carbonyl compounds?
    a. aldehydes and ketones   b. amides
    c. carboxylic acids   d. all of the above

18. Which of the following places the molecules in order from smallest to largest?
    a. protein, tripeptide, peptide   b. amino acid, tripeptide, dipeptide
    c. amino acid, dipeptide, protein   d. peptide, protein, amino acid

19. Complex carbohydrates are
    a. nonpolar   b. long chains
    c. all digestible by humans   d. a and b

20. Lipids
    a. are soluble in nonpolar solvents.
    b. are important components of cell membranes
    c. often contain fatty acids bonded to a glycerol backbone
    d. all of the above

Fill-in-the-Blank

21. Because of carbon's flexible _____ there are more than 40 million organic

    compounds. These are organized into families according to the _____ they contain.

22. _____ contain only hydrogen and _____. _____

    have only single bonds, _____ have at least one double bond, and

    _____ have at least one _____ bond.

23. When six-membered rings alternate single and _____ bonds, the result is a(n)

    _____ ring.

24. _____ are both structural components of plants and important sources of

    _____ for animals.

25. Proteins are made up of long chains of _____ _____. When they catalyze reactions in biological systems, they are called _____.

26. Nucleic acids are an essential part of biological systems. In _____, copies of the molecules are made. In transcription, _____ molecules are made that are then _____ into _____.

Matching

27. Alcohol

28. Aldehyde

29. Alkane

30. Alkene

31. Alkyne

32. Amino acid

33. Amide

34. Amine

35. Anabolism

36. Aromatic

37. Carbonyl group

38. Catabolism

39. Cycloalkane

40. Enzyme

41. Ester

a. protein monomer containing an –NH$_2$ group and a –COOH group

b. metabolic processes that build larger molecules from smaller ones

c. metabolic reactions that break larger molecules into smaller molecules, releasing energy

d. carbonyl group with hydrogen on one side and an alkyl group on the other

e. hydrocarbon containing only single bonds that contains a ring structure

f. C=O

g. molecule that contains two alkyl groups bonded to a single-bonding oxygen

h. organic derivative of ammonia

i. hydrocarbon containing at least one double bond

j. carboxylic acid with a long hydrocarbon tail

k. organic molecule containing an –OH group

l. molecule containing two alkyl groups on either side of a carbonyl group

m. organic molecule containing an alkyl group and a nitrogen bonded to a carbonyl group

n. total of all the reactions that go on in a cell

o. carbohydrate that cannot be broken down by acid hydrolysis

42. Ether

p. organic molecule that contains a $-\overset{\overset{\displaystyle O}{\|}}{C}-O-R$

43. Fatty acid

q. containing a benzene ring

44. Hydrocarbon

r. hydrocarbon containing only single bonds

45. Ketone

s. protein that catalyzes biological reactions

46. Lipid

t. molecule that contains only hydrogen and carbon

47. Metabolism

u. making RNA molecules from DNA templates

48. Monosaccharide

v. hydrocarbon containing at least one triple bond

49. Peptide

w. having as many hydrogens as possible, containing only single bonds

50. Replication

x. naturally occurring molecule that is soluble in non-polar solvents

51. Saturated

y. mRNA to protein

52. Transcription

z. not having as many hydrogens as possible: containing one or more multiple bonds

53. Translation

aa. process by which DNA molecules are exactly copied

54. Unsaturated

bb. chain of two to 100 amino acids

Problems
55. Draw the following compounds:
   a. *meta*-dibromobenzene
   c. 2-hexanone
   e. 4-ethyl-6-methyl-1-heptene

   b. 2,3,4-trimethyl-1-hexanol
   d. 3-ethyl-2,4-dimethylpentanal

56. Name the following compounds:

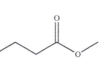

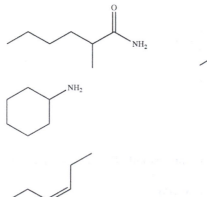

57. Identify the functional groups in the following molecules:

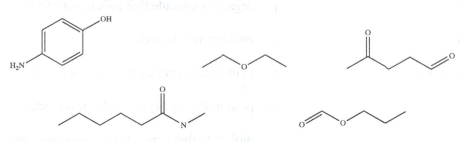

58. Predict the results of the following reactions giving all possible products:
    a. benzene plus nitric acid in the presence of sulfuric acid
    b. ethylbenzene plus chlorine in the presence of FeCl₃
    c. 3-hexene plus water
    d. 2-butene plus hydrogen
    e. ethene and chlorine
    f. dehydration of ethanoic acid and ethanol
    g. *N*-methylacetamide plus acid

N-methylacetamide

59. Briefly describe the four stages of catabolism.

60. Draw the two possible dipeptides that can be formed by alanine and serine.

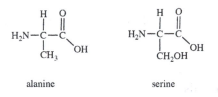

alanine   serine

61. List the six possible tripeptides that can be formed by alanine (ala), leucine (leu), and histidine (his).

62. Classify the following monosaccharides:

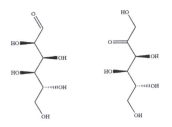

63. Draw the basic structure of a saturated triglyceride.

64. From what DNA sequence was the following mRNA transcribed? For how many amino acids does it code?

UAGCACUGAUCAGUAAGC

Challenge Problem

65. Citric acid is an unsaturated, tricarboxylic acid. When 5.00 g of the acid undergoes combustion, 6.87 g of CO_2 and 1.87 g of H_2O are produced. When a 0.312 g sample of citric acid is titrated with 0.100 M sodium hydroxide, 48.8 mL are needed for complete titration. Determine the molecular formula of the acid.

APPENDIX

Solutions to Study Guide Problems

Chapter 1 – Chemical Tools: Experimentation and Measurement

Workbook Problems

Workbook Problem 1.1

Strategy: Consider the differences in the degree size and zero-point adjustments for the Fahrenheit and Celsius scales.

Step 1:   A Fahrenheit degree is smaller than a Celsius degree. Just think about the melting (freezing) point of water. Water melts at 0° on the Celsius scale and at 32° on the Fahrenheit scale. Therefore, when converting from Fahrenheit to Celsius, the temperature should be lower.

Step 2:   Apply the formula for the conversion of °F to °C. (Remember, for this conversion, you do a zero-point correction followed by a size correction.)

$$°C = \frac{5}{9} \times (102 - 32) = 38.9\ °C$$

Step 3:   Does your answer in step 2 agree with your answer in step 1?
               We predicted that the Celsius temperature would be lower, and it is.

Step 4:   Apply the formula for the conversion of °C to K.

$$K = 38.9\ °C + 273.15\ \frac{K}{°C} = 312\ K$$

Workbook Problem 1.2

Strategy: Think about the information you have been given. Set up a mathematical equation that allows you to solve for the mass of lead. Remember that the definition of density is mass per unit volume (or mass divided by volume).

$$Density\left(\frac{g}{mL}\right) = \frac{Mass\ of\ sample\ (g)}{Volume\ of\ sample\ (mL)}$$

Step 1:   Substitute the information given in the problem into the mathematical equation you set up.

$$11.34\frac{g}{mL} = \frac{mass\ of\ lead\ (g)}{53.43\ mL}$$

Step 2:   Solve for the mass of lead.

$$\left(11.34\,{}^{g}\!\!\diagup\!\!{}_{mL}\right)(53.43\,mL) = 605.9\,g\,Pb$$

Workbook Problem 1.3

Strategy: Following the example and Workbook Problem 1.2 set up an equation to determine the volume of the cylinder.

$$Density\left(\frac{g}{mL}\right) = \frac{Mass\,of\,sample\,(g)}{Volume\,of\,sample\,(mL)}$$

Step 1:   Solve for the volume of the cylinder.

$$1.977\frac{g}{L} = \frac{4.84365\,g}{volume\,of\,sample\,(L)}$$

$$volume\,of\,sample\,(mL) = \frac{4.84365\,g}{1.977\frac{g}{L}} = 2.451365706\,L$$

Step 2:   Apply the rules for multiplication and division to determine the number of significant numbers in your answer.

When carrying out either multiplication or division, your answer cannot have more significant figures than either of the original numbers. 1.977 has four significant figures; 4.84365 has six significant figures. Your answer can't have more than four significant figures.

Step 3:   If necessary, use the rules for rounding-off numbers.

We must round off 2.451 365 706 L to four significant figures. The first digit we must remove is a 3; therefore the number can simply be truncated at four digits The final answer is 2.451 L.

Workbook Problem 1.4

Strategy: Using the conversion table on the back cover of your book, determine the conversion factor for meters to miles and convert time to one unit.

$$1\,mi = 1.6093\,km$$

$$27\,min \times \frac{60\,s}{1\,min} = 1620\,s$$

$$1620\,s + 46\,s = 1666\,s$$

Step 1: Calculate the amount of time it will take for the athlete to run that distance, and convert back to minutes and seconds.

$$10 \text{ km} \times \frac{1 \text{ mi}}{1.6093 \text{ km}} \times \frac{1666 \text{ s}}{5 \text{ mi}} = 2070 \text{ s}$$

$$2070 \text{ s} \times \frac{1 \text{ min}}{60 \text{ s}} = 34.5 \text{ mins}$$

Self–Test

True–False
1. F. A theory is formed only after a hypothesis has been tested several times and is shown to be correct.
2. T. Remember that leading zeroes only place the decimal—they are not significant—but internal zeroes *are* significant.
3. F. Mass is the amount of matter in an object, while weight is the gravitation pull on that object. Mass and weight are often used interchangeably because it is rare to be making measurements in any gravitational field but Earth's, but it is important to understand the distinction.
4. F. The units most commonly used in the laboratory to measure volume are the liter (dm^3) and the milliliter (cm^3).
5. T
6. T
7. T
8. F. The SI unit for temperature is degree Celsius.
9. T
10. F. Density describes an object's mass and its volume.

Multiple Choice

11. c	12. d	13. a	14. d
15. c	16. a		

Matching

17. d	18. f	19. a	20. c
21. b	22. e		

Fill-in-the-Blank
23. theory
24. kilogram, kelvin, meter
25. kelvin, Celsius, Fahrenheit
26. density
27. accurate, precise
28. scientific notation
29. round, significant figures
30. conversion factor
31. the number with the fewest digits to the right of the decimal
32. giga, G
33. mass
34. volume

Problems

35. $246\,cm \times \dfrac{10\,mm}{1\,cm} = 2460\,mm$ $2460\,mm \times \dfrac{1\,m}{1000\,mm} = 2.46\,m$

$2.46\,m \times \dfrac{1\,km}{1000\,m} = 2.46 \times 10^{-3}\,km$

36. a. $\dfrac{8.3}{2.793} = 3.0$ (The least amount of significant figures, two, is found in the numerator.)

      b. $23.945\ 62 \times 8.3421 = 199.76$ (The least amount of significant figures, five, is found in 8.3421.)

      c. $3.2 + 5.3875 = 8.6$ (There is only one digit to the right of the decimal point in the number 3.2, so the answer must be rounded off to one digit after the decimal point.)

      d. $9434.21 - 4.199 = 9430.01$ (There are only two digits to the right of the decimal point in 9434.21.)

37. $°F = \left(\dfrac{9}{5} \times 56\right) + 32 = 130\,°F$

38. The conversion factor needed to solve this problem is 1 mile = 1.6093 km

$$290,587\,mi^2 \times \left(\dfrac{1.6093\,km}{1\,mi}\right)^2 = 752,576\,km^2$$

39. The conversion factor needed to solve this problem is 1×10^9 nm = 1 m.

$$6.73 \times 10^{-7}\,m \times \dfrac{1 \times 10^9\,nm}{1\,m} = 673\,nm$$

40. $78.5\,°C + 273.15 = 351.6\,K$ $°F = \left(\dfrac{9}{5} \times 78.5\right) + 32 = 173°F$

41. The conversion factors needed to solve these problems are 1 mile = 1.6093 km and 1 hour = 3600 s

Sprinter: $\dfrac{100\,m}{9.58\,s} \times \dfrac{3600\,s}{1\,hr} \times \dfrac{1\,km}{1000\,m} \times \dfrac{1\,mi}{1.6093\,km} = 23.4\,{mi}/{hr}$

Swimmer: $\dfrac{100\,m}{47.85\,s} \times \dfrac{3600\,s}{1\,hr} \times \dfrac{1\,km}{1000\,m} \times \dfrac{1\,mi}{1.6093\,km} = 4.68\,{mi}/{hr}$

42. density = mass/volume

density = $\dfrac{9.634\ g}{(16.2\ mL\ -\ 11.5\ mL)} = 2.049787$ g/mL = 2.05 g/mL

43. It is important to realize that the speed of light, 3.00×10^8 m/s, can also be used as a conversion factor. 3.00×10^8 m = 1 s

$$13.6 \times 10^9 \text{ km} \times \frac{1000 \text{ m}}{1 \text{ km}} \times \frac{1 \text{ s}}{3.00 \times 10^8 \text{ m}} \times \frac{1 \text{ hour}}{3600 \text{ s}} = 12.6 \text{ hours}$$

44. The first problem is figuring out the volume of the pycnometer based on the mass of the water.

37.82 g – 23.75 g = 14.07 g water. The mass of this water is 14.07g/0.9982g/mL = 14.095 mL of water. This is the volume of the pycnometer.

When the beryllium is put in, we need to solve for the mass of the remaining water:
Total mass – mass of pycnometer – mass of beryllium = mass of remaining water
40.72 g – 23.75 g – 6.345 g = 10.625 g water.

The volume of this water will be 10.625 g/0.9982 g/mL = 10.644 mL

The difference between this volume and the total volume of the apparatus is the volume of the metal.

Beryllium volume = 14.095 mL – 10.644 mL = 3.451 mL beryllium.

Density = mass/volume = 6.345 g/3.451mL = 1.84 g/mL (At this point, the answer is rounded off to the correct number of significant figures.)

45. Volume = mass/density. 10.67g/4.55 g/mL = 2.35 mL

46. This problem requires a system of equations. Let x = grams of oil, and y = grams of trichloroethane.

Assume 100 g total solution: $x + y = 100$ g

If there is 100 g of total solution, and the liquids have been mixed so as to give the same density as water, it is possible also to determine the volume of the 100 g of solution:

100g/0.997 g/mL = 100.3mL

It is also possible to solve for the volumes of each of the components.
$x/0.918 + y/1.3492 = 100.3$ mL

Multiply through both sides by 1.3492: $1.4697x + y = 135.32$

Set up a system of equations:
$$\begin{array}{r} 1.4697x + y = 135.32 \\ - \quad \underline{x + y = 100} \\ 0.4697x \quad = 35.32 \end{array}$$

$x = 75.2$ g $y = 24.8$ g

47. 0 °C is the freezing point of water, while 0 K is the coldest possible temperature, also known as absolute zero.

48. The numbers in a calculation are not rounded or truncated until the final step. When a number is rounded because of significant figures, it can be changed slightly. If each number along the way is rounded, error will be introduced at each step. For example, if a box is measured as being 2.57 cm long, 3.634 cm wide, and 1 cm tall, the volume would properly be calculated 2.57 cm × 3.634 cm × 1 cm = 9.339 cm³, which then is rounded to 9 cm³. If an overzealous student were instead to round all the numbers to one significant figure before beginning the calculation, the answer would be 12 cm³, a significant error.

Challenge Problem

49. On a mercury thermometer, the liquid range of gallium is 2373.22 °C (the difference in temperature between the melting and boiling points). On the gallium thermometer, the liquid range of gallium is 1000 °C. The size adjustments for these two scales are

$$1°C \times \frac{1000\,°Ga}{2373.22\,°C} = 0.421\,°Ga \qquad 1°Ga \times \frac{2373.22\,°C}{1000\,°Ga} = 2.37\,°C$$

Therefore, 1 °C is 0.421 times smaller than 1 °Ga, and 1 °Ga is 2.37 times larger than 1°C. The melting point of gallium is higher by 29.78° on the Celsius scale, so we would first do the size adjustment followed by the zero-point adjustment when converting from °Ga to °C.

$$°C = (2.37 \times °Ga) + 29.78$$

Now convert the melting point of copper from °Ga to °C:

$$°C = (2.37 \times 447\,°Ga) + 29.78 = 1091\,°C$$

We can determine the accuracy of the thermometer by comparing the difference between the converted and reported value to the reported value.

$$\frac{(|1085° - 1091°|)}{1085°} \times 100\% = 0.52\%$$

Chapter Two – Atoms, Molecules, and Ions

Workbook Problems

Workbook Problem 2.1

Strategy: Determine the carbon-to-hydrogen mass ratios for both compounds, and compare these ratios to each other.

Step 1:   Determine the carbon-to-hydrogen ratio for the first compound.

$$\text{First compound: } C:H \text{ mass ratio} = \frac{3.43\,g\,C}{0.857\,g\,H} = 4.00$$

Step 2:   Determine the carbon-to-hydrogen ratio for the second compound.

$$\text{Second compound: } CH \text{ mass ratio} = \frac{4.80\,g\,C}{0.400\,g\,H} = 12.0$$

Step 3:   Divide the ratio found for the first compound by the ratio for the second compound.

$$\frac{C:H \text{ mass ratio in 1st compound}}{C:H \text{ mass ratio in 2nd compound}} = \frac{4.0}{12.0} = 0.33$$

Step 4:   Can your answer be converted to a small whole-number ratio?

$$\text{Yes. } 0.33 = \frac{33}{100} = \frac{1}{3}$$

Workbook Problem 2.2

Strategy: Use the periodic table to determine the atomic number and the number of neutrons in each isotope. Give the standard symbol for each.

Step 1:   First, it's necessary to know the chemical symbol for oxygen. The chemical symbol for oxygen is O.

Step 2:   Use the periodic table to determine the atomic number for oxygen. From the periodic table, we find that the atomic number for O is 8.

Step 3:   The number of neutrons can be determined from the definition for mass number.

A (mass number) = Z (atomic number) + number of neutrons.
number of neutrons = $A - Z$

For oxygen-16: number of neutrons = 16 – 8 = 8
For oxygen-17: number of neutrons = 17 – 8 = 9
For oxygen-18: number of neutrons = 18 – 8 = 10

Step 4:   The standard symbol is written with the mass number as a superscript and the atomic number as a subscript, both to the left of the symbol.

$$^{16}_{8}O \qquad\qquad ^{17}_{8}O \qquad\qquad ^{18}_{8}O$$

Workbook Problem 2.3

Strategy: Determine the atomic number from the information given.

We are told that element X has 51 protons; therefore, the atomic number of the element is 51.

Step 1:   Now you know the atomic number, identify the element by using the periodic table.

From the periodic table, we find that the element with $Z = 51$ is Sb (antimony).

Step 2:   The standard symbol is written with the mass number as a superscript and the atomic number as a subscript, both to the left of the symbol.

The mass number is determined by adding the number of protons and the number of neutrons.

$A = 51 + 70 = 121$. The standard symbol is $^{121}_{51}\text{Sb}$.

Workbook Problem 2.4

Strategy: Determine the number of grams of gold in the ring and find the conversion factors needed to convert from grams of sample to number of atoms.

If the ring is 14/24 gold, and weighs a total of 3.65 grams, there are 14/24 × 3.65 grams of gold present:

$$\frac{14}{24} \times 3.65 = 2.13 \, g \, gold$$

We are beginning with grams and want to end up with atoms. The first conversion factor we need is one that will allow us to cancel out the unit grams.

$$\frac{1 \, amu}{1.66054 \times 10^{-24} \, g}$$

Using this conversion factor leaves us with the unit, amu. We now need a conversion factor that will cancel out the unit, amu.

$$\frac{1 \, atom \, Au}{196.7 \, amu}$$

(196.7 amu is the atomic mass for gold from the periodic table; one atom of gold weighs this much.)   This conversion factor leaves us with the desired unit.

Step 1:   Set up a mathematical equation such that all of the units except number of atoms cancel.

$$2.13 \, g \times \frac{1 \, amu}{1.66054 \times 10^{-24} \, g} \times \frac{1 \, atom \, Au}{196.7 \, amu} = 6.52 \times 10^{21} \, atoms \, of \, Au$$

Workbook Problem 2.5

Strategy: Remember, the atomic mass of an element = Σ(mass of each isotope × the abundance of the isotope), and the abundances must add up to 100%.

Step 1: Use the information given to determine the abundance of ^{204}Pb.

The abundance of ^{204}Pb is not given, but the abundances of the other isotopes are, making it a straightforward process to find the missing abundance:

Abundance ^{204}Pb + 52.4% + 22.1% + 24.1% = 100%
Abundance ^{204}Pb = 1.4%

Step 2: Use the information given to determine the average atomic mass.

$$(0.524 \times 207.977\,amu) + (0.221 \times 206.976\,amu) + (0.241 \times 205.974\,amu) +$$
$$(0.014 \times 203.973\,amu) = 207.217\,amu$$

Workbook Problem 2.6

Step 1: Determine whether the formula contains both metals and nonmetals or only nonmetals.
  a) HI – two nonmetals   b) NO_2 – two nonmetals
  c) $Ba(OH)_2$ – metal and nonmetal   d) PCl_3 –two nonmetals

Step 2: Identify the compounds as either molecular or ionic, based on your conclusions in Step 1.
  a) , b), and d) – molecular compounds   c) – ionic compound

Step 3: Determine if the substance is capable of providing either an H^+ or OH^- ion in water.
  a) HI – produces H^+ when dissolved in water.
  b) NO_2 – cannot produce either H^+ or OH^- when dissolved in water.
  c) $Ba(OH)_2$ – produces OH^- when dissolved in water.
  d) PCl_3 – cannot produce either H^+ or OH^- when dissolved in water.

Step 4: Identify the compounds as either acids or bases (if applicable) based on your conclusions in step 3.
  a) HI – acid   c) $Ba(OH)_2$ – base

Workbook Problem 2.7

Step 1: When naming compounds, determine if the metal is a main-group metal or a transition metal. If it is a transition metal, you must indicate the charge on the metal when naming the compound. The charge on the metal is determined from the number and charge on the anion. (Remember, you must maintain electrical neutrality.)

MgCl₂   Magnesium chloride. (Mg^{2+} is a group 2A metal and has only one charge.)

CaO   Calcium oxide. (Ca^{2+} is a group 2A metal and has only one charge.)

TiI₄   Titanium is a transition metal; therefore, we need determine and note the charge on the metal ion. Iodine is a group 7A nonmetal, so it carries a charge of –1. Because there are four iodide ions, the name of the compound is titanium(IV) iodide.

CoF₃   Cobalt is a transition metal. We have three fluorides for a total charge of –3. To maintain electrical neutrality, the total charge on the cobalt ions must be +3. The name of the compound is cobalt(III) fluoride.

CuSe   Copper (II) selenide. (Se is a group 6A metalloid and carries a charge of –2.)

Step 2:   When writing formulas, use the periodic table or the name of the compound to determine the charge on the metal. Use the periodic table to determine the charge on the anion. (Remember to make sure that all charges add up to zero.)

sodium chloride	Sodium is a group 1A metal and forms only Na^+. Chloride has a –1 charge. For the charges to balance, one of each is needed: NaCl.
chromium (III) oxide	The name tells us that the charge on the chromium cation is +3. Using the periodic table, we find that the charge on the oxide anion is –2. The lowest common multiple of these charges is 6, which means that there will need to be six positive charges and six negative charges. These requirements are fulfilled by two chromium ions and three oxygen ions: Cr_2O_3.
bismuth (II) nitride	Bismuth has a charge of +2, nitride has a charge of –3. Again, the lowest common multiple is 6, so that six positive and six negative charges are needed: Bi_3N_2.
aluminum sulfide	Aluminum is a group 3A metal and forms Al^{3+} cations. The sulfide anion has a –2 charge. Al_2S_3.

Step 3:   Make sure the formula contains the smallest whole-number ratio of cation to anion. The formulas above all have the smallest whole-number ratios of cation to anion.

Workbook Problem 2.8

Step 1:   When naming the molecules, remember that the element that is more anion-like uses the –*ide* suffix. Also, remember to use numerical prefixes to indicate the number of each atom.

PF$_6$ dinitrogen tetroxide   selenium dioxide   NBr$_3$ HF (g)

PF$_6$ phosphorus hexafluoride.
NBr$_3$ nitrogen tribromide.
HF (g) The formula was written to indicate that this binary hydrogen
          compound is a gas. Therefore, it is named as hydrogen fluoride gas.

Step 2:   Use the correct numerical prefixes:

dinitrogen tetroxide   N$_2$O$_4$
selenium dioxide   SeO$_2$

Workbook Problem 2.9

Step 1:   When naming the compounds, refer to your textbook for the names of
polyatomic ions, but begin memorizing them. Also, don't forget that you must
indicate the charge on certain metal cations.

Fe(NO$_3$)$_3$ NO$_3$ is the nitrate ion and has a charge of –1. Since we have
three nitrate ions, we have an overall charge of –3. Therefore,
the charge on the iron must be +3. Iron is a transition metal, and
we must include the charge on the metal when naming the
compound. The name of the compound is iron(III) nitrate.

Al$_2$(SO$_4$)$_3$ SO$_4$ is the sulfate ion and has a charge of –2. Since we have
three sulfate ions, we have an overall charge of –6. Therefore,
the overall charge on the metal must be +6. There are two
aluminum ions, which must each carry a charge of +3.
Aluminum is a main-group metal, so a charge of +3 is expected;
it need not be denoted with a Roman numeral. This is aluminum
sulfate.

Step 2:   When writing formulas, use the periodic table or the name of the compound to
determine the charge on the metal. To determine the formula for a polyatomic
ion, refer to your textbook. Remember, you must maintain electrical neutrality.

potassium chromate   The formula for chromate is CrO$_4^{2-}$. Potassium is a
group 1A metal and has a charge of +1. To maintain
electrical neutrality, we need to have two potassium
cations for every one CrO$_4^{2-}$ anion. The formula is
K$_2$CrO$_4$.

magnesium phosphate   The formula for phosphate is PO$_4^{3-}$ (this is the only
polyatomic ion with a charge of –3 that you need to
be concerned about). Magnesium is a group 2A metal
and has a charge of +2. To balance the charges, three
magnesium ions are needed for every two phosphate
ions. The formula is Mg$_3$(PO$_4$)$_2$

ammonium nitrate	The formula for ammonium is NH_4^+. The formula for nitrate is NO_3^-. There is no reason in the world that two polyatomic ions cannot bind together to form an ionic compound, but ammonium is the only polyatomic ion you need to worry about that carries a positive charge. The formula for the compound is NH_4NO_3.

Self–Test

True–False
1. F. *Ions* are the same element carrying a different charge; *isotopes* are the same element with differing mass numbers.
2. T. This is the law of conservation of mass.
3. T. Broken atoms no longer retain their identity as elements.
4. F. Nonmetals have a strong tendency to form multiple compounds—think of carbon monoxide and carbon dioxide—and so must be named using prefixes.
5. F. Protons and neutrons are found in the nucleus, electrons in a cloud around the nucleus.
6. T. If the number of protons in a nucleus is changed, the identity of the atom also changes.
7. F. Atoms of the same element can have different masses.
8. F. The ratio of elements in an ionic compound is identified by charge balance.
9. F. Transition metals can carry a variety of charges. Main group metals have invariant charges.

Multiple Choice

10. c	11. d	12. a	13. b
14. d	15. c	16. b	17. a
18. c	19. b		

Fill-in-the-Blank
20. atoms, protons, neutron, and electrons   21. protons, atomic number
22. neutrons, isotope, electrons, ion   23. compounds, mixtures
24. covalent, shared   25. metal, nonmetal, ionic, opposite charges
26. polyatomic ion, nitrate, ammonium, sulfate   27. di-, tri-, penta-
28. reactants, law of conservation of mass

Matching

29. g	30. i	31. b	32. m
33. a	34. q	35. j	36. c
37. l	38. o	39. r	40. k
41. d	42. p	43. n	44. e
45. h	46. g		

Problems
47. mass of oxygen $3.2 \text{ L oxygen} \times \dfrac{1.43 \text{ g}}{\text{L}} = 4.58 \text{g oxygen}$

hydrogen + 4.58 g oxygen → 5.15 g water

mass of hydrogen = 5.15 g – 4.58 g = 0.57 g hydrogen.

volume of hydrogen $= \dfrac{0.57 \text{ g}}{0.0893 \text{ g/L}} = 6.4 \text{ L}$

48. Compound 1: $O:P = \dfrac{11.4 \text{ g O}}{4.3 \text{ g P}} = 2.65$ Compound 2: $O:P = \dfrac{5.33 \text{ g O}}{3.34 \text{ g P}} = 1.59$

$$\dfrac{O:P \text{ ratio in compound 2}}{O:P \text{ ratio in compound 1}} = \dfrac{1.59}{2.65} = 0.6 = \dfrac{6}{10} = \dfrac{3}{5}$$

49. a) $p = 42$, $n = 55$, $e = 42$ b) $p = 92$, $n = 143$, $e = 86$
    c) $p = 93$, $n = 142$, $e = 93$ d) $p = 80$, $n = 122$, $e = 76$

50. Abundances must all add up to 100%:
    Abundance of Neon-22 = 100% – 90.48% – 0.27% = 9.25%
    Solve for mass of Neon-22
    $= (19.992 \, amu \times 0.9048) + (20.993 \, amu \times 0.0027) + (x \, amu \times 0.0925) = 20.18 \, amu$

51. Average atomic mass
    $= (0.7899 \times 23.985 \text{ amu}) + (0.1000 \times 24.986 \text{ amu}) + (0.1101 \times 25.983 \text{ amu}) = 24.305 \text{ amu}$

52. Naphthalene = C_8H_{10}

53. a) LiCl – lithium chloride   b) Na_2S – sodium sulfide
    c) CaO – calcium oxide   d) Mg_3N_2 – magnesium nitride

54. a) carbon monoxide – covalent   b) strontium chloride – ionic
    c) carbon tetrachloride – covalent   d) platinum (IV) oxide – ionic

55. $9500 \text{ atoms} \times \dfrac{196.97 \text{ amu}}{1 \text{ atom}} \times \dfrac{1.6605 \times 10^{-24} \text{ g}}{1 \text{ amu}} = 3.11 \times 10^{-18} \text{ g}$

56. In dichloromethane, there are two chlorine atoms, two hydrogen atoms, and one carbon atom. From the periodic table, the masses of these atoms add up to

$$(2 \times 35.5) + (2 \times 1.00) + 12.01 = 85.01 \text{ amu}$$

The percentage of chlorine will be $(2 \times 35.5)/85.01 \times 100 = 83.5\%$ Cl
The percentage of hydrogen will be $(2 \times 1.00)/85.01 \times 100 = 2.35\%$ H
The percentage of carbon will be $12.01/85.01 \times 100 = 14.1\%$ C

57. a. CaF_2 b. $Ca_3(PO_4)_2$ c. $Ca(NO_3)_2$ d. $CaCO_3$

58. a. tin (IV) chloride   b. phosphoric acid   c. magnesium acetate
    d. hydrogen chloride   e. rubidium sulfide   f. carbon dioxide
    g. chromium (III) phosphate h. difluorine monoxide   i. phosphorous pentoxide
    j. nickel (II) chloride   k. iron (III) hydroxide   l. hydrochloric acid
    m. lead (II) iodide   n. nitric acid   o. barium bromide
    p. aluminum nitrate

59. a. $Ca(NO_2)_2$ b. NaCl   c. Cl_2O d. H_2O
    e. NiH_2 f. BBr_3 g. $CuCl_2$ h. XeF_4
    i. NO_2 j. Ba_3N_2 k. $H_3PO_4(aq)$ l. HF (aq)
    m. $LiMnO_4$ n. H_2CO_3 (aq) o. PCl_3 p. CBr_4

Challenge Problem

60. To calculate the volume of the sphere, it is necessary first to find the radius. It is also a good idea to convert the radius into centimeters, as the final desired unit is cm^3.

Diameter = 0.2 mm
Radius = 0.2 mm/2 × 1cm/10mm = 0.01cm
Volume of a sphere = $4/3\pi r^3$ = 4.19×10^{-6} cm^3

With the volume and the densities of the two different metals, it is possible to calculate the masses of the spheres, then convert the masses to amu and the amu to atoms:

$$Osmium = 22.61 \text{ g/cm}^3 \text{ x } 4.19 \text{ x } 10^{-6} \text{ cm}^3 \times \frac{1 \text{ amu}}{1.6605 \times 10^{-24} \text{ g}} \text{ x } \frac{1 \text{ atom}}{190.23 \text{ amu}} = 2.999 \times 10^{17} \text{ atoms}$$

$$Aluminum = 2.70 \text{ g/cm}^3 \text{ x } 4.19 \text{ x } 10^{-6} \text{ cm}^3 \times \frac{1 \text{ amu}}{1.6605 \times 10^{-24} \text{ g}} \text{ x } \frac{1 \text{ atom}}{26.98 \text{ amu}} = 2.525 \times 10^{17} \text{ atoms}$$

If these atoms were the size of sand grains, 1 mm square, what volume would they take up?

Volume per "atom" = (0.1 mm × 1 cm/10 mm)^3 = (0.01 cm)^3 = 1×10^{-6} cm^3
Osmium: 2.999×10^{17} atoms ×1 × 10^{-6} cm^3 = 2.999×10^{11} cm^3 = 299,900,000 liters
Aluminum: 2.525×10^{17} atoms ×1 ×10^{-6} cm^3 = 2.525×10^{11} cm^3 = 299,900,000 liters

To put this in perspective, an Olympic-sized swimming pool holds about 2,500,000 liters, so if the atoms in this miniscule sphere were expanded to the size of small grains of sand, they would be equivalent in volume to around 120 Olympic-sized swimming pools. Atoms are really, really tiny.

Chapter Three – Mass Relationships in Chemical Reactions

Workbook Problems

Workbook Problem 3.1

Step 1:   Remember, the term *combustion* is used to indicate reaction with oxygen. When hydrocarbons (compounds containing primarily C and H) undergo a combustion reaction, carbon dioxide and water are the products.

$$C_6H_{14} + O_2 \rightarrow CO_2 + H_2O$$

Step 2:   Use coefficients to balance the equation. (Remember, it helps to save oxygen for last.)

Begin with carbon. There are 6 carbons on the reactant side, but only 1 carbon on the product side. Place a 6 in front of CO_2.

$$C_6H_{14} + O_2 \rightarrow 6\,CO_2 + H_2O$$

There are 14 hydrogens on the reactant side, but only 2 hydrogens on the product side. Place a 7 in front of H_2O.

$$C_6H_{14} + O_2 \rightarrow 6\,CO_2 + 7\,H_2O$$

Now balance the oxygens. There are 2 oxygens on the reactant side and a total of 19 on the product side. This odd number of oxygens will require a doubling of everything we have done so far to make it even:

$$2\,C_6H_{14} + O_2 \rightarrow 12\,CO_2 + 14\,H_2O$$

Now there are 2 oxygens on the reactant side, and 38 oxygens on the product side. Place a 19 in front of the oxygen on the reactant side:

$$2\,C_6H_{14} + 19\,O_2 \rightarrow 12\,CO_2 + 14\,H_2O$$

Step 3:   Reduce the coefficients to their smallest whole-number ratio.

The ratio is 2:19:12:14. This cannot be reduced, as 19 is a prime number.

Step 4:   Check your answer.

Reactant side	*Product side*
12 C	12 C
28 H	28 H
38 O	38 O

Workbook Problem 3.2

Part A

Step 1: Write an unbalanced chemical equation based on the information given in the problem.

$$H_2O_2 \ (aq) \rightarrow H_2O \ (l) + O_2 \ (g)$$

Step 2: Use coefficients to balance the equation.

$$2 \ H_2O_2 \ (aq) \rightarrow 2 \ H_2O \ (l) + O_2 \ (g)$$

Step 3: Reduce the coefficients to their smallest whole-number ratio if necessary.

The ratio is 2:2:1 and cannot be reduced further.

Step 4: Check your answer.

Reactant side	Product side
4 H	4 H
4 O	4 O

Part B

Step 1: Determine the mole ratio for hydrogen peroxide and oxygen from your balanced chemical equation.

For every two moles of hydrogen peroxide, two moles of water will be produced: this is a 1:1 ratio.

For every two moles of hydrogen peroxide, one mole of oxygen gas will be produced: this is a 2:1 ratio

Step 2: Use the mass and molecular weight of the hydrogen peroxide solution to determine the moles of hydrogen peroxide present.

$$60 \ g \ x \ \frac{1 \ mol}{34.02 \ g} = 1.76 \ mol$$

Step 3: Use the mole ratio from above as a conversion factor, and calculate the number of moles of oxygen generated by the decomposition reaction.

$$1.76 \ mol \ H_2O_2 \ x \ \frac{1 \ mol \ O_2}{2 \ mol \ H_2O_2} = 0.88 \ mol \ O_2$$

Step 4: Use the mole ratio from above as a conversion factor, and calculate the number of moles of water generated by the decomposition reaction.

$$1.76 \ mol \ H_2O_2 \ x \ \frac{1 \ mol \ H_2O}{1 \ mol \ H_2O_2} = 1.76 \ mol \ H_2O$$

Ballpark check: The number of moles of water generated should be two times the number of moles of oxygen generated based on the ratio found in the balanced equation.

Workbook Problem 3.3

Step 1:   Write an unbalanced chemical equation from the information given in the problem:

$$Fe + O_2 \rightarrow Fe_2O_3 \text{ (this provides a little nomenclature review!)}$$

Step 2:   Balance the equation.

$$4Fe + 3O_2 \rightarrow 2Fe_2O_3$$

Step 3:   From the balanced chemical equation, determine the mole-to-mole ratio of iron to iron (III) oxide.

4 mol Fe : 2 mol Fe_2O_3 can be reduced (or not) to 2 mol Fe : 1 mol Fe_2O_3

Step 4:   Find the atomic mass of iron.

From the periodic table, Fe = 55.85 g/mol

Step 5:   Determine the molecular mass for iron (III) oxide.

$$(55.85 \text{ g/mol} \times 2) + (16.00 \times 3) = 159.7 \text{ g/mol}$$

Step 6:   From the information in Steps 3, 4, and 5, create conversion factors. Using these conversion factors, set up a dimensional-analysis problem to convert from grams of iron to moles of iron, to moles of iron oxide, to grams of iron oxide.

$$14 \text{ g Fe} \times \frac{1 \text{ mol Fe}}{55.85 \text{ g}} \times \frac{2 \text{ mol Fe}_2O_3}{4 \text{ mol Fe}} \times \frac{159.7 \text{ g Fe}_2O_3}{\text{mol Fe}_2O_3} = 20.0 \text{ g Fe}_2O_3$$

Step 7:   Using the law of mass conservation, how many grams of oxygen were added to the iron?

There are two ways to solve this: the easy way, as indicated in the question, and the hard way. The easy way to solve this is to see that on the process of rusting, the iron has gained 6.0 grams, and that must have come from the only other reagent, oxygen.

The hard way is to do another stoichiometry problem like the one above:

$$14 \text{ g Fe} \times \frac{1 \text{ mol Fe}}{55.85 \text{ g}} \times \frac{3 \text{ mol O}_2}{4 \text{ mol Fe}} \times \frac{32.0 \text{ g O}_2}{\text{mol O}_2} = 6.0 \text{ g O}_2$$

As you continue through chemistry, you will often find situations like this, where there are numerous ways to get to the same answer. Always look for the easiest: $20 - 14 = 6$ beats the alternative any day.

Workbook Example 3.4

Step 1:   Balance the equation.

$$3H_2 + N_2 \rightarrow 2NH_3$$

Step 2:   Calculate the theoretical yield using dimensional analysis.

$$18.0 \text{ g N}_2 \times \frac{1 \text{ mol N}_2}{28.02 \text{ g}} \times \frac{2 \text{ mol NH}_3}{1 \text{ mol N}_2} \times \frac{17.01 \text{g NH}_3}{\text{mol NH}_3} = 21.85 \text{ g}$$

Step 3:   Calculate the percent yield using the actual yield stated in the problem and the theoretical yield just calculated.

$$\% \text{ yield } = \frac{\text{actual yield}}{\text{theoretical yield}} \times 100 = \frac{14.6 \text{ g}}{21.85 \text{ g}} \times 100 = 66.8\%$$

Workbook Problem 3.5

Step 1:   Using the information given, write an unbalanced chemical equation.

$$Al + HCl \rightarrow AlCl_3 + H_2$$

Step 2:   Balance the chemical equation.

$$2Al + 6HCl \rightarrow 2AlCl_3 + 3H_2$$

Step 3:   Determine the number of moles of each reactant present.

$$5.3 \text{ g Al} \times \frac{1 \text{ mol Al}}{27.0 \text{ g Al}} = 0.196 \text{ mol Al}$$

$$9.2 \text{ g HCl} \times \frac{1 \text{ mol HCl}}{36.5 \text{ g HCl}} = 0.252 \text{ mol HCl}$$

Step 4:   Determine which reactant gives the smallest number of moles of product. This gives the limiting reagent and the theoretical yield.

$$0.196 \text{ mol Al} \times \frac{2 \text{ mol AlCl}_3}{2 \text{ mol Al}} = 0.196 \text{ mol AlCl}_3$$

$$0.252 \text{ mol HCl} \times \frac{2 \text{ mol AlCl}_3}{6 \text{ mol HCl}} = 0.084 \text{ mol AlCl}_3$$

The limiting reagent is HCl, and the theoretical yield is 0.084 mol $AlCl_3$.

Step 5:   Determine how many moles of the excess reagent were used.

$$0.252 \text{ mol HCl} \times \frac{2 \text{ mol Al}}{6 \text{ mol HCl}} = 0.084 \text{ mol Al used}$$

Step 6:   Determine how many moles of the excess reagent remain.

$$\underline{0.196 \text{ mol Al} - 0.084 \text{ mol Al} = 0.112 \text{ mol Al remaining}}$$

Step 7:   Determine how many grams of the excess reagent are left over.

$$0.112 \text{ mol Al} \times \frac{27.0 \text{ g Al}}{\text{mol Al}} = 3.02 \text{ g Al left over}$$

Step 8:   Determine the percent yield of the reaction.

First, convert the theoretical yield to grams:

$$0.084 \text{ mol AlCl}_3 \times \frac{133.5 \text{ g AlCl}_3}{\text{mol AlCl}_3} = 11.2 \text{ g AlCl}_3$$

$$\% \text{ yield} = \frac{8.3 \text{ g AlCl}_3}{11.2 \text{ g AlCl}_3} \times 100 = 74\%$$

NOTE: The same answer could be obtained by converting the actual yield to moles:

$$8.3 \text{ g AlCl}_3 \times \frac{1 \text{ mol AlCl}_3}{133.5 \text{ g AlCl}_3} = 0.062 \text{ mol AlCl}_3$$

$$\% \text{ yield} = \frac{0.062 \text{ mol AlCl}_3}{0.084 \text{ mol AlCl}_3} \times 100 = 74\%$$

Workbook Problem 3.6

Step 1:   Convert the grams of each element into moles of each element.

$$18.9\% \text{ Li} = 18.9 \text{ g Li} \times \frac{1 \text{ mol Li}}{6.94 \text{ g Li}} = 2.72 \text{ mol Li}$$

$$16.2\% \text{ C} = 16.2 \text{ g C} \times \frac{1 \text{ mol C}}{12.01 \text{ g C}} = 1.35 \text{ mol C}$$

$$64.9\% \text{ O} = 64.9 \text{ g O} \times \frac{1 \text{ mol O}}{16.0 \text{ g O}} = 4.06 \text{ mol O}$$

Step 2:   Find the mole ratios.

There are no obvious ratios here—nothing is 1:1—so it makes sense to determine how many there are of each element relative to carbon, the element of which there is the least. Assume there is one carbon, and go from there:

$$\frac{2.72 \text{ mol Li}}{1.35 \text{ mol C}} = 2\frac{\text{Li}}{\text{C}}$$

$$\frac{4.06 \text{ mol O}}{1.35 \text{ mol C}} = 3\frac{\text{O}}{\text{C}}$$

Step 3:   If necessary, convert the ratios to whole numbers, and write the empirical formula.

Li_2CO_3 —this is lithium carbonate.

Workbook Problem 3.7

Step 1:   Determine the formula of the compound.

NaOCl – this is common household bleach

Step 2:   Determine the molecular weight and the weights of each component

Na	=	23
O	=	16
Cl	=	35.5
--		
Molecular weight		74.5

Step 3:   Determine the percent composition by dividing the mass of each element present by the total mass of the compound and multiplying by 100.

$$\%Na = \frac{23}{74.5} \times 100 = 30.9\%$$

$$\%O = \frac{16}{74.5} \times 100 = 21.5\%$$

$$\%Cl = \frac{35.5}{74.5} \times 100 = 47.6\%$$

Workbook Problem 3.8

Step 1:   Find the molar amounts of C and H in CO_2 and H_2O.

$$5.70 \text{ g CO}_2 \text{ x} \frac{1 \text{ mol CO}_2}{44 \text{ g CO}_2} = 0.129 \text{ mol CO}_2 \text{ x} \frac{1 \text{ mol C}}{\text{mol CO}_2} = 0.129 \text{ mol C}$$

$$2.31 \text{ g H}_2\text{O x} \frac{1 \text{ mol H}_2\text{O}}{18 \text{ g H}_2\text{O}} = 0.128 \text{ mol H}_2\text{O x} \frac{2 \text{ mol H}}{\text{mol H}_2\text{O}} = 0.257 \text{ mol H}$$

Step 2:   Carry out mole-to-gram conversions to find the number of grams of C and H in the original sample.

$$0.129 \text{ mol C x} \frac{12 \text{ g C}}{\text{mol C}} = 1.55 \text{ g C}$$

$$0.257 \text{ mol H x} \frac{1.008 \text{ g H}}{\text{mol H}} = 0.259 \text{ g H}$$

Step 3:   Subtract the masses of C and H from the mass of the starting sample to determine the mass of S.

$$2.85\text{g} - 1.55\text{g} - 0.259\text{g} = 1.04\text{g}$$

Step 4:   Convert the mass of S to moles of S.

$$1.04 \text{ g S x} \frac{1 \text{ mol S}}{32 \text{ g S}} = 0.0325 \text{ mol S}$$

Step 5:   Find the mole ratios.

$$\frac{0.129 \text{ mol C}}{0.0325 \text{ mol S}} = 4\frac{C}{S}$$

$$\frac{0.257 \text{ mol H}}{0.0325 \text{ mol S}} = 8\frac{H}{S}$$

Step 6:   Use the ratios above to write the empirical formula of the compound.

$$C_4H_8S$$

Step 7:   Compare the empirical formula weight to the molecular formula weight to see if a conversion is needed.

The empirical formula weight is 88 amu—no conversion is needed.

Step 8:   Write the molecular formula.

The molecular formula is C_4H_8S. Phew!

Self–Test

True–False
1. F. The number of atoms of each type must be the same, but the number of molecules is often different.
2. T
3. T
4. F. The coefficients in a balanced chemical equation show the *mole* ratios of the components.
5. F. Theoretical yields of chemical reactions are determined through mathematical calculation, not experiment.
6. T
7. F. Combustion analysis can only be used to determine the empirical formula of a compound.
8. T

Matching

9. g	10. c	11. i	12. d
13. j	14. a	15. b	16. f
17. e	18. j		

Fill-in-the-Blank
19. balanced
20. coefficients, subscripts
21. molecular mass or formula mass
22. limiting reactant
23. excess reactant
24. theoretical yield, actual yield, percent yield
25. empirical formula, molecular formula, molecular mass
26. carbon dioxide, water
27. Avogadro's number, 6.02×10^{23}

Problems

28. a. $2N_2 + 5O_2 \rightarrow 2N_2O_5$
    b. $C_3H_8 + 5O_2 \rightarrow 3CO_2 + 4H_2O$
    c. $H_3PO_4 + 3NaOH \rightarrow Na_3PO_4 + 3H_2O$
    d. $Ba(NO_3)_2 + 2NaOH \rightarrow Ba(OH)_2 + 2NaNO_3$

29. a. 45 g calcium chloride – formula is $CaCl_2$, molecular mass = $40 + (2 \times 35.5) = 111$ g/mol
    b. 23 g sucrose ($C_{12}H_{22}O_{11}$) – molecular mass = $(12 \times 12) + (22 \times 1.008) + (11 \times 16) = 342$ g/mol
    c. 67 g phosphorus triflouride – formula is PF_3, molecular mass = $31 + (19 \times 3) = 88$ g/mol
    d. 5 grams ammonium carbonate – formula is $(NH_4)_2CO_3$, molecular mass is $(18 \times 2) + 60 = 96$ g/mol

$$5 \text{ g } (NH_4)_2CO_3 \times \frac{1 \text{ mol } (NH_4)_2CO_3}{96 \text{ g } (NH_4)_2CO_3} = 0.052 \text{ mol } (NH_4)_2CO_3$$

30. To determine whether there is a limiting reactant, run the reaction mathematically using both reagents.

$$2N_2 + 5O_2 \rightarrow 2N_2O_5$$

$$4 \text{ mol } N_2 \times \frac{2 \text{ mol } N_2O_5}{2 \text{ mol } N_2} = 4 \text{ mol } N_2O_5$$

$$10 \text{ mol } O_2 \times \frac{2 \text{ mol } N_2O_5}{5 \text{ mol } O_2} = 4 \text{ mol } N_2O_5$$

The reactants are present in stoichiometric ratios, so the theoretical yield is 4 mol of N_2O_5. The molecular mass of dinitrogen pentoxide is 108 g/mol, so the theoretical yield in grams is

$$4 \text{ mol } N_2O_5 \times 108 \frac{g}{mol} = 432 \text{ g}$$

If the actual yield is 376 grams, the percent yield is calculated

$$\frac{376 \text{ g}}{432 \text{ g}} \times 100 = 87\%$$

31. Balance the equation.

$$2NaHCO_3 \rightarrow Na_2CO_3 + CO_2 + H_2O$$

Convert the 20 g of sodium hydrogen carbonate to moles: the formula weight is 84.

$$20 \text{ g NaHCO}_3 \times \frac{1 \text{ mol NaHCO}_3}{84 \text{ g NaHCO}_3} = 0.238 \text{ mol NaHCO}_3$$

Set up a stoichiometric ratio to determine the theoretical yield of water:

$$0.238 \text{ mol NaHCO}_3 \times \frac{1 \text{ mol } H_2O}{2 \text{ mol NaHCO}_3} = 0.119 \text{ mol } H_2O$$

$$0.119 \text{ mol } H_2O \times \frac{18 \text{ g } H_2O}{\text{mol } H_2O} = 2.14 \text{ g } H_2O$$

If the reaction has a typical percent yield of 87%, then only 87% of the theoretical yield can be expected.

$$2.14 \text{ g } H_2O \times 0.87 = 1.86 \text{ g } H_2O$$

32. Write and balance the equation.

$$CH_4 + 2O_2 \rightarrow CO_2 + 2H_2O$$

Convert the quantities given to moles:

$$6.7 \text{g CH}_4 \times \frac{1 \text{ mol CH}_4}{16 \text{ g CH}_4} = 0.419 \text{ mol CH}_4$$

$$20 \text{ g O}_2 \times \frac{1 \text{mol O}_2}{32 \text{g O}_2} = 0.625 \text{ mol O}_2$$

Run the reaction with each quantity to determine the limiting reagent and the theoretical yields.

$$0.419 \text{ mol CH}_4 \times \frac{1 \text{ mol CO}_2}{\text{mol CH}_4} = 0.419 \text{ mol CO}_2 \times \frac{44 \text{ g CO}_2}{\text{mol CO}_2} = 18.4 \text{ g CO}_2$$

$$0.419 \text{ mol CH}_4 \text{ x} \frac{2 \text{ mol H}_2\text{O}}{\text{mol CH}_4} = 0.838 \text{ mol H}_2\text{O x} \frac{18 \text{ g H}_2\text{O}}{\text{mol H}_2\text{O}} = 15.1 \text{ g H}_2\text{O}$$

$$0.625 \text{ mol O}_2 \text{x} \frac{1 \text{ mol CO}_2}{2 \text{ mol O}_2} = 0.313 \text{ mol CO}_2 \text{x} \frac{44 \text{ g CO}_2}{\text{mol CO}_2} = 13.8 \text{ g CO}_2$$

$$0.625 \text{ mol O}_2 \text{x} \frac{2 \text{ mol H}_2\text{O}}{2 \text{ mol O}_2} = 0.625 \text{ mol H}_2\text{O x} \frac{18 \text{ g H}_2\text{O}}{\text{mol H}_2\text{O}} = 11.3 \text{ g H}_2\text{O}$$

The lower theoretical yields come from the 20 g of oxygen, so this is the limiting reagent, and the theoretical yields are 13.8 g of carbon dioxide and 11.3 g of water.

To determine how many grams of methane will be left over, set up another stoichiometry problem to determine how many moles are consumed.

$$0.625 \text{ mol O}_2 \times \frac{1 \text{ mol CH}_4}{2 \text{ mol O}_2} = 0.313 \text{ mol CH}_4 \text{ consumed}$$

0.419 moles of methane were available, 0.313 were consumed, leaving 0.106 moles.

$$0.106 \text{ moles CH}_4 \text{ x} \frac{16 \text{ g CH}_4}{\text{mol CH}_4} = 1.7 \text{ g CH}_4 \text{ left over}$$

33. Carbon dioxide
CO_2 molecular weight = 44 amu

$$\%C = \frac{12}{44} \text{x} 100 = 27.3\% \qquad\qquad \%O = \frac{32}{44} \text{x} 100 = 72.7\%$$

Sodium phosphate
Na_3PO_4 molecular weight = 164 amu

$$\%Na = \frac{69}{164} \text{x} 100 = 42.1\% \qquad \%P = \frac{31}{164} \text{x} 100 = 18.9\% \qquad \%O = \frac{64}{164} \text{x} 100 = 39.0\%$$

CH_3COOH
Acetic acid molecular weight = 60 amu

$$\%C = \frac{24}{60} \text{x} 100 = 40\% \qquad \%H = \frac{4}{60} \text{x} 100 = 6.7\% \qquad \%O = \frac{32}{60} \text{x} 100 = 53.3\%$$

Adenosine triphosphate: $C_{10}H_{16}N_5O_{13}P_3$
Molecular weight = 507 amu

$$\%C = \frac{120}{507} \text{x} 100 = 23.7\% \qquad \%H = \frac{16}{507} \text{x} 100 = 3.16\% \qquad \%N = \frac{70}{507} \text{x} 100 = 13.8\%$$

$$\%O = \frac{208}{507} \text{x} 100 = 41.0\% \qquad \%P = \frac{93}{507} \text{x} 100 = 18.3\%$$

34. Convert percentages to mole ratios assuming a 100 g sample of the unknown compound.

$$40\% \text{ C} = 40 \text{ g C x} \frac{1 \text{ mol C}}{12 \text{ g C}} = 3.33 \text{ mol C}$$

$$6.7\% \text{ H} = 6.7 \text{ g H x} \frac{1 \text{ mol H}}{1.008 \text{ g H}} = 6.65 \text{ mol H}$$

$$53.3\% \text{ O} = 53.3 \text{ g O x} \frac{1 \text{ mol O}}{16 \text{ g O}} = 3.33 \text{ mol O}$$

Carbon and oxygen have a 1:1 ratio, and carbon and hydrogen are 1:2, making the empirical formula CH_2O (this empirical formula is the origin of the term "carbohydrate"—you know from this that you are looking at a sugar molecule). The empirical formula weight is 30, so there are six empirical formula units in the molecular formula, which must be $C_6H_{12}O_6$. For the record, this is glucose.

35. Convert grams of water and carbon dioxide to moles, then grams of carbon and hydrogen.

$$3.71 \text{ g CO}_2\text{x} \frac{1 \text{ mol CO}_2}{44 \text{ g CO}_2} \text{ x } \frac{1 \text{ mol C}}{\text{mol CO}_2} = 0.084 \text{ mol Cx} \frac{12 \text{ g C}}{\text{mol C}} = 1.01 \text{ g C}$$

$$1.51 \text{ g H}_2\text{O x} \frac{1 \text{ mol H}_2\text{O}}{18 \text{ g H}_2\text{O}} \text{ x } \frac{2 \text{ mol H}}{\text{mol H}_2\text{O}} = 0.1667 \text{ mol H x} \frac{1.008 \text{ g H}}{\text{mol H}} = 0.169 \text{ g H}$$

From this, it is possible to solve for the grams of oxygen in the sample.

$$1.57\text{g} - 1.01\text{g} - 0.169\text{g} = 0.391 \text{ g O}$$

$$0.391 \text{ g O x} \frac{1 \text{ mol O}}{16 \text{ g O}} = 0.024 \text{ mol O}$$

It is now possible to develop mole ratios for the formula.

$$\frac{0.084 \text{ mol C}}{0.024 \text{ mol O}} = 3.5\frac{\text{C}}{\text{O}} \qquad\qquad \frac{0.167 \text{ mol H}}{0.024 \text{ mol O}} = 7\frac{\text{H}}{\text{O}}$$

From this, an initial formula can be developed: $C_{3.5}H_7O$. This can now be refined to the proper formula: $C_7H_{14}O_2$.

Chapter Four – Reactions in Aqueous Solution

Workbook Problems

Workbook Problem 4.1

Step 1:   First, determine the number of moles found in 500 mL of a 0.30 M solution.

$$\text{moles} = M \times L = 0.3 \ \frac{\text{moles}}{L} \times 0.500L = 0.150 \text{ moles}$$

Step 2:   Determine the number of grams of sodium chloride from the number of moles

$$0.150 \text{ mole NaCl} \times \frac{58.5 \text{ g}}{\text{mol}} = 8.8 \text{ g NaCl}$$

Workbook Problem 4.2

Step 1:   Rearrange the equation $M_i \times V_i = M_f \times V_f$ to solve for the final molarity.

$$M_f = \frac{M_i \times V_i}{V_f} = \frac{50 \text{ mL} \times 10 \text{ M}}{1000 \text{ mL}} = 0.5 \text{ M}$$

Note that here, the conversion of volumes *was* needed, so that they would match and cancel out.

Workbook Problem 4.3

Strategy: $AlCl_3$ is a strong electrolyte and completely dissociates in water.

Step 1:   Determine the total number of moles of ions formed when $AlCl_3$ completely dissociates in water.

$$AlCl_3 \rightarrow Al^{3+} + 3Cl^-$$

Step 2:   Create a conversion factor comparing the total number of moles of ions in solution to 1 mol of $AlCl_3$.

Four ions are formed in the dissociation of aluminum chloride, so the conversion factor is

$$\frac{4 \text{ mol ions}}{\text{mol AlCl}_3}$$

Step 3:   Use the conversion factor to calculate the molar concentration of ions in solution.

$$\frac{0.75 \text{ mol AlCl}_3}{L} \times \frac{4 \text{ mol ions}}{\text{mol AlCl}_3} = 3.0 \text{ M ions}$$

Workbook Problem 4.4

Strategy: From the information given, write a molecular equation and determine if any of the reactants or products are strong electrolytes.

Step 1: Write and balance the molecular equation.

$$HNO_3 \, (aq) + KOH \, (aq) \rightarrow H_2O \, (l) + KNO_3 \, (aq)$$

Step 2: Write the strong electrolytes as free ions for the ionic equation.

Ionic Equation:
$$H^+ \, (aq) + NO_3^- \, (aq) + K^+ \, (aq) + OH^- (aq) \rightarrow H_2O \, (l) + K^+ \, (aq) + NO_3^- \, (aq)$$

Step 3: Eliminate any spectator ions to write the net ionic equation.

spectator ions: K^+ and NO_3^-
Net ionic equation: $H^+ \, (aq) + OH^- \, (aq) \rightarrow H_2O \, (l)$

This net ionic equation is characteristic of acid–base neutralization reactions.

Workbook Problem 4.5

Strategy: Using the solubility guidelines and solution stoichiometry, determine suitable reactants to use in the preparation of lead chloride and the amount of reactants needed.

Step 1: Determine reactants that are soluble and will produce the insoluble lead (II) chloride and another soluble product.

Using the guidelines in the study guide, an alkali metal will bring just about anything into solution, as will ammonium, so two easy ways to get the chloride ion into solution would be NH_4Cl or $NaCl$, both of which are highly soluble.

To bring a lead (II) ion into solution, nitrate will always work, but the textbook indicates that acetate and perchlorate can also be used to bring lead into solution. Therefore, some options for the lead (II) ion are $Pb(NO_3)_2$, $Pb(CH_3CO_2)_2$, and $Pb(ClO_4)_2$. Note that the ratios of ions are the same in all the compounds: the chloride ion is being brought into solution that has ratio MCl, and can be generically represented that way. The lead (II) cation will be brought into solution using a compound with two anions, which can be generically represented PbA_2. This will make it possible to devise a generic "recipe" for making lead (II) chloride.

Step 2: Write a balanced chemical equation.

$$2MCl \, (aq) + PbA_2 \, (aq) \rightarrow PbCl_2 \, (s) + 2MA \, (aq)$$

Step 3: Use solution stoichiometry to determine the volume of each reactant needed. First it is necessary to determine how many moles of lead (II) chloride need to be made.

$$1.3 \text{ g PbCl}_2 \times \frac{1 \text{ mol PbCl}_2}{278 \text{ g PbCl}_2} = 4.67\times10^{-3} \text{ mol or } 4.67 \text{mmol PbCl}_2$$

From this, determine the moles of each component that are needed:

$$4.67\times10^{-3} \text{ mol PbCl}_2 \times \frac{1 \text{ mol PbA}_2}{\text{mol PbCl}_2} = 4.67\times10^{-3} \text{ mol PbA}_2$$

$$4.67\times10^{-3} \text{ mol PbCl}_2 \times \frac{2 \text{ mol MCl}}{\text{mol PbCl}_2} = 9.35\times10^{-3} \text{ mol MCl}$$

From this, the volume of solution can be determined using the definition of molarity: $M = \text{mol}/V$, so $V = \text{mol}/M$

$$V \text{ PbA}_2 = \frac{4.67\times10^{-3} \text{ mol PbA}_2}{1M \text{ PbA}_2} = 4.67\times10^{-3} \text{ L PbA}_2 \text{ or } 4.67 \text{ mL PbA}_2$$

$$V \text{ MCl} = \frac{9.35\times10^{-3} \text{ mol MCl}}{1M \text{ MCl}} = 9.35\times10^{-3} \text{ L MCl or } 9.35 \text{ mL MCl}$$

Regardless of the exact composition of the solutions, as long as they follow the generic ratios, 4.67 mL of the lead solution combined with 9.35 mL of chloride-containing solution will have a theoretical yield of 1.3 g lead (II) chloride. If you really need 1.3 g of lead chloride, you'll want to add a little more, remembering that you always need twice as much of the chloride solution as the lead solution.

Workbook Problem 4.6

Determine the reaction type by examining the reactants—are they acids or bases or neutral salts?—then write and balance the equations for the reactions.

Step 1: Write the balanced molecular equation for each reaction.

$$2\text{LiOH} + \text{H}_2\text{SO}_4 \rightarrow \text{Li}_2\text{SO}_4 + 2\text{H}_2\text{O}$$
$$\text{NH}_4\text{Cl} + \text{AgNO}_3 \rightarrow \text{NH}_4\text{NO}_3 + \text{AgCl}$$

Step 2: Determine the presence of any strong electrolytes. Write the ionic equation, showing the strong electrolytes in terms of their free ions.

$$2\text{Li}^+ + 2\text{OH}^- + 2\text{H}^+ + \text{SO}_4^{2-} \rightarrow 2\text{Li}^+ + \text{SO}_4^{2-} + 2\text{H}_2\text{O}$$
$$\text{NH}_4^+ + \text{Cl}^- + \text{Ag}^+ + \text{NO}_3^- \rightarrow \text{NH}_4^+ + \text{NO}_3^- + \text{AgCl}$$

Step 3:   Write the net ionic equation by removing spectator ions.

$$2OH^- + 2H^+ \rightarrow 2H_2O \text{ – This is an acid–base reaction}$$
$$Cl^- + Ag^+ \rightarrow AgCl \text{ – This is a precipitation reaction}$$

Workbook Problem 4.7

Step 1:   Begin by writing a balanced chemical equation.

$$Zn\,(s) \ + \ 2\,HCl\,(aq) \ \rightarrow \ ZnCl_2\,(aq) \ + \ H_2\,(g)$$

Step 2:   Determine the number of moles of zinc present.

$$\text{mol Zn} \ = \ 4.75 \text{ g Zn x } \frac{1 \text{ mol Zn}}{65.4 \text{ g}} = 0.073 \text{ mol Zn}$$

Step 3:   Determine the number of moles of HCl needed.

$$0.073 \text{ mol Zn x } \frac{2 \text{ mol HCl}}{\text{mol Zn}} = 0.145 \text{ mol HCl needed}$$

Step 4:   Calculate the volume of 5.0 M HCl needed for this reaction.

$$V = \frac{0.145 \text{ mol}}{5.0 \dfrac{\text{mol}}{\text{L}}} \text{ x } \frac{1000 \text{ mL}}{\text{L}} = 29 \text{ mL HCl}$$

Step 5:   To calculate the volume of H_2 gas produced, we first need to calculate the moles, then grams of H_2 gas produced.

$$0.073 \text{ mol Zn x } \frac{1 \text{ mol H}_2}{\text{mol Zn}} \text{ x } \frac{2.016 \text{ g H}_2}{\text{mol H}_2} = 0.147 \text{ g H}_2$$

Step 6:   Using the density of H_2 gas, determine the volume produced:

$$V = \frac{0.147 \text{ g H}_2}{0.0899 \dfrac{\text{g}}{\text{L}}} = 1.64 \text{ L H}_2$$

Workbook Problem 4.8

Step 1:   Write a balanced chemical equation for the reaction.

$$2NaOH + H_2SO_4 \rightarrow Na_2SO_4 + 2H_2O$$

Step 2:   Calculate the number of moles of NaOH present.

$$0.153 \text{ g NaOH x } \frac{1 \text{ mol NaOH}}{40.0 \text{ g NaOH}} = 0.00383 \text{ mol}$$

Step 3: Determine the number of moles of sulfuric acid needed to react with that quantity of NaOH.

$$\text{moles } H_2SO_4 = 0.00383 \text{ mol NaOH} \times \frac{1 \text{ mol } H_2SO_4}{2 \text{ mol NaOH}} = 0.00191 \text{ mol } H_2SO_4$$

Step 4: Calculate the molarity of the acid solution.

$$\text{Molarity} = \frac{0.00191 \text{ mol } H_2SO_4}{0.01637 \text{ L}} = 0.117 \text{ M } H_2SO_4$$

Workbook Problem 4.9

Strategy: Follow the rules for determining oxidation states found above:

Step 1: Identify the elements that have a fixed oxidation state.

VO_4^{3-} – oxygen carries an oxidation state of -2

$NaVO_3$ – oxygen carries an oxidation state of -2, Na carries an oxidation state of $+1$

Step 2: Determine the oxidation number of the remaining elements, keeping in mind that the sum of all oxidation numbers must be equal to the ionic charge or to zero for a neutral molecule.

VO_4^{3-} – oxygen carries an oxidation state of -2
The overall charge is -3, so $(4 \times -2) + V = -3$, vanadium has an oxidation state of $+5$

$NaVO_3$ – oxygen carries an oxidation state of -2, Na carries an oxidation state of $+1$. This is a neutral molecule, so $+1 + (3 \times -2) + V = 0$, vanadium is again $+5$.

Workbook Problem 4.10

Strategy: Determine the oxidation number of all species present.

$$\overset{0}{Cu} (s) + 2\overset{+1}{Ag^+} (aq) + 2\overset{-1}{NO_3^-} (aq) \rightarrow 2\overset{0}{Ag} (s) + \overset{+2}{Cu^{2+}} (aq) + 2\overset{-1}{NO_3^-} (aq)$$

Step1: Identify the species that have a change in oxidation number.
Copper goes from an oxidation number of 0 to an oxidation number of 2+.
Silver goes from an oxidation number of 1+ to and oxidation number of 0.

Step 2: Based on the change in oxidation number, identify the species oxidized, the species reduced, and the oxidizing and reducing agent.
Copper loses electrons—it is oxidized and is the reducing agent in this reaction.
Silver gains electrons—it is reduced and is the oxidizing agent in this reaction.

Workbook Problem 4.11

Strategy: Remember that any element higher in the activity series will react with the ion of any element lower in the activity series. Also remember, metals above the H^+ ion in the activity series will displace the hydrogen ion from an acid to form H_2 gas.

Step 1:   Predict the outcome of these reactions.

a. $Cu\ (s)\ +\ HCl\ (aq)\ \rightarrow$ no reaction: copper is below H^+ in the activity series. Most metals will dissolve in acid, but mercury, platinum, and what are sometimes called the coinage metals—silver, copper, and gold—will not.

b. $Mn\ (s)\ +\ ZnCl_2\ (aq)\ \rightarrow MnCl_2\ (aq) + Zn\ (s)$

c. $3Mg\ (s)\ +\ 2Al(NO_3)_3\ (aq)\ \rightarrow 3Mg(NO_3)_2\ (aq) + 2Al\ (s)$

Workbook Problem 4.12

Step 1:   Determine the number of moles of thiosulfate ion that were required to react with the iodine solution to completely reduce the iodine.
Using the definition of molarity $M = mol/V$, so $mol = V \times M$

$$\text{moles } S_2O_3^{2-} \text{ ion} = 0.15\frac{mol}{L} \times 0.0157\ L = 2.36 \times 10^{-3}\ mol \text{ or } 2.36\ mmol$$

Step 2:   Balance the equation for the acid solution.

$$I_2 + S_2O_3^{2-} \rightarrow I^- + S_4O_6^{2-}$$
$$\text{Reduction: } I_2 + 2e \rightarrow 2I^-{}^-$$

$$\text{Oxidation: } 2\ S_2O_3^{2-} \rightarrow S_4O_6^{2-} + 2e^-$$
$$I_2 + 2S_2O_3^{2-} \rightarrow S_4O_6^{2-} + 2I^-$$

Step 3:   Use mole ratios to determine the amount of iodine present in the reaction mixture.

$$2.36 \times 10^{-3}\ mol\ S_2O_3^{2-} \text{ ion} \times \frac{1\ mol\ I_2}{2\ mol\ S_2O_3^{2-}\ \text{ion}} = 1.18 \times 10^{-3}\ mol\ I_2$$

Step 4:   Determine the original concentration of I_2 in solution.

$$\frac{1.18 \times 10^{-3}\ mol\ I_2}{0.050\ L\ solution} = 0.024\ M\ I_2$$

Self–Test

True–False

1. F. An acid and a base react to form water and a salt.
2. F. Both sodium and nitrate indicate solubility.
3. T
4. T
5. T
6. F. Acids release hydrogen ions or hydronium ions into solution.
7. T

8. F. Oxidation numbers can be equal to ionic charge, but are not always equal.
9. F. An activity series makes it possible to predict the results of redox reactions.
10. T

Matching

11. h	12. f	13. l	14. a
15. b	16. o	17. j	18. n
19. d	20. m	21. e	22. k
23. I	24. c	25. g	26. r
27. p	28. q		

Fill-in-the-Blank

29. soluble, insoluble, precipitation
30. acid, base, water, salt
31. redox or oxidation/reduction
32. strong electrolytes, weak electrolytes
33. molecular, ionic, net ionic
34. hydrogen or hydronium, hydroxide
35. loses, gains

Problems

36. a. HNO_3 (aq) + KOH (aq) $\rightarrow$ KNO_3 (aq) + H_2O (l).

    Neutralization: net ionic equation is $H^+ + OH^- \rightarrow H_2O$

    b. $Cu(NO_3)_2$ (aq) + 2NaOH (aq) $\rightarrow$ $Cu(OH)_2$ (aq) + $2NaNO_3$ (aq)

    Precipitation: two soluble components generate and insoluble product.

    c. $Cu(NO_3)_2$ (aq) + Zn (s) $\rightarrow$ $Zn(NO_3)_2$ (aq) + Cu (s)

    Redox: Cu and Zn are transferring electrons – Zn is oxidized, Cu is reduced.

37. Convert grams to moles.

$$15 \text{ g } Na_3PO_4 \times \frac{1 \text{ mol}}{164 \text{ g}} = 0.0915 \text{ mol}$$

Convert to molar concentration of ions.

$$\frac{0.0915 \text{ mol } Na_3PO_4}{0.250 \text{ L}} \times \frac{4 \text{ ions}}{Na_3PO_4} = 1.46 \text{ M}$$

38. a. $Pb(NO_3)_2$ (aq) + 2NaCl (aq) $\rightarrow$ $PbCl_2$ (s) + $2NaNO_3$ (aq)
    Pb^{2+} (aq) + $2NO_3^-$ (aq) + $2Na^+$ (aq) + $2Cl^-$ (aq) $\rightarrow$ $PbCl_2$ (s) + $2Na^+$ (aq) + $2NO_3^-$ (aq)
    Pb^{2+} (aq) + 2Cl (aq) $\rightarrow$ $PbCl_2$ (s)

    b. H_2SO_4 (aq) + 2LiOH (aq) $\rightarrow$ Li_2SO_4 (aq) + $2H_2O$ (l)
    $2H^+$ (aq) + SO_4^{2-} (aq) + $2Li^+$ (aq) + $2OH^-$ (aq) $\rightarrow$ $2Li^+$ (aq) + SO_4^{2-} (aq) + $2H_2O$ (l)
    H^+ (aq) + OH^- (aq) $\rightarrow$ H_2O (l)

    c. KCl (aq) + $AgCH_3CO_2$ (aq) $\rightarrow$ KCH_3CO_2 (aq) + AgCl (s)
    K^+ (aq) + Cl^- (aq) + Ag^+ (aq) + $CH_3CO_2^-$ (aq) $\rightarrow$ K^+ (aq) + $CH_3CO_2^-$ (aq) + AgCl (s)
    Ag^+ (aq) + Cl^- (aq) $\rightarrow$ AgCl (s)

39. Write a balanced net ionic equation.

$$Ca^{2+} (aq) + 2OH^- (aq) \rightarrow Ca(OH)_2 (s)$$

Determine the number of moles of calcium hydroxide required:

$$1.8g\ Ca(OH)_2 \times \frac{1\ mol}{74.1\ g} = 0.243\ mol$$

From the balanced equation above, 0.243 mol of Ca ions are required, and 0.0486 mol of hydroxide ions are required.

From the definition of molarity, solve for the volumes needed.

$$M = \frac{mol}{L} \therefore V = \frac{mol}{M}$$

$$V_{OH} = \frac{0.0486\ mol}{1\ M} = 0.0486\ L\ or\ 48.6\ mL$$

$$V_{Ca} = \frac{0.0243\ mol}{1\ M} = 0.0243\ L\ or\ 24.3\ mL$$

48.6 mL 1 M $Ca(NO_3)_2$ solution would need to be combined with 24.3 mL 1 M KOH to produce 1.8 g $Ca(OH)_2$.

40. a. No reaction. Possible products are lead acetate (soluble) and copper (II) nitrate (soluble).
    b. No reaction. Possible products are sodium sulfate (soluble) and ammonium nitrate (soluble).
    c. $Ag^+ (aq) + Cl^- (aq) \rightarrow AgCl (s)$

41. a. Cl_2 Elemental forms, even in molecules, always have oxidation numbers of 0
    b. NaCl   Na = +1, Cl = −1
    c. CO_3^{2-} O = −2, C = +4 $4 + (3 \times -2) = -2$, which is the charge on the ion.
    d. Cr_2O_3 O = −2, Cr = +3 $(3 \times -2) + (2 \times +3) = 0$
    e. ClO_3^- O = −2, Cl = +5 $(3 \times -2) + 5 = -1$

42. a. $C_2H_6 + O_2 \rightarrow CO_2 + H_2O$ (Combustion is a form of redox reaction).
        Oxidation number of C goes from −3 to +4   oxidized, reducing agent
        Oxidation number of O goes from 0 to −2   reduced, oxidizing agent
    b. $Na + Cl_2 \rightarrow NaCl$
        Oxidation number of Na goes from 0 to +1   oxidized, reducing agent
        Oxidation number of Cl goes from 0 to −1   reduced, oxidizing agent
    c. $Mg + Fe^{2+} \rightarrow Mg^{2+} + Fe$
        Oxidation number of Mg goes from 0 to +2   oxidized, reducing agent
        Oxidation number of Fe goes from +2 to 0   reduced, oxidizing agent
    d. $Ca + H^+ \rightarrow Ca^{2+} + H_2$
        Oxidation number of Ca goes from 0 to +2   oxidized, reducing agent
        Oxidation number of H goes from +1 to 0   reduced, oxidizing agent

43. a. chromium and copper (II) nitrate
$$2Cr \ (s) + 3Cu(NO_3)_2 \ (aq) \rightarrow 2Cr(NO_3)_3 \ (aq) + 3Cu \ (s)$$
b. tin and silver (I) nitrate
$$Sn \ (s) + 2AgNO_3 \rightarrow Sn(NO_3)_2 + 2Ag$$
c. copper (II) nitrate and cobalt
$$2Co \ (s) + 3Cu(NO_3)_2 \ (aq) \rightarrow 2Co(NO_3)_3 \ (aq) + 3Cu \ (s)$$
d. silver and HCl – no reaction.

44. Balance the equation.

$$SO_3^{2-} \ (aq) + Cr_2O_7^{2-} \ (aq) \rightarrow SO_4^{2-} \ (aq) + 2Cr^{3+}(aq)$$

Balance half-reactions in H, O, and e^-.

$$SO_3^{2-} \ (aq) + H_2O \ (l) \rightarrow SO_4^{2-} \ (aq) + 2H^+ \ (aq) + 2e^-$$
$$Cr_2O_7^{2-}(aq) + 14H^+ \ (aq) + 6e^- \rightarrow 2Cr^{3+} \ (aq) + 7H_2O \ (l)$$

Equalize numbers of electrons.

$$3SO_3^{2-} \ (aq) + 3H_2O \ (l) \rightarrow 3SO_4^{2-} \ (aq) + 6H^+ \ (aq) + 6e^-$$
$$Cr_2O_7^{2-}(aq) + 14H^+ \ (aq) + 6e^- \rightarrow 2Cr^{3+} \ (aq) + 7H_2O \ (l)$$

Add equations.

$$3SO_3^{2-} \ (aq) + 3H_2O \ (l) + Cr_2O_7^{2-}(aq) + 14H^+ \ (aq) + 6e^- \rightarrow$$
$$3SO_4^{2-} \ (aq) + 6H^+ \ (aq) + 6e^- + 2Cr^{3+} \ (aq) + 7H_2O \ (l)$$

Cancel.

$$3SO_3^{2-} \ (aq) + Cr_2O_7^{2-}(aq) + 8H^+ \ (aq) \rightarrow 3SO_4^{2-} \ (aq) + 2Cr^{3+} \ (aq) + 4H_2O \ (l)$$

Determine the moles of potassium chromate.

$$0.0237 \ L \times 0.124 \ M = 0.00294 \ mole \ K_2Cr_2O_7$$

Determine moles of Na_2SO_3 using the balanced equation.

$$0.00294 \ mol \ K_2Cr_2O_7 \times \frac{3 \ mol \ Na_2SO_3}{mol \ K_2Cr_2O_7} = 0.00882 \ mol \ Na_2SO_3$$

$$M = \frac{0.00882 \ mol \ NaSO_3}{0.015 \ L} = 0.588 \ M$$

45. a. K_2SO_4 (aq) + $Pb(NO_3)_2$ → $2KNO_3$ (aq) + $PbSO_4$ (s)
    $2K^+$ (aq) + SO_4^{2-} (aq) + Pb^{2+} (aq) + $2NO_3^-$ (aq) → $2K^+$ (aq) + $2NO_3^-$ (aq) + $PbSO_4$ (s)
    Pb^{2+} (aq) + SO_4^{2-} (aq) → $PbSO_4$ (s)

 b.   Determine the number of moles of each reactant.

$$5.3 \text{ g K}_2SO_4 \times \frac{1 \text{mol}}{174 \text{ g K}_2SO_4} = 0.0304 \text{ mol K}_2SO_4$$

$$4.9 \text{ g Pb(NO}_3)_2 \times \frac{1 \text{mol}}{331 \text{ g Pb(NO}_3)_2} = 0.0148 \text{ mol Pb(NO}_3)_2$$

Reactants react in a 1:1 ratio: $Pb(NO_3)_2$ is the limiting reagent.

$$0.0148 \text{ mol Pb(NO}_3)_2 \times \frac{1 \text{ mol PbSO}_4}{\text{mol Pb(NO}_3)_2} \times 303 \frac{g}{\text{mol PbSO}_4} = 4.48 \text{ g PbSO}_4$$

 c.   Subtract the moles of limiting reagent used from the moles of excess reagent:

$$0.0304 \text{ mol} - 0.0148 \text{ mol} = 0.0156 \text{ mol K}_2SO_4$$

$$0.0156 \text{ mol K}_2SO_4 \times \frac{174 \text{ g K}_2SO_4}{1 \text{ mol}} = 2.71 \text{ g K}_2SO_4$$

 d.   percent yield$= \dfrac{4.37}{4.48} \times 100 = 98\%$

 e.   For this, the number of moles of K_2SO_4 in excess and the moles of KNO_3 need to be converted to ions and diluted.

$$0.0156 \text{ mol K}_2SO_4 \times \frac{3 \text{ mol ions}}{\text{mol K}_2SO_4} = 0.0468 \text{ mol ions}$$

$$0.0148 \text{ mol Pb(NO}_3)_2 \times \frac{2 \text{ mol KNO}_3}{\text{mol Pb(NO}_3)_2} = 0.0296 \text{ mol KNO}_3$$

$$0.0296 \text{ mol KNO}_3 \times \frac{2 \text{ mol ions}}{\text{mol KNO}_3} = 0.0592 \text{ mol ions}$$

$$M_{ions} = \frac{0.0592 \text{ mol} + 0.0468 \text{ mol}}{.200 \text{ L}} = 0.53 \text{ M}$$

Challenge Problem
46. To approach this problem, first solve for the mole quantity of hydroxide generated by the reaction.

$$0.0597 \text{ L HCl} \times 7.5 \frac{\text{mol}}{\text{L}} \times \frac{1 \text{mol OH}}{\text{mol HCl}} = 0.448 \text{ mol OH}$$

This is also equivalent to the number of moles of the alkali metal. There is now enough information to set up a system of equations.

$$Li + Na = 4 \text{ g}$$
$$\frac{Li}{6.9} + \frac{Na}{23} = 0.448 \text{mol}$$

This second equation is equivalent to $0.145\text{Li} + 0.0435\text{Na} = 0.448$. To solve by substitution,

$$\text{Li} = 4 - \text{Na, so}$$

$$0.145(4 - \text{Na}) + 0.0435\text{Na} = 0.448$$

$$0.58 - 0.145\text{Na} + 0.0435\text{Na} = 0.448$$

$$-0.145\text{Na} + 0.0435\text{Na} = 0.448 - 0.58$$

$$-0.102\text{Na} = -0.132$$

$$\text{Na} = 1.3 \text{ g}$$

$$\text{Na} + \text{Li} = 4$$

$$\text{Li} = 2.7 \text{ g}$$

47. As always, the first step is to write and balance an equation that will describe what is happening and allow for stoichiometric determinations. The trick to this problem is realizing that tin that is oxidized to +2 can be further oxidized to +4.

$$\text{Sn}^{2+} (aq) + \text{NO}_3^- (aq) \rightarrow \text{Sn}^{4+} (aq) + \text{NO} (g)$$

Separate and balance the half-reactions.

$$3 \text{ e}^- + 4\text{H}^+ (aq) + \text{NO}_3^- (aq) \rightarrow \text{NO} (g) + 2\text{H}_2\text{O} (l)$$
$$\text{Sn}^{2+} (aq) \rightarrow \text{Sn}^{4+} (aq) + 2\text{e}^-$$

Equalize the numbers of electrons, add, and cancel.

$$8\text{H}^+ (aq) + 3\text{Sn}^{2+} (aq) + 2\text{NO}_3^- (aq) \rightarrow 2\text{NO} (g) + 4\text{H}_2\text{O} (l) + 3\text{Sn}^{4+} (aq)$$

Determine the number of moles of NO_3^- required in titration.

$$0.0276 \text{ L} \times 0.0563 \text{ M} = 0.00155 \text{ mol NO}_3^-$$

Determine the grams of tin titrated.

$$0.00155 \text{ mol NO}_3^- \times \frac{3 \text{ mol Sn}}{2 \text{ mol NO}_3^-} \times \frac{118.7 \text{ g Sn}}{\text{mol Sn}} = 0.276 \text{ g Sn}$$

Determine the percentage of tin in the sample.

$$\% \text{ Sn} = \frac{0.276 \text{ g Sn}}{5.36 \text{ g sample}} \times 100 = 5.14\% \text{ Sn}$$

Chapter Five – Periodicity and the Electronic Structure of Atoms

Workbook Problems

Workbook Problem 5.1

Step 1:   Choose the appropriate form of the equation, and solve:

$$\lambda = \frac{c}{\upsilon} = \frac{3.00 \times 10^8 \, \text{m/s}}{88.5 \times 10^6 \, /\text{s}} = 3.39 \, \text{m}$$

As this demonstrates, wavelengths can be quite long. Although visible light is in the nanometer range, it would not make sense to convert this wavelength to nanometers.

Workbook Problem 5.2

Strategy: Use the equation for the energy of a photon, and multiply by Avogadro's number. Then convert frequency to wavelength.

Step 1:   Determine the energy of one photon, then convert to one mole of photons.

$$E = h\nu = 6.626 \times 10^{-34} \, \text{J} \bullet \text{s} \times 4.4 \times 10^{14} \, /\text{s} = 2.91 \times 10^{-19} \, \text{J}$$

$$2.91 \times 10^{-19} \, \text{J} \times 6.022 \times 10^{23} \, /\text{mol} = 176,000 \, \text{J/mol} = 176 \, \text{kJ/mol}$$

Step 2:   Convert the frequency to wavelength.

$$c = \lambda\nu \therefore \lambda = \frac{c}{\nu} = \frac{3.00 \times 10^8 \, \text{m/s}}{4.4 \times 10^{14} \, /\text{s}} = 6.82 \times 10^{-7} \, \text{m} = 682 \, \text{nm}$$

Step 3:   Consult your textbook for the color of this wavelength of light.

Light of 682 nm will be reddish-orange.

Workbook Problem 5.3

Strategy: Determine the values of n that will make λ the longest and shortest. Remember that the value of λ is greatest when the value of n is smallest and the value λ is smallest when the value of n is greatest.

Step 1:   Determine the values of n that will make λ the shortest and longest.

The longest wavelength line will be at the smallest possible value of n. If $m = 5$, and n must be greater than 5, the longest wavelength will occur when $n = 6$. The shortest wavelength will occur with the largest possible value of n. If $n = \infty$, $1/n = 0$.

Step 2:   Use the Balmer–Rydberg equation with $m = 5$ and solve for the shortest wavelength.

$$\frac{1}{\lambda} = R\left[\frac{1}{m^2} - \frac{1}{n^2}\right] = 1.097 \times 10^{-2}\,\text{nm}^{-1}\left[\frac{1}{25} - \frac{1}{\infty}\right] = 4.39 \times 10^{-4}\,\text{m}$$

$$\frac{1}{\lambda} = \frac{1}{4.39 \times 10^{-4}\,\text{m}} = 2280\,\text{nm}$$

Step 3:   Solve for λ using the Balmer–Rydberg equation and the values of m that make λ the longest.

$$\frac{1}{\lambda} = R\left[\frac{1}{m^2} - \frac{1}{n^2}\right] = 1.097 \times 10^{-2}\,\text{nm}^{-1}\left[\frac{1}{25} - \frac{1}{36}\right] = 1.34 \times 10^{-4}\,\text{nm}^{-1}$$

$$\frac{1}{\lambda} = \frac{1}{1.34 \times 10^{-4}\,\text{nm}^{-1}} = 7460\,\text{nm}$$

Workbook Problem 5.4

Strategy: Identify n and m, solve for wavelength, then energy, and convert to kJ/mol.

Step 1:   Identify n and m for this problem.

As the intention is to completely remove the electron and make a H^+ ion, $n = 3$, and $m = \infty$

Step 2:   Solve the Balmer–Rydberg equation to get the wavelength of the energy needed:

$$\frac{1}{\lambda} = 1.097 \times 10^{-2}\,/\text{nm}\left[\frac{1}{3^2}\right] = 0.00122\,/\text{nm}$$

$$\lambda = 820\,\text{nm}$$

Step 3:   Solve for the energy of one photon of this energy:

$$E = \frac{hc}{\lambda} = \frac{6.626 \times 10^{-34}\,\text{J} \bullet \text{s} \times 3.00 \times 10^8\,\text{m/s}}{820 \times 10^{-9}\,\text{m}} = 2.42 \times 10^{-19}\,\text{J}$$

Step 4:   Convert to kJ/mol:

$$2.42 \times 10^{-19}\,\frac{\text{J}}{\text{photon}} \times 6.022 \times 10^{23}\,\frac{\text{photons}}{\text{mol}} \times \frac{1\,\text{kJ}}{1000\,\text{J}} = 146\,\frac{\text{kJ}}{\text{mol}}$$

Workbook Problem 5.5

Strategy: Determine the subshell associated with a value of $l = 0$.

Step 1:   Identify the subshell with the value of n and the letter designation for $l = 0$.

When $l = 0$, the subshell is s. This is a $3s$ subshell.

Step 2:   Determine the number of orbitals in this subshell.

The number of orbitals in a subshell is determined by the magnetic quantum number, m_l. For $l = 0$, only one value of m_l is available, so there is only one orbital.

Workbook Problem 5.6

Step 1:   Determine the number of electrons in calcium.

Calcium has atomic number 20, so it has 20 electrons.

Step 2:   Use the Aufbau principle to determine the ground-state electronic configuration.

Begin building up the electron configuration from $n = 1$

Ca: $1s^2\,2s^2 2p^6 3s^2 3p^6 4s^2$

Step 3:   Determine the noble gas in the previous row, and remove the electrons associated with it to a shorthand notation. Specifically name the electrons in unfilled subshells.

The noble gas with fewer electrons than calcium is argon. With the argon atoms delineated using shorthand, the electron configuration becomes:

Ca: $[Ar]4s^2$

Step 4:   Draw the orbital-filling diagram:

$$\underset{1s}{\uparrow\downarrow}\quad \underset{2s}{\uparrow\downarrow}\quad \underset{2p}{\uparrow\downarrow\ \uparrow\downarrow\ \uparrow\downarrow}\quad \underset{3s}{\uparrow\downarrow}\quad \underset{3p}{\uparrow\downarrow\ \uparrow\downarrow\ \uparrow\downarrow}\quad \underset{4s}{\uparrow\downarrow}$$

Self–Test

True–False
1. F. Amplitude is the height of a wave from the midline.
2. T
3. T
4. F. Above the atomic level, de Broglie wavelengths are irrelevant.
5. F. The Bohr model only holds for hydrogen.
6. T
7. F. The angular momentum quantum number must be less than the principal quantum number.

8. F. f-orbitals are only found in the lanthanide and actinide metals.
9. T
10. F. Electrons that share an orbital must have opposite spins.

Multiple Choice

11. d	12. b	13. c	14. a
15. b	16. c	17. b	18. a
19. d	20. c		

Matching

20. b	21. g	22. e	23. d
24. c	25. I	26. a	27. f
28. h			

Fill-in-the-Blank

29. wavelength, frequency	30. line spectra
31. photons	32. emission
33. quantum numbers	34. s-orbitals
35. p-orbitals, two, six	36. degenerate, parallel spins
37. valence electrons	38. atomic radius

Problems

39. First, convert from frequency to wavelength.

$$c = \lambda \nu, \text{ so } \lambda = \frac{c}{\nu} = \frac{3.00 \times 10^8 \, \text{m/s}}{835.6 \times 10^6 \, \text{/s}} = 0.359 \text{ m or } 359 \text{ mm}$$

Then calculate the energy of 1 mol of these photons.

$$E = h\nu = 6.626 \times 10^{-34} \, \text{J} \bullet \text{s} \times 835.6 \times 10^6 \, \text{/s} = 5.54 \times 10^{-25} \, \text{J/photon}$$

$$5.54 \times 10^{-25} \, \text{J/photon} \times 6.022 \times 10^{23} \, \text{photons/mole} = 0.333 \text{ J/mol}$$

40. First, convert from wavelength to frequency, taking care with the units of wavelength.

$$c = \lambda \nu, \text{ so } \nu = \frac{c}{\lambda} = \frac{3.00 \times 10^8 \, \text{m/s}}{2.15 \times 10^{-14} \, \text{m}} = 1.40 \times 10^{22} \, \text{/s}$$

Then calculate the energy of 1 mol of these photons.

$$E = h\nu = 6.626 \times 10^{-34} \, \text{J} \bullet \text{s} \times 1.40 \times 10^{22} \, \text{/s} = 9.28 \times 10^{-12} \, \text{J/photon}$$

$$9.28 \times 10^{-12} \, \text{J/photon} \times 6.02 \times 10^{23} \, \text{photons/mole} = 5.59 \times 10^{12} \, \text{J/mol}$$

41. First, determine the energy of one photon at this wavelength.

$$E = \frac{hc}{\lambda} = \frac{6.626 \times 10^{-34} \, \text{J} \bullet \text{s} \times 3.00 \times 10^8 \, \text{m/s}}{432 \times 10^{-9} \, \text{m}} = 4.60 \times 10^{-19} \, \text{J}$$

From this, determine the number of photons in 2.5 mJ of energy.

$$\frac{2.5 \times 10^{-3}\,\text{J}}{4.60 \times 10^{-19}\,\text{J/photon}} = 5.43 \times 10^{15}\,\text{photons}$$

42. First, calculate the energy needed to dislodge one electron. This will be the same energy found in a photon capable of dislodging an electron.

$$495.8\,\frac{\text{kJ}}{\text{mol}} \times \frac{1\,\text{mol}}{6.022 \times 10^{23}\,\text{photons}} \times \frac{1000\,\text{J}}{\text{kJ}} = 8.233 \times 10^{-19}\,\frac{\text{J}}{\text{photon}}$$

A photon with this much or more energy can dislodge an electron from sodium. From this, the threshold frequency and wavelength can be calculated.

$$E = h\nu,\, \nu = \frac{E}{h} = \frac{8.233 \times 10^{-19}\,\text{J}}{6.626 \times 10^{-34}\,\text{J}\cdot\text{s}} = 1.243 \times 10^{15}/\text{s}$$

$$\lambda = \frac{c}{\nu} = \frac{3.00 \times 10^{8}\,\text{m/s}}{1.243 \times 10^{15}/\text{s}} = 2.414 \times 10^{-7}\,\text{m} = 241.4\,\text{nm}$$

If the threshold frequency corresponds to a wavelength of 241 nm, any wavelength that gives a higher energy photon—that is, any wavelength lower than 241 nm—will eject electrons. 540 nm is below the threshold frequency and will not eject electrons. 195 nm is above the threshold frequency and will eject photons.

43.

$$(\Delta x)(\Delta mv) \geq \frac{h}{4\pi}$$

$$\Delta v \geq \frac{h}{4\pi m \Delta x} \geq \frac{6.626 \times 10^{-34}\,\frac{\text{kg}\bullet\text{m}^2}{\text{s}}}{4 \times 3.14 \times 9.11 \times 10^{-34}\,\text{kg} \times 327 \times 10^{-12}\,\text{m}} \geq 1.77 \times 10^{8}\,\text{m/s}$$

44.

$$\frac{1}{\lambda} = R\left\lfloor \frac{1}{1} - \frac{1}{6^2} \right\rfloor = 1.097 \times 10^{-2}/\text{nm} \left\lfloor 1 - \frac{1}{36} \right\rfloor = 0.0107/\text{nm}$$

$$\lambda = 93.8\,\text{nm}$$
$$E = \frac{hc}{\lambda} = \frac{6.626 \times 10^{-34}\,\text{J}\bullet\text{s} \times 3.00 \times 10^{8}\,\text{m/s}}{93.8 \times 10^{-9}\,\text{m}} = 2.12 \times 10^{-18}\,\text{J}$$

45. $Y = 1s^2 2s^2 2p^6 3s^2 3p^6 4s^2 3d^{10} 4p^6 5s^2 4d^1$

46. Neon has electrons in two energy levels, $n = 1$ and $n = 2$. It has both s and p electrons and a total of 10 electrons. They will have the following quantum numbers:

1	0	0	$+\frac{1}{2}$	$1s^1$
1	0	0	$-\frac{1}{2}$	$1s^2$
2	0	0	$+\frac{1}{2}$	$2s^1$
2	0	0	$-\frac{1}{2}$	$2s^2$
2	1	-1	$+\frac{1}{2}$	$2p^1$
2	1	-1	$-\frac{1}{2}$	$2p^2$
2	1	-0	$+\frac{1}{2}$	$2p^3$
2	1	-0	$-\frac{1}{2}$	$2p^4$
2	1	$+1$	$+\frac{1}{2}$	$2p^5$
2	1	$+1$	$-\frac{1}{2}$	$2p^6$

47. The highest energy electron in bromine will be a $4p$ electron. The quantum numbers associated with a $4p$ electron are $n = 4$, and $l = 1$.

48.   Si =   ⇅   ⇅   ⇅ ⇅ ⇅   ⇅   ↑ ↑ __
           1s   2s   2p   3s   3p

       P =   ⇅   ⇅   ⇅ ⇅ ⇅   ⇅   ↑ ↑ ↑
           1s   2s   2p   3s   3p

       Cl =   ⇅   ⇅   ⇅ ⇅ ⇅   ⇅   ⇅ ⇅ ↑_
           1s   2s   2p   3s   3p

Phosphorus has three unpaired electrons, making it the most paramagnetic. Chlorine has one unpaired electron and is therefore the least paramagnetic element.

49. Solve for the wavelength of the photon, then solve the Balmer–Rydberg equation for n.

$$E = \frac{hc}{\lambda}, \quad \lambda = \frac{hc}{E} = \frac{6.626 \times 10^{-34}\,\text{J} \bullet \text{s} \times 3.00 \times 10^{8}\,\text{m/s}}{4.58 \times 10^{-19}\,\text{J}} = 434\ \text{nm}$$

$$\frac{1}{434\,\text{nm}} = 1.097 \times 10^{-2}\,/\text{nm} \left[\frac{1}{2^2} - \frac{1}{n^2} \right]$$

$$\frac{1}{434\,\text{nm} \times 1.097 \times 10^{-2}/\text{nm}} = \left[\frac{1}{4} - \frac{1}{n^2} \right]$$

$$0.210 = .25 - \frac{1}{n^2}$$

$$.04 = \frac{1}{n^2}$$

$$n^2 = 25$$

$$n = 5$$

50.

$$\lambda = \frac{h}{mv} = \frac{6.626 \text{ x } 10^{-34} \frac{\text{kg} \bullet \text{m}^2}{\text{s}}}{9.109 \text{ x } 10^{-31} \text{kg x } 2.57 \text{ x } 10^6 \text{m/s}} = 2.83 \text{ x } 10^{-10} \text{m} = 0.283 \text{ nm}$$

51. $[Ar]4s^1 3d^{10}$ = 29 electrons – Cu $[Xe]6s^2 4f^{14} 5d^{10} 6p^5$ = 85 electrons – At
    $[Kr] 4d^{10}$ = 46 electrons – Pd $[Ar]4s^1 3d^5$ = 24 electrons – Cr

52. These elements are in different periods, so they need to be ranked from lowest to highest in the periodic table.

    Rn > Os > Ru > Fe

53. These elements are in different groups, so they need to be ranked from right to left in the periodic table.

    Mg > S > Cl > Ne

Challenge Problem
54. First, solve for the energy of one photon at 550 nm.

$$E = \frac{hc}{\lambda} = \frac{6.626 \text{x} 10^{-34} \text{ J} \bullet \text{s x } 3.00 \text{ x } 10^8 \text{m/s}}{550 \text{ x } 10^{-9} \text{m}} = 3.61 \text{ x } 10^{-19} \text{ J/photon}$$

The light bulb is emitting 40 J/s, and is allowed to do so for 4 hours. The energy emitted is

$$40 \frac{\text{J}}{\text{s}} \text{ x 4 hours x } \frac{3600 \text{ s}}{\text{h}} = 576,000 \text{ J}$$

Solve for photons emitted.

$$\frac{576,000 \text{ J}}{3.61 \text{ x } 10^{-19} \text{ J/photon}} = 1.60 \text{ x } 10^{24} \text{ photons}$$

For the second part of the problem, solve for the energy needed to heat the water.

$$\text{energy} = 200 \text{ g x } 4.184 \frac{\text{J}}{\text{g} \bullet {}^{\circ}\text{C}} \text{ x } (95 \text{ }^{\circ}\text{C} - 20 \text{ }^{\circ}\text{C}) = 62,760 \text{ J}$$

If the light bulb is emitting 576,000J of energy in 4 hours, and 95% of this is heat, the bulb is emitting 547,000 J of heat in 4 hours, or 137,000 J/hr.

If all of this heat went into the 200 g of water, solve for the time.

$$\text{time} = \frac{62,760 \text{J}}{137,000 \text{ J/hr}} = 0.459 \text{ hours x } \frac{60 \text{ min}}{\text{hr}} = 27.5 \text{ min}$$

55. 15 W = 15 J/s.

If the laser were 100% efficient, 2.73×10^{27} photons would equal 15 J. The energy of each photon would therefore be

$$\frac{15 \text{ J}}{2.73 \text{ x } 10^{19} \text{ photons}} = 5.49 \text{x} 10^{-19} \text{ J / photon}$$

$$E = \frac{hc}{\lambda} \text{ , } \lambda = \frac{hc}{E} = \frac{6.626 \text{ x } 10^{-34} \text{ J} \bullet \text{s x } 3.00 \text{ x } 10^{8} \text{ m/s}}{5.49 \text{ x } 10^{-19} \text{ J}} = 3.62 \text{ x } 10^{-7} \text{ m or } 362 \text{ nm}$$

If the laser were 80% efficient, the wavelength would be

$$\lambda = \frac{hc}{E} = \frac{6.626 \text{ x } 10^{-34} \text{ J} \bullet \text{s x } 3.00 \text{ x } 10^{8} \text{ m/s}}{5.49 \text{ x } 10^{-19} \text{ J x } 0.8} = 4.52 \text{ x } 10^{-7} \text{ m or } 452 \text{ nm}$$

If 20%

$$\lambda = \frac{hc}{E} = \frac{6.626 \text{ x } 10^{-34} \text{ J} \bullet \text{s x } 3.00 \text{ x } 10^{8} \text{ m/s}}{5.49 \text{ x } 10^{-19} \text{ J x } 0.2} = 1.81 \text{ x } 10^{-6} \text{ m or } 1810 \text{ nm}$$

To solve for the efficiency if its wavelength is 1542, rearrange the equation to solve for the percentage.

$$x\% = \frac{hc}{E\lambda} = \frac{6.626 \text{ x } 10^{-34} \text{ J} \bullet \text{s x } 3.00 \text{ x } 10^{8} \text{ m/s}}{5.49 \text{ x } 10^{-19} \text{ J x } 1.542 \text{ x } 10^{-6} \text{ m}} = 0.235$$

The laser is 23.5% efficient.

Chapter Six – Ionic Compounds: Periodic Trends and Bonding Theory

Workbook Problems

Workbook Problem 6.1

Step 1: The largest jump in ionization energies is between the fourth and fifth ionization energies. This shows that there are five valence electrons, so this must be a group 5A element. In fact, it is phosphorus. ($P \rightarrow P^{+} \rightarrow P^{2+} \rightarrow P^{3+} \rightarrow P^{4+} \rightarrow P^{5+}$)

Workbook Problem 6.2

Step 1: Identify the group number of each element and identify them as metals or nonmetals.

Rb:	Rubidium	Group 1A metal
Sc:	Scandium	Group 3B transition metal
Se:	Selenium	Group 6A nonmetal
As:	Arsenic	Group 5A nonmetal

Step 2:   Find the nearest noble gas.

> Rb: Krypton (36 electrons)
> Sc: Argon   (18 electrons)
> Se: Krypton (36 electrons)
> As: Krypton (36 electrons)

Step 3:   Identify the number of electrons that must be gained/lost to achieve the electron configuration of that noble gas.

> Rb: 37 electrons → Krypton (36 electrons) One electron must be lost
> Sc: 21 electrons → Argon (18 electrons)   Three electrons must be lost
> Se: 34 electrons → Krypton (36 electrons) Two electrons must be gained
> As: 33 electrons →Krypton (36 electrons)   Three electrons will be gained

Step 4:   Write the ionized form of the element.

> Rb^+ Sc^{3+} Se^{2-} As^{3-}

Workbook Problem 6.3

Strategy: Write and sum the energies of each of the steps in the reaction to calculate the overall energy change of the formation of rubidium chloride from its elements.

Step 1:   Write each step of the reaction with its energy required or produced per mole.

Sublimation of sodium	$Na\ (s) \rightarrow Na\ (g)$	+107.3 kJ/mol
Vaporization of bromine	$½Br_2\ (l) \rightarrow ½Br_2\ (g)$	+ 15.4 kJ/mol
Splitting of bromine molecules	$½Br_2\ (g) \rightarrow Br\ (g)$	+112 kJ/mol
Ionization of sodium	$Na\ (g) \rightarrow Na^+\ (g) + e^-$	+495.8 kJ/mol
Ionization of bromine	$Br\ (g) + e^- \rightarrow Br^-\ (g)$	–325 kJ/mol
Formation of ionic solid	$Na^+\ (g) + Br^-\ (g) \rightarrow NaBr\ (s)$	–747 kJ/mol

Step 2:   Sum the energy terms to find the overall energy change of the reaction.

> Total = –341.5 kJ/mol

Note: the values given for bond breaking and vaporizing bromine are divided in half. As given, they are for the following reactions: $Br_2\ (l) \rightarrow Br_2\ (g)$ and $Br_2\ (g) \rightarrow 2Br\ (g)$.

Workbook Problem 6.4

Strategy: Determine the relative sizes and charges for the ions and assign relative strengths based on charge and size.

Step 1:   Identify the relative charges in the two sets of ions and how they change from one set to the other.

NaCl, KCl	same charges
AlF$_3$, MgCl$_2$	Al – 3+, Mg – 2+, aluminum will provide stronger lattice energy.
CuBr$_2$, CuCl	Cu (II) – 2+, Cu (I) – 1+, copper (II) will provide stronger lattice energy

Step 2:   If necessary, identify the relative ion sizes in the two sets of ions and how they change from one to the other.

NaCl, KCl same charges. Sodium is smaller than potassium, so will provide stronger lattice energy.

Step 3:   Determine the relative lattice energies of the ion pairs.

$$NaCl > KBr \quad AlF_3 > MgCl_2 \quad CuBr_2 > CuCl$$

Self–Test

True–False
1. T
2. F. Nonmetals gain electrons to obtain noble gas electron structures.
3. F. Transition metals will sometimes form more than one ion.
4. T
5. T
6. F. Ionization energies decrease down the periodic table.
7. F. Electron affinity applies primarily to nonmetals and is difficult to determine for nonmetals.
8. T
9. F. The Born–Haber cycle provides information about ionic bonding.
10. F. Lattice energies are related to atomic size and charge.

Multiple Choice

11. d	12. b	13. c	14. b
15. a	16. c	17. d	18. a

Matching

19. e	20. d	21. a	22. b
23. h	24. g	25. f	26. c

Fill-in-the-Blank

27. argon	28. decreases	29. core, easy
30. increase, valence	31. anions	32. charge, Coulomb's
33. eight, octet rule		

Problems

34. P: ⇅ ⇅ ⇅ ⇅ ⇅ ⇅ ↿ ↿ ↿
     1s 2s   2p   3s   3p

     P³⁻: ⇅ ⇅ ⇅ ⇅ ⇅ ⇅ ⇅ ⇅ ⇅
     1s 2s   2p   3s   3p
     Electron structure is that of argon

     Ti: ⇅ ⇅ ⇅ ⇅ ⇅ ⇅ ⇅ ⇅ ⇅ ⇅ ↿ ↿
     1s   2s   2p   3s   3p 4s   3d

     Ti²⁺: ⇅ ⇅ ⇅ ⇅ ⇅ ⇅ ⇅ ⇅ ⇅ __ ↿ ↿
     1s   2s   2p   3s   3p 4s   3d
     most common

     Ti⁴⁺: ⇅ ⇅ ⇅ ⇅ ⇅ ⇅ ⇅ ⇅ ⇅
     1s   2s   2p   3s   3p 4s   3d
     also possible

     Mn: ⇅ ⇅ ⇅ ⇅ ⇅ ⇅ ⇅ ⇅ ⇅ ⇅ ↿ ↿ ↿ ↿ ↿
     1s 2s   2p   3s   3p   4s   3d

     Mn²⁺: ⇅ ⇅ ⇅ ⇅ ⇅ ⇅ ⇅ ⇅ ⇅ __ ↿ ↿ ↿ ↿ ↿
     1s 2s   2p   3s   3p   4s   3d

     K: ⇅ ⇅ ⇅ ⇅ ⇅ ⇅ ⇅ ⇅ ⇅ ↿
     1s 2s   2p   3s   3p   4s

     K⁺: ⇅ ⇅ ⇅ ⇅ ⇅ ⇅ ⇅ ⇅ ⇅
     1s 2s   2p   3s   3p
     Electron structure is that of argon

     Se: [Ar] ⇅ ⇅ ⇅ ⇅ ⇅ ⇅ ⇅ ↿ ↿
              4s   3d   4p

     Se²⁻: [Ar] ⇅ ⇅ ⇅ ⇅ ⇅ ⇅ ⇅ ⇅ ⇅
              4s   3d   4p
     Electron structure is that of krypton

35. Remembering that cations are smaller than neutral atoms, anions are larger than neutral atoms, and that atoms decrease in size from left to right across the periodic table and increase in size down the periodic table:

$$Li < Li^+ < Al < Al^{3+} < Cl < Cl^-$$

36. There is a very large jump (~4x) between the third and fourth ionization energies. The fourth electron removed is a core electron, meaning that there are three valence electrons. In the third period of the periodic table, this is aluminum.

37. Ga: [Ar] ⇅ ⇅ ⇅ ⇅ ⇅ ⇅ ↿ __ __
              4s   3d   4p

     Ge: [Ar] ⇅ ⇅ ⇅ ⇅ ⇅ ⇅ ↿ ↿ __
              4s   3d   4p

     Br: [Ar] ⇅ ⇅ ⇅ ⇅ ⇅ ⇅ ⇅ ⇅ ↿
              4s   3d   4p

Based on its electron structure, gallium is far more likely to lose an electron to form a +1 ion than to gain an electron. Looking at the periodic table, this is also confirmed by the position of gallium, which is a main-group metal. Germanium can adopt a half-filled shell, which is also a fairly stable arrangement, but it does not have the stability of a noble gas configuration. Bromine is the closest to having a noble gas configuration, so it will have the highest electron affinity.

38. Use the Born–Haber cycle. However, this time the net energy change is known, and we need to solve for the heat of sublimation of aluminum.

$Al\ (s) \rightarrow Al\ (g)$	? kJ/mol	
$Al\ (g) \rightarrow Al^+\ (g) + e^-$	578 kJ/mol	578 kJ/mol
$Al^+\ (g) \rightarrow Al^{2+}\ (g) + e^-$	1,817 kJ/mol	1,817 kJ/mol
$Al^{3+}\ (g) \rightarrow Al^{3+}\ (g) + e^-$	2,745 kJ/mol	2,745 kJ/mol
$3/2\ Cl_2\ (g) \rightarrow 3\ Cl\ (g)$	3/2(244 kJ/mol)	366 kJ/mol
$3\ Cl\ (g) \rightarrow 3\ Cl^-\ (g)$	3(−348.6 kJ/mol)	−1046 kJ/mol
$Al^{3+}\ (g)\ +\ 3\ Cl^-\ (g) \rightarrow AlCl_3\ (s)$	−5,492 kJ/mol	−5,492 kJ/mol

Overall net energy change   −704.2 kJ/mol

−704.2 = ? + 578 kJ/mol + 1,817 kJ/mol + 2,745 kJ/mol + 366 kJ/mol − 1046 kJ/mol − 5,492 kJ/mol

? = 328 kJ/mol

39. $Mg\ (s) + Cl_2\ (g) \rightarrow MgCl_2\ (s)$
$2\ Mg\ (s) + O_2\ (g) \rightarrow 2\ MgO\ (s)$
$3\ Mg\ (s) + 2\ P\ (s) \rightarrow Mg_3P_2\ (s)$

$2\ Li\ (s) + Cl_2\ (g) \rightarrow 2\ LiCl\ (s)$
$4\ Li\ (s) + O_2\ (g) \rightarrow 2\ Li_2O\ (s)$
$3\ Li\ (s) + P\ (s) \rightarrow LiP_3\ (s)$

$2Al\ (s) + 3Cl_2\ (g) \rightarrow 2\ AlCl_3\ (s)$
$4Al\ (s) + 3O_2\ (g) \rightarrow 2\ Al_2O_3\ (s)$
$Al\ (s) + P\ (s) \rightarrow AlP\ (s)$

40.
$K\ (s) \rightarrow K\ (g)$	+89 kJ/mol
$K\ (g) \rightarrow K^+\ (g) + e^-$	+418.8 kJ/mol
$½\ F_2\ (g) \rightarrow F\ (g)$	½(+158 kJ/mol)
$F\ (g) \rightarrow F\ (g)$	−328 kJ/mol
$K^+\ (g)\ +\ F^-\ (g) \rightarrow KF\ (s)$	−821 kJ/mol
Total:	−562 kJ/mol

41. $AlCl_3 > MgCl_2 > LiCl > KCl > CsCl$
Using Coulomb's law, an increase in ionic charge on the cation will dominate other factors, so the cation with the highest charge will have the highest lattice energy. With ions that have charges of 1, size becomes a factor: smaller ions can get closer together and will thus have a larger lattice energy than larger ions, so these can be ranked in order of cation size, smallest to largest.

Challenge Problem
42. The chemical equation for the reaction of the alkaline-earth chloride with silver nitrate is

$$MCl_2 \;+\; 2\,AgNO_3 \;\rightarrow\; M(NO_3)_2 \;+\; 2\,AgCl$$

a. The mass of chlorine can be determined from the mass of silver chloride produced.

$$18.8g\ AgCl \times \frac{1\ mol\ AgCl}{143.3g} \times \frac{1\ mol\ Cl}{1\ mol\ AgCl} \times \frac{35.5g\ Cl}{mol\ Cl} = 4.66g\ Cl$$

$$mass\%\ Cl\ in\ metal\ chloride \;=\; \frac{4.66g\ Cl}{5.25g\ MCl_2} = 88.7\%$$

b. The identity of the metal can be determined by calculating the molar mass of the metal chloride, then subtracting out the mass of the chloride.

$$mol\ MCl_2 = 18.8g\ AgCl \times \frac{1mol\ AgCl}{143.3g\ AgCl} \times \frac{1mol\ MCl_2}{2mol\ AgCl} = 0.0656mol\ MCl_2$$

$$molar\ mass\ MCl_2 = \frac{5.25\ g}{0.0656\ mol} = 80\frac{g}{mol}$$

$$Molar\ mass\ of\ M \;=\; 80\ g/mol - 71\ g/mol = 9\ g/mol$$

From the periodic table, the metal must be beryllium.

c. For the reaction of beryllium with chlorine, $Be\ (s)\ +\ Cl_2\ (g)\ \rightarrow\ BeCl_2\ (s)$
For the reaction of $BeCl_2$ with $AgNO_3$, $BeCl_2\ +\ 2\,AgNO_3\ \rightarrow\ Be(NO_3)_2\ +\ 2\,AgCl$

d. To determine which was the limiting reactant, the moles of beryllium and the moles of chlorine gas must be calculated.

$$1.53g\ Be \times \frac{1\ mol}{9.01g} = 0.170\ mol$$

$$4.75L\ Cl_2 \times 3.17\frac{g}{L} \times \frac{1mol}{71g} = 0.212\ mol$$

Chlorine was in excess: the liters left unreacted can now be calculated.
0.212 mol Cl_2 present – 0.170 mol Cl_2 reacted = 0.042 mol Cl_2 unreacted.

$$0.042\ mol\ Cl_2 \times \frac{71g}{mol} \times \frac{1\ L}{3.17g} = 0.94\ L$$

Chapter Seven – Covalent Bonding and Electron-Dot Structures

Workbook Problems

Workbook Problem 7.1

Step 1:   Predict the polarity of the bonds based on their locations, then check with electronegativity values from Figure 7.4 in the textbook:

H – F – nonmetals, far apart: polar covalent
F – Cl – nonmetals, close together: non polar
C – O – nonmetals, close together: non polar
K – Cl – metal and non-metal: ionic
Ge – Br – semi-metal and nonmetal: polar covalent

Step 2:   Put in the numbers to check

H – F : 4.0 – 2.1 = 1.9, polar covalent
F – Cl : 4.0 – 3.0 = 1.0, polar covalent (estimate is off—fluorine is *really* electronegative!)
C – O : 3.0 – 2.0 = 1.0, polar covalent
K – Cl : 3.0 – 0.8 = 2.2, ionic
Ge – Br : 2.8 – 1.8 = 1.0, polar covalent

Note: what this exercise shows is that the farther apart on the periodic table two elements are, the more polar/ionic their bond. Also, almost all nonmetal—nonmetal bonds are polar.

Workbook Problem 7.2

Strategy: Determine valences of compounds, then combine them to form complete octets.

Step 1:   Determine the number of electrons on each atom based on its position on the periodic table:

H – 1 valence electron   F – 7 valence electrons   C – 4 valence electrons
Br – 7 valence electrons   Cl – 7 valence electrons

Step 2:   Combine the atoms to complete octets, substituting bond lines for paired electrons.

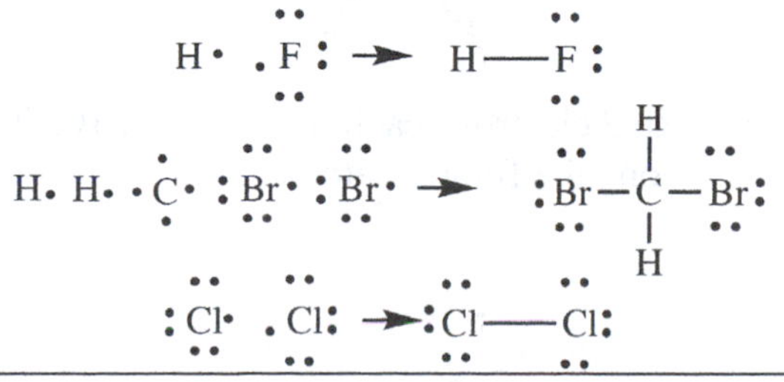

Workbook Problem 7.3

Step 1: Determine the number of electrons on each atom based on its position on the periodic table:

H: one valence electron   C: four valence electrons   O: six valence electrons

Step 2: Combine the atoms to complete octets, substituting bond lines for paired electrons.

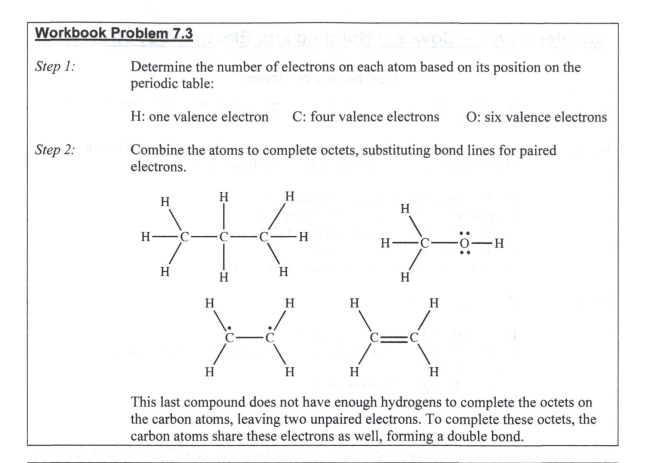

This last compound does not have enough hydrogens to complete the octets on the carbon atoms, leaving two unpaired electrons. To complete these octets, the carbon atoms share these electrons as well, forming a double bond.

Workbook Problem 7.4

Step 1: Determine the total number of valence-shell electrons.

5 (from P) + 35 (from 5 F) = 40

Step 2: Determine the connections.

P is the central atom. (This leads to the most symmetrical structure.)

Step 3: Draw the bonds, and subtract the number of electrons used from the total number of valence electrons available.

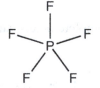

The five bonds use 2 electrons each for a total of 10. That leaves 30 electrons to be distributed among the fluorine atoms.

Step 4:   Complete the octets of the outer atoms.

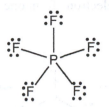

Step 5:   Place any remaining electrons on the central atom.

There are no remaining electrons, but the central phosphorus atom has an expanded octet of 10.

Workbook Problem 7.5

Step 1:   Determine the total number of valence-shell electrons.

Total number of valence-shell electrons = 4 (from C) + 18 (from 3 O) + 2 (from the 2– charge) = 24

Step 2:   Determine the connections.

Carbon is the central atom for symmetry's sake, as well as it being the most electronegative atom.

Step 3:   Draw the bonds, and subtract the number of electrons used from the total number of valence electrons available.

This uses 6 valence electrons, leaving 18.

Step 4:   Complete the octets of the outer atoms.

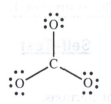

Step 5:   Place any remaining electrons on the central atom, looking for multiple possibilities.

There are no electrons remaining, but carbon has an incomplete octet and will need to share a pair of electrons from one of the oxygens. This leads to a number of resonance structures:

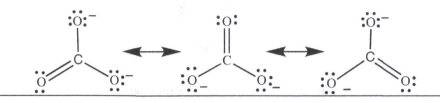

Workbook Problem 7.6

Step 1: Determine the number of valence electrons in the ion

Total number of valence-shell electrons = 4 (from C) + 6 (from O) + 5 (from N) + 1 (from the 1– charge) = 16

Step 2: Determine three reasonable electron-dot structures with carbon central and this number of valence electrons.

Step 3: Calculate the formal charge on each of the atoms in each of the structures.

Structure 1:	Structure 2:	Structure 3:
N = 5 – 2 – 4 = –1	N = 5 – 3 – 2 = 0	N = 5 – 1 – 6 = –2
C = 4 – 4 – 0 = 0	C = 4 – 4 – 0 = 0	C = 4 – 4 – 0 = 0
O = 6 – 2 – 4 = 0	O = 6 – 1 – 6 = –1	O = 6 – 3 – 2 = +1

Step 4: Choose the most stable structure.

Notice that all three have a net charge of –1, appropriate to this ion. Structure 3 (which is already violating rules about how many bonds oxygen likes to have) has formal charges on two atoms, and is therefore unstable. Structures 1 and 2 both have single negative charges on one atom. To choose between these two, choose the structure in which the negative charge is assigned to the most electronegative atom. This is structure 1.

Self–Test

True–False
1. F. Covalent bonds involve sharing of electrons.
2. T
3. T
4. T
5. T
6. F. Valences of main group elements can be easily predicted using the periodic table.
7. F. Multi-atom structures can be predicted using electron-dot models.
8. F. The structures of charged species can be predicted by removing or adding electrons as needed.

9. T
10. T

Fill-in-the-Blank
11. bond lengths, bond, nuclei
12. bond-dissociation energy, positive, released
13. polar covalent, ionic
14. nonbonding, bonding
15. octet rule, expanded octets
16. formal charges

Matching
17. j   18. a   19. i   20. b
21. d   22. g   23. f   24. e
25. c

Problems
26. LiBr – ionic (metal/nonmetal)   HCl – covalent (nonmetal/nonmetal)
    MgO – ionic (metal/nonmetal)   NF$_3$ – covalent (nonmetal/nonmetal)

27.

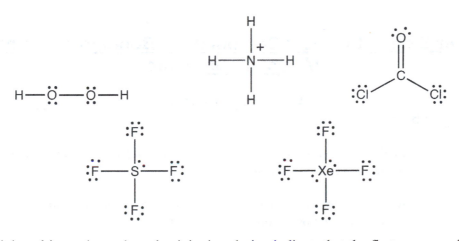

28. The high melting point and conductivity in solution indicate that the first compound is ionic. The second compound, with its low melting point and nonconducting solution, is covalent. The solubilities of the compounds provide no information about bonding.

29.

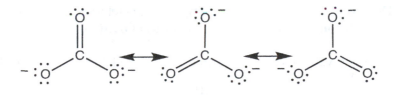

The three resonance structures are all equivalent.
Formal charges: Central C $= 4 - 4 - 0 = 0$
                    Double-bonded O $= 6 - 2 - 4 = 0$
                    Single-bonded O $= 6 - 1 - 6 = -1$

The formal charges match the ionic charge.

30. Li-Cl > Al-F > H-Cl > C-H > Cl-Cl. The atoms that are farther apart on the periodic table will have more-polar bonds.

31. Cs < Fe < O < F. Fluorine is the most electronegative element: proximity to fluorine gives a good approximation of electronegativity of an element.

32. The most symmetrical structure is usually the most stable, and formal charges bear this out.

Leftmost structure:
- Leftmost nitrogen = 5 – 1 – 6 = –2
- Central nitrogen = 5 – 4 – 0 = –1
- Oxygen = 6 – 3 – 2 = +1

Center structure:
- Leftmost nitrogen = 5 – 2 – 4 = –1
- Central nitrogen = 5 – 4 – 0 = –1

Rightmost structure:
- Leftmost nitrogen = 5 – 3 – 2 = 0
- Central nitrogen = 5 – 4 – 0 = –1
- Oxygen = 6 – 1 – 6 = –1

None of these structures is perfectly satisfactory, but the rightmost structure is the only one in which there is a negative formal charge on oxygen, the most electronegative element in the structure.

Chapter Eight – Covalent Compounds: Bonding Theories and Molecular Structure

Workbook Problems

Workbook Problem 8.1

Strategy: Draw the electron-dot structure for each of the molecules and use the list above to determine the molecular shape.

Step 1: Draw the electron-dot structure for the molecules, using the number of valence electrons and the most symmetrical structure.

CS_2 has 4 valence electrons from carbon and 6 from each of the sulfurs, for a total of 16.

$$:\!\ddot{S}\!=\!\!=\!C\!=\!\!=\!\ddot{S}\!:$$

Double bonds were needed to complete the octet around carbon. Although sulfur will form an expanded octet, here it does not need to.

SF_4 has 6 valence electrons from the sulfur, and 28 from the four fluorines, for a total of 34.

$$\begin{array}{c} :\!\ddot{F}\!: \\ | \\ :\!\ddot{F}\!-\!\ddot{S}\!-\!\ddot{F}\!: \\ | \\ :\!\ddot{F}\!: \end{array}$$

Step 2: Determine the number of charge clouds around the central atom. How many are used for bonding and how many are nonbonding?
CS_2 has two electron clouds around the central atom.
SF_4 has five electron clouds around the central atom, four bonding, and one nonbonding pair.

| Step 3: | Based on the chart, determine the shape. |
| | CS$_2$ is linear.   SF$_4$ is seesaw shaped. |

Workbook Problem 8.2

Given the following structure:

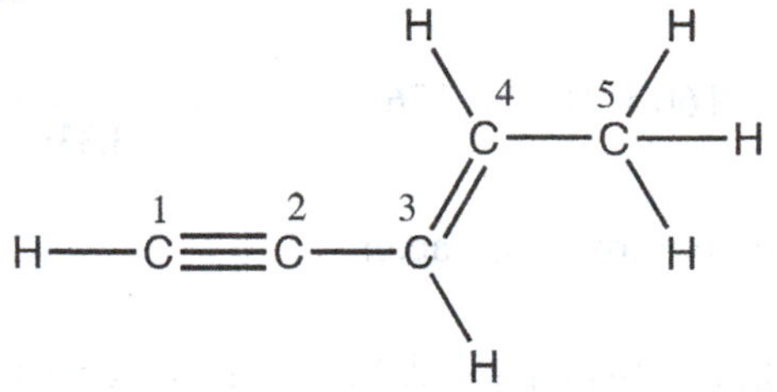

Describe the hybridization and bonding of each of the carbon atoms.

| **Strategy:** | Determine the number of electron clouds around each carbon to determine hybridization, then identify the number of σ and π bonds. |

Step 1:	Determine the number of electron clouds around each carbon and its hybridization.
	Carbon 1: two electron clouds – *sp* hybridization
	Carbon 2: two electron clouds – *sp* hybridization
	Carbon 3: three electron clouds – *sp*^{2} hybridization
	Carbon 4: three electron clouds – *sp*^{2} hybridization
	Carbon 5: four electron clouds – *sp*^{3} hybridization

Step 2:	Determine the bond types present.
	Carbon 1: σ bond to hydrogen, σ and 2π bonds to carbon 2
	Carbon 2: σ and 2π bonds to carbon 1, σ bond to carbon 3
	Carbon 3: σ bond to carbon 2, σ bond to hydrogen, σ and π to carbon 4
	Carbon 4: σ and π to carbon 3, σ to hydrogen and σ to carbon 5
	Carbon 5: σ bonds to carbon 4, and three hydrogens

Workbook Problem 8.3

| **Strategy:** | Look for asymetrical molecules with electronegativity differences. |

| Step 1: | Determine which molecules are asymetrical: |

| | All of these molecules are asymetrical, even trimethylamine, which has a lone pair of electrons on the nitrogen atom. |
| Step 2: | Determine which element is most electronegative and thus the area of highest electron concentration: |

H$_2$CO – oxygen   N(CH$_3$)$_3$ – nitrogen   HBr – bromine

All of these molecules will have dipoles.

Workbook Problem 8.4

Strategy: Calculate the dipole if it were purely ionic, then calculate the percent ionic character.

Step 1: Calculate ionic dipole:

$$\mu = Q \times r = 1.160 \times 10^{-19} \times 176 \times 10^{-12}\, m \times \frac{1\, D}{3.336 \times 10^{-30}\, C \bullet m} = 6.12\, D$$

Step 2: Determine percent ionic character.

$$\% \text{ ionic character} = \frac{1.42\, D}{6.12\, D} \times 100 = 23.2\% \text{ ionic}$$

Workbook Problem 8.5

Strategy: Examine the structures of the molecules and determine what interactions are present, then rank in order of strength.

Step 1: List the interactions experienced by each molecule:

CH_3OH: hydrogen bonding, dipole–dipole, and dispersion forces.

Xe: dispersion forces only

CH_3Cl: dipole–dipole and dispersion forces.

Step 2: Rank the molecules in order of increasing strength of interactions.

$$Xe < CH_3OH < CH_3Cl$$

Putting It Together

Strategy: Determine the empirical formula of the compound and the molecular formula. Determine the structure.

Step 1: Solve for the empirical formula:
26.68 g C = 2.221 mol C   71.09 g O = 4.443 mol O
2.224 g H = 2.222 mol H

Empirical formula = CO_2H

Step 2: Solve for the molecular formula:

Empirical formula weight = 45.01 g/mol
Molecular weight = 90.03 g/mol
Molecular formula = $C_2O_4H_2$

Step 3:   Determine the structure of the molecule from the information given:

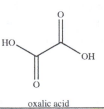

oxalic acid

Self–Test

True–False

1. T
2. T
3. T
4. F. Methane molecules are tetrahedral.
5. T
6. F. When orbitals overlap, π-bonds are formed.
7. T
8. T
9. F. The polarity of individual bonds sometimes cancels out, leaving a molecule with polar bonds without a dipole.
10. F. Intermolecular forces range from very weak dispersion forces to fairly strong.
11. T

Fill-in-the-Blank

12. tetrahedral, bent, trigonal pyramidal
13. sp^2
14. σ, π
15. Electronegativity
16. dipole–dipole, London dispersion forces, hydrogen bonds

Matching

17. i	18. h	19. g	20. f
21. e	22. d	23. c	24. b
25. a			

Problems

26. NF_3 – four electron clouds provides tetrahedral electron geometry. Three bonds lead to trigonal pyramidal molecular geometry.

    H_2S – four electron clouds again provides tetrahedral electron geometry, with two bonds leading to bent molecular geometry.

    IF_4^+ – here, there is an expanded octet to consider with 34 electrons spread over 5 atoms. There are six electron clouds in this molecule, with two nonbonding pairs on the iodine. This molecule will have octahedral electron geometry and square planar molecular geometry.

    PF_5 – here, there is another expanded octet. There are five electron clouds in this molecule, with no nonbonding pairs. Both the electronic and molecular geometries are trigonal bipyramidal.

27. The carbon with four bonds has tetrahedral geometry, which indicates sp^3 hybridization and σ binding. The carbon bonded to the oxygen has three electron groups, which indicates trigonal planar geometry and sp^2 hybridization. This atom has σ bonds to carbon and hydrogen, and one σ and one π bond to oxygen.

28.

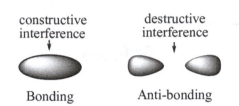

constructive
interference

destructive
interference

Bonding   Anti-bonding

Challenge Problem
29. Determine the empirical formula of the acid.

26.1 g C = 2.18 mol C
69.7 g O = 4.36 mol O
4.35 g H = 4.35 mol H

Empirical formula = CO_2H_2

Determine the molecular formula.

The moles of NaOH needed to neutralize the 0.36 grams of acid are

0.07824 L × 0.10 mol/L = 0.007824 mol

The molecular weight of the compound is 0.360 g/0.007824 mol = 46.01

The empirical formula weight is 46.01, so the empirical formula is the same as the molecular formula.

Lewis structure:

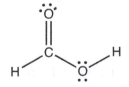

The central carbon atom is sp^2 hybridized, with trigonal planar geometry

The oxygen bonded to the C and H atoms is sp^3 hybridized, with tetrahedral geometry.

C–H = σ   C–O = σ   O–H = σ   C=O = π

Chapter Nine – Thermochemistry: Chemical Energy

Workbook Problems

Workbook Problem 9.1

Step 1:   Use $w = -(P \times \Delta V)$:

$$w = -(13 \text{ atm} \times [7.4 \text{ L} - (7.4 \text{ L} + 3.7 \text{ L})]) = 48.1 \text{ L} \cdot \text{atm}$$

Step 2:   Convert from L • atm to joules to kilojoules.

$$48.1 \text{ L} \cdot \text{atm} \times 101 \text{ J/L} \cdot \text{atm} = 4858 \text{ J} = 4.9 \text{ kJ}$$

Step 3:   What does the sign tell you? Was work done by or on the system?

The positive value of the work shows that the system gained: work was done on the system. The same thing can be determined by looking at the initial and final volume of the reactions: 11.1 L of reactants provided 7.4 L of products. The overall system contracted, so work was done on it.

Workbook Problem 9.2

Step 1:   Use $\Delta E = q - P\Delta V$

$$\Delta E = -79.9 \text{ kJ} - 4.9 \text{ kJ} = -84.8 \text{ kJ}$$

Workbook Problem 9.3

Step 1:   Determine the ΔH for the reaction given the amount of carbonic acid that dissociated.

$$0.5 \text{ mol} \times -20.7 \text{ kJ/mol} = -10.4 \text{ kJ}$$

Step 2:   Determine $P\Delta V$ and convert from L • atm to kJ.

$$1 \text{ atm} \times 11.2 \text{ L} = 11.2 \text{ L} \cdot \text{atm} \times 101 \text{ J/L} \cdot \text{atm} = 1131 \text{ J} = 1.13 \text{ kJ}$$

Step 3:   Determine ΔE from the equation: $\Delta H = \Delta E + P\Delta V$. Rearrange the equation to $\Delta E = \Delta H - P\Delta V$.

$$\Delta E = -10.4 - (-1.13) = -11.5 \text{ kJ}$$

Think it through to make sure this makes sense. Heat was released from the exothermic equation, leading to a negative ΔH, but work was done on the system when the volume decreased, which lessens the magnitude of the overall energy change. Therefore, it makes sense.

Workbook Problem 9.4

Step 1:   Convert the grams of water to moles.

$$50 \text{ g} \times \frac{1 \text{ mol}}{18 \text{ g}} = 2.8 \text{ mol}$$

Step 2:   Multiply by the heat of vaporization.

$$2.8 \text{ mol} \times 40.7 \frac{\text{kJ}}{\text{mol}} = 113 \text{ kJ}$$

Step 3:   Make sure the sign of the heat transfer makes sense: the water is the system.

Vaporizing water is the same as boiling it on the stove. This is a way of adding heat, so it is expected that heat will need to be added to the system to vaporize the water. In fact, lots of heat will need to be added, because vaporizing water requires breaking all of the hydrogen bonds that hold it together as a liquid. The sign on the energy transfer is positive, so heat is being added to the system, as expected.

Workbook Problem 9.5

Step 1:   Determine the number of moles of calcium.

$$3.24 \text{ g} \times \frac{1 \text{mol}}{40.08 \text{ g}} = 0.0808 \text{ mol}$$

Step 2:   Using the stoichiometric ratio, determine the heat involved in the reaction.

$$0.0808 \text{ mol Ca} \times \frac{2 \text{ mol CaO}}{2 \text{ mol Ca}} \times -634.9 \frac{\text{kJ}}{\text{mol CaO}} = -51.3 \text{ kJ}$$

Step 3:   Based on the sign of the heat transfer, determine whether the reaction is exothermic or endothermic.

Based on the negative sign for the energy transfer, this is an exothermic reaction.

Workbook Problem 9.6

Strategy: Determine the heat absorbed by the water, then reverse the sign to determine the number of joules evolved by the dissociation. Convert the grams of NaOH to moles, and divide the heat evolved by the number of moles of NaOH to determine the molar heat of dissociation.

Step 1:   Determine the amount of heat gained by the water using the equation $q = mc\Delta T$.

$$q = 50 \text{ g} \times 4.184 \text{ J/g} \degree\text{C} \times (30.70 \degree\text{C} - 22.73 \degree\text{C}) = 1667\text{J}$$

Step 2:   Determine the amount of heat released by the reaction.

As the dissociation released the energy absorbed by the water, it will have the same magnitude, but be opposite in sign: –1667 J.

Step 3:   Determine the number of moles of NaOH involved.

$$1.5 \text{ g} \times \frac{1 \text{mol}}{40 \text{ g}} = 0.0375 \text{ mol}$$

Step 4:   Determine the heat of dissociation per mole of NaOH.

$$\frac{-1667 \text{ J}}{0.0375 \text{ mol}} = 44\,453 \frac{\text{J}}{\text{mol}} \times \frac{1 \text{ kJ}}{1000 \text{ J}} = -44.5 \frac{\text{kJ}}{\text{mol}}$$

Workbook Problem 9.7

Strategy: Get the reactants on the right and the products on the left, flipping reactions around if needed, and multiplying them through if necessary to get the equation as written:

Step 1:   Reactants on the left, multiplying through if necessary.

Fe_2O_3 (s) → 2 Fe (s) + 3/2 O_2 (g) $\Delta H° = +824.2$ kJ (turned around, and multiplied by ½)

3 CO (g) + 3/2O_2 (g) → 3 CO_2 (g) $\Delta H° = -848.1$ kJ (multiplied by 3/2)

Step 2:   Products on the right, multiplying through if necessary.
Products are already on the right as a result of the manipulations above.

Step 3:   Stack up the rewritten reactions, and add them together.

Fe_2O_3 (s)   → 2 Fe (s) + 3/2 O_2 (g) $\Delta H° = +824.2$ kJ
3 CO (g) + 3/2O_2 (g) → 3 CO_2 (g) $\Delta H° = -848.1$ kJ
Fe_2O_3 (s) + 3 CO (g) + 3/2O_2 (g) → 2 Fe (s) + 3/2 O_2 (g) + 3 CO_2 (g) $\Delta H° = -23.9$ kJ

Step 4:   Cancel terms that appear on both sides.

Fe_2O_3 (s) + 3 CO (g) → 2 Fe (s) + 3 CO_2 (g) $\Delta H° = -23.9$ kJ

Workbook Problem 9.8

Strategy: Write a balanced equation for the reaction, and use the standard heats of formation to determine the overall heat of reaction per mole of glucose, then convert grams to moles and determine the heat released by 10 g of glucose.

Step 1:   Write and balance the equation.

$C_6H_{12}O_6$ (s) + 6 O_2 (g) → 6 H_2O (l) + 6 CO_2 (g)

Step 2:	Determine the overall heat of reaction using $\Delta H^\circ = \sum \Delta H^\circ_{products} - \sum \Delta H^\circ_{reactants}$.

$$\Delta H^\circ = [(6 \times -285.8 \text{ kJ/mol}) + (6 \times -393.5 \text{ kJ/mol})] - (-1273.3 \text{ kJ/mol})$$
$$= -2800 \text{ kJ/mol}$$

Step 3:   Convert the grams of glucose to moles.

$$10 \text{ grams} \times \frac{1 \text{ mol}}{180 \text{ g}} = 0.056 \text{ mol glucose}$$

Step 4:   Determine the energy released by 10 g of glucose.

$$0.056 \text{ mol} \times -2800 \text{ kJ/mol} = -157 \text{ kJ}$$

Step 5:   Using $q = mc\Delta T$, determine the temperature change of 500 g of water.

$$157{,}000 \text{ J} = 500 \text{ g} \times 4.184 \text{ J/g} \cdot {}^\circ\text{C} \times \Delta T$$

$$\Delta T = \frac{157{,}000 \text{ J}}{500 \text{g} \times 4.184 \text{ J/g} \cdot {}^\circ\text{C}} = 75 \text{ }^\circ\text{C}$$

Workbook Problem 9.9

Predict the energy change for combustion of methane (CH_4) using bond-dissociation energies. Use the following information:

	Bond Energy
C–H	414 kJ/mol
O=O	498 kJ/mol
O–H	464 kJ/mol
C=O	799 kJ/mol

Strategy:	Write a balanced equation for the reaction, and determine the bond energies for both products and reactants. Subtract products for reactants to determine the overall energy change of the reaction.

Step 1:   Write and balance the equation.

$$CH_4 \text{ } (g) + 2 \text{ } O_2 \text{ } (g) \rightarrow CO_2 \text{ } (g) + 2 \text{ } H_2O \text{ } (l)$$

Step 2:   Determine the energy of the bonds on the reactant side.

$$(4 \times 414 \text{ kJ/mol}) + (2 \times 498 \text{ kJ/mol}) = 2652 \text{ kJ/mol}$$

Step 3:   Determine the energy of the bonds on the product side.

$$(2 \times 799 \text{ kJ/mol}) + (4 \times 464 \text{ kJ/mol}) = 3454 \text{ kJ/mol}$$

Step 4:   Determine the energy released by the reaction.

$$\text{Reactants} - \text{products} = 2652 \text{ kJ/mol} - 3454 \text{ kJ/mol} = -802 \text{ kJ/mol}$$

Workbook Problem 9.10

Given the following data, is the reaction of hydrogen and chlorine gas spontaneous at room temperature?

ΔH_f HCl $(g) = -92.3$ kJ/mol $S°$ H$_2$ $(g) = 130.6$ J/K•mol
$S°$ Cl$_2$ $(g) = 223$ J/K•mol $S°$ HCl $(g) = 186.8$ J/K•mol

Strategy: Write a balanced equation for the reaction, and determine ΔH and ΔS for the reaction. Using the Gibbs free-energy equation, determine ΔG for the reaction at 25°C, not forgetting to convert to kelvin first.

Step 1:   Write and balance the equation.

$$H_2 (g) + Cl_2 (g) \rightarrow 2HCl (g)$$

Step 2:   Determine ΔH.

$$\Delta H = \text{products} - \text{reactants} = 2 \times -92.3 \text{ kJ/mol} = -184.6 \text{ kJ/mol}$$

Step 3:   Determine ΔS.

$$\Delta S = 2 \times 186.8 \text{ J/K•mol} - (130.6 \text{ J/K•mol} + 223 \text{ J/K•mol}) = 20 \text{ J/ K•mol}$$

Step 4:   Determine the free energy of the reaction at 25°C, first converting the temperature to K.

$$\Delta G = -184.6 \text{ kJ/mol} - 298K(0.020 \text{ kJ/ K•mol}) = -191 \text{ kJ/mol}$$

Step 5:   Determine the spontaneity of the reaction.

The negative sign on the free-energy term indicates that this reaction is spontaneous at 25°C. If the value here is written per mole of HCl formed, the value is –95.3 kJ/mol. This matches the value for the standard free energy of formation of HCl.

Putting It Together

Balance the decomposition equation:

$$N_2H_4 (l) \rightarrow N_2 (g) + 2H_2 (g)$$

Convert the grams of hydrazine to moles:

$$3g \times \frac{1 \text{ mol}}{32 \text{ g}} = 0.0938 \text{ mol hydrazine}$$

Calculate ΔH:

$$\Delta H = \text{products} - \text{reactants} = 0 - 95.4 \text{ kJ/mol} = -95.4 \text{ kJ/mol}$$

Calculate ΔS:

$$\Delta S = \text{products} - \text{reactants} = [191.5 + 2(130.6)] - 121.2 = 331.5 \text{ J/mol}\bullet\text{K}$$

Calculate ΔG:

$$\Delta G = \Delta H - T\Delta S = -95.4\text{kJ/mol} - 298\text{K}(0.3315 \text{ kJ/mol}\bullet\text{K}) = -194 \text{ kJ/mol}$$

$$-194 \text{ kJ/mol} \times 0.0938 \text{ mol} = -18.2 \text{ kJ}$$

Self–Test

True–False

1. T
2. F. Changes in energies are always calculated final – initial.
3. F. State functions are path independent.
4. T
5. F. Standard state is measured at 25 °C , which is room (or lab) temperature.
6. F. It is possible for a reaction to both transfer heat and do work.
7. F. Melting is an endothermic process—heat must be added.
8. F. Specific heat is an intensive property: it can be used to identify materials.
9. T
10. T
11. T
12. F. A positive change in entropy—an increase in disorder—is favorable.
13. F. Gases have higher entropy than liquids, because they are more disordered.
14. T
15. T

Matching

16. e	17. h	18. d	19. m
20. g	21. l	22. q	23. n
24. p	25. i	26. c	27. k
28. j	29. o	30. a	31. b
32. f			

Fill-in-the-Blank

33. kinetic energy, potential energy	34. temperature, heat
35. negative, exothermic	36. reversible
37. volume, work	38. thermodynamic standard state
39. specific heat	40. surroundings, system, positive
41. bond-dissociation energy, positive	42. entropy, increase
43. Gibbs free-energy, equilibrium, negative, positive	

Problems

44. To easily solve this problem, set the two energies equal to one another to solve for the needed velocity of the marble. Watch your units: they must match for the answer to come out correctly.

$$\frac{1}{2} \text{ x } 1.3 \text{ kg x } \left(5\frac{\text{m}}{\text{s}}\right)^2 = \frac{1}{2} \text{ x } 0.002 \text{ kg x } \left(v\frac{\text{m}}{\text{s}}\right)^2$$

$$16.25 \ \frac{kg \cdot m^2}{s^2} = 0.001 \ v^2 \ kg$$

$$16250 \ \frac{m^2}{s^2} = v$$

$$v = 127 \ \frac{m}{s}$$

45. Work is being done on the system, since the change in volume is negative. Knowing q and w makes it possible to calculate ΔE.

$$w = -P\Delta V = -8 atm \times (0.550L - 0.950L) = 3.2 \ L\bullet atm \times 101 \ J/(L\bullet atm) = 323 \ J = 0.323 \ kJ$$
$$q = 12 \ kJ \ (positive, \ as \ it \ is \ being \ absorbed \ by \ the \ system)$$

For the system, $\Delta E = q + w$; $\Delta E = 12 \ kJ + 0.323 \ kJ = 12.3 \ kJ$.
For the surroundings, $\Delta E = -12.3 \ kJ$

46. Using $q = m \times c \times \Delta T$:

$$150 \ J = 15 \ g \times 0.140 \ J/g\bullet{}^\circ C \times \Delta T$$
$$\Delta T = 150 \ J/(15 \ g \times 0.140 \ J/g\bullet{}^\circ C)$$
$$\Delta T = 71.4 \ {}^\circ C$$

47. For this reaction, both w and q need to be calculated to be able to determine the overall energy change of the system.

$$w = -P\Delta V = -1 atm \times (1L - 3L) = 2 \ L\bullet atm \times 101 \ J/(L\bullet atm) = 202 \ J = 0.202 \ kJ$$

To calculate q it is necessary to know how many moles of chlorine gas were involved in the reaction:

$$3.17 \ \frac{g}{L} \ x \ 1.00 \ L \ = 3.17 \ g$$

$$3.17 \ gx \ \frac{1mol}{71 \ g} = 0.0446 \ mol$$

Multiply by the heat involved per mole of chlorine.

$$0.00446 \ mol \times 80.3 \ kJ/mol = 3.58 \ kJ$$
$$\Delta E = q + w; \ \Delta E = 3.58 \ kJ + 0.202 \ kJ = 3.79 \ kJ.$$

48. From these data, it is possible to calculate ΔS for the reaction, and thus the free energy of the reaction.

$$\Delta S = products - reactants = 266.1 \ J/mol\bullet K - [223.0 \ J/mol\bullet K + 0.5(205 \ J/mol\bullet K)]$$
$$= -59.4 \ J/mol\bullet K$$

This is the entropy change for the overall reaction. For our reaction, which involved 0.0446 mole of chlorine, the entropy change is

$$-59.4 \ J/mol\bullet K \times 0.0446 \ mol = -2.65 \ J/K$$

Free energy is calculated $\Delta G = \Delta H - T\Delta S$.

$$\Delta G = 3.58 \text{ kJ} - 298\text{K} \bullet (-0.00265 \text{ kJ/K}) = 4.37 \text{ kJ}$$

This is not a spontaneous reaction at 25° C.

Because the enthalpy terms and the entropy terms are both unfavorable, this will not be a spontaneous reaction at any temperature.

49. Heat is being absorbed from the water by the salt as it dissolves (indicating an entropy-driven process). The ΔH for the solution is per mole, not per gram, so it is first necessary to convert the grams of Epsom salts to moles.

$$15\text{g} \times \frac{1 \text{ mol}}{246.46 \text{ g}} = 0.0609 \text{ mol} \times 16.11 \frac{\text{kJ}}{\text{mol}} = 0.980 \text{ kJ} = 980 \text{ J}$$

$$q_{rxn} = -q_{soln} = -[(\text{sp. heat}) \times (\text{mass of soln}) \times (\Delta T)]$$

$$980 \text{ J} = -4.184 \frac{\text{J}}{\text{g} \bullet °\text{C}} \times 100 \text{ g} \times \Delta T$$

$$\Delta T = -2.34 \text{ °C}$$

The final temperature will be 24.36 °C + (–2.34 °C) = 22.0 °C

50. A number of calculations are required for this problem: the ice and water must be warmed and two phase changes completed.

$$50 \text{ g } H_2O = 2.8 \text{ moles } H_2O$$

–10 °C to 0 °C:
$$q = mc\Delta T = 50 \text{ g} \times 2.108 \text{ J/g°C} \times 10 \text{ °C} = 1054 \text{ J} = 1.054 \text{ kJ}$$

Ice to water:
$$q = 2.8 \text{ moles} \times 6.01 \text{ kJ/mol} = 16.8 \text{ kJ}$$

0 °C to 100 °C:
$$q = 50 \text{ g} \times 4.184 \text{ J/g°C} \times 100 \text{ °C} = 20,900 \text{ J} = 20.9 \text{ kJ}$$

Water to steam:
$$q = 2.8 \text{ moles} \times 40.7 \text{ kJ/mol} = 114 \text{ kJ}$$

Total energy = 153 kJ.

Of this energy, 75% is used in the phase change from liquid to gas. This makes sense, because this phase transition requires a breaking of all intermolecular attractions. Warming ice or water is simply adding kinetic energy, and the phase change from solid to liquid loosens, but does not break, the interactions between the molecules.

51. This problem requires a bit of algebra. Let the amount of copper = Cu and the amount of gold = Au. The specific heat of the metal chunk can be determined by seeing how much heat it released to the water.

$$q = 25.0 \text{ g} \times 4.184 \text{ J/g} \bullet °\text{C} \times (26.65 \text{ °C} - 25.00 \text{ °C}) = 173 \text{ J}$$

Switching to the metal

$$-173 \text{ J} = 6.73 \text{ g} \times c \times (26.65 \text{ °C} - 100 \text{ °C})$$

$$c = 0.350 \text{ J/g•°C}$$

To solve for the amount of each, it is necessary to set up a system of equations.

$$\text{Cu} + \text{Au} = 6.73\text{g}; \text{ Cu} = 6.73\text{g} - \text{Au}$$

$$(6.73 - \text{Au})\left(0.385 \frac{\text{J}}{\text{g} \cdot \text{°C}}\right) + \frac{\text{Au}}{6.73\text{g}}\left(0.129 \frac{\text{J}}{\text{g} \cdot \text{°C}}\right) = 0.350 \frac{\text{J}}{\text{g} \cdot \text{°C}}$$

Substituting:

$$\frac{6.73\text{g} - \text{Au}}{6.73\text{g}}\left(0.385 \frac{\text{J}}{\text{g} \cdot \text{°C}}\right) + \frac{\text{Au}}{6.73\text{g}}\left(0.129 \frac{\text{J}}{\text{g} \cdot \text{°C}}\right) = 0.350 \frac{\text{J}}{\text{g} \cdot \text{°C}}$$

$$6.73\text{g} - \text{Au}\left(0.385 \frac{\text{J}}{\text{g} \cdot \text{°C}}\right) + \frac{\text{Au}}{6.73\text{g}}\left(0.129 \frac{\text{J}}{\text{g} \cdot \text{°C}}\right) = 0.350 \frac{\text{J}}{\text{g} \cdot \text{°C}} \times 6.73\text{g}$$

$$2.59 \frac{\text{J}}{\text{°C}} - 0.385\text{Au} \frac{\text{J}}{\text{g} \cdot \text{°C}} + 0.129\text{Au} \frac{\text{J}}{\text{g} \cdot \text{°C}} = 2.36 \frac{\text{J}}{\text{°C}}$$

$$2.59 \frac{\text{J}}{\text{°C}} - 2.36 \frac{\text{J}}{\text{°C}} = 0.385\text{Au} \frac{\text{J}}{\text{g} \cdot \text{°C}} - 0.129\text{Au} \frac{\text{J}}{\text{g} \cdot \text{°C}}$$

$$0.23 \frac{\text{J}}{\text{°C}} = 0.256\text{Au} \frac{\text{J}}{\text{g} \cdot \text{°C}}$$

$$\text{Au} = 0.90\text{g}$$

$$\text{Cu} = 5.83\text{g}$$

52. $2 \text{ Fe (s)} + 3/2 \text{ O}_2 \text{ (g)} \rightarrow \text{Fe}_2\text{O}_3 \text{ (s)}$

Given the following:

$$\text{Fe}_2\text{O}_3 \text{ (s)} + 3\text{CO (g)} \rightarrow 2 \text{ Fe (s)} + 3\text{CO}_2 \text{ (g)} \qquad \Delta H° = -26.7 \text{ kJ}$$

$$\text{CO} + \tfrac{1}{2} \text{ O}_2 \rightarrow \text{CO}_2 \text{ (g)} \qquad \Delta H° = -283.0 \text{ kJ}$$

The first reaction needs to be turned so as to give the proper product. Once this is done, the second reaction is fine in the direction it is written, but needs to be multiplied through by 3 to give the proper number of CO_2 molecules.

$$2 \text{ Fe (s)} + 3\text{CO}_2 \text{ (g)} \rightarrow \text{Fe}_2\text{O}_3 \text{ (s)} + 3\text{CO (g)} \qquad\qquad \Delta H° = +26.7 \text{ kJ}$$
$$\underline{3\text{CO} + 3/2 \text{ O}_2 \rightarrow 3\text{CO}_2 \text{ (g)} \qquad\qquad\qquad\qquad\qquad \Delta H° = -849.0 \text{ kJ}}$$
$$2 \text{ Fe (s)} + 3\text{CO}_2 \text{ (g)} + 3\text{CO} + 3/2 \text{ O}_2 \rightarrow \text{Fe}_2\text{O}_3 \text{ (s)} + 3\text{CO (g)} + 3\text{CO}_2 \text{ (g)} \quad \Delta H° = -822.3 \text{ kJ}$$

Cancel terms that appear on both sides.

$$2 \text{ Fe (s)} + 3/2 \text{ O}_2 \rightarrow \text{Fe}_2\text{O}_3 \text{ (s)} \qquad \Delta H° = -822.3 \text{ kJ}$$

53. $2 \text{ NaHCO}_3 \text{ (s)} \rightarrow \text{Na}_2\text{CO}_3 \text{ (s)} + \text{H}_2\text{O (l)} + \text{CO}_2 \text{ (g)}$

Given the following heats of formation, what is the approximate ΔH of this decomposition reaction?

$$\Delta H_f^\circ \text{ NaHCO}_3 = -947.7 \text{ kJ/mol}$$

$$\Delta H_f^\circ \text{ Na}_2\text{CO}_3 = -1131 \text{ kJ/mol}$$

$$\Delta H_f^\circ \text{ H}_2\text{O} \quad = -285.9 \text{ kJ/mol}$$

$$\Delta H_f^\circ \text{ CO}_2 \quad = -393.5 \text{ kJ/mol}$$

The approximate Δ*H* is calculated by subtracting the sum of the heats of formation of the reactants from the sum of the heats of formation of the products.

[(–1131 kJ/mol) + (–285.9 kJ/mol) + (–393.5 kJ/mol)] – (2 × –947.7 kJ/mol) = 85.00 kJ/mol

54. Using the bond-dissociation energies given, calculate the enthalpy of the following reaction:

$$2 \text{ Cl}_2 \text{ (g)} + \text{CH}_4 \text{ (g)} \rightarrow 2 \text{ H}_2 \text{ (g)} + \text{CCl}_4 \text{ (g)}$$

Cl–Cl = 243 kJ/mol	H–H = 436 kJ/mol
C–H = 410 kJ/mol	C–Cl = 330 kJ/mol

Products: Reactants:

2 H–H bonds:	872 kJ/mol	2 Cl–Cl bonds:	486 kJ/mol
4 C–Cl bonds:	1320 kJ/mol	4 C–H bonds:	1640 kJ/mol
	2192 kJ/mol bonds formed		2126 kJ/mol bonds broken

reactants – products = –66 kJ/mol

55. $\Delta G^\circ = \Delta H^\circ - T\Delta S^\circ$, when $\Delta G^\circ = 0$, $T = \Delta H / \Delta S$

ΔH = +66 kJ/mol for the reverse reaction, solve for Δ*S* as follows:

S° Cl$_2$ = 223 J/mol•K	S° H$_2$ = 130.6 J/mol•K
S° CH$_4$ = 188 J/mol•K	S° CCl$_4$ = 214 J/mol•K

Products – reactants =
  [2(223 J/mol•K) + 188 J/mol•K)] – [2(130.6 J/mol•K) + 214 J/mol•K] = 158.8 J/mol•K
                                                                    = 0.1588 kJ/mol•K

T = 66,000/158.8 = 415.6 K = 143 °C

56. First, write and balance the equation for the combustion of pentane.

$$\text{C}_5\text{H}_{12} \text{ (l)} + 8 \text{ O}_2 \text{ (g)} \rightarrow 5 \text{ CO}_2 \text{ (g)} + 6 \text{ H}_2\text{O (l)}$$

$$\Delta H_f^\circ \text{ pentane} \quad = -146.3 \text{ kJ/mol}$$

$$\Delta H_f^\circ \text{ H}_2\text{O} \qquad = -285.9 \text{ kJ/mol}$$

$$\Delta H_f^\circ \text{ CO}_2 \qquad = -393.5 \text{ kJ/mol}$$

$$[5(-393.5 \text{ kJ/mol}) + 6(-285.9 \text{ kJ/mol})] - [-146.3 \text{ kJ/mol}] = -3537 \text{ kJ/mol}$$

$$-3537 \, \frac{\text{kJ}}{\text{mol}} \times \frac{1 \, \text{mol}}{72 \, \text{g}} = -49 \, \frac{\text{kJ}}{\text{g}}$$

$$-49 \, \frac{\text{kJ}}{\text{g}} \times 0.626 \, \frac{\text{g}}{\text{mL}} = -30.8 \, \frac{\text{kJ}}{\text{mL}}$$

Chapter Ten – Gases: Their Properties and Behavior

Workbook Problems

Workbook Problem 10.1

Strategy: Determine the pressure of the gas in mm Hg; convert to atmospheres.

Step 1:   Determine the pressure of the gas in mm Hg.

The pressure is higher inside the manometer. Pressure = 760 mm Hg + 73 mm Hg = 833 mm Hg

Step 2:   Convert from mm Hg to atmospheres.

$$833 \text{ mm Hg} \times \frac{1 \text{ atm}}{760 \text{ mm Hg}} = 1.10 \text{ atm}$$

Workbook Problem 10.2

Strategy: Determine which gas law is involved, convert temperatures, solve for the final volume.

Step 1:   Determine the gas law to use.

This problem involves only volume and temperature: use Charles's law.

Step 2:   Convert the temperatures to K.

$$37 \, ^\circ\text{C} + 273.15 = 310 \text{ K}$$
$$350 \, ^\circ\text{C} + 273.15 = 623 \text{ K}$$

Step 3:   Solve for the final volume.

$$\frac{V_1}{T_1} = \frac{V_2}{T_2} \qquad \frac{27 \text{ L}}{310 \text{ K}} = \frac{V_2}{623 \text{ K}}$$

$$V_2 = \frac{27 \text{ L} \times 623 \text{K}}{310 \text{ K}} = 54.3 \text{ L}$$

Workbook Problem 10.3

Strategy: Rearrange the ideal gas law to solve for pressure, convert grams to moles, and °C to K, solve.

Step 1:   Rearrange the ideal gas law.

$$PV = nRT \text{ becomes } P = \frac{nRT}{V}$$

Step 2:   Convert grams to moles and °C to K.

$$6.7 \text{ g } F_2 \times \frac{1 \text{ mol } F_2}{38 \text{ g } F_2} = 0.176 \text{ mol } F_2$$

$$25.0 \,°C + 273.15 = 298 \text{ K}$$

Step 3:   Solve for the final pressure.

$$P = \frac{nRT}{V} = \frac{0.176 \text{ mol} \times 0.08206 \frac{L \cdot atm}{mol \cdot K} \times 298 \text{ K}}{3.0 \text{ L}} = 1.43 \text{ atm}$$

Workbook Problem 10.4

If 12.3 g of neon were compressed from 1.32L at 1 atm and 25 °C to 18 atm and 750 °C, what would the final pressure be in the container?

Strategy: Determine what form of the gas law is needed, convert temperatures and moles if necessary, and solve for final pressure.

Step 1:   Determine the form of gas law needed.

Although this problem again requires solving for one variable, it is a very different type of problem. Whenever a change is indicated, this is an indication that the full ideal gas law is not needed. Instead, a version of the simple gas law will be required.

Here, the temperature, pressure, and volume are all changing, but the number of moles is not.

$$\frac{P_1 V_1}{T_1} = \frac{P_2 V_2}{T_2}$$

Step 2:   Convert grams to moles and °C to K.
There is no need to convert to moles of fluorine.

$$25 \,°C + 273.15 = 298 \text{ K}$$
$$750 \,°C + 237.15 = 1023 \text{ K}$$

Step 3: Solve for the final pressure.

$$\frac{1 \text{ atm} \cdot 3.2 \text{ L}}{298 \text{ K}} = \frac{18 \text{ atm} \cdot V_2}{1023 \text{ K}}$$

$$V_2 = \frac{1 \text{ atm} \cdot 3.2 \text{ L} \cdot 1023 \text{ K}}{298 \text{ K} \cdot 18 \text{ atm}} = 0.61 \text{ L}$$

Workbook Problem 10.5

Strategy: Determine which gas law is involved; solve for the volumes of oxygen and nitrogen dioxide.

Step 1: Determine the gas law to use.

This is Avogadro's law, or at least a corollary of it. Since the volume of a molar amount of gas is constant, gases can be made to react in equivalent volumes. The volumes will be in stoichiometric ratios.

Step 2: Solve for the volume of oxygen.

$$7.45 \text{ L NO} \times \frac{1 \text{ L O}_2}{2 \text{ L NO}} = 3.73 \text{ L O}_2$$

Step 3: Solve for the volume of nitrogen dioxide.

$$7.45 \text{ L NO} \times \frac{2 \text{ L NO}_2}{2 \text{ L NO}} = 7.45 \text{ L O}_2$$

Workbook Problem 10.6

Strategy: Convert temperatures and pressures to make it possible to solve for the volume of 1 mol under these conditions. Divide the mass of 1 mol of xenon by the volume to determine the density.

Step 1: Convert the temperature and pressure.

$$100 \text{ °C} + 273.15 = 373.15 \text{ K}$$

$$300 \text{ mm Hg} \times \frac{1 \text{ atm}}{760 \text{ mm Hg}} = 0.395 \text{ atm}$$

Step 2: Use the ideal gas law to calculate the volume of the gas.

$$V = \frac{nRT}{P} = \frac{1 \text{ mol} \times 0.08206 \frac{\text{L} \cdot \text{atm}}{\text{mol} \cdot \text{K}} \times 373 \text{K}}{0.395 \text{ atm}} = 77.5 \text{ L}$$

Step 3:   Determine the density of the gas.

$$\text{Mass of 1 mol of xenon} = 131.3 \text{ g}$$

$$\text{Density} = \frac{131.3 \text{ g}}{77.5 \text{ L}} = 1.69 \frac{\text{g}}{\text{L}}$$

Workbook Problem 10.7

Strategy: Decide on a volume of gas, then solve for *n*. Using the density makes it possible to solve for the mass for that volume and then for the molecular weight.

Step 1:   Determine the number of moles in 1 L of gas.

$$n = \frac{PV}{RT} = \frac{3.75 \text{ atm x 1 L}}{0.08206 \dfrac{\text{L} \cdot \text{atm}}{\text{mol} \cdot \text{K}} \text{x}(700°\text{C} + 273.15)\text{K}} = 0.0470 \text{ mol}$$

Step 2:   Solve for the mass in 1 L of gas using the density.

$$\text{Mass} = \text{density} \times \text{volume} = 0.798 \text{ g/L} \times 1 \text{ L} = 0.798 \text{ g}$$

Step 3:   Determine the molecular weight by dividing g/moles.

$$MW = 0.798\text{g}/0.0470 \text{ mol} = 17.0 \text{ g/mol}.$$

Ammonia is a gas that fits this molecular weight.

Workbook Problem 10.8

Strategy: Determine the number of moles of each gas, the total number of moles, and the mole fraction of each component. Using these, calculate the total pressure and the partial pressures using the ideal gas law and Dalton's law.

Step 1:   Determine the number of moles of each gas:

$$18.3 \text{ g } CO_2 \text{ x } \frac{1 \text{ mol}}{44.0 \text{ g}} = 0.416 \text{ mol } CO_2$$

$$23.1 \text{ g Ne x } \frac{1 \text{ mol}}{20.2\text{g}} = 1.14 \text{ mol Ne}$$

$$5.26 \text{ g } H_2 \text{ x } \frac{1 \text{ mol}}{2.02 \text{ g}} = 2.61 \text{ mol } H_2$$

Step 2:   Use the total number of moles to calculate the total pressure:

$$P = \frac{nRT}{V} = \frac{(0.416 \text{ mol} + 1.14 \text{ mol} + 2.61 \text{ mol})\, 0.08206 \dfrac{L \bullet atm}{mol \bullet K}\, (25°C + 273.15)K}{30 \text{ L}}$$

$$= \frac{4.17 \text{ mol} \times 0.08206 \dfrac{L \bullet atm}{mol \bullet K} \times 298 \text{ K}}{30 \text{ L}} = 3.40 \text{ atm}$$

Step 3:   Use the mole fraction (X) of each component to determine the partial pressures:

$$X_{CO_2} = \frac{0.416 \text{ mol}}{4.17 \text{ mol}} = 0.100 \qquad\qquad X_{Ne} = \frac{1.14 \text{ mol}}{4.17 \text{ mol}} = 0.273$$

$$X_{H_2} = \frac{2.61 \text{ mol}}{4.17 \text{ mol}} = 0.626$$

$$P_{CO_2} = 0.100 \times 3.40 \text{ atm} = 0.340 \text{ atm} \qquad P_{Ne} = 0.273 \times 3.40 \text{ atm} = 0.928 \text{ atm}$$

$$P_{H_2} = 0.626 \times 3.40 \text{ atm} = 2.13 \text{ atm}$$

These problems are particularly easy to check because the mole fractions must add up to 1, and the partial pressures must add up to the total pressure, as they do here.

Workbook Problem 10.9

Strategy: Determine the speed of H_2 molecules at 25 °C, then solve the equation again for temperature needed for the much larger I_2 molecules.

Step 1:   Determine the speed of H_2 at 25 °C

$$u = \sqrt{\frac{3RT}{M}} = \sqrt{\frac{3 \times 8.314\, \dfrac{kg \bullet \frac{m^2}{s^2}}{K \bullet mol} \times (25 + 273)K}{0.00202 \dfrac{kg}{mol}}} = \sqrt{3{,}680{,}000 \frac{m^2}{s^2}} = 1918 \frac{m}{s}$$

Step 2:   Determine the temperature required for the iodine molecules to achieve the same speed.

$$1918 \frac{m}{s} = \sqrt{\frac{3 \times 8.314\, \dfrac{kg \bullet \frac{m^2}{s^2}}{K \bullet mol} \times T}{0.2538 \dfrac{kg}{mol}}} = \sqrt{98.27 \frac{\frac{m^2}{s^2}}{K} \times T}$$

Square both sides to solve for T:

$$3{,}680{,}000 \frac{m^2}{s^2} = 98.27 \frac{\frac{m^2}{s^2}}{K} \times T$$

$$T = 37{,}450 \text{ K}$$

Workbook Problem 10.10

Oxygen-16 has an atomic mass of 15.995, and oxygen-18 has an atomic mass of 17.999. What will be their relative rates of effusion?

Strategy: Solve using Graham's law.

Step 1:   Solve using Graham's law:

$$\frac{\text{Rate of effusion of } {}^{16}O}{\text{Rate of effusion of } {}^{18}O} = \sqrt{\frac{35.998 \text{ g } {}^{18}O/\text{mol}}{31.990 \text{ g } {}^{16}O/\text{mol}}} = 1.061$$

Workbook Problem 10.11

Strategy: Solve using the ideal gas law and the van der Waals equation.

Step 1:   Solve using the ideal gas law:

$$P = \frac{nRT}{V} = \frac{1 \text{ mol x } 0.08206 \frac{L \cdot atm}{mol \cdot K} \text{ x } 300 \text{ K}}{0.100 \text{ L}} = 246 \text{ atm}$$

Step 2:   Solve using the van der Waals equation:

$$P = \frac{nRT}{V-nb} - \frac{an^2}{V^2} = \frac{1 \text{ mol x } 0.08206 \frac{L \cdot atm}{mol \cdot K} \text{ x } 300 \text{ K}}{0.100 \text{ L } - (1 \text{ mol x } 0.0318 \text{ L/mol})} - \frac{(1 mol)^2 \text{ x } 1.38 \text{ } (L^2 \cdot atm)/mol^2}{(0.100 \text{ L})^2}$$

$$= \frac{24.6 \text{ L} \cdot atm}{0.0682 \text{ L}} - \frac{1.38 \text{ L}^2 \cdot atm}{0.0100 \text{ L}^2} = 361 \text{ atm} - 138 \text{ atm} = 223 \text{ atm}$$

Step 3:   Compare the two values:

The ideal and real values are reasonably close, given the large pressures involved. Whether the real value would be acceptable would depend on the precision required in the work.

Putting It Together

Strategy: Determine the empirical formula, the number of moles present, and the molecular weight, then determine the molecular formula.

40.56 g Br = 0.5076 mol	18.02 g Cl = 0.5083 mol
28.93 g F = 1.522 mol	0.51 g H = 0.506 mol
12.19 g C = 1.015 mol	

Empirical formula = C_2BrClF_3H, empirical formula weight = 197.4 g/mol

Solve for moles:

$$n = \frac{PV}{RT} = \frac{3.0 \text{ atm x } 13.67 \text{ L}}{0.08206 \dfrac{\text{L} \cdot \text{atm}}{\text{mol} \cdot \text{K}} x 500 \text{ K}} = 1.00 \text{ mol}$$

The empirical formula and the molecular formulas are the same.

Self–Test

True–False
1. F. The atmosphere is approximately 70% nitrogen.
2. T
3. T
4. T
5. T
6. F. The smaller a gas particle is, the faster it travels.
7. F. The ratio is inverse.
8. F. The ideal gas law begins to fail at high pressures.
9. F. We breathe troposphere.
10. T

Matching

11. j	12. a	13. l	14. i
15. h	16. o	17. m	18. b
19. e	20. n	21. c	22. f
23. d	24. p	25. r	26. q
27. g	28. k		

Fill-in-the-Blank
29. pressure, collisions
30. pressure, volume, temperature, moles
31. mole fraction, partial pressure
32. volume, negligible, pressure
33. low, faster

Problems
34. To determine the mm of silicone oil that the pressure will support, first determine the mm of Hg that it will support by converting from atmospheres to mmHg.

$$0.983 \text{ atm x } \frac{760 \text{ mm Hg}}{1 \text{ atm}} = 747 \text{ mm Hg}$$

Then convert from mm Hg to mm silicone oil using the relative densities as a conversion factor.

$$747 \text{ mm Hg x } \frac{13.6 \text{ mm oil}}{0.970 \text{ mm Hg}} = 10,500 \text{ mm oil}$$

35. a) 340 mm ethyl alcohol greater than atmospheric pressure.

$$\text{Pressure in bulb} = 340 \text{ mm EtOH x} \frac{0.789 \text{ mm Hg}}{13.6 \text{ mm EtOH}} = 19.7 \text{ mm Hg above atmospheric pressure.}$$

Total pressure in the bulb = 760 mm Hg + 19.7 mm Hg = 780 mmHg

b) 257 mm ethyl alcohol less than atmospheric pressure.

$$\text{Pressure in bulb} = 257 \text{ mm EtOH} \times \frac{0.789 \text{ mm Hg}}{13.6 \text{ mm EtOH}} = 14.9 \text{ mm Hg below atmospheric pressure.}$$

Total pressure in the bulb = 760 mm Hg – 14.9 mm Hg = 745 mmHg

36. $n = \dfrac{5.78 \text{ atm} \times 0.050 \text{ L}}{0.08206 \dfrac{\text{L} \cdot \text{atm}}{\text{mol} \cdot \text{K}} \times (\text{-20 °C} + 273.15)\text{K}} = 0.0139 \text{ mol}$

If the mass of 0.0139 mol is 0.281 g, the molecular weight is 0.281 g/0.0139 mol = 20.2 g/mol. The gas is neon.

37. Assume that you have 1 L of the gas and determine the number of moles in that liter. Alternatively, you can do the reverse: assume that you have 1 mole of gas, and determine the number of liters. Both will get you the correct answer.

$$n = \frac{4.00 \text{ atm} \times 1.00 \text{ L}}{0.08206 \dfrac{\text{L} \cdot \text{atm}}{\text{mol} \cdot \text{K}} \times (100 \text{ °C} + 273.15)\text{K}} = 0.131 \text{ mol}$$

$$\text{density} = \frac{0.131 \text{ mol}}{\text{L}} \times \frac{16.0 \text{ g}}{\text{mol}} = 2.10 \frac{\text{g}}{\text{L}}$$

38. a. Assume a 100 g sample, and convert 92.3 g C and 7.69 g H to moles.

$$92.3 \text{ g C} \times \frac{1 \text{ mol C}}{12.0 \text{ g C}} = 7.69 \text{ mol C}; \ 7.69 \text{ g H} \times \frac{1 \text{ mol H}}{1.0 \text{ g H}} = 7.69 \text{ mol H}$$

This gives rise to a 7.69:7.69 C to H mole ratio.

There is one carbon atom for every hydrogen atom in the molecule, so the empirical formula is CH.

b. To solve for molar mass, we need to determine the moles of gas.

$$n = \frac{\dfrac{763}{760} \text{ atm} \times 0.250 \text{ L}}{0.08206 \dfrac{\text{L} \cdot \text{atm}}{\text{mol} \cdot \text{K}} \times (35 + 273) \text{ K}} = 0.00993 \text{ mol} \qquad \frac{0.258 \text{ g}}{0.00993 \text{ mol}} = 26.0 \text{ g/mol}$$

c. The empirical formula weight is 13 g/mol, so the molecular formula is two empirical formula units: C_2H_2.

39. Determine the number of moles in 1 L under these conditions.

$$n = \frac{15 \text{ atm} \times 1.00 \text{ L}}{0.08206 \dfrac{\text{L} \cdot \text{atm}}{\text{mol} \cdot \text{K}} \times 573 \text{ K}} = 0.319 \text{ mol}$$

14.04 g/0.319 mol = 44.0 g/mol

40. Write and balance the equation.

$$2 \, C_4H_{10} + 13 \, O_2 \, (g) \rightarrow 8 \, CO_2 \, (g) + 10 \, H_2O \, (g)$$

Determine the number of moles of butane burned.

$$25mL \times 0.6014 \frac{g}{mL} \times \frac{1 \, mol}{58.12 \, g} = 0.26 \, mol$$

Determine the moles of oxygen consumed, then convert to liters.

$$0.26 \, mol \, C_4H_{12} \times \frac{13 \, mol \, O_2}{2 \, mol \, C_4H_{12}} = 1.7 \, mol \, O_2 \, ;$$

$$V = \frac{nRT}{P} = \frac{1.7 \, mol \times 0.08206 \, \frac{L \cdot atm}{mol \cdot K} \times (30 + 273) \, K}{1 \, atm} = 42.2 \, L$$

Determine the number of moles of carbon dioxide produced, then convert to liters.

$$0.26 \, mol \, C_4H_{12} \times \frac{8 \, mol \, CO_2}{2 \, mol \, C_4H_{12}} = 1.04 \, mol \, CO_2 \, ;$$

$$V = \frac{nRT}{P} = \frac{1.04 \, mol \times 0.08206 \, \frac{L \cdot atm}{mol \cdot K} \times (30 + 273) \, K}{1 \, atm} = 25.9 \, L$$

Determine the number of moles of water vapor produced, then convert to liters.

$$0.26 \, mol \, C_4H_{12} \times \frac{10 \, mol \, H_2O}{2 \, mol \, C_4H_{12}} = 1.30 \, mol \, H_2O \, ;$$

$$V = \frac{nRT}{P} = \frac{1.30 \, mol \times 0.08206 \, \frac{L \cdot atm}{mol \cdot K} \times (30 + 273) \, K}{1 \, atm} = 32.3 \, L$$

Determine the overall energy change of the reaction.

Volume of products – volume of reactants = $(32.3 \, L + 25.9 \, L) - (42.2 \, L + 0.26 \, L) = +15.97 \, L$

Work is done by this reaction.

41. Find the number of moles of each gas.

$$7.03 \, g \, Ne \times \frac{1 \, mol}{20.2 \, g} = 0.348 \, mol \qquad 11.7 \, g \, N_2 \times \frac{1 \, mol}{28.0 \, g} = 0.418 \, mol \qquad 15.7 \, g \, He \times \frac{1 \, mol}{4.00 \, g} = 3.93 \, mol$$

The total moles of gas = $0.348 \, mol + 0.418 \, mol + 3.93 \, mol = 4.70 \, moles$

Find the total pressure.

$$P = \frac{nRT}{V} = \frac{(0.348 \text{ mol} + 0.418 \text{ mol} + 3.93 \text{ mol}) \times 0.08206 \frac{L \cdot atm}{mol \cdot K} \times 298 \text{ K}}{5.000 \text{ L}} = 22.97 \text{ atm}$$

Find the mole fraction and partial pressure of each component.

$$Ne : \frac{0.348 \text{ mol}}{4.70 \text{ mol}} \times 23.0 \text{ atm} = 1.70 \text{ atm} \quad N_2 : \frac{0.418 \text{ mol}}{4.70 \text{ mol}} \times 23.0 \text{ atm} = 2.05 \text{ atm}$$

$$He : \frac{3.93 \text{ mol}}{4.70 \text{ mol}} \times 23.0 \text{ atm} = 19.2 \text{ atm}$$

42. $\dfrac{\text{rate He}}{\text{rate H}_2} = \dfrac{\sqrt{1.00794 \times 2}}{\sqrt{4.002602}} = \dfrac{1.4198}{2.00065} = 0.7096$ or $\dfrac{\text{rate H}_2}{\text{rate He}} = \dfrac{\sqrt{4.002602}}{\sqrt{1.00794 \times 2}} = \dfrac{2.00065}{1.4198} = 1.409$

43. $\dfrac{P_1 V_1}{T_1} = \dfrac{P_2 V_2}{T_2}$ $\qquad T_2 = \dfrac{P_2 V_2 T_1}{P_1 V_1} = \dfrac{1 \text{ atm} \times 0.375 \text{ L} \times 298 \text{ K}}{7.8 \text{ atm} \times 0.050 \text{ L}} = 287 \text{ K} = 13.4 \text{ °C}$

Chapter Eleven – Liquids, Solids, and Phase Changes

Workbook Problems

Workbook Problem 11.1

Strategy: Rearrange the equation $\Delta G = \Delta H - T\Delta S$ to solve for the entropy change.

Step 1:   Rearrange to solve for ΔS.

$$0 = \Delta H - T\Delta S$$

$$\Delta S = \frac{\Delta H}{T}$$

Step 2:   Solve for ΔS, being careful to maintain the correct sign.

$$\Delta S = \frac{\Delta H}{T} \approx \frac{-5970 \frac{J}{mol}}{(-107 + 273.15)\text{K}} = -35.9 \frac{J}{mol \cdot K}$$

It is important to remember that in freezing, heat is released and order increases. This means using $-\Delta H_{fus}$. It will be immediately apparent if this was not done correctly, because an increase in entropy will result, while freezing decreases entropy.

Workbook Problem 11.2

Strategy: Use the Clausius–Clapeyron equation to solve for ΔH.

Step 1:   Convert temperatures to K.

$$78.4\ °C + 273.15 = 351.55\ K \qquad\qquad 63.5\ °C + 273.15 = 336.65\ K$$

Step 2:   Solve for ΔH.

$$\Delta H_{vap} = \frac{(\ln P_2 - \ln P_1)R}{\left(\dfrac{1}{T_1} - \dfrac{1}{T_2}\right)} = \frac{(\ln 760 - \ln 400)8.314\,\dfrac{J}{mol \bullet K}}{\left(\left[\dfrac{1}{336.65K}\right] - \left[\dfrac{1}{351.55K}\right]\right)} = \frac{(6.63 - 5.99)8.314\,\dfrac{J}{mol \bullet K}}{0.000126\,\dfrac{1}{K}} = 42.3\ kJ/mol$$

Workbook Problem 11.3

Strategy: Use the Clausius–Clapeyron equation to solve for the vapor pressure at 25 °C.

Step 1:   Use the Clausius–Clapeyron equation to solve for the vapor pressure:

$$\ln P_1 + \frac{42,300\ J/mol}{8.314\,\dfrac{J}{mol \bullet K}(25 + 273.15)K} = \ln 760 + \frac{42,300\ J/mol}{8.314\,\dfrac{J}{mol \bullet K}(351.55K)}$$

$$\ln P_1 + 17.06 = 21.10$$
$$\ln P_1 = 4.035$$
$$P_1 = e^{4.035} = 56.8\ mm\ Hg$$

Workbook Problem 11.4

Polonium has a density of 9.3 g/cm^3 and a simple cubic.

Strategy: Use the number of atoms per unit cell and the molecular weight to determine the mass of a unit cell, then use the density to solve for the volume of a cell. With the volume, determine the length of one side, then determine the atomic radius.

Step 1:   Determine the mass of a unit cell:

A simple cubic structure has a primitive cubic unit cell, which contains one atom.

$$1\ atom \times \frac{1\ mol}{6.022 \times 10^{23}\,atoms} \times 208.98\,\frac{g}{mol} = 3.470 \times 10^{-22}\ g$$

Step 2:   Solve for the volume of a unit cell:

$$\frac{3.470 \times 10^{-22}\,g}{9.3\,\dfrac{g}{cm^3}} = 3.73 \times 10^{-23}\ cm^3$$

Step 3:   Determine the edge length of a unit cell:

$$\text{Edge length} = \sqrt[3]{3.731 \times 10^{-23}\,cm^3} = 3.34 \times 10^{-8}\ cm \times \frac{1\ m}{100\ cm} \times \frac{1 \times 10^{12}\ pm}{m} = 334\ pm$$

Step 4:   Determine the atomic radius:

For a primitive cubic unit cell, the radius of an atom is half the edge length.

Atomic radius = 0.5 × 334 pm = 167 pm.

Putting It Together

Strategy: Draw the electron-dot structure of SO_2 to determine the molecular shape and the intermolecular forces present, then use the information provided to calculate the heats of reaction.

Sulfur and oxygen both have 6 valence electrons, but a total of 18 valence electrons for the total molecule. These can be distributed to provide two valence structures, in which the double bond can occur on either side of the sulfur. The result of this is that, unlike carbon dioxide, sulfur dioxide is bent, and it will have dipole–dipole forces:

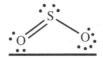

The equation for the reaction of sulfur dioxide with atmospheric oxygen to form sulfur trioxide is

$$SO_2\,(g) + \tfrac{1}{2}\,O_2 \rightarrow SO_3\,(g)$$

Using the provided heats of formation, the heat of reaction is

$$\Delta H_{rxn} = (-395.7\ \text{kJ/mol}) - (-296.8\ \text{kJ/mol} + 0.0\ \text{kJ/mol}) = -98.9\ \text{kJ/mol}$$

When sulfur trioxide reacts with water to form sulfuric acid, the reaction is

$$SO_3\,(g) + H_2O\,(l) \rightarrow H_2SO_4\,(l)$$

Using the provided heats of formation, the heat of this reaction is

$$\Delta H_{rxn} = (-814.0\ \text{kJ/mol}) - (-395.7\ \text{kJ/mol} + -285.8\ \text{kJ/mol}) = -132.5\ \text{kJ/mol}$$

Self–Test

True–False
1. T
2. F. Vapor pressure is temperature dependent. The closer the liquid is to its boiling point, the higher the vapor pressure will be.
3. T
4. F. X ray crystallography is used to examine the structure of crystalline solids.
5. F. The coordination number is the number of other atoms each atom is touching.
6. F. Fullerene and graphite are both allotropes of carbon.
7. T

Matching

8. f	9. c	10. b	11. h
12. g	13. d	14. j	15. a
16. e	17. i		

Multiple Choice

18. d	19. c	20. c	21. a
22. b	23. b	24. a	

Fill-in-the-Blank

25. enthalpy   26. phase changes

27. sublimation   28. decreases, increase

29. ionic solid, molecular solid, covalent network solid

Problems

30. $CH_3CH_2CH_2OH$ this molecule is capable of hydrogen bonding. Because hydrogen bonding is the strongest intermolecular force, this intermolecular force will predominate.

   $CH_3(CH_2)_4CH_3$ this molecule is nonpolar, so it will be unable to interact except with London dispersion forces.

   PCl_3 this is a polar molecule with a dipole. The most important interactions will be dipole–dipole.

31. a. **Br_2** or Cl_2 – Bromine is larger, so will have more dispersion forces and a higher boiling point.

   b. CH_3OH or **CH_3CH_2OH** – both will hydrogen bond, but the second will also have higher dispersion forces.

   c. CH_4 or **CH_3OH** – the second molecule can hydrogen bond, while the first cannot: it will have the highest boiling point.

32. $T = \dfrac{\Delta H_{vap}}{\Delta S_{vap}}$; $T = \dfrac{565,300 \text{ J/mol}}{150.8 \text{ J/K} \cdot \text{mol}} = 3749 \text{ K} = 3476\,^\circ C$

33. $0.237 \text{ g} \times \dfrac{1 \text{ mol}}{18.0 \text{ g}} \times -40.7 \dfrac{\text{kJ}}{\text{mol}} = -0.536 \text{ kJ}$ released by the condensing water.

$$536 \text{ J} = 50.0 \text{ g} \times 0.449 \dfrac{\text{J}}{\text{g} \cdot \,^\circ C} \times \Delta T$$

$$\Delta T = 23.9\ ^\circ C$$

   The final temperature is $25.0\ ^\circ C + 23.9\ ^\circ C = 48.9\ ^\circ C$

34. To heat the ice cube from $-20\ ^\circ C$ to $0.0\ ^\circ C$:

$$q = 45 \text{g} \times 1 \text{mol}/18 \text{g} \times 36.57 \text{ J/g} \bullet\,^\circ C \times 20\ ^\circ C = 1830 \text{ J} = 1.83 \text{ kJ}$$

To melt the ice cube at 0.0 °C:

$$45 \text{ g} \times \frac{1 \text{ mol}}{18 \text{ g}} \times 6.01 \frac{\text{kJ}}{\text{mol}} = 15 \text{ kJ}$$

To warm the water to body temperature:

$$q = 45 \text{ g} \times 4.184 \text{ J/g} \bullet °\text{C} \times 37 °\text{C} = 7.0 \text{ kJ}$$

Total heat absorbed by water/lost by victim = 23.8 kJ.

35. $\ln P_2 = \ln P_1 + \dfrac{\Delta H_{vap}}{R}\left(\dfrac{1}{T_1} - \dfrac{1}{T_2}\right)$

$$\ln P_2 = \ln(760 \text{ mm Hg}) + \frac{30{,}800 \dfrac{\text{J}}{\text{mol}}}{\left(8.3145 \dfrac{\text{J}}{\text{mol} \cdot \text{K}}\right)}\left(\frac{1}{(80.1 + 273.15) \text{ K}} - \frac{1}{(75.0 + 273.15) \text{ K}}\right) = 6.48;$$

$$P_2 = e^{6.48} = 653 \text{ mm Hg}$$

36. Face-centered cubic unit cells contain four total atoms.

Mass of unit cell:

$$4 \text{ atoms} \times \frac{1 \text{ mol}}{6.022 \times 10^{23} \text{ atoms}} \times \frac{196.97 \text{ g}}{\text{mol}} = 1.308 \times 10^{-21} \text{ g}$$

Volume of unit cell:

$$\frac{1.308 \times 10^{-21} \text{ g}}{19.3 \dfrac{\text{g}}{\text{cm}^3}} = 6.779 \times 10^{-23} \text{ cm}^3$$

Edge length of unit cell =

$$\sqrt[3]{6.779 \times 10^{-23} \text{ cm}^3} = 4.147 \times 10^{-8} \text{ cm} = 415 \text{ pm}$$

A face-centered cubic cell has an edge diagonal equal to 4r. Using the Pythagorean theorem:

$$(4r)^2 = 2(415 \text{ pm})^2$$
$$r = 146 \text{ pm}$$

37. A face-centered cubic unit cell has four total atoms, and the diagonal of a side is equal to 2x the atomic diameter.

$$\text{Mass of unit cell} = 4 \text{ atoms} \times \frac{1 \text{ mol}}{6.022 \times 10^{23} \text{ atoms}} \times \frac{106.42 \text{ g}}{\text{mol}} = 7.069 \times 10^{-22} \text{ g}$$

Volume of unit cell:

Diagonal = 137 pm × 4 = 548 pm = (548 × 10^{-10} cm)

Edge length:

$$2l^2 = (548 \times 10^{-10} \text{ cm})^2$$
$$l = 390 \times 10^{-10} \text{ cm}$$
$$\text{Volume} = (390 \times 10^{-10} \text{ cm})^3 = 5.81 \times 10^{-23} \text{ cm}^3$$

$$\text{Density of metal} = \frac{7.069 \times 10^{-22} \text{ g}}{5.81 \times 10^{-23} \text{ cm}^3} = 12.15 \frac{\text{g}}{\text{cm}^3}$$

38. normal freezing point = 197.5 K   triple point = 216.6 K
critical point = 304.25 K   boiling = line C
sublimation = line A   melting = line B

The forward slope of the central line (the freezing/melting line) shows that the solid form of the substance is more dense than the liquid form. As pressure increases, it is more likely to be a solid.

Challenge Problem
39. To solve this problem, first determine the number of joules available.

$$320 \text{ g} \times \frac{1 \text{ mol}}{44.0 \text{ g}} \times 2220 \frac{\text{kJ}}{\text{mol}} = 16,145 \text{ kJ}$$

Now set up an algebraic equation to solve for the mass of water through the phase changes.

x = mass of water

$x/18$ = moles of water

16145 kJ =

$$\left(0.00203 \frac{\text{kJ}}{\text{g} \cdot {}^\circ\text{C}}\right) 20\,^\circ\text{C}x + \left(6.01 \frac{\text{kJ}}{\text{mol}}\right) \frac{x}{18 \frac{\text{g}}{\text{mol}}} + \left(0.004184 \frac{\text{kJ}}{\text{g} \cdot {}^\circ\text{C}}\right) 100\,^\circ\text{C}x + \left(40.67 \frac{\text{kJ}}{\text{mol}}\right) \frac{x}{18 \frac{\text{g}}{\text{mol}}} + \left(0.00208 \frac{\text{kJ}}{\text{g} \cdot {}^\circ\text{C}}\right) 15\,^\circ\text{C}x =$$

$$16145 \text{kJ} = 0.406 \frac{\text{kJ}}{\text{g}} x + 0.334 \frac{\text{kJ}}{\text{g}} x + 0.414 \frac{\text{kJ}}{\text{g}} x + 2.26 \frac{\text{kJ}}{\text{g}} x + 0.0312 \frac{\text{kJ}}{\text{g}} x$$

$$16145 \text{kJ} = 3.445 \frac{\text{kJ}}{\text{g}} x$$

$$x = 4690 \text{g}$$

Chapter Twelve – Solutions and Their Properties

Workbook Problems

Workbook Problem 12.1

Strategy: Determine which is the major component and which is the minor component.

In a 3% solution of hydrogen peroxide, hydrogen peroxide is the solute and the water, which makes up the remaining 97%, is the solvent.

Workbook Problem 12.2

Strategy: Charge and molecular size determine the hydration energy.

Smaller, more highly charged ions have the greatest hydration energy:

$$Cl^- < Li^+ < Mg^{2+}$$

Workbook Problem 12.3

Strategy: Calculate mass percent of solution, volume of solution, and moles of solute to provide all units.

Step 1:   Determine mass percent.

$$\text{mass \%} = \frac{15.43 \text{ g}}{315.43 \text{ g}} \times 100 = 4.89\% \text{ glucose}$$

Step 2:   Calculate the moles of solute and the molality of the solution.

$$15.43 \text{ g} \times \frac{1 \text{ mole}}{180 \text{ g}} = 0.0857 \text{ mol}$$

$$m = \frac{0.0857 \text{ mol}}{0.31543 \text{ kg}} = 0.272 \text{ m}$$

Step 3:   Calculate the volume of the solution and the molarity of the solution.

$$\frac{315.43 \text{ g}}{1.09 \dfrac{\text{g}}{\text{mL}}} = 289 \text{ mL}$$

$$M = \frac{0.0857 \text{ mol}}{0.289 \text{ L}} = 0.296 \text{ M}$$

Workbook Problem 12.4

Strategy: Calculate the initial and final amounts of CO_2 in a saturated solution; determine the moles and then the liters of CO_2.

Step 1: Calculate the moles of dissolved CO_2 under pressure.

$$Solubility = 0.0032\frac{mol}{L \cdot atm} \times 2.3 \; atm = 0.00736 \; \frac{mol}{L}$$

$$Dissolved \; CO_2 = 0.00736 \; mol/L \times 0.500 \; L = 0.00368 \; mol$$

Step 2: Calculate the moles of dissolved CO_2 after pressure is released and equilibrium with the atmosphere reached:

$$Solubility = 0.0032\frac{mol}{L \cdot atm} \times 0.30 \; mm \; Hg \times \frac{1 \; atm}{760 \; mm \; Hg} = 0.00000126 \; \frac{mol}{L}$$

$$0.00000126 \; \frac{mol}{L} \times 0.5 \; L = 0.000000632 \; mol$$

Step 3: Calculate the change in dissolved CO_2—this is the CO_2 that has been released.

$$0.00368 \; mol - 0.000000632 \; mol = 0.00368 \; mol.$$

For all practical purposes, all of the CO_2 has bubbled away.

Step 4: Determine the volume of gas at 25°C and 1 atm pressure.

$$V = \frac{nRT}{P} = \frac{0.00368 \; mol \times 0.08602 \; \frac{L \cdot atm}{mol \cdot K} \times (25 + 273.15)K}{1 \; atm} = 0.0944L = 94.4 \; mL$$

Workbook Problem 12.5

Strategy: Determine the mole fraction of ions needed to lower the vapor pressure by this amount, then convert to grams of each ionic substance.

Step 1: Calculate the mole fraction of water in a solution with this vapor pressure lowering:

$$230.7 \; mm \; Hg = 233.7 \; mm \; Hg \times X_{H_2O}$$

$$0.987 = X_{H_2O}$$

Step 2: Calculate the moles of dissolved ions that will lead to this mole fraction:

$$500 \; g/18.0 \; g/mol = 27.78 \; mol \; H_2O$$

$$X_{H_2O} = \frac{27.78 \text{ mol}}{(27.78 \text{ mol} + x \text{ mol})} = 0.987$$

$$27.78 \text{ mol} = 0.987x(27.78 \text{ mol} + x \text{ mol})$$

$$27.78 - 27.42 = 0.987x$$

$$x = 0.3659 \text{ mol ions}$$

Step 3:   Convert to moles, then grams of $CaCl_2$:

$$0.3659 \text{ mol ions} \times \frac{1 \text{ } CaCl_2}{3 \text{ ions}} = 0.1220 \text{ mol } CaCl_2$$

$$0.1220 \text{ mol } CaCl_2 \times 111 \text{ g/mol} = 13.6 \text{ g } CaCl_2$$

Step 4:   Convert to moles, then grams of LiCl:

$$0.3659 \text{ mol ions} \times \frac{1 \text{ LiCl}}{2 \text{ ions}} = 0.1830 \text{ mol LiCl}$$

$$0.1830 \text{ mol LiCl} \times 42.5 \text{ g/mol} = 7.77 \text{ g LiCl}$$

Workbook Problem 12.6

Strategy: Determine the mole fraction of each component, and calculate the partial pressure of each.

Step 1:   Solve for the mass and moles of each component:

$$0.13 \times 150 \text{ g} = 19.5 \text{ g ethanol}/46 \text{ g/mol} = 0.424 \text{ mol ethanol}$$

$$0.87 \times 150 \text{ g} = 130.5 \text{ g water}/18.0 \text{ g/mol} = 7.25 \text{ mol water}$$

Step 2:   Determine the mole fraction of each component:

$$X_{H_2O} = \frac{7.25 \text{ mol}}{7.25 \text{ mol} + 0.424 \text{ mol}} = 0.945$$

$$X_{EtOH} = \frac{0.424 \text{ mol}}{7.25 \text{ mol} + 0.424 \text{ mol}} = 0.0552$$

Step 3:   Determine the partial pressure of each component and the total pressure:

$$P_{H_2O} = 0.945 \times 23.8 \text{ mm Hg} = 22.5 \text{ mm Hg}$$

$$P_{EtOH} = 0.0552 \times 61.2 \text{ mm Hg} = 3.38 \text{ mm Hg}$$

$$P_{total} = 22.5 \text{ mm Hg} + 3.38 \text{ mm Hg} = 25.9 \text{ mm Hg}$$

Workbook Problem 12.7

Strategy: Using the equation for freezing-point depression, solve for the molality of ions needed, then convert to moles and grams of ammonium sulfate.

Step 1:   Determine the molality of ions needed to generate this freezing-point depression.

$$15.0 \,°C = 1.86 \,°C/m \times X \,m$$

$$X = 8.06 \,m$$

Step 2:   Determine the moles of $(NH_4)_2SO_4$ needed to generate this quantity of ions.

$$8.06 \,m = \frac{x \,\text{moles}}{0.300 \,\text{kg}}$$

$$8.06 \frac{\text{moles}}{\text{kg}} \times 0.300 \,\text{kg} = 2.42 \,\text{mol ions}$$

$$2.42 \,\text{mol ions} \times \frac{1 \,\text{mol salt}}{4 \,\text{mol ions}} = 0.605 \,\text{mol salt}$$

Step 3:   Convert to grams of ammonium sulfate needed.

$$0.605 \,\text{mol} \times \frac{150 \text{g}}{\text{mol}} = 90.8 \,g$$

Workbook Problem 12.8

Strategy: Using the equation for osmotic pressure; solve for the molarity of the solution:

$$7.34 \,\text{atm} = x \,M \times 0.08206 \frac{\text{L} \cdot \text{atm}}{\text{mol} \cdot \text{K}} \times 298 \,K$$

$$x = 0.300 \,M$$

Workbook Problem 12.9

Strategy: Using the equation for osmotic pressure, calculate the molarity of the solution, and from that the molecular weight of the compound.

Step 1:   Calculate the molarity of the solution:

$$\Pi = M \times R \times T$$

$$8.428 \,\text{mm Hg} \times \frac{1 \,\text{atm}}{760 \,\text{mm Hg}} = M \times 0.08206 \frac{\text{L} \cdot \text{atm}}{\text{mol} \cdot \text{K}} \times 298 \text{K}$$

$$M = 0.000453 \,M$$

Step 2:   Calculate the number of moles in the solution:

$$0.00453\frac{mol}{L} \times 0.100 \text{ L} = 0.0000453 \text{ mol or } 45 \text{ } \mu\text{mol}$$

Step 3:   Determine the molecular weight:

$$23 \times 10^{-3}g = 0.0000453 \text{ mol} \times \text{M.W.}$$

$$\text{M. W.} = 507.18 \text{ } \frac{g}{\text{mol}}$$

Putting It Together

Calculate the empirical formula of the compound from the combustion analysis data:

$$2.026 \text{ g} \times \frac{1 \text{ mol}}{44.0 \text{ g}} \times \frac{1 \text{ C}}{CO_2} = 0.0460 \text{ mol C}$$

$$0.0460 \text{ mol C} \times 12\frac{g}{\text{mol}} = 0.553 \text{ g C}$$

$$0.8288 \text{ g} \times \frac{1 \text{ mol}}{18.0 \text{ g}} \times \frac{2 \text{ H}}{H_2O} = 0.0921 \text{ mol H}$$

$$0.0921 \text{ mol H} \times 1.008\frac{g}{\text{mol}} = 0.0928 \text{ g H}$$

Grams oxygen: $0.83 \text{ g} - 0.0928 \text{ g H} - 0.553 \text{ g C} = 0.184 \text{ g O}$

Moles oxygen: $\dfrac{0.184 \text{ g O}}{16.0\frac{g}{\text{mol}}} = 0.0115 \text{ mol O}$

Empirical formula:
oxygen has the smallest number of moles:

$$\frac{0.0115 \text{ mol O}}{0.0115 \text{ mol O}} = 1 \text{ oxygen}$$

$$\frac{0.0921 \text{ mol H}}{0.0115 \text{ mol O}} = 8 \text{ hydrogens per oxygen}$$

$$\frac{0.0460 \text{ mol C}}{0.0115 \text{ mol O}} = 4 \text{ carbons per oxygen}$$

Empirical formula = C_4H_8O

Determine the molarity of the solution formed:

$$1.02 \text{ atm} = x \text{ M} \times 0.08206 \frac{\text{L} \cdot \text{atm}}{\text{mol} \cdot \text{K}} \times 298 \text{ K} \qquad x = 0.0417 \text{ M}$$

$$0.0416 \frac{\text{mol}}{\text{L}} \times 0.250L = 0.0104 \text{ mol} \qquad \frac{1.50 \text{ g}}{0.0104 \text{ mol}} = 144.2 \frac{\text{g}}{\text{mol}}$$

Determine the molecular formula:

Empirical formula weight = 72 g/mol
Molecular weight = 2 × empirical formula weight
Molecular formula = 2 × empirical formula: $C_8H_{16}O_2$

Self–Test

True–False
1. T
2. T
3. F. ppm or ppb are the most convenient units for environmental work.
4. F. Potassium nitrate will dissolve readily in water (polar solvent).
5. T
6. F. Supersaturated solutions are always unstable.
7. F. Increasing pressure causes gases to become more soluble.
8. T
9. F. Concentrated solutions have a lower freezing point.
10. T

Multiple Choice
11. a	12. d	13. c	14. b
15. c			

Matching
16. k	17. f	18. a	19. m
20. d	21. l	22. b	23. g
24. c	25. n	26. h	27. i
28. e	29. j		

Fill-in-the-Blank
30. solvent–solvent, solute–solute   31. positive, increased disorder of both solvent and solute
32. density   33. more
34. drop, rise, drop, osmotic pressure 35. molecular mass
36. solvated, hydrated   37. exothermic, endothermic

Problems
38. The rankings are from most polar (least soluble) to most nonpolar (most soluble). Iodine is completely nonpolar, so will be readily soluble in benzene. Butanol $CH_3(CH_2)_2CH_2OH$ is primarily nonpolar, but has a polar oxygen group that reduces its solubility. NaCl is ionic, and so will be essentially insoluble in benzene.

$$NaCl < CH_3(CH_2)_2CH_2OH < I_2$$

39. a. $\text{mass \%} = \dfrac{28.4 \text{ g}}{(28.4 \text{ g} + 350 \text{ g})} \times 100 = 7.50\%$

b.   mole fraction:

   moles sucrose = 28.4 g/342.3 g/mol = 0.0830 mol

   moles water = 350 g/ 18.0 g/mol = 19.4 mol

   $\text{mol fraction} = \dfrac{0.0830 \text{ mol}}{(19.4 \text{ mol} + 0.0830 \text{ mol})} = 0.00425$

c. $\text{molality} = \dfrac{0.0830 \text{ mol}}{0.350 \text{ kg}} = 0.237 \text{ m}$

d. $\text{molarity} = \dfrac{0.0830 \text{ mol}}{0.374 \text{ L}} = 0.222 \text{ M}$

40. 100 g of this solution will contain 3 g of hydrogen peroxide and have a volume of 100 g/1.01 g/mL = 99 mL

$$\text{molarity} = \dfrac{3 \text{ g} \times \dfrac{1 \text{ mol}}{34 \text{ g}}}{0.099 \text{ L}} = 0.891 \text{ M}$$

41. $15 \text{ppb} = \dfrac{0.015 \text{ mg}}{\text{L}} \times \dfrac{1 \text{ g}}{1000 \text{ mg}} \times \dfrac{1 \text{ mol As}}{74.9 \text{ g}} = 2.00 \times 10^{-7} \text{M}$

   $15 \text{ppb} = \dfrac{0.015 \text{ mg}}{\text{L}} \times \dfrac{1 \text{ g}}{1000 \text{ mg}} \times 100{,}000 \text{ L} = 1.5 \text{ g}$

42. $\text{solubility} = 1.32 \times 10^{-3} \dfrac{\text{mol}}{\text{L} \cdot \text{atm}} \times 0.21 \text{ atm} = 2.77 \times 10^{-4} \text{ M}$

43. $\text{solubility} = 6.25 \times 10^{-4} \dfrac{\text{mol}}{\text{L} \cdot \text{atm}} \times 976 \text{ mm Hg} \times \dfrac{1 \text{ atm}}{760 \text{ mm Hg}} = 8.03 \times 10^{-4} \text{ M}$

44. moles of glucose = 4.5 g/180 g/mol = 0.025 mol
    moles of water = 55 g/18 g/mol = 3.06 mol

$$P_{\text{soln}} = 42.175 \text{ mm Hg} \times \dfrac{3.06 \text{ mol}}{3.06 \text{ mol} + 0.025 \text{ mol}} = 41.9 \text{ mm Hg}$$

45. moles of $FeCl_3$ = 5.8 g /162.2 g/mol = 0.36 mol $FeCl_3$
    moles of ions = 0.36 mol $\times$ 3.4 = 0.122 mol ions
    moles of water = 72.1 mol /18 g/mol = 4.01 mol

$$P_{\text{soln}} = 23.8 \text{ mm Hg} \times \dfrac{4.01 \text{ mol}}{4.01 \text{ mol} + 0.122 \text{ mol}} = 23.1 \text{ mm Hg}$$

46. moles of benzene: $10g/78.0$ g/mol $= 0.128$ mol
    moles of toluene: $10g/92.0$ g/mol $= 0.109$ mol

$$\text{mole fraction benzene} = \frac{0.128 \text{ mol}}{(0.128 \text{ mol} + 0.109 \text{ mol})} = 0.540$$

$$\text{mole fraction tolulene} = \frac{0.109 \text{ mol}}{(0.128 \text{ mol} + 0.109 \text{ mol})} = 0.460$$

$$P_{total} = (0.540 \times 93.4 \text{ mm Hg}) + (0.460 \times 26.9 \text{ mm Hg}) = 62.8 \text{ mm Hg}$$

47. $\Delta T_b = 80.0 \text{ °C} - 76.7 \text{ °C} = 3.3 \text{ °C} = 5.03 \text{ °C/m} \times X$
    $X = 0.656$ m

$$0.656 \text{ m} = \frac{x \text{ mol napthalene}}{1.0 \text{ kg}} = 0.656 \text{ mol naphthalene} \times \frac{128 \text{ g}}{\text{mol}} = 84.0g$$

48. $$m = \frac{10.0 \text{ g NaCl} \times \dfrac{1 \text{ mol}}{58.5 \text{ g}}}{0.090 \text{ kg}} = 1.90 \text{ m} \times \frac{2 \text{ ions}}{\text{mol}} = 3.80 \text{ m ions}$$

$$\Delta T_b = 0.51 \frac{\text{°C}}{\text{m}} \times 3.80 \text{ m} = 1.94 \text{ °C}$$

$$\Delta T_f = 1.86 \frac{\text{°C}}{\text{m}} \times 3.80 \text{ m} = 7.07 \text{ °C}$$

Boiling point $= 101.94$ °C   Freezing point $= -7.07$ °C

49. $\Delta T_f = |3.5 - 5.5| = 2.0 \text{ °C}$

$2.0 \text{ °C} = 5.12 \dfrac{\text{°C}}{\text{m}} \times X$

$X = 0.390$ m

$0.390 \text{ m} = x \text{ mol}/0.075 \text{ kg}$ $x = 0.0292$ mol

MW $= 2.50 \text{ g}/0.0292 \text{ mol} = 85.33$ g/mol

50. moles glucose $= 18.0g/180$ g/mol $= 0.10$ mol

molarity $= 0.10 \text{ mol}/1.2 \text{ L} = 0.0833$ M

$$\Pi = 0.0833 \text{ M} \times 0.08206 \frac{\text{L} \cdot \text{atm}}{\text{K} \cdot \text{mol}} \times (273.15 + 15)\text{K} = 1.97 \text{atm}$$

51. $16.4 \text{ mm Hg} \times \dfrac{1 \text{ atm}}{760 \text{ mm Hg}} = X \text{ M} \times 0.08206 \dfrac{\text{L} \cdot \text{atm}}{\text{K} \cdot \text{mol}} \times (273.15 + 15) \text{K}$

$$X = 9.13 \times 10^{-4} \text{ M} \times 1.00 \text{ L} = 9.13 \times 10^{-4} \text{ mol}$$

$$\dfrac{0.250 \text{ g}}{9.13 \times 10^{-4} \text{ mol}} = 274 \dfrac{\text{g}}{\text{mol}}$$

Challenge Problem

52. First, determine the mole fraction of pentane in the vapor.

100 g vapor will be 43.2 g pentane, 56.8 g hexane

43.2 g pentane /72 g/mol = 0.60 mol   56.8 g hexane /86 g/mol = 0.66 mol

$$X_{pentane} = \dfrac{0.60 \text{ mol}}{0.60 \text{ mol} + 0.66 \text{ mol}} = 0.476$$

$$X_{hexane} = \dfrac{0.66 \text{ mol}}{0.60 \text{ mol} + 0.66 \text{ mol}} = 0.524$$

Now set up a two-variable system to solve for the mole fractions in the solution.

$0.476 P_{total} = 425\ x$, where x = mole fraction of pentane in solution

$0.524 P_{total} = 151\ (1 - x)$, where $1 - x$ = mole fraction of hexane in solution

$$P = 425\ x/0.476 = 893\ x$$
$$-\quad \underline{P = [151\ (1 - x)]/0.524 = 288 - 288x}$$
$$0 = -288 + 1181\ x$$

$x = 0.243$ = mole fraction of pentane $1 - x = 0.757$ = mole fraction of hexane

$$P_{total} = (151 \text{ mmHg} \times 0.757) + (425 \text{ mmHg} \times 0.243) = 218 \text{ mmHg}$$

Chapter Thirteen – Chemical Kinetics

Workbook Problems

Workbook Problem 13.1

Step 1:   Balance the equation:

$$2 \text{ HBr } (g) \rightarrow \text{H}_2\ (g) + \text{Br}_2\ (g)$$

Step 2:   Write the relative reaction rates:

$$\text{General rate} = -\dfrac{\Delta \text{HBr}}{2\Delta t} = \dfrac{\Delta \text{H}_2}{\Delta t} = \dfrac{\Delta \text{Br}_2}{\Delta t} \quad \text{or}$$

Step 3:   Determine the rate of formation of bromine from the rate of disappearance of HBr.

$$-\frac{-3.26 \times 10^{-3} \text{ M/s}}{2} = 0.00163 \text{ M/s}$$

This is also the rate of formation of hydrogen.

Workbook Problem 13.2

Step 1:   Zeroth-order reaction:

Rate = k – units will be M/s.

Step 2:   First-order reaction:

Rate = $k[A]$, so M/s = $k \cdot$ M, $k = s^{-1}$

Step 3:   Second-order reaction:

Rate = $k[A]^2$, so M/s = $k \cdot M^2$, $k = M^{-1}s^{-1}$

Workbook Problem 13.3

Step 1:   Determine the order of the reaction with respect to NO_2.

In reactions 1 and 2, the concentration of NO_2 is doubled, while the concentration of F_2 is held constant. Doubling $[NO_2]$ causes the rate to double, so the reaction is first order in NO_2.

Step 2:   Determine the order of the reaction with respect to H_2.

In reactions 2 and 3, the concentration of NO_2 is held constant while the concentration of F_2 is varied. Doubling $[F_2]$ causes the rate of the reaction to quadruple, so the reaction is second-order in F_2.

Step 3:   Write the equation for the rate law.

$$\text{Rate} = k[NO_2][F_2]^2$$

Step 4:   Calculate the value of k.

We can calculate the value of k using the data from any one of the experiments. Using the data in experiment 1 gives

$$0.026\frac{M}{s} = k(0.10 \ M)(0.10 \text{ M})^2$$

$$k = \frac{0.026\frac{M}{s}}{1.0 \times 10^{-3} M^3} = 26.0\frac{1}{M^2 \cdot s}$$

Workbook Problem 13.4

Step 1:   Solve for k:

$$\ln \frac{0.03 \text{ M}}{0.10 \text{ M}} = -k(73 \text{ min})$$

$$k = 0.0165 \text{ min}^{-1}$$

Step 2:   Determine the concentration of the reactant at 80% completion:

$$0.10 \text{ M} \times 0.80 = 0.08 \text{ M}$$
$$0.10 \text{ M} - 0.08 \text{ M} = 0.02 \text{ M}$$

Step 3:   Solve for t at 80% completion:

$$\ln \frac{0.02 \text{ M}}{0.10 \text{ M}} = -0.0165 \text{ min}^{-1} \times t$$

$$t = 97.5 \text{ min}$$

Workbook Problem 13.5

Strategy: Substitute into the half-life equation.

$$t_{1/2} = \frac{0.693}{0.0165 \text{ min}^{-1}} = 42 \text{ min}$$

Workbook Problem 13.6

Step 1:   Solve for k:

$$5730 \text{ years} = \frac{\ln 2}{k}$$

$$k = 1.21 \times 10^{-4} \text{ year}^{-1}$$

Step 2:   Substitute into the integrated rate law and solve for t after 23% decay:

$$\ln (1.0-0.23) = -1.21 \times 10^{-4}/\text{year} \times t$$

$$t = 2160 \text{ years old}$$

Workbook Problem 13.7

Step 1:   Solve for k

$$\frac{1}{[A]_t} = kt + \frac{1}{[A]_0}$$

$$\frac{1}{0.0495 \text{ M}} = k \times 400 \text{ s} + \frac{1}{0.1000 \text{ M}}$$

$$k = 0.0255 \text{ M}^{-1}s^{-1}$$

Step 2: Use the half-life equation to determine the time required for 50% completion.

$$t_{1/2} = \frac{1}{0.0255 \text{ M}^{-1}s^{-1} \times 0.100 \text{ M}} = 392 \text{ s}$$

Step 3: Determine the concentration of NO_2 at 95% completion (5% of NO_2 remaining).

$$0.100 \text{ M} \times 0.05 = 0.00500 \text{ M}$$

Step 4: Substitute the concentrations into the integrated rate law, and solve for t.

$$\frac{1}{0.00500 \text{ M}} = 0.0255 \text{ M}^{-1}s^{-1} \times t + \frac{1}{0.100 \text{ M}}$$

$$t = 7450 \text{ s} \times \frac{1 \text{ hour}}{3600 \text{ s}} = 2.07 \text{ hours, or 2 hours 4 minutes}$$

Workbook Problem 13.8

Step 1: Solve for k.

$$\text{Rate} = k = 0.026 \text{ M} \cdot s^{-1}$$

Step 2: Substitute into the integrated rate law and solve for [A] after 20 seconds:

$$[A] = -0.026 \text{ M} \cdot s^{-1} \times 20 \text{ s} + 1.0 \text{ M} = 0.48 \text{ M}$$

Workbook Problem 13.9

Step 1: Write the expected rate law:

$$\text{Rate} = k[A]^2$$

Step 2: Sum for the overall reaction:

$$2 \text{ A} + A_2 + \text{B} \rightarrow A_2 + A_2\text{B, cancels to}$$

$$2 \text{ A} + \text{B} \rightarrow A_2\text{B}$$

Workbook Problem 13.10

If a reaction has a rate constant of 0.0123 s^{-1} at 700 K and 0.00342 s^{-1} at 250 K, what is the activation energy of the reaction, and what will the rate constant be at 500 K?

Strategy: Solve for E_a using the two-point Arrhenius equation, then use the same equation to solve for the rate constant at 500 K.

Step 1: Solve for E_a using the two-point Arrhenius equation:

$$\ln\left(\frac{0.00342\ \text{s}^{-1}}{0.0123\ \text{s}^{-1}}\right) = \left(\frac{-E_a}{8.314\ \text{J/mol} \cdot \text{K}}\right)\left(\frac{1}{250\ \text{K}} - \frac{1}{700\ \text{K}}\right)$$

$$E_a = 4138\ \text{J/mol}$$

Step 2: Solve for k at 500 K:

$$\ln\left(\frac{k_2}{0.0123\ \text{s}^{-1}}\right) = \left(\frac{-4138\ \text{J/mol}}{8.314\ \text{J/mol} \cdot \text{K}}\right)\left(\frac{1}{500\ \text{K}} - \frac{1}{700\ \text{K}}\right)$$

$$\ln\left(\frac{k_2}{0.0123\ \text{s}^{-1}}\right) = -0.284$$

$$\left(\frac{k_2}{0.0123\ \text{s}^{-1}}\right) = e^{-0.284}$$

$$k_2 = 0.00926\ \text{s}^{-1}$$

Putting It Together

a. The first step has a molecularity of 1. The second step is bimolecular.

b. The reaction is first order in $Ni(CO)_4$. Because changing the concentration of reactants in the second step has no effect on the overall rate, the first step is rate-determining.

The rate law is

$$\text{Rate} = k[Ni(CO)_4]$$

c. Substitute into the integrated first-order rate law. $[Ni(CO)_4]_0 = 0.30$ M, $[Ni(CO)_4]_t = 0.30$ M $\times$ 0.40 = 0.12 M.

$$\ln\frac{0.12\ \text{M}}{0.30\ \text{M}} = -9.3 \times 10^{-3}\text{s}^{-1} \times t$$

$$t = 98.5\ \text{s}$$

d. Determine the overall reaction by summing the reaction steps:

$$Ni(CO)_4 + P(C_6H_5)_3 \rightarrow Ni(CO)_3[P(C_6H_5)_3] + CO$$

Determine the number of moles of nickel:

$$0.237 \text{ g Ni(CO)}_4 \text{ x } \frac{1 \text{ mol}}{170.69 \text{ g}} = 1.39 \text{ x } 10^{-3} \text{mol}$$

If the reaction goes to completion, the number of moles of triphenylphosphine will be equal to the starting moles of $Ni(CO)_4$, because the ratio is 1:1.

The amount of $Ni(CO)_4$ remaining after 2 minutes can be calculated as follows:

$$\ln \frac{x \text{ mol}}{1.39 \times 10^{-3} \text{ mol}} = -9.3 \times 10^{-3} \text{s}^{-1} \times 120 \text{ s}$$

$$\frac{x \text{ mol}}{1.39 \times 10^{-3} \text{ mol}} = e^{-1.116}$$

$$x = 4.55 \times 10^{-4} \text{ mol}$$

The amount of $Ni(CO)_4$ consumed is equal to the amount of triphenylphosphine produced:

$$(1.39 \times 10^{-3} \text{ mol} - 4.55 \times 10^{-4} \text{ mol}) \times 433 \text{ g/mol} = 0.405 \text{ g}$$

The reaction is 32.7% complete.

Self–Test

True–False

1. T
2. F. The exponents must be experimentally determined.
3. F. The order of a reaction is determined by adding the exponents in the rate law.
4. T
5. T
6. F. The half-life is the time required for the concentration to drop to half its initial level.
7. T
8. F. Zeroth-order reactions are quite uncommon.
9. F. Reaction steps can occur at all different rates.
10. T
11. F. Activation energy is always positive, although reactions can be net exothermic.
12. T
13. F. Biological reactions almost invariably involve catalysts.
14. T

Multiple Choice

15. b	16. c	17. a	18. a
19. c	20. d	21. d	22. c
23. b	24. d		

Matching

25. k	26. d	27. n	28. b
29. f	30. a	31. m	32. c
33. o	34. i	35. p	36. g
37. h	38. j	39. l	40. e

Fill-in-the-Blank

41. reaction rate, decrease, increase, product
42. rate constant, k
43. 2, 4, 8, 1
44. reverse reaction
45. first-order
46. elementary reactions, rate-determining step, slowest
47. activation energy, transition state
48. lower energy
49. double
50. adsorb, release/desorb

Problems

51. $\dfrac{\Delta[C]}{\Delta t}$ can be calculated arithmetically.

$$\text{initial rate} = \frac{\Delta[C]}{\Delta t} = \frac{0.005 - 0.000}{15 - 0} = 3.3 \times 10^{-4} \text{ M/s}$$

$$\text{rate from 30 to 75s} = \frac{\Delta[C]}{\Delta t} = \frac{0.019 - 0.010}{75 - 30} = 2.0 \times 10^{-4} \text{ M/s}$$

52. Rate of appearance of $H_2 = \left[1.67 \times 10^{-2} \text{ mol } PH_3/(L \cdot s)\right] \times \dfrac{6 \text{ mol } H_2}{4 \text{ mol } PH_3} = 2.50 \times 10^{-2} \text{ mol } H_2/(L \cdot s)$

53. When the value of [NO] is doubled, the reaction rate increases by a factor of four. The reaction is second order in NO. When the concentration of Cl_2 is doubled, the reaction rate doubles. The reaction is first order in Cl_2.

The rate law is

$$\text{Rate} = k[NO]^2[Cl_2]$$

Using the data in the second experiment:

$$63.6 \text{ mol/(L} \cdot \text{s)} = k(0.050 \text{ mol/L})^2 (0.025 \text{ mol/L})$$

$$k = \frac{63.6 \text{ mol/(L} \cdot \text{s)}}{6.25 \times 10^{-5} \text{ mol}^3/L^3} = 1.02 \times 10^6 \text{ L}^2/(\text{mol}^2 \cdot \text{s})$$

54. When the concentration of A is doubled, the reaction rate increases by a factor of eight, so the reaction is third order in A. When the concentration is B is doubled, the rate is unaffected: the reaction is zero order in B.

The reaction is third order overall.

The rate law is

$$\text{Rate} = k[A]^3$$

Using the data in second experiment:

$$0.134 \ mol/(L\cdot s) = k(0.300 \ mol/L)^3$$

$$k = \frac{0.134 \ mol/(L \cdot s)}{(0.027 \ mol^3/L^3)} = 4.96 \ L^2/(mol^2 \cdot s)$$

$$Rate = 4.96 \ L^2/(mol^2 \cdot s) \times (0.542 \ mol/L)^3 = 0.790 \ mol/(L \cdot s)$$

55. $\dfrac{1}{[O_3]} = \left[1.40 \times 10^{-2} L/(mol \cdot s)\right]\left(12 \ h \times \dfrac{3600 \ s}{1 \ h}\right) + \dfrac{1}{2.37 \ M} = 605 \ L/mol$

$$[O_3] = 1.65 \times 10^{-3} \ M$$

56. The reaction order can be determined graphically by plotting the data as [HI] versus time, ln [HI] versus time, and 1/[HI] versus time. The graph that provides a straight line provides the order of the reaction.

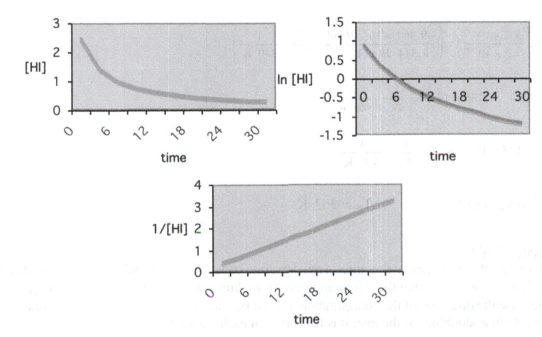

The plots show that the reaction is second order. Using the integrated rate law and the initial rate,

$$\frac{1}{1.45 \ M} = k(3 \ min) + \frac{1}{2.50 \ M}$$

$$k = \frac{\dfrac{1}{1.45 \ M} - \dfrac{1}{2.50 \ M}}{3 \ min} = 9.66 \times 10^{-2} \ L/(mol \cdot min)$$

$$t_{\frac{1}{2}} = \frac{1}{\left[9.66 \times 10^{-2} \ L/(mol \cdot min)\right] \times 2.50 \ mol/L} = 4.14 \ min$$

57. The overall reaction is

$$2\ NO_2\ (g)\ +\ Cl_2\ (g)\ \rightarrow\ 2\ NO_2Cl\ (g)$$

Both steps are bimolecular, and Cl (g) is a reaction intermediate: it does not appear in the final equation.

The expected rate is

$$\text{Rate} = k[NO_2][Cl_2].$$

This is a second-order reaction.

58. $$k\ =\ Ae^{\frac{-E_a}{RT}}$$

$$A = \frac{k}{e^{\frac{-E_a}{RT}}} = \frac{3.79 \times 10^{-5}\,L/mol \bullet s}{e^{\frac{-1.18 \times 10^4\,J/mol}{8.314\,\frac{J}{mol \bullet K} \times 381\,K}}} = \frac{3.79 \times 10^{-5}\,L/mol \bullet s}{e^{-3.73}} = 1.57 \times 10^{-3}$$

59. $$\ln\left(\frac{3.2 \times 10^{-3}}{7.2 \times 10^{-5}}\right) = \left(\frac{-9.862 \times 10^4\ J/mol}{8.314\ J/(mol \cdot K)}\right)\left(\frac{1}{T_2} - \frac{1}{340\ K}\right)$$

$$3.79 = -1.186 \times 10^4\ K\left(\frac{1}{T_2} - \frac{1}{340\ K}\right)$$

$$-3.195 \times 10^{-4}\ K^{-1} = \frac{1}{T_2} - \frac{1}{340\ K}$$

$$\frac{1}{T_2} = 3.26 \times 10^{-3}\ K^{-1} \qquad T_2 = 307\ K$$

Challenge Problem

60. To calculate the initial rate requires a rate law, rate constant, and the initial concentrations of C_2H_5I and/or OH^-. From the information given, we know that the rate = $k[C_2H_5I][OH^-]$, because the doubling of the concentration of each reactant while the other is held constant results in a doubling of the rate: it is first order in each reactant.

The rate constant, k, is calculated using the Arrhenius equation, $k = Ae^{-E_a/RT}$.

$$k = 1.09 \times 10^9\ M^{-1}s^{-1}e^{-\left(7.97 \times 10^4\ J/mol\right)/\left(8.314\ J/mol \cdot K\right)\left(318.15\,K\right)} = 8.956 \times 10^{-5}\ M^{-1}s^{-1}.$$

Remembering that the two 250 mL solutions are going to be combined, and assuming that they sum to 500 mL, the molar concentrations of the reactants are

$$\frac{0.475\ g\ KOH}{0.500\ L} \times \frac{1\ mol}{56.1\ g} = 1.69 \times 10^{-2}\ M\ OH$$

$$\frac{1.378 \text{ g } C_2H_5I}{0.500 \text{ L}} \times \frac{1 \text{ mol}}{156 \text{ g}} = 1.76 \times 10^{-2} \text{ M } C_2H_5I$$

$$\text{Rate} = \left(8.956 \times 10^{-5} \text{ M}^{-1}\text{s}^{-1}\right)\left(1.692 \times 10^{-2} \text{ M}\right)\left(1.768 \times 10^{-3} \text{ M}\right) = 2.68 \times 10^{-8} \text{ M} \cdot \text{s}^{-1}$$

Chapter Fourteen – Chemical Equilibrium

Workbook Problems

Workbook Problem 14.1

Step 1:   Write the equation for the equilibrium constant:

$$K_c = \frac{[H][A]}{[HA]}$$

Step 2:   Determine the concentration of each species:

If the solution is 1.37% ionized, the concentrations of H and A will be 0.0137 × 0.250 M = 3.43 × 10^{-3} M.   The concentration of HA will be 0.247 M

Step 3:   Solve for K_c:

$$K_c = \frac{[3.43 \times 10^{-3}][3.43 \times 10^{-3}]}{[0.247]} = 4.77 \times 10^{-5}$$

Workbook Problem 14.2

Strategy: Write the equilibrium equation for K_c.

$$K_c = \frac{[PCl_3][Cl_2]}{[PCl_5]}$$

Step 1:   Solve for K_c, using the molar concentrations given.

$$K_c = \frac{(1.56)(1.56)}{(2.35)} = 1.04$$

Step 2:   Determine Δn.

$$\Delta n = 2 - 1 = 1$$

Step 3: Solve for K_p, using the equation that expresses the relationship between K_c and K_p.

$$K_p = (1.04)\left[\left(\frac{0.0821\ L\cdot atm}{mol\cdot K}\right)(250\ K)\right]^1 = 21.3$$

Workbook Problem 14.3

Step 1: Write the balanced equation for the reaction.

$$C\ (s)\ +\ H_2O\ (g)\ \leftrightarrows\ CO\ (g)\ +\ H_2\ (g)$$

Step 2: Make a table listing the initial concentration, the change in concentration, and the equilibrium concentration. (Let x = the amount of substance that reacts.)

Principal Reaction	$C\ (s)$	+	$H_2O\ (g)$	$\leftrightarrows$	$CO\ (g)$	+	$H_2\ (g)$
Initial Concentration (M)			11.75 M		0		0
Change (M)			$-x$		$+x$		$+x$
Equilibrium Conc. (M)			$11.75-x$		$+x$		$+x$

Step 3: From the balanced equation, write the equilibrium equation. Substitute the algebraic expressions for the equilibrium concentrations into the equilibrium equation, and solve for x. Use the quadratic equation if necessary.

$$K_c = 3.0\times10^{-2} = \frac{[CO][H_2]}{[H_2O]} = \frac{(x)(x)}{(11.75-x)}$$

$$3.0\times10^{-2}(11.75-x) = (x)^2$$

$$x^2 + (3.0\times10^{-2})x - 0.353 = 0$$

$$x = \frac{-(3.0\times10^{-2})^2 \pm \sqrt{(3.0\times10^{-2}) - 4(-0.353)}}{2}$$

$x = 0.579$ or $x = -0.609$. The meaningful solution is $x = 0.579$. Substituting the value of x back into the equilibrium concentrations, we have

$[CO] = [H_2] = 0.579\ M$ $[H_2O] = 11.75 - 0.579 = 11.17\ M$

Workbook Problem 14.4

Step 1: Determine the direction of reaction needed to minimize each stress.

BrNO added: reaction will shift away from BrNO to the right, toward products

BrNO removed: reaction will shift toward BrNO to the left, toward reactants

NO added: reaction will shift away from NO, to the left, toward reactants

Br_2 removed: reaction will shift toward Br_2, to the right, toward products.

Workbook Problem 14.5

Step 1: Determine the direction of reaction needed to minimize each stress.

pressure increased: reaction will shift toward N_2O_4; to the left, toward reactants.

pressure decreased: reaction will shift away from N_2O_4; to the right, toward products.

Ne added: equilibrium will be unaffected.

Workbook Problem 14.6

Step 1: Determine the direction of reaction needed to minimize each stress and its effect on $[CO_2]$.

Increased pressure: increased $[CO_2]$ (3 mol of gas on left, 2 mol on right)

Decreased pressure: decreased $[CO_2]$

Removed catalyst: decreased $[CO_2]$

Lowered temperature: increased $[CO_2]$

Added O_2:   increased $[CO_2]$

Putting It Together

Dalton's law of partial pressures can be used to determine the pressure of all species present at equilibrium. As with any equilibrium problem, the first step is to set up a chart of initial values, changes, and final values, using the stoichiometry of the balanced equation.

First, convert to atmospheres:

$$NO = \frac{78.4 \text{ mm Hg}}{760 \ \frac{\text{mm Hg}}{\text{atm}}} = 0.103 \text{ atm} ;$$

$$Br_2 = \frac{51.3 \text{ mm Hg}}{760 \ \frac{\text{mm Hg}}{\text{atm}}} = 0.0675 \text{ atm} ;$$

$$\text{final pressure} = \frac{120.5 \text{ mm Hg}}{760 \ \frac{\text{mm Hg}}{\text{atm}}} = 0.159 \text{ atm}$$

The total pressure at equilibrium is equal to the sum of the pressures of each species at equilibrium. The total pressure at equilibrium is 0.159 atm.

$$0.159 = (0.103 - 2x) + (0.0675 - x) + 2x = -x + 0.1705$$

$$x = 0.0115 \text{ atm}$$

	2 NO (g) +	Br$_2$ (g) →	2 NOBr (g)
Initial Pressure	0.103	0.0675	0
Change	$-2x$	$-x$	$+2x$
Eq Pressure	$0.103 - 2x$	$0.0675 - x$	$2x$

At equilibrium:

$$P(NO) = 0.103 \text{ atm} - 2(0.0115 \text{ atm}) = 0.0800 \text{ atm}$$

$$P(Br_2) = 0.0675 \text{atm} - 0.0115 \text{ atm} = 0.0560 \text{ atm}$$

$$P(NOBr) = 2(0.0115 \text{ atm}) = 0.0230 \text{ atm}$$

$$K_p = \frac{(0.0230)^2}{(0.0800)^2 (0.0560)} = 1.48$$

Self–Test

True–False
1. T
2. F. Reaction needs to be run at different concentrations, not temperatures.
3. F. Equilibrium constants are unitless.
4. T
5. F. There will be more products than reactants.
6. F. The reaction will go in the direction of fewer moles of gas.
7. T
8. T
9. T
10. T

Multiple Choice

11. d	12. b	13. c	14. a
15. d	16. a	17. d	18. c
19. a	20. b		

Matching

21. g	22. d	23. a	24. j
25. b	26. h	27. c	28. i
29. f	30. e		

Fill-in-the-Blank
31. dynamic, concentration, constant
32. heterogeneous, liquids, solids
33. minimize, products/right, reactants/left, temperature, kinetic, catalyst, increases
34. direction, equilibrium, products, reactants, reactants, products
35. lowering

Problems

36. a. $K_p = \dfrac{[O_2][ClO]}{[O_3][ClO]}$ b. $K_p = \dfrac{[NOBr]^2}{[NO]^2[Br_2]}$

c. $K_p = \dfrac{1}{[O_2]}$

d. $K_p = \dfrac{[N_2]^2[O_2]}{[N_2O]^2}$

37. $K_p = \dfrac{(p_{NO})^2}{(p_{N_2})(p_{O_2})}$ $K_p = \dfrac{(0.045)^2}{(0.27)(0.187)} = 4.0 \times 10^{-2}$

38. $K_c = \dfrac{K_p}{(RT)^{\Delta n}}$ $\Delta n = 2 - 2 = 0;$ $K_c = K_p = 5.1 \times 10^{-2}$

39. a. mainly products
    b. slightly more products, reactants also present
    c. mainly reactants

1. To determine if a system is at equilibrium, you must first solve for the value of Q_c and then compare it to the value of K_c.

$$Q_c = \dfrac{[PCl_3][Cl_2]}{[PCl_5]}$$ $$Q_c = \dfrac{(0.83)(0.83)}{(2.43 \times 10^{-3})} = 2.83 \times 10^{2}$$

Since $Q_c > K_c$, the system is not at equilibrium. The reaction will shift from right to left to achieve equilibrium.

41. To solve this problem, make a table of initial, change, and final concentrations. The equilibrium concentrations of N_2O and O_2 can be determined by allowing x to be the concentration of N_2O that reacts on going to the equilibrium state. Since the mole ratios are 2:3:4, if $2x$ mol/L of N_2O reacts, then $3x$ mol/L of O_2 reacts and $4x$ mol/L of NO_2 are produced.

This can be summarized in the following table.

Principal Reaction	2 N_2O (g) +	3 O_2 (g) →	4 NO_2 (g)
Initial Concentration (M)	0.0342	0.0415	0
Change (M)	$-2x$	$-3x$	$+4x$
Eq Concentration (M)	$0.0342 - 2x$	$0.0415 - 3x$	$+4x$

We know that the equilibrium concentration of NO_2 is 0.0298 M and we can, therefore, solve immediately for the value of x. $4x = 0.0298; x = 7.45 \times 10^{-3}$

The value of x can be substituted back into the expressions for the equilibrium concentrations of N_2O and O_2.

$$[N_2O] = 0.0342\ M - (2 \times 7.45 \times 10^{-3}\ M) = 1.93 \times 10^{-2}\ M;$$

$$[O_2] = 0.0415\ M - (3 \times 7.45 \times 10^{-3}\ M) = 1.91 \times 10^{-2}\ M$$

Now calculate K_c by substituting the equilibrium concentrations of all species into the equilibrium expression.

$$K_c = \frac{[NO_2]^4}{[N_2O]^2[O_2]^3} = \frac{\left(7.45 \times 10^{-3}\right)^4}{\left(1.93 \times 10^{-2}\right)^2\left(1.91 \times 10^{-2}\right)^3} = 1.19$$

42. The initial concentrations of SO_2 and NO_2 are 0.100 M. Let x be the concentration of SO_2 that reacts on going to the equilibrium state. Since the mole ratios are 1:1:1:1, if x mol/L of SO_2 reacts, then x mol/L of NO_2 also reacts and x mol/L of SO_3 and NO are produced. This can be summarized in the following table:

Principal Reaction	SO_2 (g)	NO_2 (g) $\rightleftarrows$	SO_3 (g) + NO (g)	
Initial Concentration (M)	0.100	0.100	0	
Change	$-x$	$-x$	$+x$	
Eq Concentration (M)	$0.100 - x$	$0.100 - x$	$+x$	

The algebraic expressions for the equilibrium concentrations from the table can be substituted into this expression, and we can solve for x.

$$K_c = \frac{[NO][SO_3]}{[SO_2][NO_2]}$$

$$85.0 = \frac{(x)(x)}{(0.100 - x)(0.100 - x)} = \left(\frac{x}{0.100 - x}\right)^2$$

Taking the square root of both sides gives

$$\pm 9.22 = \frac{x}{0.100 - x}$$

Solving for x, there are two possible solutions. The equation with the positive square root of 85 gives

$$+9.22(0.100 - x) = x \qquad +0.922 = x + 9.22x \qquad x = \frac{0.922}{10.22} = 9.02 \times 10^{-2}$$

The equation with the negative square root of 85 gives

$$-9.22(0.100 - x) = x; \qquad -0.922 = x - 9.22x \qquad x = \frac{-0.922}{-8.22} = 0.112$$

Because the initial concentrations of SO_2 and NO_2 are 0.100, x cannot exceed 0.100. Therefore, $x = 0.112$ can be discarded as unreasonable and $x = 9.02 \times 10^{-2}$ is the solution. The equilibrium concentrations are

$$[SO_2] = [NO_2] = 0.100 - (9.02 \times 10^{-2}) = 0.0098 \text{ M}; \qquad [SO_3] = [NO] = 0.0902 \text{ M}$$

43. a. reaction shifts to the left
    b. shifts left (reaction that absorbs heat, reverse reaction generates heat)
    c. shifts to the left (greater number of moles of gas)
    d. no effect
    e. shifts to the right

44. $K_c = \dfrac{k_f}{k_r}$ $\qquad$ $K_c = \dfrac{6.8 \times 10^{-5}}{7.2 \times 10^{-7}} = 94.4$

45. The initial concentration of $HCONH_2$ is 0.125 M. Let x be the concentration of $HCONH_2$ that reacts on going to the equilibrium state. Since the mole ratios are 1:1:1, if x mol/L of $HCONH_2$ reacts, then x mol/L of NH_3 and CO are produced. This can be summarized in the following table:

Principal Reaction	$HCONH_2$ (g)	$\rightleftarrows$	NH_3 (g)	+	CO (g)
Initial Concentration (M)	0.125		0		0
Change	$-x$		$+x$		$+x$
Eq Concentration (M)	$0.125 - x$		$+x$		$+x$

The equilibrium expression for the reaction is $K_c = \dfrac{[NH_3][CO]}{[HCONH_2]}$. The equilibrium concentrations from the table can be substituted into this expression.

$$4.84 = \frac{(x)(x)}{(0.125 - x)} = \frac{x^2}{0.125 - x}$$

To solve for x, we need to rearrange this expression and use the quadratic equation.

$$4.84(0.125 - x) = x^2 \quad x^2 + 4.84x - 0.605 = 0;$$

$$x = \frac{-4.84 \pm \sqrt{(4.84)^2 - 4(-0.605)}}{2} \qquad x = -4.96 \text{ or } 0.120$$

The mathematical solution that makes chemical sense is 0.120. This can be used to calculate the equilibrium concentrations.

$$[HCONH_2] = 0.125 - 0.120 = 0.005 \text{ M} \qquad [NH_3] = [CO] = 0.120 \text{ M}$$

46. The position of equilibrium is to the left, given the very small value of K_c. We can calculate the value of Q_c from the concentrations given.

$$Q_c = \frac{[H_2]^4[CS_2]}{[H_2S]^2[CH_4]} = \frac{(0.450)^4(0.085)}{(2.00)^2(1.75)} = 4.98 \times 10^{-4}$$

$Q_c > K_c$; therefore, the reaction mixture is not at equilibrium. The reaction will proceed from right to left to reach equilibrium.

Challenge Problem

47. The equilibrium expression is

$$K_c = \frac{[N_2O][O_2]}{[NO_2][NO]}$$

The initial concentrations (mol/L) for each species are

$[NO_2] = 0.100$ M $\qquad$ $[NO] = 0.050$ M $\qquad$ $[N_2O] = 0.0125$ M $\qquad$ $[O_2] = 0.00125$ M

The mole ratio of NO_2 to NO is 1:1. Therefore, the amount of NO_2 that reacts equals the amount of NO that reacts. Define this amount as x. We can summarize the initial and equilibrium concentrations in the following table:

Principal Reaction	NO_2 (g) +	NO (g) $\rightleftharpoons$	N_2O (g) +	O_2 (g)
Initial Concentration (M)	0.100	0.050	0.0125	0.00125
Change (M)	$-x$	$-x$	$+x$	$+x$
Eq. Concentration (M)	$0.100 - x$	$0.050 - x$	$0.0125 + x$	$0.00125 + x$

The algebraic equations for the equilibrium concentrations can be substituted back into the equation for the equilibrium constant:

$$0.914 = \frac{(0.0125 + x)(0.00125 + x)}{(0.100 - x)(0.050 - x)}$$

$$0.914 = \frac{\left(1.56 \times 10^{-5} + 0.0138x + x^2\right)}{\left(0.0050 - 0.150x + x^2\right)}$$

$$0.00457 - 0.137x + 0.914x^2 = 1.56 \times 10^{-5} + 0.0138x + x^2$$

$$0 = 0.086x^2 + 0.151x - 0.00455$$

$$x^2 + 1.76x - 0.0529 = 0$$

$$x = \frac{-1.76 \pm \sqrt{1.76^2 - 4(-0.0529)}}{2}$$

$$x = 0.0296, -1.60$$

Substituting the value of x into the equilibrium concentrations leads to

$[NO_2]_e = 0.100 - 0.0296 = 0.0704$ M $\qquad\qquad$ $[NO]_e = 0.050 - 0.0296 = 0.0204$ M

$[N_2O]_e = 0.0125 + 0.0296 = 0.0421$ M $\qquad\qquad$ $[O_2]_e = 0.00125 + 0.0296 = 0.0308$ M

Chapter Fifteen – Aqueous Equilibria: Acids and Bases

Workbook Problems

Workbook Problem 15.1

Strategy: Write the reactions with water in which the acid protonates the water and the base deprotonates the water.

Step 1: Write the reactions in the forward direction.

$$HNO_2 + H_2O\ (l) \leftrightharpoons NO_2^- + H_3O^+\ (aq)$$

$$SO_4^{-2}\ (aq) + H_2O\ (l) \leftrightharpoons HSO_4^- + OH^-\ (aq)$$

Step 2: Identify the conjugate acid and conjugate base.

Reaction 1: acid = HNO_2, conjugate base = NO_2^-

Reaction 2: base = SO_4^{-2}, conjugate acid = HSO_4^-

Workbook Problem 15.2

Strategy: Determine which species is the stronger acid and which is the stronger base, then transfer protons accordingly.

Step 1: Write the reactions, transferring protons from the stronger acid to the stronger base:

$$HCl\ (aq) + CO_3^{2-}\ (aq) \leftrightharpoons Cl^-\ (aq) + HCO_3^-\ (aq)$$

$$CN^-\ (aq) + H_3PO_4\ (aq) \leftrightharpoons HCN\ (aq) + H_2PO_4^-\ (aq)$$

Workbook Problem 15.3

Strategy: Use the water dissociation constant to determine the concentrations. A basic solution will have $[OH^-] > [H_3O^+]$ and an acid solution will have $[H_3O^+] > [OH^-]$.

Step 1: Use K_w to calculate the ion concentrations:

$$1 \times 10^{-14} = [OH^-](2.35 \times 10^{-5})$$

$$[OH^-] = \frac{1 \times 10^{-14}}{2.35 \times 10^{-5}} = 4.26 \times 10^{-10}$$

This solution is acidic: the concentration of hydronium ions is much larger than the concentration of hydroxide ions.

$$1\times10^{-14} =[H_3O^+](4.26\times10^{-10})$$

$$[H_3O^+]=\frac{1\times10^{-14}}{4.26\times10^{-10}}=2.35\times10^{-5}$$

This solution is also acidic.

Workbook Problem 15.4

Strategy: Calculate the $[H_3O^+]$; determine pH.

Step 1: Calculate the $[H_3O^+]$, using the expression for the dissociation of water.

$$1\times10^{-14} =[H_3O^+](2.35\times10^{-6})$$

$$[H_3O^+]=\frac{1\times10^{-14}}{2.35\times10^{-6}}=4.25\times10^{-9}$$

Step 2: Calculate the pH.

$$pH = - \log (4.25 \times 10^{-9}) = 8.37$$

Step 3: Determine if the solution is acidic, basic, or neutral.

The pH is higher than 7, so the solution is basic.

Workbook Problem 15.5

Step 1: Determine the concentration of $[OH^-]$.

KOH is a strong base and exists in aqueous solution as K^+ and OH^-. Therefore, $[OH^-]$ = initial concentration of KOH.

$$[OH^-] = 7.2 \times 10^{-3} \text{ M}$$

Step 2: Calculate the $[H_3O^+]$.

The H_3O^+ concentration is calculated from the OH^- concentration.

$$[H_3O^+]=\frac{K_w}{[OH^-]}=\frac{1.0\times10^{-14}}{7.2\times10^{-3}}=1.4\times10^{-12} \text{ M}$$

Step 3: Calculate the pH.

$$pH = -\log (1.4 \times 10^{-12}) = 11.85$$

Workbook Problem 15.6

Step 1: Determine the species present initially.

$$HC_3H_5O_3 \text{ and } H_2O$$

Step 2: Write the proton-transfer reaction.

$$HC_3H_5O_3 \ (aq) \ + \ H_2O \ (l) \rightleftharpoons H_3O^+ \ (aq) \ + \ C_3H_5O_3^- \ (aq) \qquad K_a = 1.4 \times 10^{-4}$$

Step 3: Make a table showing the principal reaction and the initial and equilibrium concentrations.

Principal Reaction	$HC_3H_5O_3 \ (aq)$	$\rightleftharpoons$	$H_3O^+ \ (aq)$	$+$	$C_3H_5O_3^- \ (aq)$
Initial Concentration	1.63		0		0
Change	$-x$		$+x$		$+x$
Eq. Concentration	$1.63 - x$		$+x$		$+x$

Step 4: Substitute the equilibrium concentrations into the equilibrium equation, and solve for x. (Since $K_a < 1.0 \times 10^{-3}$, assume x is negligible and that $1.63 - x \cong 1.63$)

$$K_a = \frac{[H_3O^+][C_3H_5O_3^-]}{[HC_3H_5O_3]} = \frac{(x)(x)}{(1.63-x)}$$

$$1.4 \times 10^{-4} = \frac{x^2}{1.63} \qquad x = 1.51 \times 10^{-2}$$

Step 5: Calculate the equilibrium concentrations.

$$[H_3O^+]_e = [C_3H_5O_3^-]_e = 1.51 \times 10^{-2} \ M$$

$$[HC_3H_5O_3]_e = 1.63 - (1.51 \times 10^{-2}) = 1.61 \ M$$

Step 6: Calculate the pH.

$$pH = -\log (1.51 \times 10^{-2}) = 1.82$$

Workbook Problem 15.7

Strategy: To determine the percent dissociation we need to first determine the concentration of the dissociated HA. Therefore, we need to calculate the equilibrium concentration of H_3O^+ and $C_6H_5COO^-$.

Step 1: Determine the species present initially, and write the proton-transfer reaction.

The species present initially are C_6H_5COOH and H_2O (acid or base).

The proton-transfer reaction is:

$$C_6H_5COOH \ (aq) \ + H_2O \ (l) \leftrightarrows H_3O^+ \ (aq) \ + \ C_6H_5COO^- \ (aq) \qquad K_a = 6.5 \times 10^{-5}$$

Step 2: Make a table showing the principal reaction, initial concentration, change in concentration, and the equilibrium concentration.

Principal Reaction	$C_6H_5COOH \ (aq)$	$\leftrightarrows$	H_3O^+	$+$	$C_6H_5COO^-$
Initial Concentration	0.79		0		0
Change	$-x$		$+x$		$+x$
Equilibrium Concentration	$0.79 - x$		$+x$		$+x$

Step 3: Substitute the equilibrium concentrations into the equilibrium expression, and solve for x.

$$K_a = 6.5 \times 10^{-5} = \frac{[H_3O^+][C_6H_5COO^-]}{[C_6H_5COOH]} = \frac{(x)(x)}{(0.79 - x)}$$

Given the value of K_a, assume that x is negligible compared with the initial concentration of the acid; therefore, $0.79 - x \approx 0.79$. Using this value in the denominator, solve for x.

$$x^2 \approx (6.5 \times 10^{-5})(0.79)$$
$$x \approx 7.2 \times 10^{-3}$$

Step 4: Calculate the equilibrium concentrations of the major species present.

The equilibrium concentrations of H_3O^+ and $C_6H_5COO^-$ are $[H_3O^+] = [C_6H_5COO^-] = 0.0072$ M. This concentration also represents the amount of C_6H_5COOH that dissociated.

Step 5: Calculate the percent dissociation.

The percent dissociation can now be calculated.

$$\frac{0.0072}{0.79} \times 100 = 0.91\%$$

Workbook Problem 15.8

Step 1: Write the stepwise dissociation for arsenic acid.

The stepwise dissociation for arsenic acid is

$$H_3AsO_4 \ (aq) \ + \ H_2O \ (l) \leftrightarrows H_3O^+ \ (aq) \ + \ H_2AsO_4^- \ (aq)$$
$$H_2AsO_4^- \ (aq) \ + \ H_2O \ (l) \leftrightarrows H_3O^+ \ (aq) \ + \ HAsO_4^{2-} \ (aq)$$
$$HAsO_4^{2-} \ (aq) \ + \ H_2O \ (l) \leftrightarrows H_3O^+ \ (aq) \ + \ AsO_4^{3-} \ (aq)$$

Step 2:   Determine the principal reaction.

Since $K_{a1} > K_{a2}$, K_{a3}, and K_w, we know that the principal reaction is the first dissociation step and that all of the H_3O^+ present is produced from this step. Therefore, we only need to consider the first dissociation step when calculating the pH of the solution.

Step 3:   Make a table showing the principal reaction, initial concentration, change in concentration, and equilibrium concentration of the reactant and products.

Principal Reaction	$H_3AsO_4\ (aq)$	$\rightleftharpoons$	H_3O^+	+	$H_2AsO_4^-$
Initial Concentration	1.69×10^{-3}		0		0
Change	$-x$		$+x$		$+x$
Equilibrium Concentration	$(1.69 \times 10^{-3}) - x$		$+x$		$+x$

Step 4:   Substitute the equilibrium concentrations into the equilibrium expression.

Because K_a is so close to the concentration of the acid, it is unlikely that x will be negligible.

If it is assumed to be negligible, the result is

$$5.62 \times 10^{-3} = \frac{x^2}{1.69 \times 10^{-3}} \qquad x = 3.08 \times 10^{-3}$$

The quadratic equation must be used to solve for the equilibrium concentration of H_3O^+. Using $(1.69 \times 10^{-4}) - x$ and rearranging the equilibrium expression gives

$$5.62 \times 10^{-3}[(1.69 \times 10^{-3}) - x] = x^2 \qquad x^2 + (5.62 \times 10^{-3})x - (9.50 \times 10^{-6})$$

$$x = \frac{-(5.62 \times 10^{-3}) \pm \sqrt{(5.62 \times 10^{-3})^2 - 4(-9.50 \times 10^{-6})}}{2}$$

$$x = 2.72 \times 10^{-3}$$

Step 5:   Calculate the pH of the solution.

$$pH = -\log(2.72 \times 10^{-3}) = 2.56$$

Workbook Problem 15.9

Strategy: Solve using a table of equilibrium values:

Step 1:   Write the principal reaction.

$K_b > K_w$; therefore, the principal reaction is

$$(CH_3)_3N\ (aq) + H_2O\ (l) \rightleftharpoons (CH_3)_3NH^+\ (aq) + OH^-\ (aq)$$

Step 2: Construct a table with concentrations of the reactant and products.

Principal Reaction	$(CH_3)_3N$ *(aq)*	$\rightleftarrows$	$(CH_3)_3N^+$	+	OH^-
Initial Concentration	0.975		0		0
Change	$-x$		$+x$		$+x$
Equilibrium Concentration	$0.975 - x$		$+x$		$+x$

Step 3: Substitute the equilibrium concentrations into the equilibrium expression, and solve for x.

$$K_a = 6.5 \times 10^{-5} = \frac{[(CH_3)_3NH^+][OH^-]}{[(CH_3)_3N]} = \frac{(x)(x)}{0.975 - x}$$

Assume that x is negligible compared with the initial concentration of the base; therefore, $0.975 - x \approx x$. Using this value in the denominator, we can now solve for x. Remember: x represents the equilibrium concentration of the OH^- ion.

$$6.5 \times 10^{-5} = \frac{x^2}{0.975} \qquad x = 0.0080$$

Step 4: Determine the equilibrium concentrations of the species present.

The equilibrium concentrations are $[(CH_3)_3N] = 0.975 - 0.0080 = 0.967$ M; $[OH^-] = [(CH_3)_3NH^+] = 0.0080$ M.

Step 5: Calculate the pH of the solution.

Knowing the equilibrium concentration of OH^-, we can calculate the H_3O^+ concentration from the K_W expression.

$$[H_3O^+] = \frac{1.0 \times 10^{-14}}{0.0080} = 1.3 \times 10^{-12}$$

$$pH = -\log (1.3 \times 10^{-12}) = 11.90$$

Workbook Problem 15.10

Strategy: Determine if KOCl produces an acidic, basic, or neutral aqueous solution, write the hydrolysis reaction for this salt, and determine pH.

K^+ is an inert cation and will not react. However, OCl^- is the conjugate base of a weak acid and will react with water.

Step 1: Write the hydrolysis reaction for this salt.

$$OCl^- \ (aq) + H_2O \ (l) \rightleftarrows HOCl \ (aq) + OH^- \ (aq)$$

Step 2:   Write an equilibrium expression for this reaction.

$$K_b = \frac{[\text{HOCl}]\,[\text{OH}^-]}{[\text{OCl}^-]} \qquad K_b = \frac{1.0 \times 10^{-14}}{K_a\,(\text{HOCl})} = \frac{1.0 \times 10^{-14}}{3.5 \times 10^{-8}} = 2.9 \times 10^{-7}$$

Step 3:   Construct a table.

Principal Reaction	OCl⁻ (aq)	⇌	HOCl (aq)	+	OH⁻
Initial Concentration	0.137		0		0
Change	−x		+x		+x
Equilibrium Concentration	0.137 − x		+x		+x

Step 4:   Substitute the equilibrium concentrations into the equilibrium expression.

Assume that x is negligible; therefore, $0.137 - x \approx 0.137$.

$$2.9 \times 10^{-7} = \frac{(x)(x)}{0.137} \qquad x = 2.0 \times 10^{-4}$$

Step 5:   Calculate the pH of the solution.

The equilibrium concentration of OH^- is 2.0×10^{-4}. Determine the H_3O^+ concentration to calculate pH.

$$\left[\text{H}_3\text{O}^+\right] = \frac{1.0 \times 10^{-14}}{2.0 \times 10^{-4}} = 5.0 \times 10^{-11}$$

$$\text{pH} = -\log\,(5.0 \times 10^{-11}) = 10.30$$

Putting It Together

To determine the amount of aniline used, determine the theoretical yield expected in this reaction.

$$\%\ \text{yield}\ =\ \frac{\text{Experimental yield}}{\text{Theoretical yield}} \times 100$$

$$\text{Theoretical yield} = \frac{125\ \text{g}}{0.90} \times 100 = 139\ \text{g}$$

Calculate the number of moles of sulfanilic acid in 139 g:

$$139\ \text{g} \times \frac{1\ \text{mol}}{173\ \text{g}} = 0.803\ \text{mol}$$

The mole ratio of aniline to sulfanilic acid is 1:1. Therefore, we have 0.803 mol of aniline. This needs to be converted to mass, then to volume:

$$0.803 \text{ mol} \times \frac{93 \text{ g}}{\text{mol}} \times \frac{1 \text{ mL}}{1.02 \text{ g}} = 73.2 \text{ mL}$$

$Na(C_6H_4NH_2SO_3)$, sodium sulfanilate, contains the anion of a weak acid and the cation of a strong base. Therefore, the aqueous solution of this salt should be basic. The hydrolysis reaction for this salt is

$$C_6H_4NH_2SO_3^- \ (aq) \ + \ H_2O \ (l) \ \leftrightharpoons \ HC_6H_4NH_2SO_3 \ (aq) \ + \ OH^- \ (aq)$$

$$K_b = \frac{K_w}{K_a} = \frac{1 \times 10^{-14}}{5.9 \times 10^{-4}} = 1.69 \times 10^{-11}$$

Determine the initial concentration of the sulfanilate ion:

$$\text{mol } C_6H_4NH_2SO_3^- = 5.73 \text{ g NaC}_6H_4NH_2SO_3H \times \frac{1 \text{ mol NaC}_6H_4NH_2SO_3H}{196 \text{ g NaC}_6H_4NH_2SO_3H} \times \frac{1 \text{ mol } C_6H_4NH_2SO_3^-}{1 \text{ mol NaC}_6H_4NH_2SO_3H}$$

$$= 2.92 \times 10^{-2} \text{ mol } C_6H_4NH_2SO_3^-$$

$$M = \frac{2.92 \times 10^{-2} \text{mol } C_6H_4NH_2SO_3^-}{0.250 \text{ L}} = 0.117 \text{ M}$$

We can now set up our table.

Principal Reaction	$C_6H_4NH_2SO_3^- \ (aq) \ + \ H_2O \ (l) \ \leftrightharpoons \ HC_6H_4NH_2SO_3 \ (aq) +$	$OH^- \ (aq)$
Initial Concentration	0.117	0	0
Change	$-x$	$+x$	$+x$
Eq. concentration	$0.117 - x$	$+x$	$+x$

We can assume that x is negligible, given the value of K_b. Therefore, $0.117 - x \approx x$.

$$1.69 \times 10^{-11} = \frac{x^2}{0.117}$$

$$x^2 = 1.97 \times 10^{-12}$$

$$x = [OH^-]_e = 1.41 \times 10^{-6} \text{ M}$$

Calculate $[H_3O^+]$:

$$K_w = [H_3O^+][OH^-]$$

$$\left[H_3O^+ \right] = \frac{1 \times 10^{-14}}{1.41 \times 10^{-6}} = 7.09 \times 10^{-9}$$

$$pH = -\log(7.09 \times 10^{-9}) = 8.15$$

Self–Test

True–False
1. F. The Lewis theory deals with electron transfers, but an electron acceptor is an acid, and an electron donor is a base.
2. F. A strong acid fully dissociates. A concentrated solution of a weak acid can be far more harmful than a dilute solution of a strong acid.
3. T
4. T
5. F. The strength of an acid and the strength of its conjugate base are inversely related (the stronger the acid, the weaker the conjugate base).
6. T
7. F. The principal reaction for a diprotic acid is the dissociation of H_2A.
8. T
9. T
10. F. $Mg(OH)_2$ is a strong base.

Multiple Choice

11. c	12. d	13. a	14. b
15. c	16. d	17. d	18. a

Matching

19. f	20. a	21. j	22. d
23. g	24. b	25. k	26. c
27. m	28. o	29. e	30. n
31. h	32. i	33. l	34. q
35. p			

Fill-in-the-Blank

36. weak, proton
37. hydrogen or hydronium, hydroxide, neutral, equals
38. increases, decreases
39. stepwise, smaller
40. electrons, hydrogen

Problems

41. We first need to solve for the concentration of the H_3O^+ ion, then solve for [OH⁻].

$$[H_3O^+] = \text{antilog}\,(-11.78) = 1.66 \times 10^{-12}\ M \quad [OH^-] = \frac{1.0 \times 10^{-14}}{1.66 \times 10^{-12}} = 6.0 \times 10^{-3}\ M$$

42. $Ba(OH)_2$ is a strong base and completely dissociates in water according to the reaction

$$Ba(OH)_2\ (aq) \rightarrow Ba^{2+}\ (aq)\ +\ 2\ OH^-\ (aq)$$

Determine the molar concentration of OH⁻.

$$1.83\ g\ Ba(OH)_2 \times \frac{1\ mol\ Ba(OH)_2}{171.3\ g\ Ba(OH)_2} \times \frac{2\ mol\ OH^-}{mol\ Ba(OH)_2} = 2.14 \times 10^{-2}\ mol\ OH^-;$$

$$\frac{2.14 \times 10^{-2} \text{ mol OH}^-}{0.150 \text{ L}} = 0.142 \text{ M}$$

$$[H_3O^+] = \frac{1.0 \times 10^{-14}}{0.142} = 7.02 \times 10^{-14}; \text{ pH} = -\log(7.02 \times 10^{-14}) = 13.15$$

43. $0.51\% = \frac{[\text{acid}] \text{ dissociated}}{7.35 \times 10^{-3}} \times 100 \qquad [\text{acid}] \text{ dissociated} = 3.74 \times 10^{-5}$

The dissociated acid represents the $[H_3O^+]$ and $[A^-]$ at equilibrium.

$$\text{pH} = -\log(3.74 \times 10^{-5}) = 4.427$$

The equilibrium concentration of acid is $7.35 \times 10^{-3} - 3.74 \times 10^{-5} = 7.31 \times 10^{-3} \text{ M}$.

$$K_a = \frac{(3.74 \times 10^{-5})(3.74 \times 10^{-5})}{(7.31 \times 10^{-3})} = 1.91 \times 10^{-7}$$

44. Calculate $[H_3O^+]$ at equilibrium from the pH.

$$[H_3O^+] = \text{antilog } (-6.03) = 9.33 \times 10^{-7}.$$

This concentration also represents the amount of histidine that dissociated, which can be used to calculate the equilibrium concentration of histidine.

$$(1.30 \times 10^{-3}) - (9.33 \times 10^{-7}) = 1.30 \times 10^{-3}$$

$[H_3O^+] = [A^-]$ at equilibrium. Calculate K_a.

$$K_a = \frac{(9.33 \times 10^{-7})(9.33 \times 10^{-7})}{(1.30 \times 10^{-3})} = 6.70 \times 10^{-10}$$

45. This is an equilibrium problem.

Principal Reaction	C_6H_5N (aq)	+	H_2O (l)	$\rightleftarrows$	$C_6H_5NH^+$ (aq)	+	OH^- (aq)
Initial Concentration	1.24				0		0
Change	$-x$				$+x$		$+x$
Equilibrium Concentration	$1.24 - x$				$+x$		$+x$

Substituting these values into the equilibrium expression gives

$$K_b = 1.8 \times 10^{-9} = \frac{[C_6H_5NH^+][OH^-]}{[C_6H_5N]} = \frac{(x)(x)}{(1.24 - x)}$$

Assume that $1.24 - x \approx 1.24$. The above expression then simplifies to

$$x^2 = (1.8 \times 10^{-9})(1.24); \quad x = 4.72 \times 10^{-5} \text{ M}$$

$$[OH^-] = x = 4.72 \times 10^{-5}$$

Solve for $[H_3O^+]$ from $[OH^-]$

$$[H_3O^+] = \frac{1.0 \times 10^{-14}}{4.72 \times 10^{-5}} = 2.12 \times 10^{-10} \text{ M}$$

Calculate pH.

$$pH = -\log(2.12 \times 10^{-10}) = 9.67$$

46. Oxalic acid is a diprotic acid and has two possible dissociations:

Principal Reaction	$H_2C_2O_4$ (aq) + H_2O (l) $\leftrightarrows$	H_3O^+ (aq) +	$HC_2O_4^-$ (aq)
Initial Concentration	0.125	0	0
Change	$-x$	$+x$	$+x$
Equilibrium Concentration	$0.125 - x$	$+x$	$+x$

Substituting these values into the equilibrium expression gives

$$K_{a1} = 5.9 \times 10^{-2} = \frac{[H_3O^+][HC_2O_4^-]}{[H_2C_2O_4]} = \frac{(x)(x)}{(0.125 - x)}$$

Due to the value of K_{a1}, we cannot assume that the value of x is negligible. Therefore, we must use the quadratic equation to solve for x. Rearranging the above expression gives

$$(7.375 \times 10^{-3}) - (5.9 \times 10^{-2})x = x^2$$

$$x^2 + (5.9 \times 10^{-2})x - (7.375 \times 10^{-3}) = 0$$

$$x = \frac{-(5.9 \times 10^{-2}) \pm \sqrt{(5.9 \times 10^{-2})^2 - 4(-7.375 \times 10^{-3})}}{2}$$

$$x = 6.13 \times 10^{-2}$$

$$[H_2C_2O_4] = 0.125 - 0.061 = 0.064 \text{ M}; \quad [H_3O^+] = [HC_2O_4^-] = x = 0.0613$$

To calculate the $[C_2O_4^{2-}]$, use the second dissociation step of $H_2C_2O_4$, remembering that the dissociation takes place in the presence of 0.0613 M H_3O^+ from the first dissociation.

$$HC_2O_4^- \text{ (aq)} + H_2O \text{ (l)} \leftrightarrows H_3O^+ \text{ (aq)} + C_2O_4^{2-} \text{ (aq)}$$

$$K_{a2} = 6.4 \times 10^{-5} = \frac{[H_3O^+][C_2O_4^{2-}]}{[HC_2O_4^-]}$$

Use the concentration of $HC_2O_4^-$ calculated above in the calculation of $[C_2O_4^-]$. The quantities of $C_2O_4^-$ generated will be negligible compared to the ions already present.

$$6.5 \times 10^{-5} = \frac{(0.061 + x)x}{0.061 - x} = \frac{0.061x}{0.061}$$

$$x = 6.5 \times 10^{-5}$$

$$[C_2O_4^{2-}] = x = 6.5 \times 10^{-5}\, M$$

$$pH = -\log(0.0613) = 1.21$$

47. Let the formula B represent the formula for benzylamine.

Principal Reaction	B (aq) + H$_2$O (l)	$\rightleftharpoons$	BH$^+$ (aq) +	OH$^-$ (aq)
Initial Concentration	0.350		0	0
Change	– x		+ x	+ x
Equilibrium Concentration	0.350 – x		+ x	+ x

Substituting these values into the equilibrium expression gives

$$K_b = 2.14 \times 10^{-5} = \frac{\left[BH^+\right]\left[OH^-\right]}{[B]} = \frac{(x)(x)}{(0.350 - x)}$$

Assume that $0.350 - x \approx 0.350$. The above expression then simplifies to:

$$x^2 = (2.14 \times 10^{-5})(0.350) \qquad\qquad x = 2.74 \times 10^{-3}$$

$$[OH^-] = 2.74 \times 10^{-3}$$

$$\left[H_3O^+\right] = \frac{1.0 \times 10^{-14}}{2.74 \times 10^{-3}} = 3.65 \times 10^{-12}$$

$$pH = -\log(3.65 \times 10^{-12}) = 11.44$$

48. a. Ascorbic acid: $K_b = \dfrac{1.0 \times 10^{-14}}{8.0 \times 10^{-5}} = 1.2 \times 10^{-10}$

    b. Hydrazine: $K_a = \dfrac{1.0 \times 10^{-14}}{8.9 \times 10^{-7}} = 1.1 \times 10^{-8}$

49. KCN is a salt derived from a weak acid (HCN) and a strong base (KOH). This solution will be basic. The hydrolysis reaction is

$$CN^- \text{ (aq)} + H_2O \text{ (l)} \rightleftharpoons HCN \text{ (aq)} + OH^- \text{ (aq)}$$

The K_b for this reaction can be calculated from K_w and K_a for HCN.

$$K_b = \frac{1.0 \times 10^{-14}}{4.9 \times 10^{-10}} = 2.04 \times 10^{-5}$$

To determine the HCN equilibrium concentration and the pH of the solution, determine the initial concentration of CN^-.

$$1.75 \text{ g KCN} \times \frac{1 \text{ mol KCN}}{65.1 \text{ g KCN}} \times \frac{1 \text{ mol CN}^-}{1 \text{ mol KCN}} = 0.0269 \text{ mol CN}^-$$

$$\frac{0.0269 \text{ mol CN}^-}{0.150 \text{ L}} = 0.179 \text{ M}$$

Principal Reaction	CN^- (aq) + H_2O (l) $\leftrightarrows$	HCN (aq) +	OH^- (aq)
Initial Concentration	0.179	0	0
Change	$-x$	$+x$	$+x$
Equilibrium Concentration	$0.179 - x$	$+x$	$+x$

Substituting these values into the equilibrium expression gives

$$K_b = 2.04 \times 10^{-5} = \frac{[\text{HCN}][\text{OH}^-]}{[\text{CN}^-]} = \frac{(x)(x)}{(0.179 - x)}$$

Assume that $0.179 - x \approx 0.179$. The above expression then simplifies to

$$x^2 = (2.04 \times 10^{-5})(0.179) \qquad\qquad x = 1.91 \times 10^{-3}$$

$$x = [\text{HCN}] = [\text{OH}^-] = 1.91 \times 10^{-3}$$

$$[\text{H}_3\text{O}^+] = \frac{1.0 \times 10^{-14}}{1.91 \times 10^{-3}} = 5.23 \times 10^{-12}$$

$$\text{pH} = -\log(5.23 \times 10^{-12}) = 11.281$$

50. a. Stronger acid: HI – For binary acids, acid strength increases down a group due to the increase in the size of the atom.

    b. Stronger acid: H_3PO_4—For oxoacids, acid strength increases with increasing number of oxygen atoms.

Challenge Problem
51. Assume 100 g of solution, which is 25 g of H_3PO_4. The volume of the solution can be calculated from the mass and density of the solution.

$$\text{mL of solution} = \frac{100 \text{ g}}{1.1667 \text{ g/mL}} = 85.7 \text{ mL}$$

$$\text{mol } H_3PO_4 = 25 \text{ g} \times \frac{1 \text{ mol } H_3PO_4}{98 \text{ g}} = 0.255 \text{ mol}$$

$$M = \frac{0.255 \text{ mol } H_3PO_4}{0.0857 \text{ L}} = 2.98 \text{ M}$$

Write the stepwise dissociation for phosphoric acid.

$H_3PO_4 \text{ (aq)} + H_2O \text{ (l)} \rightleftharpoons H_3O^+ \text{ (aq)} + H_2PO_4^- \text{ (aq)}$ $\qquad K_{a1} = 7.5 \times 10^{-3}$

$H_2PO_4^- \text{ (aq)} + H_2O \text{ (l)} \rightleftharpoons H_3O^+ \text{ (aq)} + HPO_4^{2-} \text{ (aq)}$ $\qquad K_{a2} = 6.3 \times 10^{-8}$

$HPO_4^{2-} \text{ (aq)} + H_2O \text{ (l)} \rightleftharpoons H_3O^+ \text{ (aq)} + PO_4^{3-} \text{ (aq)}$ $\qquad K_{a3} = 4.8 \times 10^{-13}$

Since $K_{a1} > K_{a2}, K_{a3}$, we know that the principal reaction is the first dissociation step and that all of the H_3O^+ present is produced from this step. Therefore, we need only to consider the first dissociation step when calculating the pH of the solution.

Make a table showing the principal reaction, initial concentration, change in concentration, and equilibrium concentration of the reactant and products.

Principal Reaction	$H_3PO_4 \text{ (aq)} \rightleftharpoons$	H_3O^+	$+ \ H_2PO_4^-$
Initial Concentration	2.98	0	0
Change	− x	+ x	+ x
Eq. Concentration	2.98 − x	+ x	+ x

Substitute the equilibrium concentrations into the equilibrium expression. Since $K_{a1} > 1.0 \times 10^{-3}$, you cannot assume that x is negligible.

$$7.5 \times 10^{-3} = \frac{x^2}{2.98 - x}$$

$$x^2 + \left(7.5 \times 10^{-3}\right)x - \left(2.2 \times 10^{-2}\right) = 0$$

$$x = \frac{-\left(7.5 \times 10^{-3}\right) \pm \sqrt{\left(7.5 \times 10^{-3}\right)^2 + 4\left(2.2 \times 10^{-2}\right)}}{2}$$

$x = 0.145$ M. This value represents the equilibrium concentration of H_3O^+ and $H_2PO_4^-$.

Calculate the pH from this value.

$$pH = -\log (0.145) = 0.839$$

The equilibrium concentrations are

$$[H_3PO_4]_e = 2.98 \text{ M} - 0.145 \text{ M} = 2.83 \text{ M}$$
$$[H_3O^+]_e = [H_2PO_4^-]_e = 0.145 \text{ M}$$

We now repeat the process, using the chemical equation for $H_2PO_4^-$ and the equilibrium concentration calculated above.

Principal Reaction	$H_2PO_4^-$ (aq) $\rightleftharpoons$	H_3O^+ (aq)	+ HPO_4^{2-} (aq)
Initial Concentration	0.145	0.145	0
Change	$- x$	$+ x$	$+ x$
Eq Concentration	$0.145 - x$	$0.145 + x$	$+ x$

The value of K_{a2} is small enough that x is negligible; therefore, $0.145 - x \approx 0.145$ and $0.145 + x \approx 0.145$.

$$6.2 \times 10^{-8} = \frac{(0.145)x}{0.145}$$

$$x = 6.2 \times 10^{-8}$$

This value represents the equilibrium concentration of $[H_2PO_4^-]$.

$$[HPO_4^{2-}]_e = 6.2 \times 10^{-8} \text{ M}$$

Repeat the process to calculate the equilibrium concentration of PO_4^{2-}.

Principal Reaction	HPO_4^{2-} (aq) $\rightleftharpoons$	H_3O^+ (aq) +	PO_4^{3-} (aq)
Initial Concentration	6.2×10^{-8}	0.145	0
Change	$- x$	$+ x$	$+ x$
Eq Concentration	$(6.2 \times 10^{-8}) - x$	$0.145 + x$	$+ x$

The value of K_{a3} is small enough that x is negligible; therefore $(6.2 \times 10^{-8}) - x \approx 6.2 \times 10^{-8}$.

$$4.8 \times 10^{-13} = \frac{(0.145)x}{6.2 \times 10^{-8}}$$

$$x = 2.05 \times 10^{-19}$$

This value of x represents the equilibrium concentration of PO_4^{3-}.

$$[PO_4^{3-}]_e = 2.05 \times 10^{-19}$$

Chapter Sixteen – Applications of Aqueous Equilibria

Workbook Problems

Workbook Problem 16.1

Strategy: Write the net reaction; determine K_n.

Step 1:   Write the net neutralization reaction by writing individual reactions for the acid, base, and water:

$$HNO_2\ (aq) + H_2O\ (l) \rightleftharpoons NO_2^-\ (aq) + H_3O^+\ (aq) \qquad K_a = 4.5 \times 10^{-4}$$
$$H_3O^+\ (aq) + OH^-\ (aq) \rightarrow H_2O\ (aq) \qquad\qquad 1/K_w = 1.0 \times 10^{14}$$
$$\overline{\rule{0pt}{0.8em}\hspace{0.5em}}\ $$
$$HNO_2\ (aq) + OH^-\ (aq) \rightarrow NO_2^-\ (aq) + H_2O\ (aq)$$

Step 2:   Calculate K_n based on the equilibrium constants for the individual reactions in Step 1.

$$K_n = K_a\ x\ 1/K_w = (4.5 \times 10^{-4}) \times (1.0 \times 10^{14}) = 4.5 \times 10^{10}$$

Step 3:   Determine the position of equilibrium from the value of K_n.

Based on the value of K_n, the equilibrium will lie far to the right. The reaction will go to completion, leaving the conjugate base of the weak acid in solution.

Step 4:   Predict the pH.

The solution will be basic: the pH will be greater than 7.

Workbook Problem 16.2

Strategy: Write the reaction; determine the concentration of H_3O^+, $H_2PO_3^-$.

Step 1:   Write the overall reaction:

$$H_3PO_3\ (aq) + H_2O\ (l) \rightleftharpoons H_2PO_3^-\ (aq) + H_3O^+\ (aq)$$

Step 2:   Make a chart of initial and equilibrium concentrations:

Principal Reaction	H_3PO_3	$H_3O^+\ (aq)$	$H_2PO_3^-$
Initial Conc.	0.250	0	0.175
Change	$-x$	$+x$	$+x$
Equilibrium Conc.	$0.250 - x$	$+x$	$0.175 + x$

Step 3:   Solve for the equilibrium concentrations of H_3O^+, $H_2PO_2^-$:

$$1.0 \times 10^{-2} = \frac{x(0.175 + x)}{0.250 - x}$$

$$x^2 + (0.175x + 0.010x) - 0.0025 = 0$$

$$x^2 + 0.185x - 0.0025 = 0$$

$$x = \frac{-0.185 \pm \sqrt{(0.185)^2 + 4(0.0025)}}{2}$$

$$x = 0.0126$$

$$[H_3O^+] = 0.0126 \text{ M}$$

$$[H_2PO_3^-] = 0.0126 + 0.175 = 0.188 \text{ M}$$

Step 4:   Determine the pH of the solution:

$$-\log (0.0126) = 1.90$$

Step 5:   Determine the % dissociation of the acid:

$$\% \text{ dissociation} = \frac{0.126}{0.250} \times 100 = 50.4\%$$

Workbook Problem 16.3

Step 1:   Determine the principal reaction and equilibrium concentrations.

The principal reaction and equilibrium concentration for this solution are shown in the table:

Principal Reaction	HCOOH (aq) + H₂O (l) ⇌	H₃O⁺ (aq) +	HCO₂⁻ (aq)
Initial Conc.	0.35	0	0.25
Change	−x	+x	+x
Equilibrium Conc.	0.35 −x	+x	0.25 + x

Step 2:   Using the equilibrium equation, solve for $[H_3O^+]$, assuming that the change (x) is negligible.

$$K_a = \frac{[H_3O^+][HCO_2^-]}{[HCOOH]}$$

$$[H_3O^+] = K_a \frac{[HCOOH]}{[HCO_2^-]}$$

K_a for formic acid equals 1.8×10^{-4}. Substituting the given information into the equilibrium expression gives

$$\left[H_3O^+\right] = \left(1.8 \times 10^{-4}\right)\frac{(0.35)}{(0.25)} = 2.5 \times 10^{-4}$$

Step 3:   Solve for the pH.

$$pH = -\log\left(2.5 \times 10^{-4}\right) = 3.60$$

Step 4:   Write the neutralization reaction that occurs on addition of HCl, and set up a table showing the number of moles present before and after the addition of HCl.

Neutralization Reaction	HCO_2^- (aq) + H_3O^+ (l)	$\leftrightarrows$	HCOOH (aq) +	H_2O (l)
Before Reaction (mol)	0.35		0.10	0.25
Change (mol)	−0.10		−0.10	+0.10
After Reaction (mol)	0.25		0	0.35

Step 5:   Determine the concentrations of the buffer components after neutralization occurs.

Assuming that the volume of the solution does not change, the concentrations of the buffer components after neutralization are

$$[HCO_2^-] = \frac{0.25 \text{ mol}}{1.0 \text{ L}} = 0.25 \text{ M}$$

$$[HCOOH] = \frac{0.35 \text{ mol}}{1.0 \text{ L}} = 0.35 \text{ M}$$

Step 6:   Substitute the concentrations of the buffer components into the equilibrium expression and calculate the pH.

$$\left[H_3O^+\right] = \left(1.8 \times 10^{-4}\right)\frac{0.35}{0.25} = 2.5 \times 10^{-4}$$

$$pH = -\log\left(2.5 \times 10^{-4}\right) = 3.60$$

Workbook Problem 16.4

Strategy: Use the Henderson–Hasselbalch equation to calculate the [base]/[acid] ratio. This information provides the mole ratio of acetate to acetic acid.

Step 1:   Calculate the pK_a of acetic acid.

$$-\log\left(1.8 \times 10^{-5}\right) = 4.74$$

Step 2:   Use the Henderson–Hasselbalch equation to calculate the [base]/[acid] ratio.

$$\log \frac{[\text{base}]}{[\text{acid}]} = pH - pK_a$$

$$\log \frac{[base]}{[acid]} = 5.75 - 4.74 = 1.01$$

$$\frac{[base]}{[acid]} = 10^{1.01} = 10.2$$

Step 3:   Determine the mass of sodium acetate needed.

$$10.2 \times 0.01 \text{ mol} = 0.102 \text{ mol sodium acetate}$$

$$0.102 \text{ mol} \times \frac{82.03 \text{ g}}{\text{mol}} = 8.37g$$

Workbook Problem 16.5

Strategy: Using the Henderson–Hasselbalch equation, calculate the log of the [base]/[acid] ratio, then determine the pH at which this value is obtained.

$$\log \frac{25}{75} = pH - 8.8$$

$$-0.477 = pH - 8.8$$

$$pH = 8.3$$

Workbook Problem 16.6

Strategy: Determine the number of moles of H_3O^+ and OH^- present in the mixture and determine the pH based on the excess.

Step 1:   Calculate the moles of OH^- and the moles of H_3O^+.

$$\text{Moles } OH^- = 0.05 \text{ L} \times 0.1 \text{ mol/L} = 0.005 \text{ mol } OH^-$$

$$\text{Moles } H_3O^+ = 0.015 \times 1 \text{ mol/L} = 0.015 \text{ mol } H_3O^+$$

Step 2:   Determine the excess quantity and the resultant $[H_3O^+]$.

$$H_3O^+ \text{ is in excess: } 0.015 - 0.005 = 0.010 \text{ mol } H_3O^+$$

$$[H_3O^+] = 0.010 \text{ mol}/0.065 \text{ L} = 0.154 \text{ M HCl}$$

Step 3:   Determine the pH of the solution.

$$pH = -\log(0.154) = 0.813$$

Workbook Problem 16.7

Strategy: Determine the volume of NaOH required to provide equimolar amounts of base and acid, and determine the concentration of conjugate base and pH.

Step 1:   Calculate the moles of phenol present and thus the moles of OH⁻ needed:

$$0.100 \text{ L} \times 0.050 \text{ mol/L} = 0.0050 \text{ moles}$$

Step 2:   Determine the volume of sodium hydroxide solution required:

$$0.0050 \text{ moles} / 0.1 \text{ mol/L} = 0.050 \text{ L} = 50 \text{ mL}$$

Step 3:   Determine the total volume of the resultant solution:

$$100 \text{ mL original volume} + 50 \text{ mL NaOH solution} = 150 \text{ mL}$$

Step 4:   Solve for the concentration of salt:

$$0.0050 \text{ mol}/0.150 \text{ L} = 0.033 \text{ M}$$

Step 6:   Solve for K_b and [OH]:

$$K_b = \frac{K_w}{K_a} = \frac{1.0x10^{-14}}{1.3x10^{-10}} = 7.7x10^{-5}$$

$$7.7x10^{-5} = \frac{x^2}{(0.033 - x)} \quad \text{assume } 0.033\text{-x} \approx 0.033$$

$$x = 0.00159 \text{ M (assumption was correct)}$$

Step 7:   Solve for [H₃O] and pH:

$$[H_3O^+] = \frac{1x10^{-14}}{0.00159} = 6.3 \times 10^{-12}$$

$$pH = -\log 6.3 \times 10^{-12} = 11.20$$

Workbook Problem 16.8

a) milliliters of acid required to reach the equivalence point:

Step 1:   Write the neutralization reaction that occurs, and determine the number of mmoles of $(CH_3)_3N$ in the initial solution.

$$(CH_3)_3N \ (aq) \ + \ HCl \ (aq) \leftrightarrows (CH_3)_3NH^+ \ (aq) \ + \ Cl^- \ (aq)$$

$$mol \ (CH_3)_3N = 0.040 \ L \times 0.375 \frac{mol \ (CH_3)_3N}{L \ (CH_3)_3N} = 0.015 \ mol \ (CH_3)_3N$$

Step 2:   Determine the moles of HCl needed to react with the $(CH_3)_3N$ and the volume of 0.100 M HCl needed.

From the neutralization reaction, we know that the mole ratio of $(CH_3)_3N$:HCl is 1:1. Therefore,

$$Moles \ HCl = 0.015$$

$$mL \ HCl \ = \ \frac{0.015 \ mol \ HCl}{0.100 \ mol \ HCl/mL \ HCl} = 0.150 \ L = 150 \ mL \ HCl$$

b) Initial pH of the $(CH_3)_3N$ solution:

Step 1:   Determine the initial pH of the solution:

Principal Reaction	$(CH_3)_3N \ (aq)$	$+$	$H_2O \ (l)$	$\leftrightarrows$	$(CH_3)_3NH^+ \ (aq)$	$+$	$OH^- \ (aq)$
Initial Conc.	0.375 M				0		0
Change	$-x$				$+x$		$+x$
Equilibrium Conc.	0.375 M $- x$				$+x$		$+x$

$$K_b = \frac{[(CH_3)_3NH^+][OH^-]}{[(CH_3)N]}$$

$$6.5 \times 10^{-5} = \frac{(x)(x)}{(0.375 - x)} \cong \frac{x^2}{0.375} \qquad x = [OH^-] = 4.93 \times 10^{-3}$$

$$[H_3O^+] = \frac{K_w}{[OH^-]} = \frac{1.0 \times 10^{-14}}{4.93 \times 10^{-3}} = 2.02 \times 10^{-12} ; \qquad pH = -\log (2.02 \times 10^{-12}) = 11.7$$

c) pH of the solution after addition of 10.00 mL of acid:

Step 1:   Calculate the concentration of $(CH_3)_3N$ and its conjugate acid after the addition of 10.00 mL of HCl.

$$mol \ (CH_3)_3N = 0.015 \ mol - (0.100 \ M)(0.010 \ L) = 0.014 \ mmol$$

The total volume present = 40.00 mL $(CH_3)_3N$ + 10.00 mL HCl added.

$$[(CH_3)_3N] = \frac{0.014 \text{ mol } (CH_3)_3N}{0.050 \text{ L soln}} = 0.280 \text{ M}$$

The conjugate acid of $(CH_3)_3N$ is $(CH_3)_3NH^+$.

$$\text{mol } (CH_3)_3NH^+ = \text{mol HCl used} = (0.100 \text{ M})(0.010 \text{ L}) = 0.0010 \text{ mol}$$

$$[(CH_3)_3NH^+] = \frac{0.0010 \text{ mol } (CH_3)_3NH^+}{0.050 \text{ L}} = 0.020 \text{ M}$$

Step 3:   Determine the K_a and pK_a of the conjugate acid of $(CH_3)_3N$.

$$K_a(CH_3)_3NH^+ = \frac{K_w}{K_b(CH_3)_3N} = \frac{1.0 \times 10^{-14}}{6.5 \times 10^{-5}} = 1.5 \times 10^{-10}$$

$$pK_a = -\log (1.5 \times 10^{-10}) = 9.81$$

Step 4:   Using the Henderson–Hasselbalch equation, calculate the pH of the solution.

$$pH = pK_a + \log \frac{[\text{base}]}{[\text{acid}]} = 9.81 + \log \frac{0.280}{0.020} = 11.0$$

d) pH halfway to the equivalence point:

Step 1:   Use the Henderson–Hasselbalch equation to calculate the pH. Halfway to the equivalence point, [base] = [acid].

$$pH = pK_a = 9.81.$$

e) pH at the equivalence point:

Step 1:   Calculate the concentration of the conjugate acid, using the initial moles of base and the total volume of solution:

At the equivalence point, moles of $(CH_3)_3NH^+$ = moles of $(CH_3)_3N$ reacted (= amount of $(CH_3)_3N$ present initially).

$$\text{Moles } (CH_3)_3NH^+ = 0.015$$

Total volume = 40.00 mL $(CH_3)_3N$ + 150 mL HCl to reach eq point = 190 mL

$$[(CH_3)_3NH^+] = \frac{0.015 \text{ mol}}{0.190 \text{ L}} = 0.079 \text{ M}$$

Step 2:   Determine the pH of the solution:

Principal Reaction	$(CH_3)_3NH^+ (aq) +$	$H_2O\ (l) \rightleftharpoons$	$(CH_3)_3N\ (aq) +$	$H_3O^+\ (aq)$
Initial Conc.	0.079		0	0
Change	$-x$		$+x$	$+x$
Equilibrium Conc.	$0.079 - x$		$+x$	$+x$

$$K_a\ (CH_3)_3NH^+ = 1.5 \times 10^{-10}$$

$$K_a = \frac{[H_3O^+][(CH_3)_3N]}{[(CH_3)_3NH^+]} \qquad 1.5 \times 10^{-5} = \frac{(x)(x)}{0.079 - x} \cong \frac{x^2}{0.079}$$

$$x = [H_3O^+] = 1.09 \times 10^{-3} \qquad pH = -\log(1.09 \times 10^{-3}) = 2.96$$

f) pH after the addition of 200.00 mL of 0.100 M HCl:

Step 1:   Determine the number of mmoles of excess HCl.

The first 150 mL titrated the base, the remaining 50 mL are excess.

Moles HCl in excess = 0.050 L × 0.100 mol/L = 0.005 mol

Step 2:   Determine the concentration of excess HCl, and calculate the pH.

$$[HCl]_{xs} = \frac{0.005\ mol}{0.240\ L} = 0.021\ M \qquad pH = -\log(0.021) = 1.68$$

Workbook Problem 16.9

Strategy: Determine how far this amount of sodium hydroxide takes the titration: use the Henderson–Hasselbalch equation to determine pH. Calculate the isoelectric point from the pK_{a1} and pK_{a2}.

Step 1:   Determine how many moles of proline and how many moles of sodium hydroxide are present:

Moles proline = 0.100 L × 0.050 mol/L = 0.0050 mol proline

$$\text{Moles NaOH} = \frac{0.025\ g}{40\ g/mol} = 0.000625\ mol$$

Step 2:   Use the Henderson-Hasselbalch equation to determine the pH:

$$pH = pK_a + \log\frac{[\text{base}]}{[\text{acid}]} = 1.952 + \log\frac{0.000625\ mol\ /\ 0.100\ L}{(0.005 - 0.000625\ mol)\ /\ 0.100\ L} = 1.1$$

Step 3:   Calculate the isoelectric point:

$$\text{isoelectric point} = \frac{pK_{a1} + pK_{a2}}{2} = 6.30$$

Workbook Problem 16.10

Step 1:   Write the solubility equilibrium expression for $CdCl_2$.

The solubility equilibrium for $CdCl_2$ is

$$CdCl_2\ (s) \leftrightarrows Cd^{2+}\ (aq)\ +\ 2\ Cl^-\ (aq)$$

and the K_{sp} expression for this reaction is

$$K_{sp} = [Cd^{2+}][Cl^-]^2$$

Step 2:   Determine the concentration of Cd^{2+} and Cl^- and calculate the K_{sp}.

If $[Cd^{2+}] = 1.10 \times 10^{-5}$ then the $[Cl^-]$ is two times that, 2.20×10^{-5}. Substituting these values into the equilibrium expression gives

$$K_{sp} = \left(1.10 \times 10^{-5}\right)\left(2.20 \times 10^{-5}\right)^2 = 5.32 \times 10^{-15}$$

Workbook Problem 16.11

Strategy: Write the solubility equilibrium for $Cu(OH)_2$ and the equilibrium expression. Using the method described in Chapter 14, calculate the molar solubility of $Fe(OH)_3$.

The solubility equilibrium for $Cu(OH)_2$ is

$$Cu(OH)_2.\ (s) \leftrightarrows Cu^{2+}\ (aq)\ +\ 2\ OH^-\ (aq)$$

The equilibrium expression is

$$K_{sp} = [Cu^{2+}][OH^-]^2$$

Step 1:   Let x be the number of mol/L of $Cu(OH)_2$ that dissolves. The saturated solution then contains x mol/L of Cu^{2+} and $2x$ mol/L of OH^-. Solve for K_{sp}.

Substitute this into the equilibrium expression:

$$K_{sp} = 2.18 \times 10^{-20} = [Cu^{2+}]\ [OH^-]^2 = (x)(2x)^2$$

Solve for x:

$$2.18 \times 10^{-20} = 4x^3; \qquad x = 1.75 \times 10^{-7}\ M$$

Workbook Problem 16.12

Step 1:   Write the solubility equation and solubility expression.

$$Sn(OH)_2 \, (s) \rightleftarrows Sn^{2+} \, (aq) + 2 \, OH^- \, (aq)$$

The equilibrium expression is

$$K_{sp} = [Sn^{2+}][OH^-]^2$$

Step 2:   Using the pH, determine the $[OH^-]$ concentration.

$$[H_3O^+] = antilog \, (-9.45) = 3.55 \times 10^{-10}$$

$$[OH^-] = \frac{1 \times 10^{-14}}{3.55 \times 10^{-10}} = 2.82 \times 10^{-5}$$

Step 3:   Let x be the number of mol/L of $Sn(OH)_2$ that dissolves. Construct a table showing the equilibrium concentrations.

Solubility Equilibrium	$Sn(OH)_2 \, (s)$	$\rightleftarrows$	$Sn^{2+} \, (aq)$	+	$OH^- \, (aq)$
Initial Concentration			0		2.82×10^{-5}
Equilibrium Concentration			$+x$		$2.82 \times 10^{-5} + x$

Step 4:   Calculate the solubility of $Sn(OH)_2$.

Given the very small value of K_{sp}, we can assume that x is negligible. Substituting the equilibrium concentrations into the K_{sp} expression gives

$$5.4 \times 10^{-27} = (x)(2.82 \times 10^{-5}) \qquad x = 1.92 \times 10^{-22}$$

Workbook Problem 16.13

Strategy: Write a metathesis reaction for the reaction between $Zn(NO_3)_2$ and Na_2CO_3, and apply the solubility rules in Chapter 4 to determine if a precipitate will form.

$$Zn(NO_3)_2 \, (aq) + Na_2CO_3 \, (aq) \rightarrow 2 \, NaNO_3 \, (aq) + ZnCO_3 \, (s)$$

Step 1:   Write the solubility equation and IP expression for the precipitate.

$$ZnCO_3 \, (s) \rightarrow Zn^{2+} \, (aq) + CO_3^{2-} \, (aq) \quad K_{sp} = 1.2 \times 10^{-10}$$

The ion product expression is

$$IP = [Zn^{2+}][CO_3^{2-}]$$

Step 2:   Calculate the IP and compare its value to the K_{sp}. Will a precipitate form?

Calculate the IP from the concentrations of Zn^{2+} and CO_3^{2-}, remembering that two solutions were mixed for a total volume of 350 mL:

$$\left[Zn^{2+} \right] = \frac{(0.75 \text{ M})(100 \text{ mL})}{350 \text{ mL}} = 0.21 \text{ M} \qquad \left[CO_3^{2-} \right] = \frac{(1.50 \text{ M})(250 \text{ mL})}{350 \text{ mL}} = 1.07 \text{ M}$$

$$IP = (0.21)(1.07) = 0.23$$

$IP > K_{sp}$; Therefore, a precipitate will form.

Putting It Together

a. $AlCl_3 \ (s) + H_3PO_4 \ (aq) \rightarrow AlPO_4 \ (s) + 3 \ HCl \ (aq)$

b. Determine the limiting reactant: the mole ration of the reactants is 1:1.

$$mol \ AlCl_3 = 95 \ g \ AlCl_3 \times \frac{1 \ mol \ AlCl_3}{133.5 \ g \ AlCl_3} = 0.711 \ mol \ AlCl_3$$

$$mol \ H_3PO_4 = 1.50 \ L \times \frac{1 \ mol \ H_3PO_4}{1 \ L} = 1.125 \ mol \ H_3PO_4$$

The limiting reactant is $AlCl_3$.

$$g \ AlPO_4 = 0.711 \ mol \ AlCl_3 \times \frac{1 \ mol \ AlPO_4}{1 \ mol \ AlCl_3} \times \frac{122 \ g \ AlPO_4}{1 \ mol \ AlPO_4} = 86.74 \ g \ AlPO_4$$

c. $AlPO_4 \ (s) \rightleftharpoons Al^{3+} \ (aq) + PO_4^{3-} \ (aq) \ K_{sp} = 1.3 \times 10^{-20}$

Let x represent the number of moles of $AlPO_4$ that will dissolve. Therefore, the number of moles of Al^{3+} and PO_4^{3-} formed is also x:

$$K_{sp} = 1.3 \times 10^{-20} = (x)(x) = x^2; \quad x = 1.1 \times 10^{-10} \text{ M}$$

d. The solubility of $AlPO_4$ will increase on the addition of HCl: hydrogen ions can react with phosphate ions to produce HPO_4^{2-}, removing the phosphate ions from solution. Using Le Châtelier's principle, this will cause more $AlPO_4$ to dissolve.

$$H^+ \ (aq) + PO_4^{3-} \ (aq) \rightleftharpoons HPO_4^{2-} \ (aq)$$

To determine if a precipitate will form, calculate Q and then compare this value to K_{sp}. (Remember that Q, the ion product, is calculated using the same formula for K_{sp}.)

$$Q = [Al^{3+}][PO_4^{3-}]$$

e. Calculate the concentration of Al^{3+} and PO_4^{3-} from the total volume of mixing the two solutions. Remember that when you dilute the solutions $M_f \times V_f = M_i \times V_i$. The total volume of the solution is 2.00 L.

$$\left[Al^{3+}\right] = \frac{0.75\,L \times \left(4.00 \times 10^{-3}\,M\right)}{2.00\,L} = 1.50 \times 10^{-3}\,M$$

$$\left[PO_4^{3-}\right] = \frac{1.25\,L \times \left(0.700\,M\right)}{2.00\,L} = 0.438\,M$$

$$Q = (1.50 \times 10^{-3})(0.438) = 6.56 \times 10^{-4}$$

$Q \gg K_{sp}$; therefore, a precipitate will form. Given that the molar solubility of $AlPO_4$ is so small (1.1×10^{-10}), the mass of $AlPO_4$ can be calculated based on the number of moles of Al^{3+}, the limiting reactant:

$$mol\ Al^{3+} = 3.00 \times 10^{-3} \qquad mol\ PO_4^{3-} = 0.876$$

$$g\ AlPO_4 = 3.00 \times 10^{-3} \times \frac{1\,mol\ AlPO_4}{1\,mol\ Al^{3+}} \times \frac{122\,g\ AlPO_4}{1\,mol\ AlPO_4} = 0.366\,g\ AlPO_4$$

Self-Test

True–False
1. T
2. F. When a weak acid reacts with a strong base the pH of the solution is greater than 7.00.
3. T
4. F. The equilibrium shifts to more dissociated acid to neutralize the base.
5. F. Half of the acid is dissociated.
6. T
7. F. The equivalence point will be lower than 7, because there is conjugate acid left in solution.
8. F. Molar solubility and K_{sp} can be mathematically interconverted.
9. T
10. F. Na and K ions cannot be precipitated out, but can be identified by flame tests.

Matching

11. d	12. i	13. b	14. g
15. a	16. c	17. h	18. e
19. j	20. f		

Fill-in-the-Blank
21. strong, completion, protons
22. weak, conjugate, pH, equilibrium, buffers
23. metal, coordinate, molecules

Problems
24. a. HNO_3 and NaOH: This is a reaction between a strong acid and a strong base. The net ionic equation is H_3O^+ (*aq*) + OH^- (*aq*) $\rightarrow$ 2 H_2O (*l*); pH = 7.

b. HCl and NH_2NH_2: NH_2NH_2 is a weak base. The net ionic equation is H_3O^+ (aq) + $NH_2NH_2 \rightarrow H_2O$ (l) + $NH_2NH_3^+$ (aq); pH < 7.00.

c. Acetic acid (CH_3COOH) and NaOH: Acetic acid is a weak acid. The net ionic equation is $CH_3COOH + OH^-$ (aq) $\rightarrow CH_3COO^-$ (aq) + H_2O (l); pH > 7.00.

25. The salt NH_4Cl is 100% dissociated, so the species present initially are NH_3, NH_4^+, Cl^-, and H_2O. NH_3 is the strongest base present ($K_b \gg K_w$). This gives the following table:

Principal Reaction	NH_3 (aq) + H_2O (l) $\leftrightarrows$	NH_4^+ (aq) +	OH^- (aq)
Initial Concentration (M)	0.10	0.25	0
Change (M)	$-x$	$+x$	$+x$
Eq Concentration (M)	$0.10 - x$	$0.25 + x$	$+x$

The common ion in this problem is NH_4^+. The equilibrium equation for the principal reaction is

$$K_b = 1.8 \times 10^{-5} = \frac{\left[NH_4^+ \right]\left[OH^- \right]}{\left[NH_3 \right]} = \frac{(0.25 + x)(x)}{(0.10 - x)} \approx \frac{(0.25)(x)}{(0.10)}$$

x is assumed to be negligible.

$$x = [OH^-] = \frac{\left(1.8 \times 10^{-5} \right)\left(0.10 \right)}{(0.25)} = 7.2 \times 10^{-6}$$

The assumption concerning the size of x is justified. Since we now know the [OH^-], we can calculate the [H_3O^+].

$$[H_3O^+] = \frac{1.0 \times 10^{-14}}{7.2 \times 10^{-6}} = 1.4 \times 10^{-9}$$

$$pH = -\log (1.4 \times 10^{-9}) = 8.85$$

26. The principal reaction and equilibrium concentrations for this solution are

Principal Reaction	HClO (aq) + H_2O (l) $\leftrightarrows$	H_3O^+ (aq) +	ClO^- (aq)
Initial Concentration (M)	0.45	0	0.25
Change (M)	$-x$	$+x$	$+x$
Equilibrium Concentration (M)	$0.45 - x$	$+x$	$0.25 + x$

If we solve the equilibrium equation for H_3O^+, we obtain

$$K_a = \frac{\left[H_3O^+ \right]\left[ClO^- \right]}{\left[HClO \right]} \qquad \left[H_3O^+ \right] = K_a \frac{\left[HClO \right]}{\left[ClO^- \right]}$$

K_a for HClO is 3.5×10^{-8}. Substituting gives:

$$[H_3O^+] = (3.5 \times 10^{-8}) \frac{(0.45)}{(0.25)} = 6.3 \times 10^{-8}$$

$$pH = -\log(6.3 \times 10^{-8}) = 7.20$$

After addition of NaOH:

Neutralization Reaction	HClO (aq) +	OH⁻ (aq) ⇄	H₂O (l) +	ClO⁻ (aq)
Before Reaction (mol)	0.45	0.10		0.25
Change (mol)	− 0.10	−0.10		+ 0.10
After Reaction (mol)	0.35	0		0.35

Substituting these values into the expression for $[H_3O^+]$, we can then calculate the pH.

$$[H_3O^+] = (3.5 \times 10^{-8}) \frac{(0.35)}{(0.35)} = 3.5 \times 10^{-8}$$

$$pH = -\log(3.5 \times 10^{-8}) = 7.46$$

To determine the pH of the solution after addition of HCl, we must take into account the neutralization reaction that takes place.

Neutralization Reaction	ClO⁻ (aq) +	H₃O⁺ (aq) ⇄	H₂O (l) +	HClO (aq)
Before reaction (mol)	0.30	0.10		0.30
Change (mol)	−0.10	−0.10		+0.10
After Reaction (mol)	0.20	0		0.40

Substituting these values into the expression for $[H_3O^+]$, we can then calculate the pH.

$$[H_3O^+] = (3.5 \times 10^{-8}) \frac{(0.40)}{(0.20)} = 7.0 \times 10^{-8}$$

$$pH = -\log(7.0 \times 10^{-8}) = 7.15$$

27. Rearrange the Henderson–Hasselbalch equation:

$$\log \frac{[\text{base}]}{[\text{acid}]} = pH - pK_a$$

The pK_a for lactic acid is $-\log(1.4 \times 10^{-4}) = 3.85$

Substituting the pH and pK_a values gives

$$\log \frac{[\text{lactate}]}{[\text{lactic acid}]} = 4.75 - 3.85 = 0.90$$

Taking the antilog of both sides gives

$$\frac{[\text{lactate}]}{[\text{lactic acid}]} = 7.94$$

This gives the ratio of lactate ion to lactic acid. To determine the ratio of lactic acid to the lactate ion, take the reciprocal:

$$\text{lactic acid} = 0.126 \text{ x lactate}$$

$$\frac{[\text{lactic acid}]}{[\text{lactate}]} = 0.126$$

If the solution is to be 0.750 M overall, this can be rearranged to solve for the amount of each:

$$\text{lactic acid} + \text{lactate} = 0.750 \text{ mol}$$

$$x + 0.126x = 0.750$$

$$x = 0.666 \text{ mol}$$

0.666 mol lactic acid and 0.084 mol lactate are required. To check, 0.084/0.666 = 0.126.

28. Calculate the pH of the buffer solution using the Henderson–Hasselbalch equation:

$$pH = pK_a + \log\frac{[HCO_2^-]}{[HCOOH]} \qquad pH = 3.74 + \log\frac{0.25}{0.50} = 3.44$$

Account for the reaction on addition of HCl:

Neutralization Reaction	HCO_2^- (aq) +	H_3O^+ (aq) $\rightarrow$	$HCOOH$ (aq) + H_2O (l)
Before Reaction (mol)	0.250	0.300	0.50
Change (mol)	–0.250	–0.250	0.250
After Reaction (mol)	0	0.050	0.75

The limiting reactant in this problem is the HCO_2^- ion, which means that this amount of HCl exceeds the capacity of the buffer. The pH will be determined by the excess HCl present.

$$pH = -\log(0.050) = 1.30$$

29. a.   0.0 mL of NaOH: $[H_3O^+] = 0.0400$ M; pH = 1.398

   b.   10.0 mL of NaOH = 0.010 L × 0.100 M = 1 mmol

   75 mL of HCl = 0.075 L × 0.040 M = 3 mmol

2 mmol HCl will remain after the reaction.

$[H_3O^+]$ after neutralization $= \dfrac{(2.00 \text{ mmol })}{(85.0 \text{ mL })} = 0.0235$ $pH = -\log(0.0235) = 1.629$

c. 30.0 mL of NaOH = 3mmol. This is the equivalence point: the HCl will be completely neutralized, and there is no NaOH in excess. The pH will be 7.0

d. 50.0 mL of NaOH = 5 mmol of NaOH.

  2 mmol of NaOH will remain after the reaction

$$[OH^-] \text{ after neutralization} = \dfrac{2.00 \text{ mmol}}{125 \text{ mL}} = 0.016 \text{ M}$$

$$\left[H_3O^+\right] = \dfrac{1.0 \times 10^{-14}}{0.016} = 6.25 \times 10^{-13}$$ $pH = -\log(6.25 \times 10^{-13}) = 12.20$

30. a. 0.0 mL NaOH: pH is calculated in the same manner as a weak acid.

Principal Reaction	CH_3COOH (aq)+H_2O (l) $\leftrightarrows$	H_3O^+ (aq) +	CH_3COO^- (aq)
Initial concentration (M)	0.0500	0	0
Change (M)	$-x$	$+x$	$+x$
Equilibrium Concentration (M)	$0.0500 - x$	$+ x$	$+ x$

$$K_a = 1.8 \times 10^{-5} = \dfrac{[H_3O^+][CH_3COO^-]}{[CH_3COOH]} = \dfrac{(x)(x)}{(0.0500 - x)} \approx \dfrac{(x)^2}{(0.0500)}$$

$$x = 9.48 \times 10^{-4}$$

$$pH = -\log (9.48 \times 10^{-4}) = 3.02$$

b. 10.0 mL of NaOH = 0.010 L × 0.100 M = 1 mmol

  50 mL of acetic acid = 0.050 L × 0.050 M = 2.5 mmol

  1.5 mmol acetic acid will remain after the reaction, and 1 mmol of sodium acetate will have been formed.

  Using the Henderson–Hasselbalch equation and using the pK_a for acetic acid (4.74):

$$\dfrac{\left(\left|1085\,° - 1091\,°\right|\right)}{1085\,°} \times 100 = 0.52\,\%$$ $3.2 \text{ L oxygen} \times \dfrac{1.43 \text{ g}}{L} = 4.58\text{g oxygen}$

c. 25.0 mL NaOH = 2.5 mmol. This is the equivalence point of the titration. All the acetic acid is neutralized, and what remains is a solution containing 2.5 mmol sodium acetate.

  2.5 mmol/75 mL = 0.033 M acetate

$$K_b = \frac{K_w}{K_a} = \frac{1 \times 10^{-14}}{1.8 \times 10^{-5}} = 5.56 \times 10^{-10}$$

The pH is determined from the equilibrium expression for sodium acetate:

$$K_b = 5.56 \times 10^{-10} = \frac{(x)^2}{3.33 \times 10^{-2}} \qquad\qquad x = [OH^-] = 4.35 \times 10^{-6}$$

$$[H_3O^+] = K_w/4.35 \times 10^{-6} = 2.32 \times 10^{-9}.$$

$$pH = -\log(2.32 \times 10^{-9}) = 8.63$$

d.   30.0 mL of NaOH = 0.030 L × 0.100 M = 3 mmol

50 mL of acetic acid = 0.050 L × 0.050 M = 2.5 mmol

0.5 mmol NaOH will remain after the reaction. This will determine the pH of the solution.

$$[OH^-] = \frac{0.50 \text{ mmol}}{80.0 \text{ mL}} = 6.3 \times 10^{-3}$$

$$[H_3O^+] = \frac{1 \times 10^{-14}}{6.3 \times 10^{-3}} = 1.59 \times 10^{-12} \qquad\qquad pH = -\log(1.59 \times 10^{-12}) = 11.80$$

31. The solubility equilibrium for $CaCO_3$ is $CaCO_3\ (s) \leftrightarrows Ca^{2+}\ (aq) + CO_3^{2-}\ (aq)$

If 7.07×10^{-5} mol is the amount of $CaCO_3$ that dissolves in 1.0 L of solution then the $[Ca^{2+}] = 7.07 \times 10^{-5}$ M and the $[CO_3^{2-}] = 7.07 \times 10^{-5}$ M. Substituting these values into the equilibrium expression gives

$$K_{sp} = [Ca^{2+}][CO_3^{2-}] = (7.07 \times 10^{-5})^2 = 5.0 \times 10^{-9}$$

32.

Solubility Equilibrium	$Ca_3(PO_4)_2\ (s) \leftrightarrows$	$3\ Ca^{2+}\ (aq)$	$2\ PO_4^{3-}\ (aq)$
Equilibrium Concentration (M)		$+3\,x$	$+2\,x$

$$K_{sp} = 2.1 \times 10^{-33} = [Ca^{2+}]^3[PO_4^{3-}]^2 = (3x)^3(2x)^2 = 108x^5$$
$$x = 1.14 \times 10^{-7}$$

33.

Solubility Equilibrium	$CaSO_4\ (s) \leftrightarrows$	$Ca^{2+}\ (aq)$	$SO_4^{2-}\ (aq)$
Equilibrium Concentration (M)		$+\,x$	$0.50 + x$

$$K_{sp} = 7.1 \times 10^{-5} = [Ca^{2+}][SO_4^{2-}] = (x)(0.50 + x) \approx (x)(0.50)$$
$$x = 1.42 \times 10^{-4}$$

34. a. AgBr: Not more soluble. Br^- is the conjugate base of a very strong acid and is very unreactive.

    b. Na_2S: More soluble. S^{2-} is the conjugate base of a weak acid.

    c. LiCN: More soluble. CN^- is the conjugate base of a very weak acid.

    d. CaF_2: More soluble. F^- is the conjugate base of a weak acid.

35. Because K_f for $Zn(NH_3)_4^{2+}$ is large, nearly all the Zn^{2+} from $Zn(NO_3)_2$ will be converted to $Zn(NH_3)_4^{2+}$.

$$Zn^{2+} (aq) + 4 NH_3 (aq) \leftrightarrows Zn(NH_3)_4^{2+}$$

Conversion of 0.50 mol/L of Zn^{2+} to $Zn(NH_3)_4^{2+}$ consumes 2.00 mol/L of NH_3 (due to a 1:4 mole ratio of Zn^{2+} to NH_3). Assuming 100% conversion to $Zn(NH_3)_4^{2+}$ gives the following concentrations:

$[Zn^{2+}] = 0 M$ $\qquad$ $[Zn(NH_3)_4^{2+}] = 0.50 M$ $\qquad$ $[NH_3] = 4.0 - 2.00 = 2.0 M$

After conversion of Zn^{2+} to $Zn(NH_3)_4^{2+}$, assume that a small amount of the reverse reaction occurs, producing Zn^{2+}.

$$Zn(NH_3)_4^{2+} \leftrightarrows Zn^{2+} (aq) + 4 NH_3 (aq)$$

Dissociation of x mol/L of $Zn(NH_3)_4^{2+}$ produces x mol/L of Zn^{2+} and $4x$ mol/L of NH_3. The equilibrium concentrations are

$[Zn(NH_3)_4^{2+}] = 0.50 - x$ $\qquad$ $[Zn^{2+}] = x$ $\qquad$ $[NH_3] = 2.0 + 4x$

The following table summarizes this reasoning under the balanced equation

Principal Reaction	$Zn^{2+} (aq)$ +	$4 NH_3 (aq) \leftrightarrows$	$Zn(NH_3)_4^{2+} (aq)$
Initial Concentration (M)	0.50	4.0	0
After 100% Reaction (M)	0	2.0	0.50
Equilibrium Concentration (M)	x	$2.0 + 4x$	$0.50 - x$

Substitute the equilibrium concentrations into the expression for K_f and make the approximation that x is negligible compared with 0.50 M:

$$K_f = 7.8 \times 10^8 = \frac{[Zn(NH_3)_4^{2+}]}{[Zn^{2+}][NH_3]^4} = \frac{(0.50 - x)}{(x)(2.0 + 4x)^4} \approx \frac{(0.50)}{(x)(2.0)^4}$$

$$[Zn^{2+}] = x = \frac{(0.50)}{(7.8 \times 10^8)(2.0)^4} = 4.01 \times 10^{-11} M$$

$$[Zn(NH_3)_4] = 0.50 - (4.01 \times 10^{-11} M = 0.50 M$$

36. The equation for this reaction is $3\,CaCl_2\,(aq) + 2\,Na_3PO_4\,(aq) \rightarrow Ca_3(PO_4)_2\,(s) + 6\,NaCl\,(aq)$

After the two solutions are mixed, the total volume is 550 mL, making the concentrations of the relevant ions

$$[Ca^{2+}] = \frac{350 \text{ mL} \times 0.75}{550 \text{ mL}} = 0.477M \qquad [PO_4^{3-}] = \frac{200 \text{ mL} \times 1.50}{550 \text{ mL}} = 0.545M$$

To determine if a precipitate will form, calculate the ion product and compare its value to K_{sp}.

$$IP = [Ca^{2+}]^3[PO_4^{3-}]^2 \qquad\qquad IP = (0.477)^3(0.545)^2 = 0.0322$$

K_{sp} for $Ca_3(PO_4)_2 = 2.1 \times 10^{-33}$; $IP \gg K_{sp}$; therefore, a precipitate will form.

37. $K_{spa}\,(CuS) = 6 \times 10^{-16}$; $K_{spa}\,(FeS) = 6 \times 10^2$

$$Q_c = \frac{[Cu^{2+}][H_2S]^2}{[H_3O^+]^2} = \frac{(0.007)(0.10)^2}{(0.20)^2} = 1.75 \times 10^{-3}$$

Since $K_{spa}\,(CuS) < Q_c < K_{spa}\,(FeS)$, Cu^{2+} will selectively precipitate out of solution.

Challenge Problem
38. For CuS

$$CuS\,(s) \leftrightarrows Cu^{2+}\,(aq) + S^{2-}\,(aq) \qquad K_{spa} = 6.0 \times 10^{-16}$$

The acidic buffer will cause S^{2-} ions to be removed from the solution to form HS^- and $H_2S\,(aq)$:

$$H^+\,(aq) + S^{2-}\,(aq) \leftrightarrows HS^-\,(aq)$$

$$H^+\,(aq) + HS^-\,(aq) \leftrightarrows H_2S\,(aq)$$

Adding the three chemical equations together, we have:

$$CuS\,(s) + 2\,H^+\,(aq) \leftrightarrows Zn^{2+}\,(aq) + H_2S\,(aq)$$

This equation is the definition of K_{spa}

$$K_{spa} = \frac{[Cu^{2+}][H_2S]}{[H_3O^+]^2}$$

To determine the molar solubility of CuS, we need to determine the concentration of Cu^{2+} at equilibrium. The concentration of H_3O^+ can be calculated from the information given about the buffer:

$$[HCHO_2] = 0.25 \text{ M and } [CHO_2^-] = 0.35 \text{ M}; K_a = 1.8 \times 10^{-4}$$

$$pH = pK_a + \log \frac{[CH_2O^-]}{[HCH_2O]} = 3.74 + \log \frac{0.25}{0.45} = 3.48$$

$$[H^+] = 3.31 \times 10^{-4}$$

Principal Reaction	CuS (s) +	2 H^+ (aq) $\rightleftharpoons$	Cu^{2+} (aq) +	H_2S (aq)
Initial concentration		3.31×10^{-4}	0	0
Change		$-2x$	$+x$	$+x$
Equilibrium Conc.		$(3.31 \times 10^{-4}) - 2x$	$+x$	$+x$

Substituting the equilibrium concentrations into the equilibrium expression, we have

$$6.0 \times 10^{-16} = \frac{(x)(x)}{\left[(3.31 \times 10^{-4}) - 2x\right]^2}$$

The value of x can be considered negligible.

$$x = 8.11 \times 10^{-12}$$

$x = [Cu^{2+}] = 8.11 \times 10^{-12}$. $[Cu^{2+}]$ represents the molar solubility of CuS.

Chapter Seventeen – Thermodynamics: Entropy, Free Energy, and Equilibrium

Workbook Problems

Workbook Problem 17.1

Strategy: Determine the number of moles of hydrogen, calculate ΔS.

Step 1: Calculate the moles of hydrogen.

$$3 \text{ g } H_2 \times \frac{1 \text{ mol}}{2.0 \text{ g}} = 1.5 \text{ mol}$$

Step 2: Calculate ΔS:

$$\Delta S = nR \ln \frac{V_{final}}{V_{initial}} = 1.5 \text{ mol} \times 8.314 \text{ J/K } \ln \frac{20}{5} = 17 \text{ J/K}$$

Workbook Problem 17.2

Strategy: Write the formula for the reaction; calculate ΔS.

Step 1:   Write the formula:

$$2H_2\ (g) + O_2\ (g) \rightarrow 2H_2O\ (g)$$

Step 2:   Calculate ΔS:

$$2(188.7) - [2(130.6) + 205.0] = -88.8\ \text{J/mol} \cdot \text{K}$$

Workbook Problem 17.3

Step 1:   Write the net ionic equation for this reaction.

$$Ag^+\ (aq)\ +\ Br^-\ (aq)\ \rightarrow\ AgBr\ (s)$$

Step 2:   Calculate ΔS_{system}.

$$\Delta S_{system} = [S°(AgBr] - [S°(Ag^+) + S°(Br^-)]$$

$$\Delta S_{system} = [107] - [72.7 + 82.4] = -48.1\ \text{J/K}$$

Step 3:   Calculate ΔH_{rxn}

$$\Delta H^{\circ}_{rxn} = [\Delta H^{\circ}_f(AgBr)] - [\Delta H^{\circ}_f(Ag^+) + \Delta H^{\circ}_f(Br^-)]$$

$$\Delta H^{\circ}_{rxn} = [-100.4] - [105.6 + (-121.5)] = -84.5\ \text{kJ}$$

Step 4:   Calculate ΔS_{total}.

$$\Delta S_{total} = -48.1\ \text{J/K} - \frac{-84,500\ \text{J}}{298\ \text{K}} = 235\ \text{J/K}$$

The reaction is spontaneous.

Workbook Problem 17.4

Strategy: Solve for ΔS and ΔH, solve for the temperature at which $\Delta G = 0$.

Step 1:   Calculate ΔS:

$$\Delta S = 238 - 127 = 111\ \text{J/mol·K}$$

Step 2:   Calculate ΔH:

$$\Delta H = -201.2 - -238.7 = 37.5\ \text{kJ/mol}$$

Step 3:   Calculate T:

$$T = \frac{\Delta H^\circ}{\Delta S^\circ} = \frac{37,500 \ J/mol}{111 \ J/mol \cdot K} = 338 \ K = 65 \ ^\circ C$$

Workbook Problem 17.5

Strategy: From the information given, you can determine ΔG° from the equation $\Delta G^\circ = \Delta H^\circ - T\Delta S^\circ$.

Step 1:   Determine ΔH°_{rxn}

$$\Delta H^\circ_{rxn} = [(-127.1) + (-447.5)] - [(-407.3) + (-101.8)] = -65.5 \ kJ/mol$$

Step 2:   Determine ΔS°_{rxn}.

$$\Delta S^\circ_{rxn} = (96.2 + 205.4) - (219.1 + 115.5) = -33.0 \ J/mol \cdot K$$

Step 3:   Calculate ΔG°_{rxn}.

$$\Delta G^\circ = \Delta H^\circ - T\Delta S^\circ = -65,500 \ J/mol - 298K(-33 \ J/mol \cdot K) = -55.7 \ kJ/mol$$

Workbook Problem 17.6

Strategy: Determine ΔG°_{rxn}, then calculate ΔH° and ΔS° to determine the temperature at which the reaction becomes spontaneous.

Step 1:   Based on the value of ΔG°_{rxn}, determine if the reaction is spontaneous.

$$\Delta G^\circ_{rxn} = (-65.3 + 0) - [(-50.8) + (2 \times 0)] = -14.5 \ kJ/mol$$

Step 2:   Determine the temperature at which the reaction will become spontaneous.

Calculate ΔH° and ΔS° for this reaction:

$$\Delta H^\circ = [(-135.4) + (0)] - [(-74.8) + (2 \times 0)] = -60.6 \ kJ/mol$$

$$\Delta S^\circ = [(216.4 + (2 \times 130.6)] - [(186.2) + (2 \times 223)] = -154.6 \ J/mol \cdot K$$

$$T = \frac{\Delta H^\circ}{\Delta S^\circ} = \frac{-60,600 \ kJ/mol}{-154.6 \ J/mol \cdot K} = 392 \ K = 119 \ ^\circ C$$

Step 3:   Determine the spontaneity at the temperatures listed.

In Step 1, it was determined that the reaction was spontaneous at 25 °C. In Step 2, it was determined that the reaction became spontaneous at 119 °C. Therefore, the reaction is spontaneous below 119 °C, and nonspontaneous above 119 °C.

400 °C = nonspontaneous   45 °C = spontaneous   −20 °C = spontaneous

Workbook Problem 17.7

Consider the following reaction:

$$2\ NO\ (g)\ +\ O_2\ (g)\ \leftrightarrows\ 2NO_2\ (g)$$

Determine the spontaneity under standard conditions and again under the following conditions: $P_{NO} = 0.150$ atm, $P_{O_2} = 0.250$ atm, $P_{NO_2} = 0.001$ atm. Under which conditions is the reaction more spontaneous?

Strategy: Determine ΔG°_{rxn}, then calculate ΔG_{rxn} and determine which reaction is more spontaneous.

$$\Delta G^\circ \text{ (kJ/mol)}$$

	ΔG° (kJ/mol)
NO (g)	86.6
O₂ (g)	0
NO₂ (g)	51.3

Step 1:   Calculate ΔG°_{rxn}, determine if the reaction is spontaneous

$$\Delta G^\circ_{rxn} = 2(51.3) - 2(86.6) = -70.6\ kJ\ /\ mol$$

Step 2:   Calculate ΔG.

$$\Delta G = -70,600\frac{J}{mol} + 8.314\frac{J}{mol \cdot K}(298\ K)\ln\frac{(0.001)^2}{(0.150)^2(0.250)} = -91,993\frac{J}{mol} = -92.0\frac{kJ}{mol}$$

The reaction is more spontaneous under the nonstandard conditions.

Workbook Problem 17.8

Consider the following reaction:

$$I_2\ (g)\ +\ Cl_2\ (g)\ \leftrightarrows\ 2ICl\ (g)$$

The equilibrium constant, K_p, for this reaction is 81.9 at 25 °C. What is the standard free-energy change for this reaction?

Strategy: Use the equation $\Delta G^\circ = -RT\ \ln K$

Step 1:   Solve for the free energy:

$$\Delta G° = -RT \ln K = -8.314 \frac{J}{mol \cdot K} \times 298\ K\ \text{x}\ \ln 81.9 = -10{,}900\ J/mol = -10.9\ kJ/mol$$

Putting It Together

To solve for the percentage of hemoglobin molecules bound to carbon monoxide, first solve for the equilibrium constant, K, at 37 °C.

$$\Delta G° = -RT\ln K;\ \ T = (273.15 + 37) = 310.15\ K$$

$$-14{,}000\ J = -\left(8.314 \frac{}{mol \cdot K}\right)(310.15\ K)\ln K$$

$$\ln K = 5.43 \qquad K = 228.1$$

With the equilibrium constant and the proportions of carbon monoxide and oxygen, calculate the proportions of hemoglobin bound to each molecule:

$$K = \frac{[Hb \cdot CO]\ [O_2]}{[Hb \cdot O_2]\ [CO]} = \frac{[Hb \cdot CO]\ [0.90]}{[Hb \cdot O_2]\ [0.10]} = 228.1$$

$$\frac{[Hb \cdot CO]}{[Hb \cdot O_2]} = 25.3$$

There are 25.3 hemoglobin molecules bound to carbon monoxide for every 1 hemoglobin molecule bound to oxygen. This proportion, 25.3:1 can be converted to a percentage:

$$25.3\ Hb\text{-}CO/26.3\ \text{total Hb} = 96.4\%\ \text{bound to carbon monoxide.}$$

Self–Test

True–False
1. F. A reaction can be spontaneous and very slow.
2. T
3. T
4. F. Standard entropies of reaction can be calculated from standard entropies of formation.
5. T
6. T. A lower temperature will reduce the contribution of the entropy term, allowing the enthalpy to dominate.
7. T
8. F. Standard enthalpy of formation for a pure element is always zero.
9. F. ΔG determines the spontaneity of a reaction.
10. T

Matching
11. c	12. a	13. f	14. e
15. d	16. h	17. b	18. g

Fill-in-the-Blank

19. spontaneous, increases
20. increases, decreases
21. temperature, motion, entropy
22. larger, smaller, gases, solids
23. entropy/enthalpy, entropy/enthalpy, free energy, temperature
24. positive, unstable.
25. reverse/leftward

Problems

26. a. The toy box has the higher entropy. A dishwasher can be arranged in a finite number of ways that could be considered orderly, but toys can be tumbled into a toybox in a multitude of ways.

    b. The sugar dissolved in tea has the higher entropy. The crystals in the bowl are relatively pure and very constrained in their arrangement.

    c. The 1 mol of O_2 gas in 35.5 has the higher entropy. The 32 g of oxygen gas will have a volume of 22.4 L, because they are at STP, and more volume means less order.

27. a. ΔS is positive; an increase in the volume of a gas at constant temperature leads to a decrease in the pressure, which leads to an increase in the entropy

    b. ΔS is negative; solids have lower entropy than aqueous ions.

    c. ΔS is positive; production of a gas increases disorder.

    d. ΔS is negative; the reactant side of the reaction has more moles of gas

    e. ΔS is negative; separating the isotopes decreases the randomness of the system.

28. a. $2 NH_4NO_3 (s) \rightarrow 2 N_2 (g) + O_2 (g) + 4 H_2O (g)$

$$\Delta S^\circ_{rxn} = [2 \times S^\circ(N_2) + S^\circ(O_2) + 4 \times S^\circ(H_2O)] - [2 \times S^\circ(NH_4NO_3)]$$
$$\Delta S^\circ_{rxn} = [(2 \times 191.5) + 205.0 + (4 \times 188.7)] - (2 \times 151.1)$$
$$\Delta S^\circ_{rxn} = 1040.6 \text{ J/K}$$

    b. $2 S (s) + 3 O_2 (g) \rightarrow 2 SO_3 (g)$

$$\Delta S^\circ_{rxn} = [2 \times S^\circ(SO_3)] - [2 \times S^\circ(S) + 3 \times S^\circ(O_2)]$$
$$\Delta S^\circ_{rxn} = (2 \times 256.6) - [(2 \times 31.8) + (3 \times 205.0)]$$
$$\Delta S^\circ_{rxn} = -165.4 \text{ J/K}$$

29. Use the equation $\Delta S^\circ_{surr} = -\dfrac{\Delta H^\circ_{rxn}}{T}$.

    Solve for ΔS°_{surr} :

$$\Delta S^\circ_{surr} = \Delta S^\circ_{total} - \Delta S^\circ_{rxn} = 1957 - 515.7 = 1441 \text{ J/K}$$

$$\Delta H^\circ_{rxn} = -T\Delta S_{surr} = -298 \times 1441 \frac{J}{K} = -4.295 \times 10^5 \frac{J}{mol} = -429.5 \frac{kJ}{mol}$$

30. Use the equation $\Delta S = R \ln \dfrac{V_{final}}{V_{initial}}$, we can calculate the value of ΔS.

$$\Delta S = 8.314 \frac{J}{mol \cdot K} \ln \frac{45.3 \text{ L}}{22.4 \text{ L}} = 5.85 \frac{J}{mol \cdot K} \times 1 \text{ mol} = 5.85 \frac{J}{K}$$

31. The temperature at which ethanol spontaneously goes from liquid to gas (and back again) can be calculated using the formula:

$$T = \frac{\Delta H^{\circ}}{\Delta S^{\circ}} = \frac{-235.1 \text{ kJ/mol} - (-277.7 \text{ kJ/mol})}{282.6 \text{ J/mol} \bullet \text{K} - 161 \text{ J/mol} \bullet \text{K}} = \frac{42.6 \text{ kJ/mol}}{121.6 \text{ J/mol} \bullet \text{K}}$$

Take care with the units!

$$T = \frac{42.6 \text{ kJ/mol}}{121.6 \text{ J/mol} \bullet \text{K}} = \frac{42,600 \text{ J/mol}}{121.6 \text{ J/mol} \bullet \text{K}} = 350 \text{ K} = 77.2 \text{ }^{\circ}\text{C}$$

32. ΔS_{sys} is the same as ΔS_{rxn}, and can be calculated using the standard enthalpies given:

$$\Delta S^{\circ}_{rxn} = [2 \text{ x } S^{\circ}(\text{HCl})] - [S^{\circ}(\text{H}_2) + S^{\circ}(\text{Cl}_2)]$$

$$\Delta S^{\circ}_{rxn} = [2 \text{ x } 186.8] - [130.6 + 223.0] = 20.0 \text{ J/(mol} \bullet \text{K)}$$

The entropy of the surroundings can be calculated using the formula:

$$\Delta S^{\circ}_{surr} = -\frac{\Delta H^{\circ}_{rxn}}{T} = -\frac{-92.3 \text{ kJ/mol}}{298 \text{K}} = 310 \text{ J/(mol} \bullet \text{K)}$$

Once again, when going from enthalpy values to entropy values, be careful with units.

$$\Delta S^{\circ}_{total} = \Delta S^{\circ}_{surr} + \Delta S^{\circ}_{rxn} = 330 \text{ J/(mol} \bullet \text{K)}$$

33. $\Delta G^{\circ}_{rxn} = [2 \text{ x } \Delta G^{\circ}_f(\text{CO}_2) + 2 \text{ x } \Delta G^{\circ}_f(\text{H}_2\text{O})] - [\Delta G^{\circ}_f(\text{C}_2\text{H}_4) + 3 \text{ x } \Delta G^{\circ}_f(\text{O}_2)]$

$\Delta G^{\circ}_{rxn} = [(2 \text{ x } -394.4) + (2 \text{ x } -237.2)] - [68.1 + 0] = -1331 \text{ kJ/mol}$

34. $\Delta G^{\circ} = [2 \text{ x } \Delta G^{\circ}_f(\text{CO}_2)] - [2 \text{ x } \Delta G^{\circ}_f(\text{CO}) + \Delta G^{\circ}_f(\text{O}_2)]$

$\Delta G^{\circ} = (2 \text{ x } -394.4) - [(2 \text{ x } -137.2) + 0] = -514.4 \text{ kJ/mol}$

$\Delta G = \Delta G^{\circ} + RT \ln Q = -514,400 \text{ J/mol} + 8.314 \text{ J/K} \ln \frac{(15)^2}{(0.5)^2(0.1)} = -514,300 \text{ J/mol} = -514.3 \text{ kJ/mol}$

35. $\Delta G^{\circ} = -RT \ln K$

$$\Delta G^{\circ}_{rxn} = [\Delta G^{\circ}_f(\text{CO}_2) + \Delta G^{\circ}_f(\text{H}_2\text{O})] - [\Delta G^{\circ}_f(\text{H}_2\text{CO}_3)]$$

$$\Delta G^{\circ}_{rxn} = [(-386.0) + (-237.2)] - (-623) = -0.2 \text{ kJ/mol}$$

$$\ln K = -\frac{\Delta G^{\circ}}{RT} = -\frac{-200 \text{ J/mol}}{8.314 \text{ J/K x } 310 \text{ K}} = 7.76 \text{x} 10^{-2}$$

$$K = e^{7.76 \text{x} 10^{-2}} = 1.08$$

36. To solve this problem, you need to use the data found in Appendix B in your textbook.

$$\Delta G^{\circ}_{rxn} = [0 + 0] - [(2 \times -384.2)] = 768.4 \text{ kJ/mol}$$

To calculate the temperature at which the reaction becomes spontaneous, we need to first calculate the ΔS°_{rxn} and ΔH°_{rxn}.

$$\Delta S^{\circ}_{rxn} = [223.0 + (2 \times 51.2)] - [(2 \times 72.1)] = 181.2 \text{ J/mol·K}$$

$$\Delta H^{\circ}_{rxn} = [0 + 0] - [(2 \times -411.2)] = 822.4 \text{ kJ/mol}$$

Solve for the temperature (*watch your units*).

$$T = \frac{\Delta H}{\Delta S} = \frac{822,400 \dfrac{J}{mol}}{181.2 \dfrac{J}{mol \cdot K}} = 4538 \text{ K}$$

$$\ln K = \frac{\Delta G^{\circ}_{rxn}}{-RT} = \frac{822,400 \dfrac{J}{mol}}{-\left[\left(8.314 \dfrac{J}{mol \cdot K}\right)(4538 \text{ K})\right]} = -21.8$$

$$K = e^{-21.8} = 3.41 \times 10^{-10}$$

Challenge Problem
37. $\Delta G^{\circ} = [(2 \times 86.6 \text{ kJ/mol})] = 173.2 \text{ kJ}$

$\Delta G^{\circ} = -RT \ln K;$

$$\ln K = \frac{1.732 \times 10^5 \text{ J}}{-\left(8.314 \dfrac{J}{mol \cdot K}\right)(298 \text{ K})} = -69.9$$

$$K = 4.39 \times 10^{-31}$$

$\Delta H^{\circ} = [2 \text{ mol NO} \times 90.2 \text{ kJ/mol}] = 180.4 \text{ kJ}$

$\Delta S^{\circ} = [2 \text{ mol NO} \times 210.7 \text{ J/mol·K}] - [(1 \text{ mol N}_2 \times 191.5 \text{ J/mol·K}) + (1 \text{ mol O}_2 \times 205.0 \text{ J/mol·K})]$

$\Delta S^{\circ} = 24.9 \text{ J/mol}$

$$\Delta G^{\circ} = \Delta H - T\Delta S = 1.804 \times 10^{5} \text{ J} - (773 \text{ K})\left(24.9 \frac{\text{J}}{\text{K}}\right) = 1.61 \times 10^{5} \text{ J}$$

$$\Delta G^{\circ} = -RT \ln K$$

$$\ln K = \frac{-1.21 \times 10^{4} \text{ J}}{-\left(8.314 \frac{\text{J}}{\text{mol} \cdot \text{K}}\right)(773 \text{ K})} = -25.05$$

$$K = e^{-25.05} = 1.32 \times 10^{-11}$$

To determine the partial pressure of N_2, O_2, and NO at equilibrium, we need to first calculate the initial pressures of N_2 and O_2 using the ideal gas law.

$$P_{N_2} = P_{O_2} = \frac{(3.00 \text{ mol})\left(\frac{0.08206 \text{ L} \cdot \text{atm}}{\text{mol} \cdot \text{K}}\right)(723 \text{ K})}{15.0 \text{ L}} = 11.9 \text{ atm}$$

Principal Reaction	N_2 (g)	+	O_2 (g)	$\rightleftarrows$	2 NO (g)
Initial Pressure	11.9		11.9		0
Change	$-x$		$-x$		$+2x$
Equilibrium Pressure	$11.9 - x$		$11.9 - x$		$+2x$

Using the equilibrium constant calculated above, we can now solve for the equilibrium pressures.

$$1.32 \times 10^{-11} = \frac{P_{NO}^{2}}{P_{N_2} P_{O_2}} = \frac{2x^2}{(11.9 - x)(11.9 - x)} = \frac{2x^2}{(11.9 - x)^2}$$

Taking the square root of both sides gives

$$3.63 \times 10^{-6} = \frac{2x}{11.9 - x}$$

$$x = 2.16 \times 10^{-5}$$

The equilibrium partial pressures are $P_{N_2} = P_{O_2} = 11.9 \text{ atm}$ and $P_{NO} = 2.16 \times 10^{-5}$ atm.

Chapter Eighteen – Electrochemistry

Workbook Problems

Workbook Problem 18.1

Strategy: Follow the steps in the preceding worked example.

Step 1:   Write two unbalanced half-reactions.

        Oxidation: $Na_2SO_3 \rightarrow Na_2SO_4$ (S goes from +4 to +6)

        Reduction: $KMnO_4 \rightarrow MnO_2$ (Mn goes from +7 to +4)

Step 2:   Balance each half-reaction for atoms other than H and O.

$$Na_2SO_3 \rightarrow Na_2SO_4$$

$$KMnO_4 \rightarrow MnO_2 + K^+$$

Step 3:   Add H_2O to balance in oxygen, and H^+ to balance in hydrogen.

$$H_2O + Na_2SO_3 \rightarrow Na_2SO_4 + 2H^+$$

$$4H^+ + KMnO_4 \rightarrow MnO_2 + K^+ + 2H_2O$$

Step 4:   Balance each reaction for charge.

$$H_2O + Na_2SO_3 \rightarrow Na_2SO_4 + 2H^+ + 2e^-$$

$$3e^- + 4H^+ + KMnO_4 \rightarrow MnO_2 + K^+ + 2H_2O$$

Step 5:   Make the electron count the same in both reactions.

$$(H_2O + Na_2SO_3 \rightarrow Na_2SO_4 + 2H^+ + 2e^-) \text{ x } 3$$

$$(3e^- + 4H^+ + KMnO_4 \rightarrow MnO_2 + K^+ + 2H_2O) \text{ x } 2$$

$$3H_2O + 3Na_2SO_3 \rightarrow 3Na_2SO_4 + 6H^+ + 6e^-$$

$$6e^- + 8H^+ + 2KMnO_4 \rightarrow 2MnO_2 + 2K^+ + 4H_2O$$

Step 6:   Add the two half-reactions together, canceling anything that appears on both sides of the equation.

$$3H_2O + 3Na_2SO_3 + 6e^- + 8H^+ + 2KMnO_4 \rightarrow 3Na_2SO_4 + 6H^+ + 6e^- + 2MnO_2 + 2K^+ + 4H_2O$$

$$3Na_2SO_3 + 2H^+ + 2KMnO_4 \rightarrow 3Na_2SO_4 + 2MnO_2 + 2K^+ + H_2O$$

Step 7:   Make the solution basic by adding 1 OH^- to each side for every H^+.

$$2OH^- + 3Na_2SO_3 + 2H^+ + 2KMnO_4 \rightarrow 3Na_2SO_4 + 2MnO_2 + 2K^+ + H_2O + 2OH^-$$

$$2H_2O + 3Na_2SO_3 + 2KMnO_4 \rightarrow 3Na_2SO_4 + 2MnO_2 + 2K^+ + H_2O + 2OH^-$$

Step 8:   Cancel water molecules that appear on both sides of the equation.

$$H_2O + 3Na_2SO_3 + 2KMnO_4 \rightarrow 3Na_2SO_4 + 2MnO_2 + 2K^+ + 2OH^-$$

Step 9:   Check your answer to make sure both atoms and charges are balanced.

Reactant side: 2H	6Na	3S	18O	2K	2Mn	net charge: 0
Product Side: 2H	6Na	3S	18O	2K	2Mn	net charge: 0

Workbook Problem 18.2

Describe the construction of a galvanic cell based on the following reaction:

$$Pb^{2+}(aq) + Cd(s) \rightarrow Pb(s) + Cd^{2+}(aq)$$

Strategy: Break the overall reaction into half-reactions and design a galvanic cell, identifying the cathode and the anode.

Step 1:   Write the half-reactions, identifying the oxidation and reduction reactions:

$$Pb^{2+}(aq) + 2e^- \rightarrow Pb(s) \qquad \text{reduction – cathode}$$

$$Cd(s) \rightarrow Cd^{2+}(aq) + 2e^- \qquad \text{oxidation – anode}$$

Step 2:   Identify solutions for each half-cell and a suitable inert electrolyte for the salt bridge:

Cathode solution: $Pb(NO_3)_2$ Anode solution: $Cd(NO_3)_2$

Salt bridge electrolyte: $NaNO_3$

Any soluble salt containing the metal ions is suitable for the half-cell solutions, and a common ion in the salt bridge will prevent any side reactions.

Workbook Problem 18.3

Step 1:   Write the half-reaction occurring at the anode. Remember, the reactant in each half-cell is written first, followed by the product.

Oxidation (loss of electrons) occurs at the anode. The half-reaction is

$$Mg(s) \rightarrow Mg^{2+}(aq) + 2e^-$$

Step 2:   Write the half-reaction occurring at the cathode.

Reduction (gain of electrons) occurs at the cathode. The half-reaction is

$$Cr^{3+} (aq) + 3 e^- \rightarrow Cr (s)$$

Step 3:   Combine the two half-reactions to obtain the overall cell reaction (recall that the equation must be balanced in electrons).

For the number of electrons lost to equal the number of electrons gained, we must multiply the anode half-reaction by 3 and the cathode half-reaction by 2. The two half-reactions are then

$$3 Mg (s) \rightarrow 3 Mg^{2+} (aq) + 6 e^-$$

$$2 Cr^{3+} (aq) + 6 e^- \rightarrow 2 Cr (s)$$

Adding the two half-reactions together gives the overall equation:

$$3 Mg (s) + 2 Cr^{3+} (aq) \rightarrow 3 Mg^{2+} (aq) + 2 Cr (s)$$

Step 5:   Describe the cell indicated in the shorthand notation.

This cell would consist of a strip of magnesium, as the anode, dipping into an aqueous solution containing the Mg^{2+} ion, such as $Mg(NO_3)_2$. The cathode would consist of a strip of chromium dipping into an aqueous solution of Cr^{3+}, such as $Cr(NO_3)_3$.

Workbook Problem 18.4

Step 1:   Write the balanced equation for the reaction:

$$Mg (s) + Zn^{2+} (aq) \rightarrow Mg^{2+} (aq) + Zn (s)$$

Step 2:   Identify n and use $\Delta G° = -nFE°$ to calculate $\Delta G°$.

In this reaction, two moles of electrons are transferred per mole of reactants.

$$\Delta G° = -2 \times 96,500 \; \frac{C}{mol} \times 1.61 \; V \times 1\frac{J}{C \cdot V} = -311 \; kJ / mol$$

Workbook Problem 18.5

Strategy: Determine the position of the reducing agent relative to the position of the oxidizing agent, and based on your results, determine the spontaneity of the reactions.

In the first reaction, the reducing agent (the species undergoing oxidation) is Cu. Cu lies above the oxidizing agent in the table of standard reduction potentials.

In the second reaction, the reducing agent is F⁻, which lies above the oxidizing agent, MnO_4^-.

In a spontaneous reaction, the reducing agent must lie below the oxidizing agent in the table of standard reduction potentials. For the first reaction, the reducing agent, Cu, lies below the oxidizing agent, Ag^+. Therefore, the first reaction is spontaneous. For the second reaction, the reducing agent, F⁻, lies above the oxidizing agent, MnO_4^-. Therefore, the second reaction is nonspontaneous.

Workbook Problem 18.6

Strategy:
Nernst

Calculate the E°_{cell} for the cell reaction, substitute the information given into the equation, and solve for $[Sn^{4+}]$.

Step 1:

Calculate E°_{cell}:

$$E^\circ_{cell} = E^\circ_{Sn^{2+} \to Sn^{4+}} + E^\circ_{Cu^{2+} \to Cu} = -0.15\ V + 0.34\ V = 0.19\ V$$

Step 2:

Using E°_{cell} and E, solve for $[Sn^{4+}]$:

$$E_{cell} = E^\circ - \frac{0.0592}{n} \log \frac{[Sn^{4+}]}{[Sn^{2+}][Cu^{2+}]}$$

$$0.17 = 0.19 - \frac{0.0592}{2} \log \frac{[Sn^{4+}]}{[0.63][0.72]}$$

$$0.02 \times \frac{2}{0.0592} = \log \frac{[Sn^{4+}]}{[0.45]}$$

$$10^{0.676} = \frac{[Sn^{4+}]}{[0.45]}$$

$$[Sn^{4+}] = 2.13\ M$$

Workbook Problem 18.7

Strategy:

Determine the half-reactions occurring and calculate E_{ref}. Substitute into the modified Nernst equation to determine pH of the solution ($Cu^{2+}(aq) + 2\ e^- \to Cu(s)$ $E^\circ = 0.34$).

Step 1:

Determine E_{ref}:

$$E_{cell} = E_{H_2 \rightarrow 2H^+} + E_{Cu^{2+} \rightarrow Cu}$$

$$0.62 \text{ V} = E_{H_2 \rightarrow 2H^+} + 0.34 \text{ V}$$

$$E_{H_2 \rightarrow 2H^+} = 0.28 \text{ V} = E_{cell} - E_{ref}$$

Step 3:   Calculate the pH of the anode solution:

$$\text{pH} = \frac{0.28}{0.0592} = 4.73$$

Workbook Problem 18.8

Find K for the reaction:

$$H_2O_2 \,(l) \ + 2 \, H^+ \,(aq) \ + 2 \, Br^- \,(aq) \rightarrow 2 \, H_2O \,(l) + Br_2 \,(l)$$

Strategy: Determine the half-reactions occurring and calculate $E°$ from the standard potentials of the half-cell reactions. Calculate K from the equation

$$\log K = E° \frac{n}{0.0592}.$$

Step 1:   Determine $E°$:

$$H_2O_2 \,(l) + 2 \, H^+ \,(aq) + 2 \, e^- \rightarrow 2 \, H_2O \,(l) \qquad E° = 1.78 \text{ V}$$

$$2 \, Br^- \,(aq) \rightarrow Br_2 \,(l) \quad + 2 \, e^- \qquad E° = -1.09 \text{ V}$$

$$E° = -1.09 \text{ V} + 1.78 \text{ V} = 0.69 \text{ V}$$

Step 2:   Calculate K from $E°$:

$$\log K = E° \frac{n}{0.0592} = 0.69 \frac{2}{0.0592} = 23.3$$

$$K = 10^{23.3} = 1.99 \times 10^{23}$$

Workbook Problem 18.9

Strategy: Determine the possible half-reactions and their standard voltages to predict the reactions at the cathode and anode, then determine the overall reaction.

Step 1:   Determine the possible reactions at the cathode and their standard voltages:

$$Li^+ \,(aq) + \ e^- \rightarrow Li \,(s) \qquad\qquad E° = -3.04 \text{ V} \quad or$$

$$2 \, H_2O \,(l) \ + \ 2 \, e^- \ \rightarrow H_2 \,(g) \ + \ 2 \, OH^- \,(aq) \qquad E° = -0.83 \text{ V}$$

Step 2: Determine the possible reactions at the anode and their standard voltages:

$$2\,SO_4^{2-}\,(aq) \rightarrow S_2O_8^{2-}\,(aq)\,+\,2e^- \qquad E° = -2.01\,V \quad or$$

$$2\,H_2O \rightarrow O_2\,(g)\,+\,4\,H^+\,(aq)\,+\,4\,e^- \qquad E° = -1.23\,V$$

Step 3: Predict which reactions will occur and calculate the overall reaction:

In both cases, the hydrolysis of water requires far less energy than the reduction or oxidation of the other species, so the overall reaction is

$$2\,H_2O\,(l)\,\rightarrow\,O_2\,(g)\,+\,2\,H_2\,(g)$$

Workbook Problem 18.10

Strategy: Write the electrolysis reaction and consider the conversion process to calculate time.

Step 1: Treat the electrons as reactants in the chemical equation, and solve for the time required to produce 12 kg. of magnesium. Remember, the conversion factor of 96,500 C (or A·s) per mole of electrons.

$$Mg^{2+}\,+\,2e^-\,\rightarrow Mg\,(s)$$

$$2\,Cl^-\,\rightarrow Cl_2\,(g)\,+\,2e^-$$

$$12\text{ kg Mg}\times\frac{1000\text{ g}}{1\text{ kg}}\times\frac{1\text{ mol Mg}}{24.3\text{ g Mg}}\times\frac{2\text{ mol e}^-}{1\text{ mol Mg}}\times\frac{96,500\text{A}\cdot\text{s}}{1\text{ mol e}^-}\times\frac{1}{15\text{ A}}\times\frac{1\text{ h}}{3600\text{ s}}\times\frac{1\text{ day}}{24\text{ h}}=73.54\text{ days}$$

Step 2: Use the information calculated above to determine the amount of Cl_2 gas.

$$15\,\frac{C}{s}\times73.54\text{ days}\times\frac{24\text{ hours}}{day}\times\frac{60\text{ min}}{hour}\times\frac{60\text{ s}}{1\text{ min}}\times\frac{1\text{ mol e}^-}{96,500\text{ C}}\times\frac{1\text{ mol Cl}_2}{2\text{ mol e}^-}\times\frac{22.4\text{ L Cl}_2}{mol}=11,062\text{ L Cl}_2$$

Putting it Together:

Write the 2 half-reactions and solving for $E° = E_{anode} + E_{cathode}$.

$$O_2\,(g)\,+\,4\,H^+\,(aq)\,+\,4\,e^-\,\rightarrow\,2\,H_2O\,(l) \qquad E° = 1.23\,V$$
$$2\,Br^-\,(aq)\,\rightarrow Br_2\,(l)\,+\,2\,e^- \qquad E° = -1.09\,V$$

$$E° = 1.23\,V\,+\,(-1.09\,V) = 0.14$$

The reaction is spontaneous, because $E°$ is positive.

To determine the value of E after addition of the buffer, determine $[H^+]$ and substitute this value into the Nernst equation.

$$K_a \, (HCHO_2) = 1.7 \times 10^{-4}; \qquad\qquad pK_a = 3.74$$

$$pH = pK_a + \log \frac{\left[CHO_2^- \right]}{\left[HCHO_2 \right]} = 3.74 + \log \frac{0.35}{0.20} = 3.98$$

$$\left[H^+ \right] = \text{antilog} \, (-3.98) = 1.05 \times 10^{-4}$$

Substitute into the Nernst equation:

$$E = E^\circ - \frac{0.0592}{n} \log \frac{1}{[Br^-][H^+]}$$

After balancing the number of electrons in the two half–reactions, $n = 4$.

$$E = 0.14 \text{ V} - \frac{0.0592}{4} \log \frac{1}{\left(1.05 \times 10^{-4} \right)} = 0.0811 \text{ V}$$

The reaction is still spontaneous under these nonstandard state conditions.

Putting It Together

Write the two half–reactions and solving for $E^\circ = E_{anode} + E_{cathode}$.

$$O_2 \, (g) \, + \, 4 \, H^+ \, (aq) \, + \, 4 \, e^- \, \to \, 2 \, H_2O \, (l) \qquad E^\circ = 1.23 \text{ V}$$
$$2 \, Br^- \, (aq) \, \to \, Br_2 \, (l) \, + \, 2 \, e^- \qquad\qquad\qquad E^\circ = -1.09 \text{ V}$$

$$E^\circ = 1.23 \text{ V} \, + \, (-1.09 \text{ V}) = 0.14$$

The reaction is spontaneous, because E° is positive.

To determine the value of E after addition of the buffer, determine $[H^+]$ and substitute that value into the Nernst equation.

$$K_a \, (HCHO_2) = 1.7 \times 10^{-4}; \qquad\qquad pK_a = 3.74$$

$$pH = pK_a = \log \frac{\left[CHO_2^- \right]}{\left[HCHO_2 \right]}$$

$$pH = 3.74 + \log \frac{[0.35]}{[0.20]} = 3.98$$

$$\left[H^+ \right] = 10^{-3.98} = 1.05 \times 10^{-4}$$

Substitute into the Nernst equation:

$$E = E° - \frac{0.0592}{n} \log \frac{1}{\left[Br^-\right]\left[H^+\right]}$$

After balancing the number of electrons in the two half–reactions, $n = 4$.

$$E = 0.14V - \frac{0.0592}{4} \log \frac{1}{1.05 \times 10^{-4}} = 0.0811V$$

The reaction is still spontaneous under these nonstandard state conditions.

Self–Test

True–False
1. T
2. F. A wire allows the transfer of electrons; the salt bridge maintains electrical neutrality.
3. F. Positive E corresponds to negative free-energy change. Both indicate spontaneity.
4. T
5. F. The Nernst equation allows calculation of E under nonstandard conditions.
6. T
7. T
8. T
9. F. Corrosion can be prevented by contact with a metal *below* the metal of interest in the table of standard reduction potentials.
10. F. Overvoltage can make electrolysis reactions in aqueous solutions difficult to predict.

Multiple Choice

11. b	12. a	13. a	14. c
15. b	16. c	17. a	18. d
19. a	20. b		

Fill-in-the-Blank
21. lead storage, alkaline dry cells
22. corrosion, galvanization, sacrificial electrode.
23. sodium or aluminum, aluminum or sodium, electricity, power plants.
24. reduction, decreasing, increasing, standard hydrogen electrode
25. equilibrium, cell potentials, small or large, small or large, completion

Matching

26. f	27. c	27. h	28. a
29. b	30. i	31. d	32. e
33. g			

Problems
34. Half-reactions:

Anode: $Mg\,(s) \rightarrow Mg^{2+}\,(aq) + 2\,e^-$
Cathode: $Ni^{2+}\,(aq) + 2\,e^- \rightarrow Ni\,(s)$

Shorthand notation:
$Mg\,(s)\,|\,Mg^{2+}\,(aq)\,\|\,Ni^{2+}\,(aq)\,|\,Ni\,(s)$

35. a. $2 \text{ Al } (s) \rightarrow 2 \text{ Al}^{3+} (aq) + 6 \text{ e}^- \quad E° = 1.66 \text{ V}$
      $3 \text{ Pb}^{2+} (aq) + 6 \text{ e}^- \rightarrow 3 \text{ Pb } (s) \quad E° = 0.13 \text{ V}$
      overall $E° = 1.79 \text{ V}$ – spontaneous as written

    b. $\text{Br}_2 (aq) + 2 \text{ e}^- \rightarrow 2 \text{ Br}^- (aq) \quad E° = 1.09 \text{ V}$
      $2 \text{ Cl}^- (aq) \rightarrow \text{Cl}_2 (g) + 2 \text{ e}^- \quad E° = -1.36 \text{ V}$
      overall $E° = -0.27 \text{ V}$ – spontaneous in reverse

    c. $\text{Ni } (s) \rightarrow \text{Ni}^{2+} (aq) + 2 \text{ e}^- \quad E° = 0.26 \text{ V}$
      $\text{Pb}^{2+} (aq) + 2 \text{ e}^- \rightarrow \text{Pb } (s) \quad E° = -0.13 \text{ V}$
      overall $E° = 0.13 \text{ V}$ – spontaneous as written

    d. $\text{Fe } (s) \rightarrow \text{Fe}^{2+} (aq) + 2 \text{ e}^- \quad E° = 0.45 \text{ V}$
      $2 \text{ H}^+ + 2 \text{ e}^- \rightarrow \text{H}_2 (g) \quad E° = 0.00 \text{ V}$
      overall $E° = 0.45 \text{ V}$ – spontaneous as written

36. The two half–reactions (balanced in e^-) and their potentials for this cell are

    Anode: $3 \text{ Mg } (s) \rightarrow 3 \text{ Mg}^{2+} (aq) + 6 \text{ e}^- \quad E° = 2.37 \text{ V}$

    Cathode: $2 \text{ Fe}^{3+} (aq) + 6 \text{ e}^- \rightarrow 2 \text{ Fe } (s) \quad E° = -0.04 \text{ V}$

    a.   At standard state, $E° = 2.33 \text{ V}$ b. $[\text{Mg}^{2+}] = 0.1 \text{ M}, [\text{Fe}^{3+}] = 2.0 \text{ M}$

$$E_{cell} = E^o_{cell} - \frac{0.0592}{6} \log \frac{[\text{Mg}^{2+}]^3}{[\text{Fe}^{3+}]^2} = 2.33 - \frac{0.0592}{6} \log \frac{[0.1]^3}{[2]^2} = 2.37 \text{ V}$$

    c. $[\text{Mg}^{2+}] = 2.0 \text{ M}, [\text{Fe}^{3+}] = 0.1 \text{ M}$

$$E_{cell} = E^o_{cell} - \frac{0.0592}{6} \log \frac{[\text{Mg}^{2+}]^3}{[\text{Fe}^{3+}]^2} = 2.33 - \frac{0.0592}{6} \log \frac{[2.0]^3}{[0.1]^2} = 2.30 \text{ V}$$

37. $\Delta G° = -nFE = -2 \times 96,500 \dfrac{\text{C}}{\text{mol e}^-} \times 0.27 \text{ V} \times \dfrac{1 \text{ J}}{\text{C} \cdot \text{V}} = -52.1 \dfrac{\text{kJ}}{\text{mol}}$

$$E° = \frac{0.0592 \text{V}}{n} \log K$$

$$\log K = \frac{2 \times 0.27 \text{V}}{0.0592 \text{ V}} = 9.12$$

$$K = 10^{9.12} = 1.32 \times 10^9$$

38. The half-reactions for the cell are

    $\text{Zn}^{2+} (aq) + 2 \text{ e}^- \rightarrow \text{Zn } (s) \quad\quad E° = -0.76 \text{ V}$

    $E^o_{cell} = E^o_{anode} + E^o_{cathode} \quad\quad +1.61 \text{ V} = E^o_{anode} + (-0.76 \text{ V}) \quad E^o_{anode} = 2.37 \text{ V}$

From the table of standard reduction potentials, we find that Mg (s) has a reduction potential of -2.37 V. The half–reaction at the anode is

$$Mg\ (s)\ \rightarrow\ Mg^{2+}\ (aq)\ +\ 2\ e^-$$

39. $E_{ref} = -0.26$ V for $Ni^{2+}\ (aq)\ +\ 2\ e^- \rightarrow Ni\ (s)$

$$pH = \frac{E_{cell} - E_{ref}}{0.0592\ V} = \frac{0.14 - (-0.26)}{0.0592\ V} = 6.75$$

40. For a metal to be able to protect iron from corrosion, it must have a reduction potential below that of Fe^{2+}. Any metal below that reaction will able to protect iron, because any iron atoms that are oxidized will be reduced again by the metal.

$$Fe^{2+}\ (aq)\ +\ 2\ e^-\ \rightarrow\ Fe\ (s) \qquad E° = -0.45$$

a. $Sn^{2+}\ (aq)\ +\ 2\ e^-\ \rightarrow\ Sn\ (s)$ $\qquad E° = -0.14$ V. This is *above* the reduction potential of iron, so would speed corrosion.

b. $Mn^{2+}\ (aq)\ +\ 2\ e^-\ \rightarrow\ Mn\ (s)$ $\qquad E° = -1.18$ V. This is *below* the reduction potential of iron, so would protect against corrosion.

c. $Cu^{2+}\ (aq)\ +\ 2\ e^-\ \rightarrow\ Cu\ (s)$ $\qquad E° = 0.34$ V. This is *above* the reduction potential of iron, so would speed corrosion.

41. $AgCl\ (s)\ \rightarrow\ Ag^+\ (aq)\ +\ Cl^-\ (aq)$

Using the table of standard reduction potentials, we can break this overall reaction into the following two half–reactions:

$$AgCl\ (s)\qquad +\ e^- \rightarrow\ Ag\ (s)\ +\ Cl^-(aq) \qquad\qquad E° = 0.22\ V$$

$$Ag\ (s)\ \rightarrow\ Ag^+(aq)\ +\ e^- \qquad\qquad\qquad\qquad E° = -0.80\ V$$

$$E^o_{cell} = 0.22\ V + (-0.80\ V) = -0.58\ V$$

$$-0.58\ V = \frac{0.0592}{1}\log K$$

$$\log K = -9.80$$

$$K = 10^{-9.80}$$

$$K_{sp} = 1.58 \times 10^{-10}$$

42. $Na^+\ (l)\ +\ e^-\ \rightarrow\ Na\ (s)$

$2\ Br^-\ (l) \rightarrow Br_2\ (g) + 2\ e^-$

To determine the current required to produce 3 g of sodium per hour, set up a unit analysis to convert 3 g Na/hr to C/s:

$$\frac{3 \text{ g Na}}{\text{hr}} \times \frac{1 \text{ mol Na}}{23.0 \text{ g}} \times \frac{96,500 \text{ C}}{\text{mol e}^-} \times \frac{1 \text{ mol e}^-}{\text{mol Na}} \times \frac{1 \text{ hr}}{3600 \text{ s}} = 3.5 \frac{\text{C}}{\text{s}} = 3.5 \text{ A}$$

43. $15 \text{ g Ag} \times \dfrac{1 \text{ mol Ag}}{108 \text{ g}} \times \dfrac{96,500 \text{ C}}{\text{mol e}^-} \times \dfrac{1 \text{ mol e}^-}{\text{mol Ag}} \times \dfrac{1 \text{ s}}{3 \text{ C}} \times \dfrac{1 \text{ min}}{60 \text{ s}} = 74 \text{ min}$

44. To begin this problem, you will need to write and balance the equation. To make it easier, take the half-reaction for the reduction of MnO_4^- straight from a table to determine the number of electrons involved:

$$5 \text{ Zn } (s) \rightarrow 5 \text{ Zn}^{2+} (aq) + 10 \text{ e}^-$$

$$2 \text{ MnO}_4^- + 16 \text{ H}^+ (aq) + 10 \text{ e}^- \rightarrow 2 \text{ Mn}^{2+} (aq) + 8 \text{ H}_2\text{O } (l)$$

The overall reaction is $5 \text{ Zn } (s) + 2 \text{ MnO}_4^- + 16 \text{ H}^+ (aq) \rightarrow 5 \text{ Zn}^{2+} (aq) + 2 \text{ Mn}^{2+} (aq) + 8 \text{ H}_2\text{O } (l)$

The problem can now be set up as a unit-analysis problem:

$$3.27 \text{ g Zn} \times \frac{1 \text{ mol Zn}}{65.4 \text{ g Zn}} \times \frac{2 \text{ mol MnO}_4^-}{5 \text{ mol Zn}} \times \frac{142 \text{ g NaMnO}_4}{\text{mol NaMnO}_4} = 2.84 \text{ g}$$

This is the minimum amount needed the reaction.

45. To calculate the volume of O_2, first calculate the moles of O_2 produced and then use the ideal gas law. The half–reaction for the electrolysis of water that is relevant here is

$$2 \text{ H}_2\text{O } (l) \rightarrow O_2 (g) + 4 \text{ H}^+ (aq) + 4 \text{ e}^-$$

$$2.97 \times 10^3 \text{ C} \times \frac{1 \text{ mol e}^-}{96,500 \text{ C}} \times \frac{1 \text{ mol O}_2}{4 \text{ mol e}^-} = 7.69 \times 10^{-3} \text{ mol O}_2$$

$$V = \frac{nRT}{P} = \frac{(7.69 \times 10^{-3})(0.0821)(298)}{\left(356 \text{ mm Hg} \times \dfrac{1 \text{ atm}}{760 \text{ mm Hg}}\right)} = 0.402 \text{ L}$$

Challenge Problem
46. Separate the overall reaction into half-reactions and calculate K for the reduction of lead. $K_{red} \times K_{sp} = K_{overall}$. From this, calculate $E°$ for the reaction.

$$\text{PbSO}_4 (s) \rightarrow \text{Pb}^{2+} (aq) + \text{SO}_4^{2-} (aq) \qquad\qquad K_{sp}$$
$$\underline{\text{Pb}^{2+} (aq) + 2 \text{ e}^- \rightarrow \text{Pb } (s) \qquad\qquad\qquad K_{red} \text{ (calculated from } E°)}$$
$$\text{PbSO}_4 (s) + 2 \text{ e}^- \rightarrow \text{Pb } (s) + \text{SO}_4^{2-} (aq) \qquad\qquad K_{overall} = K_{sp} \times K_{red}$$

Calculate $K_{overall}$, then $E_{overall}$.

$$E^o_{red} = \frac{0.0592}{n} \log K_{red}$$

$n = 2$ for this reaction.

$$\log K_{red} = \frac{2(-0.126)}{0.0592} = -4.26$$

$$K_{red} = 5.50 \times 10^{-5}$$

Calculate K_{net} for the overall reaction.

$$K_{net} = (1.8 \times 10^{-8}) \times (5.5 \times 10^{-5}) = 9.9 \times 10^{-13}$$

Now calculate $E^o_{overall}$:

$$E^o_{net} = \frac{0.0592}{n} \log K_{net}$$

$$E^o_{net} = \frac{0.0592}{2} \log(9.9 \times 10^{-13}) = -0.355 \text{ V}$$

Chapter Nineteen – Nuclear Chemistry

Workbook Problems

Workbook Problem 19.1

Step 1:   First, find the chemical symbol for oxygen.

The chemical symbol for oxygen is O.

Step 2:   Use the periodic table to determine the atomic number for oxygen.

From the periodic table, we find that the atomic number for O is 8.

Step 3:   Determine the number of neutrons from the definition for mass number.

A (mass number) = Z (atomic number) + number of neutrons.
number of neutrons = $A - Z$

For oxygen-16: number of neutrons = 16 – 8 = 8
For oxygen-17: number of neutrons = 17 – 8 = 9
For oxygen-18: number of neutrons = 18 – 8 = 10

Step 4: The standard symbol is written with the mass number as a superscript and the atomic number as a subscript, both to the left of the symbol.

Workbook Problem 19.2

Strategy: Determine the atomic number from the information given.

We are told that element X has 51 protons; therefore, the atomic number of the element is 51.

Step 1: Knowing the atomic number, identify the element by using the periodic table.

From the periodic table, we find that the element with $Z = 51$ is Sb (antimony).

Step 2: The standard symbol is written with the mass number as a superscript and the atomic number as a subscript, both to the left of the symbol.

The mass number is determined by adding the number of protons and the number of neutrons.

$A = 51 + 70 = 121.$ The standard symbol is $^{121}_{51}\text{Sb}$.

Workbook Problem 19.3

Strategy: Use the periodic table to write the nuclide symbol of polonium-214 and the alpha particle. Using this information, write an incomplete nuclear equation and solve for the missing particle.

Step 1: Write the nuclide symbols for polonium-214 and the alpha particle.

polonium-214: $^{214}_{84}Po$ alpha particle: $^{4}_{2}\alpha$

Step 2: Write an incomplete nuclear equation representing the alpha decay of polonium-214.

$$^{214}_{84}Po \rightarrow {}^{4}_{2}\alpha + \underline{\quad\quad}$$

Step 3: Determine the missing mass number.

$$214 - 4 = 210$$

Step 4: Determine the missing atomic number and the symbol of the new nucleus.

$84 - 2 = 82$ The atomic number is lead (Pb).

Step 5: Write the balanced, complete nuclear equation.

$$^{214}_{84}Po \rightarrow {}^{4}_{2}\alpha + {}^{210}_{82}Pb$$

Workbook Problem 19.4

Strategy: Use the periodic table to write the nuclide symbols given in the problem. Using this information, write an incomplete nuclear equation and solve for the missing particle.

Step 1:   Write the incomplete nuclear equation.

$$^{58}_{28}Ni + ^{1}_{1}p \rightarrow ^{4}_{2}\alpha + \underline{}$$

Step 2:   Determine the missing atomic mass.

$$(58 + 1) - 4 = 55$$

Step 3:   Determine the missing atomic number and the identity of the new nucleus.

$$(28 + 1) - 2 = 27 \qquad\qquad \text{This is Cobalt (Co)}$$

Step 4:   Write the complete balanced nuclear equation.

$$^{58}_{28}Ni + ^{1}_{1}p \rightarrow ^{4}_{2}\alpha + ^{55}_{27}Co$$

Workbook Problem 19.5

Step 1:   Solve for *k*:

$$5730 \text{ years} = \frac{\ln 2}{k}$$

$$k = 1.21 \times 10^{-4} \text{ year}^{-1}$$

Step 2:   Substitute into the integrated rate law and solve for *t* after 23% decay:

$$\ln (1.0 - 0.23) = -1.21 \times 10^{-4}/\text{year} \times t$$

$$t = 2160 \text{ years old}$$

Workbook Problem 19.6

Step 1:   Determine the number of nucleons and their mass:

First, calculate the total mass of the nucleons and the mass defect:

Xe-130 has 54 protons, and 76 neutrons

Mass of 54 protons =	(54)(1.007 28 amu)	= 54.393 12 amu
Mass of 76 neutrons =	(76)(1.008 66 amu)	= 76.658 16 amu
Mass of 54 protons and 76 neutrons		=131.051 28 amu

Mass of xenon-130	=129.903	amu
Mass of electrons = (54)(5.486 x 10⁻⁴ amu)	= 0.030	amu
Mass of xenon-130 nucleus	=129.873	amu

Mass defect = 131.051 28 amu − 129.873 amu = 1.178

Step 2: Convert the mass defect to grams:

$$1.178 \text{ amu} \times \frac{1.6605 \times 10^{-24} \text{ g}}{amu} = 1.956 \times 10^{-24} \text{ g}$$

Step 3: Determine ΔE and convert to MeV/nucleon:

$$\Delta E = 1.957 \times 10^{-24} \text{g} \times \frac{1 \text{ kg}}{1000 \text{ g}} \times (3.00 \times 10^{8} \text{m/s})^{2} = 1.761 \times 10^{-10} \text{ J}$$

Convert to MeV/nucleon:

$$1.761 \times 10^{-10} \frac{\text{J}}{\text{nucleus}} \times \frac{1 \text{ MeV}}{1.60 \times 10^{-13} \text{ J}} \times \frac{1 \text{ nucleus}}{130 \text{ nucleons}} = 8.47 \frac{MeV}{nucleon}$$

Workbook Problem 19.7

Strategy: Balance the equation, calculate the mass defect, and convert to energy using the Einstein equation.

Step 1: Write and balance the equation:

$$^{235}\text{U} + {}^{1}\text{n} \rightarrow {}^{136}\text{Xe} + {}^{99}\text{Ru} + {}^{1}\text{n}$$

Step 2: Calculate the mass defect in amu and convert to kilograms:

235.043 92 amu − (135.907 22 amu + 98.905 939 amu) = 0.230 76 amu

$$0.230 \ 76 \text{ amu} \times \frac{1.6605 \times 10^{-27} \text{ kg}}{amu} = 3.8318 \times 10^{-28} \text{ kg}$$

Step 3: Determine ΔE and convert to J/mol:

$$\Delta E = 3.8318 \times 10^{-28} \text{kg} \times (3.00 \times 10^{8} \text{m/s})^{2} = 3.35 \times 10^{-11} \text{ J}$$

$$3.35 \times 10^{-11} \frac{\text{J}}{\text{reaction}} \times 6.022 \times 10^{23} \frac{\text{reactions}}{\text{mol}} = 2.077 \times 10^{13} \text{ J/mol}$$

Workbook Problem 19.8

Strategy: Calculate the mass defect and convert to energy using the Einstein equation.

Step 1: Write and balance the equation:

$$^2H + {}^3H \rightarrow {}^4He + {}^1n$$

Step 2: Calculate the mass defect in amu and convert to kilograms:

$$(2.014\,10 \text{ amu} + 3.016\,05 \text{ amu}) - (4.002\,60 \text{ amu} + 1.008\,66 \text{ amu}) = 0.018\,89 \text{ amu}$$

$$0.01889 \text{ amu} \times \frac{1.6605 \times 10^{-27} \text{ kg}}{\text{amu}} = 3.1367 \times 10^{-29} \text{ kg}$$

Step 3: Determine ΔE and convert to J/mol:

$$\Delta E = 3.136\,7 \times 10^{-29} \text{ kg} \times (3.00 \times 10^8 \text{m/s})^2 = 2.82 \times 10^{-12} \text{ J}$$

$$2.82 \times 10^{-12} \; \frac{\text{J}}{\text{reaction}} \times 6.022 \times 10^{23} \; \frac{\text{reactions}}{\text{mol}} = 1.70 \times 10^{12} \text{ J/mol}$$

This is particularly impressive when you consider that 1 mol of ^4He has a mass of less than a nickel.

Workbook Problem 19.9

Strategy: Determine the overall mass, subtract out the neutrons, and determine the identity and mass of the element produced.

The total number of protons in the new element is 98 – 92 from uranium plus 6 from carbon.

The mass of the new element will be the masses of the uranium and the carbon, minus the mass of the neutrons: 238 + 12 – 6 = 244.

The element with 98 protons is californium, or Cf.

The final equation is: $^{238}_{92}U + {}^{12}_{6}C \rightarrow {}^{244}_{98}Cf + 6 \, {}^{1}_{0}n$

Putting It Together

The formula for sodium perchlorate is $NaClO_4$. The amount of chlorine in 78.3 mg in $NaClO_4$ is

$$0.0783 \text{ g } NaClO_4 \times \frac{1 \text{ mol } NaClO_4}{122.5 \text{ g } NaClO_4} \times \frac{1 \text{ mol } Cl}{1 \text{ mol } NaClO_4} \times \frac{35.5 \text{ g } Cl}{1 \text{ mol } Cl} = 0.0227 \text{ g } Cl$$

Of the 22.7 mg Cl present, 47.0% is ^{36}Cl, 10.7 mg.

Calculate k.

$$t_{1/2} = 3.0 \times 10^5 \, years \times \frac{365 \, days}{1 \, year} \times \frac{24 \, hours}{1 \, day} \times \frac{3600 \, sec}{1 \, hour} = 9.46 \times 10^{12} \, sec$$

$$k = \frac{0.693}{9.46 \times 10^{12} \, sec} = 7.32 \times 10^{-14} \, sec^{-1}$$

The decay rate = $k \times N$, where N is the number of radioactive nuclei. Calculate N:

$$N = 2.27 \times 10^{-8} \, g \, {}^{36}Cl \times \frac{1 \, mol \, {}^{36}Cl}{36.0 \, g \, {}^{36}Cl} \times \frac{6.022 \times 10^{23} \, nuclei \, {}^{36}Cl}{1 \, mol \, {}^{36}Cl} = 3.80 \times 10^{19} \, nuclei$$

Calculate the decay rate:

$$\text{Rate} = (7.32 \times 10^{-14} \, s^{-1}) \times 3.80 \times 10^{20} \, nuclei = 2.78 \times 10^7 \, nuclei/s \text{ or } 2.78 \times 10^7 \, disintegrations/s.$$

Self–Test

True–False
1. F. Radioactive decay is a first-order process.
2. F. Light elements have a 1:1 ratio of protons to neutrons, but heavier elements require more "glue" and thus more neutrons.
3. F. A small amount of mass is lost during its conversion to energy.
4. T
5. T
6. T
7. F. The oxidation number is always given as part of the complex name.
8. T
9. T
10. T

Multiple Choice
11. b	12. c	13. a	14. d
15. c	16. c	17. a	18. d
19. c	20. b		

Fill-in-the-Blank
21. glue, proton, proton, nucleus, mass defect, $\Delta E = \Delta mc^2$
22. maximum, fusion, fission, decay
23. transmutation, particles, nucleus, element
24. alpha, beta, internal, gamma, X, external, stopped, lead
25. in vivo, diseased, internal, external, imaging

Matching
26. e	27. g	28. d	29. a
30. c	31. b	32. f	

Problems

33. Europium-130   Mass of 63 protons = (63)(1.007 28 amu)   = 63.458 64 amu
Mass of 67 neutrons = (67)(1.008 66 amu)   = 67.580 22 amu
Mass of 63 protons and 67 neutrons   = 131.038 86 amu

Mass of europium-130 atom   = 129.963 57 amu
Mass of electrons = (63) (5.486 × 10^{-4})   =   0.034 56 amu
Mass of europium-130 nucleus   = 129.929 01 amu

Mass defect: 131.038 86 amu – 129.929 01 amu= 1.109 85 amu

Europium-153   Mass of 63 protons = (63)(1.007 28 amu)   = 63.458 64 amu
Mass of 90 neutrons = (90)(1.008 66 amu)   = 90.779 40 amu
Mass of 63 protons and 90 neutrons   = 154.238 04 amu

Mass of europium-153 atom   = 152.921 23 amu
Mass of electrons = (63) (5.486 × 10^{-4})   =   0.034 56 amu
Mass of europium-153 nucleus   = 152.886 67 amu

Mass defect: 154.238 04 amu – 152.886 67 amu= 1.351 37 amu

Europium-167   Mass of 63 protons = (63)(1.007 28 amu)   = 63.458 64 amu
Mass of 104 neutrons =(104)(1.008 66 amu)   = 104.900 64 amu
Mass of 63 protons and 104 neutrons   = 168.359 28 amu

Mass of europium-167 atom   = 166.953 21 amu
Mass of electrons = (63) (5.486 × 10^{-4})   =   0.034 56 amu
Mass of europium-167 nucleus   = 166.918 65 amu

Mass defect: 168.359 28 amu – 166.918 65 amu = 1.440 64 amu

Convert this mass to kilograms:

$$\text{Europium-130: } 1.109\ 85\ \text{amu} \times \frac{1.6605 \times 10^{-27}\ \text{kg}}{\text{amu}} = 1.842\ 91 \times 10^{-27}\ \text{kg}$$

$$\text{Europium-153: } 1.351\ 37\ \text{amu} \times \frac{1.6605 \times 10^{-27}\ kg}{amu} = 2.243\ 95 \times 10^{-27}\ kg$$

$$\text{Europium-167: } 1.440\ 64\ \text{amu} \times \frac{1.6605 \times 10^{-27}\ \text{kg}}{\text{amu}} = 2.392\ 18 \times 10^{-27}\ \text{kg}$$

Use the Einstein equation to convert the mass defect to binding energy:

$$\text{Europium-130: } \Delta E = 1.842\ 91 \times 10^{-27} kg \ x \ (3.00 \times 10^8 m/s)^2 = 1.6586 \times 10^{-10}\ J$$

$$\text{Europium-153: } \Delta E = 2.243\ 95 \times 10^{-27} kg \ x \ (3.00 \times 10^8 m/s)^2 = 2.0196 \times 10^{-10}\ J$$

$$\text{Europium-167: } \Delta E = 2.392\ 18 \times 10^{-27} kg \ x \ (3.00 \times 10^8 m/s)^2 = 2.152\ 96 \times 10^{-10}\ J$$

Convert to MeV/nucleon:

$$\text{E-130: } 1.6586\times10^{-10}\ \frac{J}{\text{nucleus}}\times\frac{1\ \text{MeV}}{1.60\times10^{-13}\ J}\times\frac{1\ \text{nucleus}}{130\ \text{nucleons}}=7.97\ \frac{\text{MeV}}{\text{nucleon}}$$

$$\text{E-153: } 2.0195\times10^{-10}\ \frac{J}{\text{nucleus}}\times\frac{1\ \text{MeV}}{1.60\times10^{-13}\ J}\times\frac{1\ \text{nucleus}}{153\ \text{nucleons}}=8.25\ \frac{\text{MeV}}{\text{nucleon}}$$

$$\text{E-167: } 2.152\,96\times10^{-10}\ \frac{J}{\text{nucleus}}\times\frac{1\ \text{MeV}}{1.60\times10^{-13}\ J}\times\frac{1\ \text{nucleus}}{167\ \text{nucleons}}=8.06\ \frac{\text{MeV}}{\text{nucleon}}$$

According to this data, from most to least stable these are

$$\text{E-153} > \text{E-167} > \text{E-130}$$

The observed half-lives of these isotopes are

E-130: 1.1 ms   E-153: stable   E-167: 200 ms

This bears out the calculated results.

34. Rearrange the equation $E = mc^2$ and solve:

$$\Delta m = \frac{\Delta E}{c^2} = \frac{-498,400\ \text{J/mol}}{(3.00\times10^8\text{m/s})^2} = -5.54\times10^{-12}\ \text{kg/mol} = -5.54\times10^{-9}\text{g/mol}$$

35. mass of products = 91.9262 amu + 140.9144 amu + 3(1.008 66 amu) = 235.867 amu

mass of reactants = 235.0439 amu + 1.008 66 amu = 236.053 amu

Δm = 236.053 amu − 235.867 amu = 0.186 amu

Convert mass to kg: $0.186\ \text{amu} \times \dfrac{1.6605\times10^{-27}\ \text{kg}}{\text{amu}} = 3.08\times10^{-28}\,\text{kg}$

Convert mass to energy: $E = 3.08\times10^{-28}kg\times(3.00\times10^8\ \text{m/s})^2 = 2.77\times10^{-11}\text{J/reaction}$

Convert to J/mol: $2.77\times10^{-11}\text{J/reaction}\times\dfrac{6.022\times10^{23}\text{reactions}}{\text{mol}} = 1.67\times10^{13}\text{J/mol}$

36. mass of products = 4.002 60 amu + 1.008 66 amu = 5.011 26 amu
mass of reactants = 2 (3.016 049 amu) = 6.032 10 amu

Δm = 6.032 10 amu − 5.011 26 amu = 1.020 08 amu

Convert mass to kg: $1.020\,08\ \text{amu}\times\dfrac{1.6605\times10^{-27}\ kg}{amu} = 1.6949\times10^{-27}kg$

Convert mass to energy: $E = 1.6949 \times 10^{-27} \text{kg} \times (3.00 \times 10^8 \text{ m/s})^2 = 1.525 \times 10^{-10}$ J/reaction

Convert to J/mol: 1.525×10^{-10} J/reaction x $\dfrac{6.022 \times 10^{23} \text{ reactions}}{\text{mol}} = 9.184 \times 10^{13}$ J/mol

37. a. $^{9}_{4}\text{Be} + ^{1}_{1}\text{H} \rightarrow ^{6}_{3}\text{Li} + ^{4}_{2}\text{He}$

    b. $^{239}_{94}\text{Pu} + ^{4}_{2}\text{He} \rightarrow ^{242}_{96}\text{X} + ^{1}_{0}\text{n}$. With an atomic number of 96, this is Curium, Cm-96.

    c. $^{208}_{82}\text{Pb} + ^{58}_{26}\text{X} \rightarrow ^{265}_{108}\text{Hs} + ^{1}_{0}\text{n}$. With an atomic number of 26, this is Fe-58.

38. The decay constant, k, can be calculated from the half-life:

$$t_{1/2} = \frac{0.693}{k}; \; k = \frac{0.693}{t_{1/2}} = \frac{0.693}{6.3 \times 10^{26} \text{ s}} = 1.1 \times 10^{-27} / s$$

39. Use the integrated first-order rate law:

$$\ln \frac{N_t}{N_0} = -0.623 \left(\frac{t}{t_{1/2}} \right)$$

Substituting:

$$\ln \frac{3.5}{15.3} = -0.693 \left(\frac{t}{5715 \text{ years}} \right) \qquad \frac{\ln 0.229 \times 5715 \text{ years}}{-0.693} = 12{,}165 \text{ years}$$

The bone fragment is approximately 12,000 years old.

40. Find the mole ratio of lead to uranium:

$$56 \text{ g Pb-207} = 0.271 \text{ mol} \qquad\qquad 27 \text{ g U-235} = 0.115 \text{ mol}$$

The total number of moles of uranium initially will be the moles of uranium remaining plus the moles of lead found in the sample.

Substitute into the rate law:

$$\ln \frac{0.115}{(0.271 + 0.115)} = -0.693 \left(\frac{t}{7.04 \times 10^8 \text{ years}} \right)$$

$$\frac{\ln 0.298 \times 7.04 \times 10^8 \text{ years}}{-0.693} = 1.23 \times 10^9 \text{ years}$$

Challenge Problem

41. The information given in the problem can be summarized by the partial equation:

$$^{10}_{5}B + {}^{1}_{0}n \rightarrow {}^{?}_{3}Li + ?$$

No protons are being added to the system, but two are being lost by boron, so the radiation produced must contain two protons. Only α particles meet this requirement.

$$^{10}_{5}B + {}^{1}_{0}n \rightarrow {}^{?}_{3}Li + {}^{4}_{2}He$$

Balancing the masses leaves Li-7.

Radiation therapies are constantly being refined so that damage to surrounding tissues can be minimized. This therapy allows a nonradioactive material to be injected, removing the need for surgery. The neutron bombardment can be tightly focused to avoid the areas around the tumor so that the α-particles produced can destroy the tumor with minimum collateral damage.

It may not be surprising that this therapy is being investigated primarily for inoperable brain cancers. Unfortunately, as elegant as it is, so far it does not appear to improve outcomes.

Chapter Twenty – Transition Elements and Coordination Chemistry

Workbook Problems

Workbook Problem 20.1

Strategy: Write the expected electron configuration for the neutral element, determine the charge, and remove electrons accordingly.

Step 1: Write the electron configuration of the neutral atom:

Ti: $[Ar]3d^2 4s^2$ Mn: $[Ar]3d^5 4s^2$ Ni: $[Ar]3d^8 4s^2$

Step 2: Determine the oxidation state of the metal:

Ti: 0 – no electrons need be removed, the answer is complete
Mn: each oxygen is –2, one negative charge remains, so oxidation state is +7
Ni: 2+

Step 3: Remove the necessary electrons:

Ti: $[Ar]3d^2 4s^2$ remove 0 electrons, $[Ar]3d^2 4s^2$
Mn: $[Ar]3d^5 4s^2$ remove 7 electrons, $[Ar]$
Ni: $[Ar]3d^8 4s^2$ remove 2 electrons, $[Ar]3d^8$

Workbook Problem 20.2

Determine the oxidation state and coordination number of the metal in the following complexes:

$$AgCl_2^- \qquad [MnCl_4]^{-2} \qquad [Fe(CN)_6]^{4-} \qquad [Co(NH_3)_4Br_2]Br$$

Strategy: With the charge and ligands, it is possible to determine the oxidation state of the metals. The coordination number is the number of ligand attachments.

Step 1:   Determine the oxidation state of the metals:

Ag + 2(–1) = –1;   Ag = +1 Mn + 4(–1) = –2;   Mn = +2 Fe + 6(–1) = –4; Fe = +2

Co + 4(0) + 2(–1) = +1 (the complex has a +1 charge, balanced by a Br⁻ ion); Co = +3

Step 2:   Determine the coordination numbers:

$AgCl_2^-$	Coordination number is 2
$[MnCl_4]^{-2}$	Coordination number is 4
$[Fe(CN)_6]^{4-}$	Coordination number is 6
$[Co(NH_3)_4Br_2]Br$	Coordination number is 6

Workbook Problem 20.3

Strategy: Using the formula $E = hc/\lambda$, determine the energy of the absorbed photons, then convert to kJ/mol.

Step 1:   Determine the energy of the photons absorbed:

$$E = 6.626 \times 0^{-34} \ J \cdot s \times \frac{3.00 \times 10^8 \ m/s}{550 \times 10^{-9} \ m} = 3.61 \times 10^{-19} J$$

Step 2:   Convert to kJ/mol:

$$3.61 \times 10^{-19} J \times \frac{1 \ kJ}{1000 \ J} \times 6.022 \times 10^{23} \ mol^{-1} = 218 \ kJ/mol$$

Workbook Problem 20.4

$[Co(Cl)_6]^{3-}$ is a high-spin complex. Draw an electron diagram to demonstrate this.

Strategy: Draw the orbital diagram and determine how to maximize unpaired electrons.

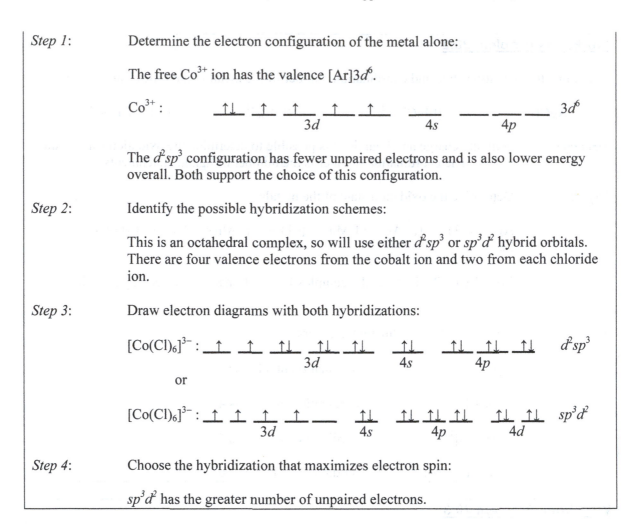

Step 1:   Determine the electron configuration of the metal alone:

The free Co^{3+} ion has the valence $[Ar]3d^6$.

Co^{3+} :   ↑↓ ↑ ↑ ↑ ↑   ___   ___ ___ ___ $3d^6$
                    3d   4s   4p

The d^2sp^3 configuration has fewer unpaired electrons and is also lower energy overall. Both support the choice of this configuration.

Step 2:   Identify the possible hybridization schemes:

This is an octahedral complex, so will use either d^2sp^3 or sp^3d^2 hybrid orbitals. There are four valence electrons from the cobalt ion and two from each chloride ion.

Step 3:   Draw electron diagrams with both hybridizations:

$[Co(Cl)_6]^{3-}$: ↑ ↑ ↑↓ ↑↓ ↑↓   ↑↓   ↑↓ ↑↓ ↑↓ d^2sp^3
                          3d   4s   4p

or

$[Co(Cl)_6]^{3-}$: ↑ ↑ ↑ ↑ ___   ↑↓   ↑↓ ↑↓ ↑↓   ↑↓ ↑↓ sp^3d^2
                          3d   4s   4p   4d

Step 4:   Choose the hybridization that maximizes electron spin:

sp^3d^2 has the greater number of unpaired electrons.

Putting It Together

Write the balanced net ionic equation for the reaction, using the half-reaction method. The unbalanced half-reactions are

$$Cr_2O_7^{2-}\ (aq)\ \rightarrow\ Cr^{2+}\ (aq) \qquad\qquad Fe^{2+}\ (aq)\ \rightarrow\ Fe^{3+}\ (aq)$$

The balanced half-reactions, after the addition of H^+ and H_2O, are

$$8\ e^-\ +\ 14\ H^+\ (aq)\ +\ Cr_2O_7^{2-}\ (aq)\ \rightarrow\ 2\ Cr^{2+}\ (aq)\ +\ 7\ H_2O\ (l)$$
$$Fe^{2+}\ (aq)\ \rightarrow\ Fe^{3+}\ (aq)\ +\ e^-$$

Balance the half-reactions in electrons:

$$8\ e^-\ +\ 14\ H^+\ (aq)\ +\ Cr_2O_7^{2-}\ (aq)\ \rightarrow\ 2\ Cr^{2+}\ (aq)\ +\ 7\ H_2O\ (l)$$
$$8\ Fe^{2+}\ (aq)\ \rightarrow\ 8\ Fe^{3+}\ (aq)\ +\ 8\ e^-$$

Add the two half–reactions together for the net ionic equation:

$$14\ H^+\ (aq)\ +\ Cr_2O_7^{2-}\ (aq)\ +\ 8\ Fe^{2+}\ (aq)\ \rightarrow\ 2\ Cr^{2+}\ (aq)\ +\ 7\ H_2O\ (l)\ +\ 8\ Fe^{3+}\ (aq)$$

With this balanced reaction and the data provided, and determine the mass of iron in the ore.

$$g\ Fe = 0.03328\ L \times \frac{0.090\ mol\ Cr_2O_7^{2-}}{1\ L} \times \frac{8\ mol\ Fe^{2+}}{1\ mol\ Cr_2O_7^{2-}} \times \frac{1\ mol\ Fe}{1\ mol\ Fe^{2+}} \times \frac{55.8\ g\ Fe}{1\ mol\ Fe} = 1.34\ g\ Fe$$

$$\%Fe = \frac{1.34\ g\ Fe}{1.513\ g\ sample} \times 100\% = 88.4\ \%$$

Self–Test

True–False

1. T
2. F. Mercury (Hg) is a liquid
3. F. Densities have a maximum in the center of the row.
4. F. Transition metals are commonly found in multiple oxidation states.
5. T
6. T
7. F. The oxidation number is always given as part of the complex name.
8. T
9. T
10. F. Crystal-field theory assumes that all interactions are electrostatic.

Multiple Choice

11. b	12. c	13. a	14. d
15. c	16. c	17. a	18. d
19. b	20. b		

Fill-in-the-Blank

21. oxidation, oxidizing, reducing
22. Coordination, ligands, coordinate covalent
23. isomers, geometric, stereoisomers
24. wavelength or frequency, energy, color
25. Valence bond theory, hybrid, crystal-field theory, electrostatic

Matching

26. h	27. m	28. A	29. F
30. b	31. g	32. c	33. k
34. n	35. e	36. q	37. i
38. p	39. j	40. l	41. D
42. o			

Questions

43. Mn^{5+}: [Ar] $4s^2$ Mo: [Kr] $5s^2\ 4d^4$

    Cr^{3+}: [Ar] $4s^2 3d^1$ Fe^{2+}: [Ar] $4s^1 3d^5$

44. Mn, Rh, La

45. Zn > Cu > Co > Mn. These metals are all better reducing agents than oxidizing agents.

46. MnO_4^- > CrO_3^- > Fe_2O_3 > CoO

47. 3, 8, 5, 2

48. a. Ligands: H_2O (donor atom—O); NH_3 (donor atom—N).
    Coordination number is 6; oxidation state of metal is +3; has the general formula MA_2B_4;
    therefore can be geometrical isomers.
    b. Ligands: NH_3 (donor atom—N); Br (donor atom—Br); H_2O (donor atom—O).
    Coordination number is 6; oxidation state of metal is +3; has the general formula
    MA_4BC; therefore can be geometrical isomers. This compound can also have ionization
    isomers were the Br and NO_3 to switch places.
    c. Ligands: NH_3 (donor atom—N), SCN^- (donor atom either N or S).
    Coordination number is 6; oxidation state of metal is +2. Because SCN^- can bond
    through either the N or S, this compound can display linkage isomerism.
    d. Ligands: ox (donor atom—O)
    Coordination number is 6 (ox is a bidentate ligand); oxidation state of metal is +3;
    charge on complex ion is –3. Because the ligands can spiral either to the right or to the
    left, this compound can exist as enantiomers.
    e. Ligands: en (donor atoms—N)
    Coordination number is 6 (en is a bidentate ligand); oxidation state of metal is +3.
    The en ligand can spiral either to the right or to the left, indicating the compound can
    exist as enantiomers.

49. a. diammineatetraaquacobalt(III) chloride
    b. tetraammineaquabromocobalt(II) nitrate
    c. tetraamminedithiocyanatochromate(II) or tetraamminediisothiocyanatochromate(II)
    depending on the linkage isomer.
    d. ammonium trioxaloferrate(III)
    e. tris(ethylenediamine)cobalt(III) iodide

50. a. $[Cr(NH_3)_4(H_2O)_4]Cl_3$ b. $[Cu(en)(SCN)_2]$
    c. $[Mn(CO)_5Cl]$ d. $K_4[Fe(CN)_4(Cl)_2] \cdot 2\ H_2O$
    e. $[Cu(H_2O)_4]SO_4$

51 a. Since $[Zn(NH_3)_4]^{2+}$ is tetrahedral, the hybridization on the metal ion is sp^3. The Zn^{2+} ion
    has the electron configuration $[Ar]\ 4s^2 3d^8$. The orbital diagram for this ion is

    [Ar] ↑↓ ↑↓ ↑↓ ↑ ↑   ↑↓

    ___ ___ ___ ___ ___   __ __ __
          3d   4s   4p

    To share the four electron pairs provided by the ammonia molecules, Zn^{2+} must use the
    empty $4s$ and $4p$ orbitals to form hybrid orbitals. The orbital diagram for the complex is

    [Ar] ↑↓ ↑↓ ↑↓ ↑↓ ↑↓   ↑↓   ↑↓   ↑↓ ↑↓

    __ __ __ __ __   ___ ___ ___ ___
          3d   4s   4p

    This complex is low spin, since it has no unpaired electrons.
    b. $[Ni(CN)_4]^{2-}$ is square planar; therefore, the hybrid orbitals being used by the Ni^{2+} ion are
    dsp^2. The Ni^{2+} ion has the electron configuration $[Ar]\ 3d^8$. The orbital diagram for this
    ion is

[Ar] ↑↓ ↑↓ ↑↓ ↑ ↑   ↑↓   ↑↓ ↑↓ ↑↓

‾‾ ‾‾ ‾‾ ‾‾ ‾‾   ‾‾   ‾‾ ‾‾ ‾‾
      3d   *4s*   *4p*

For Ni^{2+} to use dsp^2 hybrid orbitals, the two unpaired electrons must first pair up. Ni^{2+} can then accept a share in the four electron pairs provided by the CN^- ion. The orbital diagram for the complex is

[Ar] ↑↓ ↑↓ ↑↓ ↑↓   ↑↓ ↑↓ ↑↓   ↑↓

‾‾ ‾‾ ‾‾ ‾‾   ‾‾ ‾‾ ‾‾ ‾‾   ‾‾
      3d   *dsp^2*   *4p*

The complex is low spin, since there are no unpaired electrons.

52. The crystal-field splitting in tetrahedral complexes is half of the crystal-field splitting in octahedral complexes because none of the d orbitals points directly at the ligands and there are only four ligands, as opposed to six in an octahedral complex.

Challenge Problem

53. Chromium forms octahedral complexes, and there are nine possible ligands. Therefore, the compound can be a hydrate or a salt: not all nine ligands will be complexed to the chromium. The experimental data must be used to determine what is bonded and what is hydrate water or balancing ions.

In each case, the moles of compound dehydrated are the same. The molecular weight of all three is 266.45 g/mol.

$$0.500 \text{ g} \times \frac{1 \text{ mol}}{273.5 \text{ g}} = 1.88 \times 10^{-3} \text{mol}$$

The moles of compound in each solution is also the same:

$$0.125 \text{ mol/L} \times 0.050 \text{ L} = 6.25 \times 10^{-3} \text{ mol}$$

If the mass of the compound does not change when it is dehydrated, all six waters are complexed to the chromium. This is seen with compound C. Therefore, it is likely that this is $[Co(H_2O)_6]Cl_3$. This can be confirmed by determining whether 3 mol of AgCl are produced per mol of compound:

$$0.809 \text{ g AgCl} \times \frac{1 \text{ mol}}{143.4 \text{ g}} = 5.64 \times 10^{-3} \text{ mol AgCl}$$

$$\frac{5.64 \times 10^{-3} \text{mol AgCl}}{1.88 \times 10^{-3} \text{mol C}} = 3.00$$

This confirms that the chlorides are balancing ions and will separate in solution. This is $[Co(H_2O)_6]Cl_3$.

If there is a mass change on dehydration, the number of hydrate waters can be determined.

The mass change for compound A is 0.500 g – 0.432 g = 0.068 g water. The number of hydrate waters can be determined by comparing the moles of water to the moles of compound.

$$0.068 \text{ g H}_2\text{O} \times \frac{1 \text{ mol}}{18.01 \text{ g}} = 3.76 \times 10^{-3} \text{ mol H}_2\text{O}$$

$$\frac{3.76 \times 10^{-3} \text{ mol H}_2\text{O}}{1.88 \times 10^{-3} \text{ mol A}} = 2.00$$

If there are two hydrate waters, the compound will likely be $[\text{Co}(\text{H}_2\text{O})_4\text{Cl}_2]\text{Cl} \cdot 2\text{H}_2\text{O}$. This can be confirmed by the number of moles of AgCl produced per mole of compound.

$$0.269 \text{ g AgCl} \times \frac{1 \text{ mol}}{143.4 \text{ g}} = 1.88 \times 10^{-3} \text{ mol AgCl}$$

$$\frac{1.88 \times 10^{-3} \text{ mol AgCl}}{1.88 \times 10^{-3} \text{ mol A}} = 1.00$$

This confirms one balancing chloride ion. This is $[\text{Co}(\text{H}_2\text{O})_4\text{Cl}_2]\text{Cl} \cdot 2\text{H}_2\text{O}$.

The mass change for compound B is 0.500 g – 0.466 g = 0.034 g water. The number of hydrate waters can be determined by comparing the moles of water to the moles of compound.

$$0.034 \text{ g H}_2\text{O} \times \frac{1 \text{ mol}}{18.01 \text{ g}} = 1.88 \times 10^{-3} \text{ mol H}_2\text{O}$$

$$\frac{1.88 \times 10^{-3} \text{ mol H}_2\text{O}}{1.88 \times 10^{-3} \text{ mol B}} = 1.00$$

If there is one hydrate water, the compound is likely to be $[\text{Co}(\text{H}_2\text{O})_5\text{Cl}]\text{Cl}_2 \cdot \text{H}_2\text{O}$. This can be confirmed by the number of moles of AgCl produced per mole of compound.

$$0.539 \text{ g AgCl} \times \frac{1 \text{ mol}}{143.4 \text{ g}} = 3.76 \times 10^{-3} \text{ mol AgCl}$$

$$\frac{3.76 \times 10^{-3} \text{ mol AgCl}}{1.88 \times 10^{-3} \text{ mol B}} = 2.00$$

This confirms two balancing chloride ions. This is $[\text{Co}(\text{H}_2\text{O})_5\text{Cl}]\text{Cl}_2 \cdot \text{H}_2\text{O}$.

Chapter Twenty-One – Metals and Solid-State Materials

Workbook Problems

Workbook Problem 21.1

Strategy: Write and balance equations for purification of the metal from the mineral ore.

Step 1: Write the equation for the roasting of covellite:

$$2 \text{ CuS } (s) + 3 \text{ O}_2 \ (g) \rightarrow 2 \text{ CuO } (s) + 2 \text{ SO}_2 \ (g)$$

Step 2: Write the equation for the reaction of the resultant oxide with sulfuric acid:

$$\text{CuO } (s) + \text{H}_2\text{SO}_4 \ (aq) \rightarrow \text{H}_2\text{O } (l) + \text{CuSO}_4 \ (aq)$$

Step 3: Write the equation for the electrolytic purification of the solution:

$$\text{Cu}^{2+} \ (aq) + 2 \text{ e}^- \rightarrow \text{Cu } (s)$$

Workbook Problem 21.2

Strategy: Determine the number of bonding and antibonding electrons, and rank them in order by maximum bonding electrons and minimum antibonding electrons.

Step 1: Determine the numbers of electrons of each type:

Mo: $4d^4 5s^2$ 6 electrons: bonding orbitals are filled and antibonding orbitals are empty.

Cd: $4d^{10} 5s^2$ – 12 electrons: bonding and antibonding orbitals are filled.

Y: $4d^1 5s^2$ – 3 electrons: bonding orbitals half-filled, antibonding orbitals are empty.

Step 2: Rank in order of expected bonding strength:

$$\text{Cd} < \text{Y} < \text{Mo}$$

The published melting points are 321°C for Cd, 1523 °C for Y, and 2623 °C for Mo, which means this order is correct.

Workbook Problem 21.3

Strategy: Determine the number of valence electrons in the semiconductor and compare to the number of valence electrons in the dopant. Because all of the elements are main-group, their group numbers can be used.

Step 1:   Determine the numbers of valence electrons in each element:

Si: Group 4A – four valence electrons   Ge: Group 4A – four valence electrons
Ga: Group 3A – three valence electrons Sb: Group 5A – five valence electrons
As: Group 5A – five valence electrons   P: Group 5A – five valence electrons

Step 2:   Determine whether the dopant has more or fewer valence electrons than the semiconductor:

a. Si > Ga   b. Ge < Sb   c. Si < As   d. Ge < P

Step 3:   Identify the doped semiconductors as *n*- or *p*-type:

a. *p*-type   b. *n*-type   c. *n*-type   d. *n*-type

Workbook Problem 21.4

Strategy: Convert the band gap energy to photon energy, and then to wavelength.

Step 1:   Determine the energy of an individual photon:

$$178 \ \frac{kJ}{mol} \times \frac{1000 \ J}{kJ} \times \frac{1 \ mol}{6.022 \times 10^{23} \ photons} = 2.96 \times 10^{-19} \ J/photon$$

Step 2:   Convert the photon energy to wavelength:

$$E = \frac{hc}{\lambda}; \ \lambda = \frac{hc}{E} = \frac{6.626 \times 10^{-34} \ J \cdot s \times 3.00 \times 10^{8} \ m/s}{2.96 \times 10^{-19} \ J} = 673 \ nm$$

This is red light.

Putting It Together

Write and balance the equations for the reaction of Fe^{2+} with MnO_4^- and the reaction of Fe^{2+} with $Cr_2O_7^{2-}$.

The two half–reactions for the reaction of Fe^{2+} with MnO_4^- in acidic solution are

$$MnO_4^- \ (aq) \ \rightarrow \ Mn^{2+} \ (aq) \qquad\qquad Fe^{2+} \ (aq) \ \rightarrow \ Fe^{3+} \ (aq)$$

For the first half–reaction, balance the oxygens by the addition of water and H^+.

$$5 \ e^- \ + \ 8 \ H^+ \ (aq) \ + \ MnO_4^- \ (aq) \ \rightarrow \ Mn^{2+} \ (aq) \ + \ 4 \ H_2O \ (l)$$

For the second half–reaction, balance the charge.

$$Fe^{2+} \ (aq) \ \rightarrow \ Fe^{3+} \ (aq) \ + \ e^-$$

Now multiply the second half–reaction by 5 so that the number of electrons lost equals the number of electrons gained.

$$5\ e^- + 8\ H^+\ (aq) + MnO_4^-\ (aq) \rightarrow Mn^{2+}\ (aq) + 4\ H_2O\ (l)$$

$$5\ Fe^{2+}\ (aq) \rightarrow 5\ Fe^{3+}\ (aq) + 5\ e^-$$

Adding the two half–reactions together gives

$$8\ H^+\ (aq) + MnO_4^-\ (aq) + 5\ Fe^{2+}\ (aq) \rightarrow Mn^{2+}\ (aq) + 4\ H_2O\ (l) + 5\ Fe^{3+}\ (aq)$$

The balanced net ionic equation for the reaction of Fe^{2+} with $Cr_2O_7^{2-}$ is

$$14\ H^+\ (aq) + Cr_2O_7^{2-}\ (aq) + 8\ Fe^{2+}\ (aq) \rightarrow 2\ Cr^{2+}\ (aq) + 7\ H_2O\ (l) + 8\ Fe^{3+}\ (aq)$$

From the molarity and volume of $FeSO_4$, we know that we have 4.0×10^{-3} moles of Fe^{2+}. We also know that moles of Fe^{2+} consumed in the first reaction + moles of Fe^{2+} consumed in the second reaction = Total moles of Fe^{2+} = 4.0×10^{-3} moles Fe^{2+}

By determining the number of moles of Fe^{2+} consumed in the second reaction, we can determine the moles of Fe^{2+} consumed in the first reaction, which will allow us to determine the moles of manganese present.

For the reaction

$$14\ H^+\ (aq) + Cr_2O_7^{2-}\ (aq) + 8\ Fe^{2+}\ (aq) \rightarrow 2\ Cr^{2+}\ (aq) + 7\ H_2O\ (l) + 8\ Fe^{3+}\ (aq)$$

$$mol\ Fe^{2+} = 0.0224\ L \times \frac{0.0100\ mol\ Cr_2O_7^{2-}}{1\ mol\ Cr_2O_7^{2-}} \times \frac{8\ mol\ Fe^{2+}}{1\ mol\ Cr_2O_7^{2-}} = 1.79 \times 10^{-3}\ mol\ Fe^{2+}$$

The moles of Fe^{2+} consumed in the reaction

$$8\ H^+\ (aq) + MnO_4^-\ (aq) + 5\ Fe^{2+}\ (aq) \rightarrow Mn^{2+}\ (aq) + 4\ H_2O\ (l) + 5\ Fe^{3+}\ (aq)$$

mole Fe^{2+} = $(4.00 \times 10^{-3}\ mol\ Fe^{2+}) - (1.79 \times 10^{-3}\ mol\ Fe^{2+}) = 2.21 \times 10^{-3}$ moles Fe^{2+} consumed in the first reaction

We can now calculate the mass of manganese in the steel sample.

$$g\ Mn = 2.21 \times 10^{-3}\ mol\ Fe^{2+} \times \frac{1\ mol\ MnO_4^-}{5\ mol\ Fe^{2+}} \times \frac{1\ mol\ Mn}{1\ mol\ MnO_4^-} \times \frac{54.9\ g\ Mn}{1\ mol\ Mn} = 0.0243\ g\ Mn$$

$$\%\ Mn = \frac{0.0243\ g\ Mn}{0.450\ g\ steel} \times 100\% = 5.4\%$$

Self–Test

True–False
1. F. Most silicates are poor metal sources.
2. T
3. T
4. F. Reduction is by carbon monoxide.
5. F. There is no good theory for superconductivity at the present time.
6. T
7. T
8. F. Heating semiconductors increases conductivity. Heating metals decreases conductivity.
9. T
10. F. The primary drawback of ceramics is that they are brittle

Multiple Choice

11. c	12. d	13. b	14. a
15. d	16. b	17. a	18. c
19. c	20. b		

Fill-in-the-Blank
21. malleability and ductility, localized, brittle
22. current, conductance, antibonding, empty
23. below, resistance, nitrogen
24. metals, density, insulators, corrosion, malleable
25. (In any order): ceramics, metals, fibers, epoxy

Matching

26. m	27. g	28. a	29. l
29. c	30. j	31. p	32. t
33. e	34. k	35. f	36. b
37. r	38. d	39. h	40. i
41. o	42. n	43. q	44. u
45. s			

Problems

46.   Write and balance the equations:

$$2 \ MnCO_3 \ (s) + O_2 \ (g) \rightarrow 2 \ MnO_2 \ (s) + CO_2 \ (g)$$

$$3 \ MnO_2 \ (s) + 4 \ Al \ (s) \rightarrow 3 \ Mn \ (s) + 2 \ Al_2O_3 \ (s)$$

Determine the number of moles and kg of Mn in 1500 kg of $MnCO_3$:

$$1500 \text{ kg } MnCO_3 \times \frac{1000 \text{ g}}{1 \text{ kg}} \times \frac{1 \text{ mol } MnCO_3}{115 \text{ g}} = 13{,}050 \text{ mol}$$

$$13{,}050 \text{ mol } MnCO_3 \times \frac{1 \text{ mol } Mn}{\text{mol } MnCO_3} \times \frac{54.94 \text{ g}}{\text{mol } Mn} \times \frac{1 \text{ kg}}{1000 \text{ g}} = 717 \text{ kg}$$

Determine the moles, then kg of aluminum needed:

$$13{,}050 \text{ mol MnCO}_3 \times \frac{1 \text{ mol MnO}_2}{\text{mol MnCO}_3} \times \frac{4 \text{ mol Al}}{3 \text{ mol MnO}_2} \times \frac{26.98 \text{ g Al}}{\text{mol Al}} \times \frac{1 \text{ kg}}{1000 \text{ g}} = 469 \text{ kg Al}$$

47. As fourth-period elements, these all have access to $4s$ orbitals and $3d$ orbitals. Twelve electrons can be accommodated, 6 in bonding orbitals, and 6 in nonbonding orbitals. The more electrons there are in bonding orbitals, and the fewer electrons in the nonbonding orbitals, the harder the element and the higher the melting point.

    Cr: [Ar] $4s^2 3d^4$ – six electrons in bonding orbitals, no electrons in antibonding orbitals

    Co: [Ar] $4s^2 3d^7$ – six electrons in bonding orbitals, three electrons in antibonding orbitals

    Cu: [Ar] $4s^2 3d^9$ – six electrons in bonding orbitals, five electrons in antibonding orbitals

Highest to lowest melting point: Cr > Co > Cu

The observed melting points are   Cr: 1857 °C   Co: 1495 °C   Cu: 1083 °C

This confirms the determination made using molecular orbital theory.

48. Determine the number of valence electrons in each of the materials:

Si – Group 4A – four valence electrons   Ge – Group 4A – four valence electrons
P – Group 5A – five valence electrons   As – Group 5A – five valence electrons
B – Group 3A – three valence electrons

Identify the combinations and their types:

SiP   - 4A5A – n-type   SiAs   - 4A5A – n-type
SiB   - 4A3A – p-type   GeP   - 4A5A – n-type
GeAs - 4A5A – n-type   GeB   - 4A3A – p-type

Challenge Problem

49. The feasibility of using cyanidation to remove silver from argentite can be determined by calculating ΔG for the reaction. If ΔG is negative, the process may be a means of obtaining silver from the ore. Determine ΔG for the process from the K_{net} for the reaction. The two equations involved are:

   $\text{Ag}_2\text{S (s)} \rightarrow 2 \text{ Ag}^+ \text{(aq)} + \text{S}^{2-} \text{(aq)}$ $K_{sp} = 6 \times 10^{-51}$

   $2 \text{ Ag}^+ \text{(aq)} + 4 \text{ CN}^- \text{(aq)} \rightarrow 2 \text{ [Ag(CN)}_2]^-$ $K_f^2 = 1 \times 10^{42}$

(Square the value of K_f because the stoichiometry of the chemical equation is two times that of the chemical equation for the formation of $[\text{Ag(CN)}_2]^-$.)

The overall reaction and K_{net} are

$$Ag_2S \text{ (s)} + 4\,CN^- \text{ (aq)} \rightarrow 2\,[Ag(CN)_2]^- \text{ (aq)} + S^{2-} \text{ (aq)} \qquad K_{net} = 6 \times 10^{-9}$$

Calculate ΔG (assume 25°C):

$$\Delta G = -RT(\ln K)$$

$$\Delta G = -(8.314\,J/mol \cdot K)(298K)\ln 6 \times 10^{-9} = 4.69 \times 10^4\,J/mol = +46.9\,kJ/mol$$

Both the small value of K_{net} and the positive value of ΔG indicate that the cyanidation process for the removal of silver from argentite is not practical.

Use the same approach to determine if it is possible to use cyanidation to remove silver from horn silver.

$$AgCl \text{ (s)} \rightarrow Ag^+ \text{ (aq)} + Cl^- \text{ (aq)} \qquad K_{sp} = 1.8 \times 10^{-10}$$

$$Ag^+ \text{ (aq)} + 2\,CN^- \text{ (aq)} \rightarrow [Ag(CN)_2]^- \text{ (aq)} + Cl^- \text{ (aq)} \qquad K_f = 1 \times 10^{21}$$

The overall chemical equation is

$$AgCl \text{ (s)} + 2\,CN^- \text{ (aq)} \rightarrow [Ag(CN)_2]^- \text{ (aq)} + Cl^- \text{ (aq)} \qquad K_{net} = 1.8 \times 10^{11}$$

$$\Delta G = -(8.314\,J/mol \cdot K)(298K)\ln 1.8 \times 10^{11} = -6.42 \times 10^4\,J/mol = -64.2\,kJ/mol$$

Both the large value of K_{net} and the negative value of ΔG indicate that cyanidation is practical for extracting silver from horn silver.

Chapter Twenty-Two – The Main-Group Elements

Workbook Problems

Workbook Problem 22.1

Strategy: Examine the position of the elements on the periodic table.

Step 1: Predict the metallic character of the elements based on their position on the periodic table:

C or Si : Silicon is below carbon in the periodic table, so has more metallic character

Be or Li: Lithium is to the left of beryllium, so has more metallic character

Se or Br: Selenium is to the left of bromine, so has more metallic character

As or I: Astatine is above iodine, but iodine is to the right of astatine. This ambiguity is resolved by the fact that iodine is a nonmetal, while astatine is a semi-metal: astatine is more metallic than iodine.

Workbook Problem 22.2

Strategy: Calculate the molecular masses of the four molecules, calculate the percent differences, and describe the kinetic isotopic effects.

Step 1: Calculate the molecular weights for the four molecules:

$$C_6H_6 = (6 \times 12.01) + (6 \times 1.01) = 78.12 \text{ g/mol}$$

$$C_6D_6 = (6 \times 12.01) + (6 \times 2.02) = 84.18 \text{ g/mol}$$

$$H_2O = 16.00 + (2 \times 1.01) = 18.02$$

$$D_2O = 16.00 + (2 \times 2.02) = 20.04$$

Step 2: Calculate the percent difference:

$$\text{benzene \% difference} = \frac{84.18 - 78.12}{84.18} \times 100 = 7.20\%$$

$$\text{water \% difference} = \frac{20.04 - 18.02}{20.04} \times 100 = 10.1\%$$

Step 3: Predict the expected properties:

The increase in melting point from H_2O to D_2O is less than 4 °C, and the increase in boiling point is even smaller. The increases in these properties for the deuterated benzene are expected to be smaller still.

Given the increased strength of O–D bonds compared to O–H bonds, it can be expected that C–D bonds will also be stronger than C–H bonds, so more energy would be released by burning C_6D_6.

Workbook Problem 22.3

Strategy: Set up a ratio to calculate the volume, choose an appropriate unit for the answer.

$$\frac{0.083 \text{ L D}_2\text{O}}{2400 \text{ L H}_2O} = \frac{x \text{ L D}_2\text{O}}{25 \text{ L}}$$

$$2400x \text{ L} = 2.075 \text{ L}^2$$

$$x = 0.0008645 \text{ L} = 0.8645 \text{ mL or } 864.5 \text{ μL}$$

Workbook Problem 22.4

Write a balanced equation for the production of synthesis gas from ethane, C_2H_6. If each step in the process has a 60% yield, how many liters of hydrogen gas at 25 °C and 1 atm can be produced from 15 kg of ethane?

Strategy: Write and balance the equation, and determine the quantities of reactants and products at each step. Use the ideal gas law to calculate the volume of hydrogen gas.

Step 1:   Write and balance the equation for the formation of synthesis gas:

$$2\,H_2O\,(g)\; +\; C_2H_6\,(g)\; \rightarrow\; 2\,CO\,(g)\; +\; 5\,H_2\,(g)$$

Step 2:   Determine the moles of reactants entering the first step and the moles of products produced.

$$15{,}000\text{ g ethane} \times \frac{1\text{ mol}}{30.08\text{ g}} = 500\text{ moles ethane}$$

$$500\text{ mol ethane} \times \frac{2\text{ mol CO}}{\text{mol ethane}} = 1000\text{ mol CO}$$

$$500\text{ mol ethane} \times \frac{5\text{ mol H}_2}{\text{mol ethane}} = 2500\text{ mol H}_2$$

Step 3:   Multiply the moles of products by 60%, and calculate the products of the second step.

$$1000\text{ mol CO} \times 0.60 = 600\text{ mol CO}$$

$$2500\text{ mol H}_2 \times 0.60 = 1500\text{ mol H}_2$$

Only the CO continues on by the following reaction:

$$CO\,(g)\; +\; H_2O\,(g)\; \rightarrow\; CO_2\,(g)\; +\; H_2\,(g)$$

$$600\text{ mol CO} \times \frac{1\text{ mol H}_2}{\text{mol CO}} = 600\text{ mol H}_2$$

Step 4:   Multiply the moles of products by 60%, and calculate the products of the third step.

$$600\text{ mol H}_2 \times 0.60 = 360\text{ mol}$$

The third step simply removes the CO_2 into solution.

Step 5:   Multiply the moles of products by 60%, and determine the volume of the H_2 gas.

$$360 \text{ mol } H_2 \times 0.60 = 216 \text{ mol}$$

Total moles of gas = 1500 mol from Step 1 + 216 mol from Step 3 = 1716 mol H_2 gas.

$$V = \frac{nRT}{P} = \frac{1716 \text{ mol} \times 0.08206 \frac{L \cdot atm}{mol \cdot K} \times 298K}{1 \text{ atm}} = 42,000 \text{ L}$$

Workbook Problem 22.5

Strategy: Determine the initial volume, moles of each gas, and how much is left over (assume that the volume of the water produced is negligible). How many kilojoules are released in this reaction?

Step 1:   Determine the moles of each gas and the limiting reagent.

$$15 \text{ g } H_2 \times \frac{1 \text{ mol}}{2.02 \text{ g}} = 7.43 \text{ mol } H_2$$

$$100 \text{ g } O_2 \times \frac{1 \text{ mol}}{32.0 \text{ g}} = 3.13 \text{ mol } O_2$$

Oxygen is the limiting reagent.

Step 2:   Determine the moles of excess reagent remaining after the reaction.

2 mol of hydrogen react for every 1 mol of oxygen, so 6.25 mol of hydrogen will react. This leaves 0.62 mol of hydrogen unreacted.

Step 3:   Determine the volume change of the reaction.

Initial volume = 7.43 mol + 3.13 mol = 7.74 mol × 22.4 L/mol = 173 L

Final volume = 0.62 mol × 22.4 L/mol = 13.9 L

ΔV = final – initial = 13.9 L – 173 L = –159.1 L

Step 4:   Determine the kJ released by the reaction.

1 mol O_2 reacting releases 572 kJ

3.13 mol O_2 reacting releases 1790 kJ

Workbook Problem 22.6

Strategy: Write and balance the equation, determine the moles of sodium hydride, use the balanced equation to determine the moles of gas and the standard molar volume to determine liters.

Step 1:   Write and balance the equation:

$$NaH\ (s) + H_2O\ (l) \rightarrow H_2\ (g) + NaOH^-\ (aq)$$

Step 2:   Determine the moles of sodium hydride and the moles of hydrogen evolved:

$$3.65\ g\ \times \frac{1\ mol\ NaH}{24.01\ g} \times \frac{1\ mol\ H_2}{mol\ NaH} = 0.152\ mol\ H_2$$

Step 3:   Determine the volume of hydrogen evolved:

$$0.152\ mol\ H_2 \times 22.4\ L/mol = 3.4\ L\ H_2\ (g)$$

Workbook Problem 22.7

Strategy: Based on the reactivity of the metals and the expected reactions:

Step 1:   Determine what reaction, if any, will occur.

$$Sr\ (s) + O_2\ (g) \rightarrow \text{strontium oxide will form spontaneously}$$

$Ca\ (s) + H_2O\ (l) \rightarrow$ calcium hydroxide will form and hydrogen gas will be generated. Nonvigorous.

$Be\ (s) + H_2O\ (l) \rightarrow$ no reaction expected. Beryllium is the least reactive of the alkaline earth metals

Step 2:   Write balanced equations for the reactions that occur.

$$2Sr\ (s) + O_2\ (g) \rightarrow 2SrO\ (s)$$

$$Ca\ (s) + 2H_2O\ (l) \rightarrow Ca(OH)_2\ (aq) + H_2\ (g)$$

Workbook Problem 22.8

Strategy: Use the rules for assigning oxidation numbers to determine the oxidation state of carbon.

Step 1:   Determine the oxidation state of carbon.

     CO   Oxidation state for O is −2, neutral molecule, so the oxidation state for C is +2

     CO_2 Oxidation state for O is −2, neutral molecule, so the oxidation state for C is +4

C_2O_3	Oxidation state for O is –2, neutral molecule, so the oxidation state for C is +3
CO_3^{2-}	Oxidation state for O is –2, charge on molecule is –2, so the oxidation state for C is +4
CaC_2	Oxidation state for Ca is +2, neutral molecule, so the oxidation state for C is –1

Workbook Problem 22.9

Strategy: From the information provided, determine the overall silicate structure, remembering that oxygens are shared in the tetrahedral units. Balance the charge on the silicate using the ions in a correct ratio.

Step 1: Determine the silicate structure:

As the SiO_4^{4-} subunits join together, they share oxygens:

$$SiO_4^{4-} + SiO_4^{4-} = Si_2O_7^{6-} \qquad\qquad Si_2O_7^{6-} + SiO_4^{4-} = Si_3O_{10}^{8-}$$

Step 2: Determine how many of each cation is needed to balance the charge on the silicate unit:

To balance a charge of –8 with cations having charges of +3 and +2 will require two of the +3, and one of the +2. There is no other way to use both and get to +8

Step 3: Write the formula unit:

$$CaAl_2\,Si_3O_{10}$$

Workbook Problem 22.10

Step 1: Write the balanced chemical equation.

$$NH_4NO_3\,(s) \;\rightarrow\; N_2O\,(g) \;+\; 2\,H_2O\,(l)$$

Step 2: Calculate the pressure of N_2O, using the total pressure and the vapor pressure of water.

$$P_T = P_{H_2O} + P_{N_2O}$$

$$P_{N_2O} = 705 \text{ mm Hg} - 24 \text{ mm Hg} = 681 \text{ mm Hg}$$

Step 3: Calculate the number of moles of N_2O produced from 3.5 g NH_4NO_3.

$$3.5 \text{ g } NH_4NO_3 \times \frac{1 \text{ mol } NH_4NO_3}{80.0 \text{ g } NH_4NO_3} \times \frac{1 \text{ mol } N_2O}{1 \text{ mol } NH_4NO_3} = 4.38\times10^{-2} \text{ mol } N_2O$$

Step 4:   Calculate the volume of N_2O, using the ideal gas law.

$$V = \frac{nRT}{P} = \frac{\left(4.38 \times 10^{-2}\right)\left(0.0821\frac{L \cdot atm}{mol \cdot K}\right)(295.15\ K)}{681\ mm\ Hg \times \dfrac{1\ atm}{760\ mm\ Hg}} = 1.18\ L$$

Putting It Together

Write and balance the equation:

$$SO_2\ (g)\ +\ 2\ H_2S\ (aq)\ \rightarrow\ 3\ S\ (s)\ +\ 2\ H_2O\ (l)$$

To determine the volume of H_2S needed, calculate the moles of SO_2 (g) generated in the burning of the coal. Assume 100% yield.

$$mol\ SO_2 = 1.0\ tons\ coal \times \frac{0.033\ S}{coal} \times \frac{1000\ kg}{1\ kg} \times \frac{1\ mol\ S}{32.0\ g\ S} \times \frac{1\ mol\ SO_2}{1\ mol\ S} = 1031\ mol\ SO_2$$

From the balanced equation, 2 moles of H_2S are needed for every mole of SO_2, so 2062 moles H_2S are needed.

Again from the balanced equation, 3 moles of S are produced for every mole of SO_2:

$$kg\ S = 1031\ mol\ SO_2 \times \frac{3\ mol\ S}{1\ mol\ SO_2} \times \frac{32.0\ g\ S}{1\ mol\ S} \times \frac{1\ kg}{1000\ g} = 99\ kg\ S$$

Now use the ideal gas law to calculate the volume of SO_2:

$$V = \frac{\left(0.08206\frac{L \cdot atm}{K \cdot mol}\right)(1031\ mol)(423\ K)}{\left(740\ mm\ Hg \times \dfrac{1\ atm}{760\ mm\ Hg}\right)} = 3.68 \times 10^4\ L$$

Self–Test

True–False
1. F. Semimetals are on a zig-zag line from boron to astatine, not in the center of the periodic table.
2. F. Metallic character increases down a column of the periodic table.
3. T
4. F. Due to their size, it is uncommon to find double bonds in elements outside of the second row.
5. T
6. T
7. F. Silicon is too large to form double bonds: the chains consist of single bonds.
8. T

9. T

10. F. Liquid sulfur decreases in viscosity until the S_8 rings begin to break and reform into long chains. The viscosity increases at this point and decreases again when the chains begin to break as well.

Multiple Choice

11. b	12. c	13. a	14. d
15. d	16. c	17. d	18. a
19. b	20. b		

Fill-in-the-Blank

21. increases, increases, decreases, decreases   22. diamond, graphite, fullerene
23. white phosphorus, bent, red phosphorus   24. SiO_4
25. solid, liquid, S_8 rings, chains, viscosity

Matching

26. c	27. d	28. g	29. f
30. a	31. e	32. b	

Questions

33. a) Rb   b) In   c) Cs

34. a) K   b) In   c) Sb

35. Ionization energy—decreases down a group; atomic radius—increases down a group; electronegativity—decreases down a group; basicity of oxides—increases down a group

36. Each boron in diborane uses an sp^3 hybrid orbital to overlap with a terminal hydrogen $1s$ orbital. Each bridging H atom is joined to both boron atoms through a three-center, two-electron bond in which the two electrons in the B–H–B bridge are spread out over three atoms.

37. Diamond: tetrahedral bonding with sp^3 hybridization. It is nonconducting, the hardest known substance, and has the highest melting point of any element.

Graphite: trigonal planar bonding with sp^2 hybridization, with delocalized π bonds to adjacent molecules. It conducts electricity and forms sheets that can slide past one another, making it useful as a dry lubricant.

Graphene: prepared by the "Scotch tape method," graphene is a two-dimensional array of carbon molecules with a hexagonal arrangement. It is an excellent conductor of electricity.

Fullerene: this covers a number of structures from carbon nanotubes to spheres, which share a common bonding. They are molecular structures and dissolve in nonpolar solvents, as well as conducting electricity.

38. $-3 = NH_3$; $-2 = N_2H_4$; $-1 = NH_2OH$; $+1 = N_2O$; $+2 = NO$; $+3 = HNO_2$; $+4 = NO_2$; $+5 = HNO_3$

39. $HClO_4$—perchloric acid; $HClO_3$—chloric acid; $HClO_2$—chlorous acid;
$HClO$—hypochlorous acid
$HBrO_4$—perbromic acid; $HBrO_3$—bromic acid; $HBrO_2$_bromous acid;
$HBrO$—hypobromous acid
HIO_4—periodic acid; HIO_3—iodic acid; HIO_2—iodous acid; HIO—hypoiodous acid

40. $Cu\ (s)\ +\ 4\ HNO_3\ (aq)\ \rightarrow\ 2\ NO_2\ (g)\ +\ 2\ H_2O\ (l)\ +\ Cu(NO_3)_2\ (aq)$

Challenge Problem
41. $CaC_2\ (s) + H_2O\ (l) \rightarrow C_2H_2\ (g) + CaO\ (s)$

Determine the moles present of each reactant:

$$15.0\ g\ CaC_2\ x\ \frac{1\ mol\ CaC_2}{64.10\ g} = 0.234\ mol\ CaC_2$$

$$7.45\ g\ H_2O\ x\ \frac{1\ mol\ H_2O}{18.0\ g} = 0.413\ mol\ H_2O$$

Calcium carbide is the limiting reactant, as the reactants react in a 1:1 ratio.

$$0.234\ mol\ C_2H_2\ x\ \frac{26.04\ g\ C_2H_2}{mol} = 6.09\ g\ C_2H_2$$

Write the balanced reaction for the burning of acetylene:

$$C_2H_2\ (g) + 5/2\ O_2\ (g) \rightarrow 2\ CO_2\ (g) + H_2O\ (g)$$

Determine overall ΔH_{rxn} for acetylene:

$$[(-241.8\ kJ/mol) + 2\ (-393.5\ kJ/mol)\ ] - (227.4\ kJ/mol) = -1256\ kJ/mol$$

Determine the heat generated from this reaction:

$$0.234\ mol\ C_2H_2 \times -1256\ kJ/mol = -294\ kJ$$

Chapter Twenty-Three – Organic and Biological Chemistry

Workbook Problems

Workbook Problem 23.1

Strategy: Look for non-carbon atoms and multiple bonds. Compare with the chart above or the chart on Page 985 of the textbook to name the functional groups.

Step 1:   Name the functional groups:

        a. An oxygen double-bonded to a carbon that is also bonded to an alcohol group is a *carboxylic acid.*

        b. A carbon double-bonded to an oxygen and single-bonded to carbons on each side is a *ketone.*

        c. An oxygen that is double-bonded to a carbon that is also bonded to a hydrogen is an *aldehyde.*

        d. An oxygen that is single-bonded to two carbons is an *ether.*

        e. There are no non-carbon atoms here, but there are also not enough hydrogens to satisfy the bonding requirements of the carbon atoms. This is an *alkene.*

Workbook Problem 23.2

Strategy: Use the naming strategy detailed above to identify the molecules.

Step 1:   Identify and name the parent chain:

        a. pentane   b. heptane   c. hexane

Step 2:   Number the chain, beginning at the end closest to the first branch point:

        a. numbering begins on the right-hand side

        b. numbering begins on the right-hand side on any three-carbon chain

        c. numbering begins at the top of the structure.

Step 3:   Identify and number the substituents:

        a. 2-methyl, 2-methyl, 3-methyl

        b. 4-propyl, 4-ethyl

        c. 3-methyl

Step 4:   Write the name of the molecule as a single word:

        a. 2, 2, 3-trimethylpentane

        b. 4-ethyl, 4-propylheptane

        c. 3-methylhexane

Workbook Problem 23.3

Strategy: Draw the carbon backbones, and add substituents in the proper places.

Step 1:   Draw the carbon backbones:

        a. $CH_3CH_2CH_2CH_2CH_2CH_2CH_3$ (heptane)

        b. $CH_3CH_2CH_2CH_2CH_2CH_3$ (hexane)

        c. $CH_3CH_2CH_2CH_3$ (butane)

        d. $CH_3CH_2CH_2CH_2CH_2CH_2CH_2CH_3$ (octane)

Step 2:   Add the substituents, removing hydrogens as necessary to maintain four bonds per carbon:

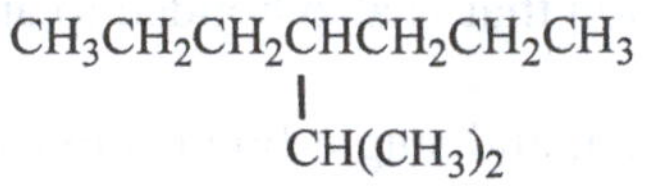

CH₃CH₂CH₂CHCH₂CH₂CH₃
                |
          CH(CH₃)₂

a.

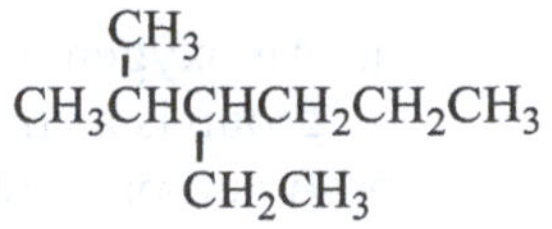

                CH₃
                |
CH₃CHCHCH₂CH₂CH₃
                |
          CH₂CH₃

b.

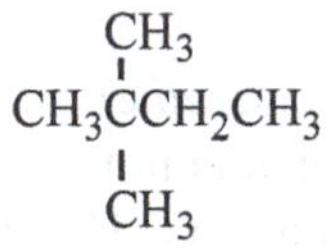

          CH₃
          |
CH₃CCH₂CH₃
          |
          CH₃

c.

                CH₃
                |
CH₃CHCHCHCH₂CH₂CH₃
          | |
          CH₃ CH₂CH₂CH₃

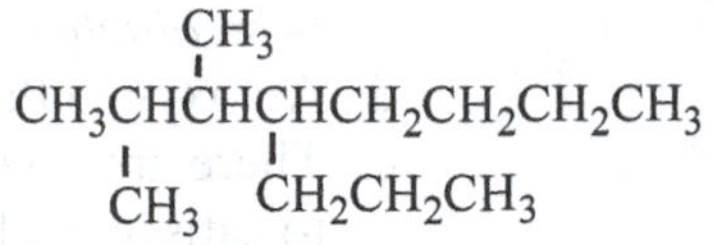

d.

Workbook Problem 23.4

Strategy: Draw the carbon backbone, and add substituents in the proper places.

Step 1:   Draw the carbon backbone:

CH₃CH₂CH₂CH₂CH₂CH₂CH₃

Step 2:   Add the substituents, subtracting hydrogens as necessary:

                    CH₃
                    |
CH₃C=CCHCH₂CHCH₃
              |
          CH₂CH₃

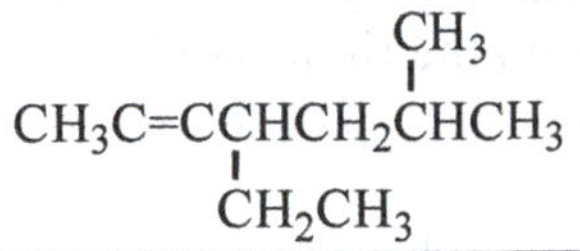

Workbook Problem 23.5

Strategy: Draw the carbon backbone, and determine the possible products of the addition reactions.

Step 1:   Draw the carbon backbone:

CH₃CH₂CH=CHCH₂CH₃

Step 2:   Add the substituents:

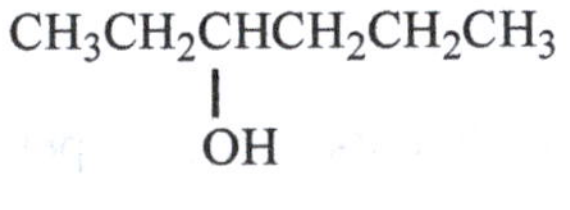

CH₃CH₂CHCH₂CH₂CH₃
              |
              OH

a.

CH₃CH₂CHCHCH₂CH₃
              | |
              Cl Cl

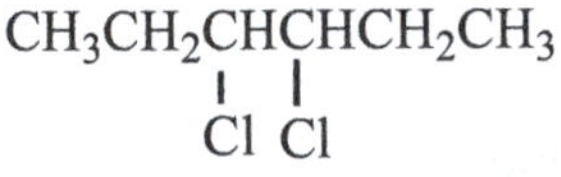

b.

c. CH₃CH₂CH₂CH₂CH₂CH₃

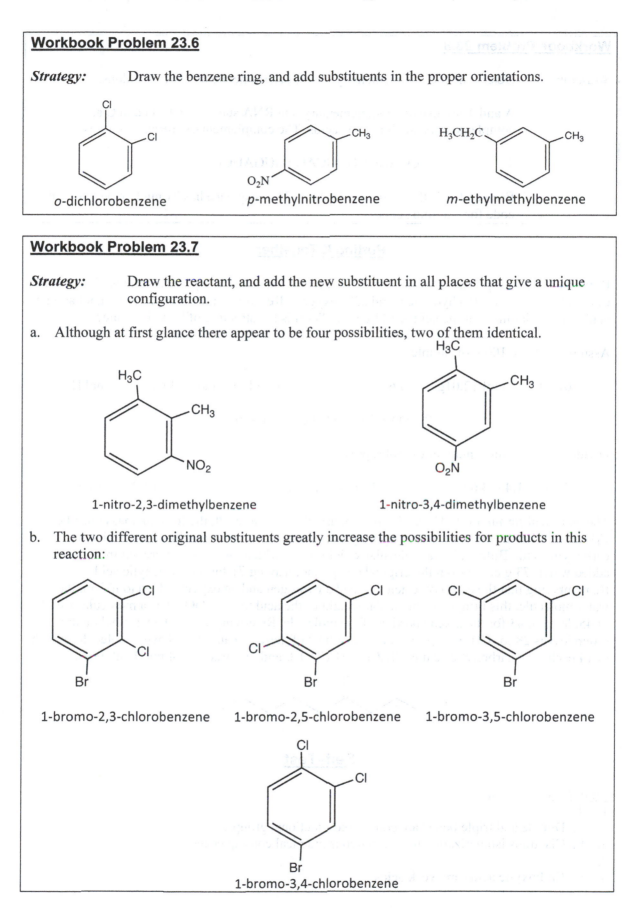

Workbook Problem 23.6

Strategy: Draw the benzene ring, and add substituents in the proper orientations.

o-dichlorobenzene   p-methylnitrobenzene   m-ethylmethylbenzene

Workbook Problem 23.7

Strategy: Draw the reactant, and add the new substituent in all places that give a unique configuration.

a.   Although at first glance there appear to be four possibilities, two of them identical.

1-nitro-2,3-dimethylbenzene   1-nitro-3,4-dimethylbenzene

b.   The two different original substituents greatly increase the possibilities for products in this reaction:

1-bromo-2,3-chlorobenzene   1-bromo-2,5-chlorobenzene   1-bromo-3,5-chlorobenzene

1-bromo-3,4-chlorobenzene

Workbook Problem 23.8

Strategy: Determine the complementary bases; determine the number of codons.

A and T are usually complementary, but RNA substitutes U. G and C are complementary in all nucleic acids. The complementary mRNA strand is:

UCGAACUGUGCACAGUGAUCU

Each codon is three nucleotides long. This sequence has 21 nucleotides, so will code for 7 amino acids.

Putting It Together

Butyl butyrate is responsible for the distinctive scent of bananas. The percent composition of this ester is 67% carbon, 11% hydrogen, and 22% oxygen. Reaction with water yields butanol and an acid. The molar mass of the acid is 144 g/mol. What is the structure of butyl butyrate?

Assume we have 100 g of sample:

$$67 \text{ g C} \times 1 \text{ mol}/12.01 \text{ g} = 5.6 \text{ mol C} \qquad 11 \text{ g H} \times 1 \text{ mol}/1.00 \text{ g} = 11 \text{ mol H}$$

$$22 \text{ g O} \times 1 \text{ mol}/16.0 \text{ g} = 1.4 \text{ mol O}$$

Divide by the smallest number of moles present:

$$5.6 \text{ C} \div 1.4 = 4.0 \text{ C} \qquad 11 \text{ H} \div 1.4 = 7.8 \text{ H} \qquad 1.4 \text{ O} \div 1.4 = 1.0 \text{ O}$$

The empirical formula is C_4H_8O. This has a mass of 72. As a result, the formula mass must be $C_8H_{16}O_2$. When an ester undergoes hydrolysis, it separates into its constituent alcohol and carboxylic acid. Butanol has a molecular weight of 74, which includes a hydrogen from the added water. 73 g come from the original compound, leaving 71 for the carboxylic acid. Remembering that the carboxylic acid has had a hydrogen and an oxygen added to it from the water molecule, this brings the molecular weight of the acid to 88. COOH has a molecular weight of 45, leaving 43 for the alkene portion of the molecule. Removing 15 for the terminal methyl group leaves 28 for CH_2 groups, so there must be 2 of these, because they have a molecular weigh of 14 each. The carboxylic acid is $CH_3CH_2CH_2COOH$, and the structure of butyl butyrate is

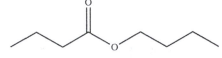

Self–Test

True–False

1. T
2. F. Double and triple bonds are considered functional groups.
3. F. Cis–trans isomerization has tremendous practical consequences.
4. T.
5. F. Carboxylic acids are weak acids.

6. T
7. T.
8. T
9. F. Cellulose is an indigestible long-chain polysaccharide.
10. F. Nucleic acids carry genetic information and provide instructions for protein production.

Multiple Choice

11. b	12. c	13. a	14. c
15. d	16. b	17. d	18. c
19. b	20. d		

Fill-in-the-Blank
21. Bonding, functional groups
22. Hydrocarbons, carbon, alkanes, alkenes, alkynes, triple, double, aromatic or benzene
23. double, benzene (or aromatic)
24. Carbohydrates, energy
25. amino acids, enzymes
26. replication, mRNA, translated, proteins

Matching

27. k	28. d	29. s	30. i
31. w	32. a	33. m	34. h
35. b	36. r	37. f	38. c
39. e	40. t	41. p	42. g
43. j	44. u	45. l	46. y
47. n	48. o	49. cc	50. bb
51. x	52. v	53. z	54. aa

Questions
55.

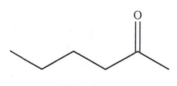

meta-dibromobenzene

2,3,4-trimethyl-1-hexanol

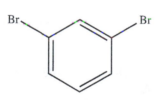

2-hexanone

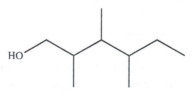

3-ethyl-2,4-dimethylpentanal

4-ethyl-6-methyl-1-heptene

56. a. 2-methyl hexanamide   b. methyl butanoate   c. cyclohexylamine
    d. cyclopentanone   e. cis-3-hexene

57.

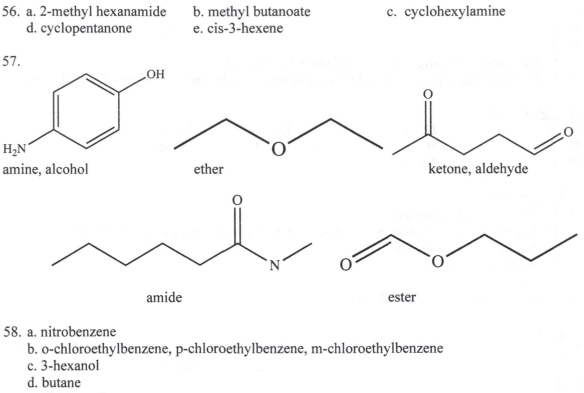

amine, alcohol   ether   ketone, aldehyde

amide   ester

58. a. nitrobenzene
    b. o-chloroethylbenzene, p-chloroethylbenzene, m-chloroethylbenzene
    c. 3-hexanol
    d. butane
    e. dichloroethane
    f. ethyl acetate or ethyl ethanoate
    g. acetic acid or ethanoic acid and methylamine

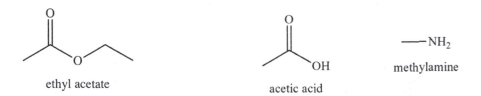

ethyl acetate   acetic acid   methylamine

59. 1. Food is digested in the stomach and small intestine to yield small molecules.
    2. Small molecules are broken down to form acetyl CoA
    3. Acetyl CoA is oxidized in the citric acid cycle to yield CO_2 and reduced coenzymes.
    4. Reduced coenzymes are oxidized by the electron transport chain and energy is released as ATP

60.

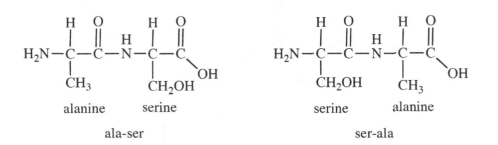

alanine   serine   serine   alanine

ala-ser   ser-ala

61. Ala-Leu-His, Ala-His-Leu, His-Leu-Ala, His-Ala-Leu, Leu-Ala-His, Leu-His-Ala

62.

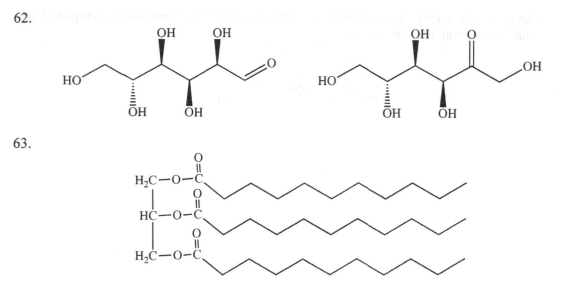

63.

64. DNA sequence = ATCGTGACTAGTCATTCG

Challenge Problem

65. 6.87 grams of CO_2 × 1 mol/12.01g = 0.156 mol of CO_2 = 0.156 mol C × 12 g/mol = 1.87 g C

1.87 grams of H_2O × 1 mol/18.01 g = 0.104 mol H_2O × 2 H/mol H_2O × 1.00 g/mol = 0.208 g H

5.00 grams of original compound – 0.208g – 1.87 g = 2.92 g left for oxygen.

2.92 g oxygen × 1 mol/16.0 g = 0.182 mol O

$$\text{moles C} = \frac{0.156}{0.156} = 1 = \frac{3}{3} = \frac{6}{6}$$

$$\text{moles H} = \frac{0.208}{0.156} = 1.3 = \frac{4}{3} = \frac{8}{6}$$

$$\text{moles O} = \frac{0.182}{0.156} = 1.17 = \frac{7}{6}$$

Empirical formula = $C_6H_8O_7$

The molecular molar mass is determined from the moles of NaOH used in the titration and the fact that citric acid is a triprotic acid. Three moles of NaOH will react for each mole of citric acid.

$$\text{mol citric acid} = 0.0488 \text{ L} \times \frac{0.100 \text{ mol NaOH}}{1 \text{ L}} \times \frac{1 \text{ mol citric acid}}{3 \text{ mol NaOH}} = 1.63 \times 10^{-3} \text{ mol citric acid}$$

$$\text{molar mass} = \frac{0.312 \text{ g citric acid}}{1.63 \times 10^{-3} \text{ mol maleic acid}} = 192 \text{ g/mol}$$

The molecular weight corresponds to the empirical formula. Therefore, the empirical and molecular formulas are the same.

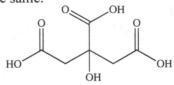